The History of Mathematics: A Source-Based Approach Volume 1

Regiomontanus *Epytoma in Almagesti Ptolemei* (Epitome of Ptolemy's *Almagest*), 1496

AMS/MAA | TEXTBOOKS

VOL 45

The History of Mathematics: A Source-Based Approach Volume 1

June Barrow-Green
Jeremy Gray
Robin Wilson

MAA PRESS
An Imprint of the American Mathematical Society
Providence, Rhode Island

2010 *Mathematics Subject Classification*. Primary 01-01, 01A05; Secondary 01A15, 01A16, 01A17, 01A20, 01A25, 01A32, 01A35, 01A45.

For additional information and updates on this book, visit
www.ams.org/bookpages/text-45

Library of Congress Cataloging-in-Publication Data

Names: Barrow-Green, June, 1953– author. | Gray, Jeremy, 1947– author. | Wilson, Robin J., author.

Title: The history of mathematics : a source-based approach / June Barrow-Green, Jeremy Gray, Robin Wilson.

Description: Providence, Rhode Island : MAA Press, an imprint of the American Mathematical Society, [2018]- | Series: AMS/MAA textbooks ; volume 45 | Includes bibliographical references and index.

Identifiers: LCCN 2018034323 | ISBN 9781470443528 (alk. paper : v. 1)| Softcover ISBN 978-1-4704-6676-3

Subjects: LCSH: Mathematics–History. | Mathematics–Study and teaching. | AMS: History and biography – Instructional exposition (textbooks, tutorial papers, etc.). msc | History and biography – History of mathematics and mathematicians – General histories, source books. msc | History and biography – History of mathematics and mathematicians – Indigenous European cultures (pre-Greek, etc.). msc | History and biography – History of mathematics and mathematicians – Egyptian. msc | History and biography – History of mathematics and mathematicians – Babylonian. msc | History and biography – History of mathematics and mathematicians – Greek, Roman. msc | History and biography – History of mathematics and mathematicians – China. msc | History and biography – History of mathematics and mathematicians – India. msc | History and biography – History of mathematics and mathematicians – Medieval. msc | History and biography – History of mathematics and mathematicians – 17th century. msc

Classification: LCC QA21 .B24 2018 | DDC 510.9–dc23

LC record available at https://lccn.loc.gov/2018034323

∞ The paper used in this book is acid-free and falls within the guidelines
established to ensure permanence and durability.
Visit the AMS home page at `https://www.ams.org/`

11 10 9 8 7 6 5 4 3 2 27 26 25 24 23 22

Contents

Acknowledgments

We would like to thank the following people for their help: Jim Bennett, Len Berggren, Henk Bos, Karine Chemla, Serafina Cuomo, Joseph Dauben, Raymond Flood, Catherine Goldstein, Niccolò Guicciardini, Jens Høyrup, Annette Imhausen, Graham Jagger, Stephen Johnston, Alex Jones, Victor Katz, Kim Plofker, Eleanor Robson, Fenny Smith, Jackie Stedall, and Glen van Brummelen. We also wish to thank our colleagues at the Open University who helped to keep the history of mathematics alive, especially Gloria Baldi, Mick Bromilow, Rebecca Browne, Giles Clark, Derek Goldrei, Sara Griffin, Tracy Johns, and Derek Richards, and for their help with TeX, Camilla Jordan and John Trapp. We also thank Gresham College, London, for a grant towards the production costs of this book.

John Fauvel was the driving force behind the Open University course upon which this book is based. Not only did he possess a wide range of knowledge about the history of mathematics, he was able to balance great complexity of material with an exquisite sensitivity to the needs and aspirations of students, to the point of making them joint explorers in the study of the subject. His emphasis on the patient reading of primary sources in translation was empowering, and his influence through the Open University, the British Society for the History of Mathematics, and in many other ways led to a remarkable growth in the appreciation of what the history of mathematics can offer. His untimely death in 2001 deprived us all of his wisdom, energy, and enthusiasm, and we hope that this book will spread his influence further.

We also thank these people at the MAA for their support: Don Albers, Beverly Ruedi, and Stephen Kennedy.

Conventions. We have used the following conventions in this book:

- Book titles are given in italics, and if the title is not in English we have followed it with a translation, which is in italics if it is the title of an English translation of the original book and in roman otherwise.
- A reference of the form '(Boyer 1959, 110)' is to page 110 of the item listed under Boyer in the Bibliography as published in 1959.
- A reference of the form 'Boyer (1959, 110) wrote' is to be read as an abbreviation for 'Boyer, on p. 110 of his book (or article) of 1959, wrote'.

Many references are given to sources in English in the form F&G 11.A1, which stands for the entry A1 in Chapter 11 of Fauvel and Gray, *The History of Mathematics: A Reader*.

Permissions & Acknowledgments

The American Mathematical Society gratefully acknowledges the following individuals for providing photographs and for permission to reproduce copyrighted materail. While every effort has been made to trace and acknowledge copyright holders, the publishers would like to apologize for any omissions and will be pleased to incorporate missing acknowledgments in any future editions.

The Asiatic Society, Kolkata

Excerpts, p. 241. H.J.J. Winter, W. Arafat, "The Algebra of Omar Khayyam" in the *Journal of the Royal Asiatic Society of Bengal: Science*, Volume 16, 1950. Reprinted by permission of The Asiatic Society, Kolkata

Bodleian Library, Oxford

Figure 7.2, p. 202. An early appearance of zero in a temple in Gwalior, 9th century. Courtesy of the Bodleian Library, Oxford (MS.+Sansk_d.14_16v).

British Library Board

Figure 11.1, p. 320. Recorde's *The Ground of Artes* (1543). ©The British Library Board.

Cambridge University Library

Figure 6.12, p. 193. Boethius, Pythagoras, Plato, and Nicomachus, from a medieval manuscript. Reproduced by kind permission of the Syndics of Cambridge University Library (N7425, li.3.12).

Figure 8.1, p. 241. A 13th-century manuscript of al-Khwarizmi's *Arithmetic*. Reproduced by kind permission of the Syndics of Cambridge University Library (N7407, li.6.5, fol 104r).

Clay Mathematics Institute

Figure 3.17, p. 81. Euclid's *Elements*, Book XII, Prop. 12, AD 888. Image courtesy of the Clay Mathematics Institute.

Florida Center for Instructional Technology

Figure 11.19, p. 358. Napier's bones: a set of rods that display the basic multiplication tables. William Dwight Whitney, *The Century Dictionary and Cyclopedia: An encyclopedic Lexicon of the English Language* (New York, NY: The Century Co., 1889), from: http://etc.usf.edu/clipart/27600/27640/napierrods_27640.htm. ClipArt courtesy FCIT, http://etc.usf.edu/clipart.

Excerpts, pp. 225–226, 230–231, and 233–234.

Green Lion Press

Excerpts, pp. 154–156. Apollonius of Perga Conics, Books I-III, ©2000 by Green Lion Press, www.greenlion.com. Used by permission.

Excerpts, pp. 370, 286–387, 389–394. Kepler's Astronomia Nova, ©2015 by Green Lion Press, www.greenlion.com. Used by permission.

The Mathematical Association of America

Figure 9.2, p. 263. An abacus of the type used by Gerbert and his followers. ©The Mathematical Association of America, 2018. All rights reserved.

Museum of the History of Science, University of Oxford

Figure 11.9, p. 343. Part of an astrolabe made by Humfrey Cole.

Pennsylvania State University Press

Figure 7.9, p. 223. Liu Hui's calculation of π. From *The Sea Island Mathematical Manual*, by Frank Swetz, ©1992 by The Pennsylvania State University Press. Reprinted by permission of The Pennsylvania State University Press.

Princeton University Press

Figure 7.8, p. 221. A Chinese diagram of the Pythagorean theorem. From: Chapter 3,"Chinese Mathematics," by Joseph Dauben in *The Mathematics of Egypt, Mesopotamia,* China, India, and *Islam: A Sourcebook*, ed. by Victor Katz. Reprinted by permission of Princeton University Press.

Excerpts, pp. 52–54, 80, 84–85, 96–97, 104, 112, 116, 466–467. Excerpted from *PROCLUS: A Commentary on the First Book of Euclid's Elements* by Glenn R. Morrow. ©1970, renewed 1998 by Princeton University Press. Reprinted by permission.

Excerpts, pp. 223–226, 228–229, and 230–232, . Excerpted from Chapter 3, "Chinese Mathematics" by Joseph Dauben in *The Mathematics of Egypt, Mesopotamia, China, India, and Islam: A Sourcebook*, ed. by Victor Katz. ©2007 by Princeton University Press, Reprinted by Permission.

Queen's College, Oxford

Figure 11.3, p. 324. The elongated "equals" sign in Record's *The Whetstone of Witte* (1557). Courtesy of the Provost and Fellows of The Queen's College, Oxford.

Figure 11.7, p. 337. Dee's *Groundplat*. Courtesy of the Provost and Fellows of The Queen's College, Oxford.

Figure 11.8, p. 340. Dee's *General and Rare Memorials*. Courtesy of the Provost and Fellows of The Queen's College, Oxford.

Springer-Verlag

Excerpts, pp. 19 and 150. Reprinted/adapted by permission from Springer Nature; Springer Verlag, DIOCLES, *On Burning Mirrors: The Arabic Translation of the Lost Greek Original* by G. J. Toomer, ©Springer-Verlag, Berlin Heidelberg 1976.

Excerpts, pp. 283–284. Reprinted/adapted by permission from Springer Nature; Springer Verlag, Nicolas Chuquet, *Renaissance Mathematician: A study with extensive translation of Chuquet's mathematical manuscript completed in 1494*, edited by Barbara Moss, Cynthia Hay, Graham Flegg, ©1985 D. Reidel Publishing Company.

Excerpts, pp. 241. Archive for History of Exact Sciences, *Greek and Arabic constructions of the regular heptagon*, Volume 30, Number 3, pp 197–330, 1984, Hogendijk, Jan P., ©Springer Verlag. With permission of Springer.

Excerpts, p. 432–433. *The History of Mathematics: A Reader*, by J. Fauvel and J. J. Gray, Macmillan, in association with the Open University, 1987. Used with permission.

University of Wisconsin Press

Excerpts, pp. 154–156. Clagett, Marshall, *Archimedes in the Middle Ages*, Volume 1, ©1964 by the Board of Regents of the University of Wisconsin System. Reprinted by permission of The University of Wisconsin Press.

Excerpts, pp. 121–124. Grant, Edward, *Nicole Oresme and the Kinematics of Circular Motion*. ©1971 by the Board of Regents of the University of Wisconsin System. Reprinted by permission of The University of Wisconsin Press.

Wikimedia Commons

Figure 2.7, p. 18. A geometry problem from the Rhind Papyrus.

Figure 2.13, p. 34. Plimpton 322.

Figure 3.13, p. 71. A Roman mosaic, c. 1st century AD, from Pompeii, showing Plato at his Academy. Used under Creative Commons License Deed Attribution-ShareAlike 3.0 Unported.

Figure 3.16, p. 81 A papyrus fragment showing a theorem from Euclid's Elements, Book II, Prop. 6.

Figure 5.8, p. 139. Part of the Archimedes palimpsest.

Figure 5.13, p. 148. How Archimedes might have set fire to the Roman fleet using parabolic mirrors, from Athanasius Kircher's *Ars Magna Lucis* (1646). Used under Creative Commons License Deed Attribution 4.0 International.

Figure 6.5, p. 176. The Ptolemaic system of planetary astronomy.

Figure 6.11, p. 184. Fermat's last Theorem.

Figure 7.11, p. 231. Zhu Shijie's table of binomial coefficients.

Figure 9.3, p. 267. Leonardo of Pisa (c.1170–c.1250). Used under Creative Commons License Deed Attribution 2.5 Generic.

Figure 9.5, p. 271. Roger Bacon (c.1212–1294). Used under Creative Commons License Deed Attribution 4.0 International.

Figure 10.3, p. 284. Luca Pacioli as a geometry teacher.

Figure 10.4, p. 285. A polyhedron by Leonardo da Vinci.

Figure 10.7, p. 302. Cardano's *Ars Magna* (Nurnberg, 1545).

Figure 10.9, p. 310. Bombelli's *Algebra*, 2nd edition (1579).

Figure 10.10, p. 314. François Viète (1540–1603).

Figure 11.4, p. 328. Recorde's *The Castle of Knowledge* (1556). Used under Creative Commons License Deed Attribution 4.0 International.

Figure 11.5, p. 334. John Dee (1527–1608). Used under Creative Commons License Deed Attribution 4.0 International.

Figure 11.6, p. 336. Euclid's *Elements*, translated by Henry Billingsley, 1570.

Figure 11.10, p. 345. The original Gresham College.

Figure 11.8, p. 357. Napier's *Mirifici Logarithmorum Canonis Descriptio* (1614).

Figure 11.22, p. 365. Some of Henry Briggs's logarithms.

Figure 11.23, p. 367. Frontispiece of Kepler's *Rudolphine Tables* (1627).

Figure 12.1, p. 377. Copernicus's heliocentric system, with Sol (the Sun) at the centre and Terra (the Earth) as one of the planets orbiting it.

Figure 12.2, p. 379. Tycho Brahe in his observatory.

Figure 12.3, p. 381. Kepler's planetary model.

Figure 12.5, p. 391. Galileo Galilei (1564–1642).

Figure 12.7, p. 396. Frontispiece to Galileo's *Dialogue Concerning the Two Chief World Systems*.

Yale Babylonian Collection

Figure 2.8, p. 22. The tablet YBC 4652 with an indication of its size. YBC 4652, courtesy of the Yale Babylonian Collection.

Figure 2.12, p. 32. YBC 7289: photograph and drawing. YBC 7289, courtesy of the Yale Babylonian Collection.

1

Introduction

Mathematics has been a major force shaping human lives for several millennia. Some of humanity's deepest thoughts are expressed in the language of mathematics, and many of its most powerful tools could not have been devised without it. At various stages in history mathematics has been highly regarded and at others largely neglected. Some people find it a source of delight, some treat it as a mere tool, some find it cold and lifeless. The study of the history of mathematics offers us a way to understand the many roles that mathematics has played, and continues to play, in the lives of societies, cultures, and individuals. We can use it to shed light on the nature of mathematics itself, and perhaps on some of the activities that make us distinctively human.

We first see mathematics at work in the creation of the great civilisations of the ancient Near East: Egypt and Mesopotamia. Then we turn to the later culture of ancient Greece, and see the emphasis placed on abstraction and proof. The classic text from this period is Euclid's *Elements*, often claimed to be one of the most published books ever written, and taught in schools in one form or another as late as the 1960s. This is a remarkable life for a book written well over 2000 years ago. What does it tell us about the work itself, and the people who used it, that it lasted so long? What are the implications of such longevity?

Later we consider some of the roles that mathematics played in the history of China and India, and in the early centuries of the world of Islam. How did mathematicians in these societies see the nature and purpose of mathematics? What did they contribute to it?

We then turn to the reawakening of intellectual life in Europe, which brought with it a renewed interest in the lost arts of mathematics, driven variously by trade, a desire to catch up with the Islamic world, and also by astronomical, scientific, and ultimately abstract intellectual issues. We conclude by looking at the work of Fermat and Descartes in the 17th century.

The topics covered here are all mainstream, important topics in the history of mathematics, but even in a book of this size some difficult choices had to be made. We have sought to include topics that carry a broad significance. So the chapter on Egyptian and Mesopotamian mathematics not only covers some of the earliest evidence we have of mathematical activity, it points firmly towards the idea that mathematics, however it is defined, is not a Greek invention. The same message is, of course, conveyed in the chapters on mathematics in India, China, and the Islamic world. Scholarship on mathematics in Egypt and Mesopotamia was largely the creation of Otto Neugebauer in the middle of the 20th century, and it has been enjoying a revival recently; comparably good work on India, China, and the Islamic world is mostly recent and growing. All these fields offer new perspectives on the nature of mathematics, its connection to proof, its uses in the societies that have sustained it, and, at least at times, the people who created it.

None of this diminishes the achievements of Greek mathematicians, either in their own day or in the centuries of the revival in the West. We have concentrated on their geometry, because that was not only a highlight but, through its emphasis on certain kinds of proof, was to have a decisive effect on the image of mathematics in later times. But we have also looked at the connections to philosophy and astronomy, and at the rather different accomplishments of Greek mathematicians in the Hellenistic period. We have also followed these stories into the Islamic world, and we look at the implications of those mathematicians for later Western work on algebra in particular. We end this account with two chapters: one on astronomy and one on the synthesis of algebra and geometry particularly associated with Descartes.

Studying the history of mathematics is by no means the same as studying mathematics itself, although some familiarity with mathematics is advisable. The questions that this book addresses are questions about the *history* of mathematics, not mathematical questions with a historical flavour (exciting though that can be, too). We are interested in understanding who did the mathematics, and why? Were they teachers — and if so, who were their students and why were they there? Was there a cultural or philosophical dimension to their mathematical work? What does it mean to discover something in mathematics? How was mathematical knowledge disseminated? Surprisingly rich answers to questions such as these can be obtained without one having to master the accompanying mathematics. What was done is interesting, but why it was done is interesting too. What was the context that made the work important in its day? Why is it still of interest to study today? These are the central questions that flow through this book.

To answer these questions, historians of mathematics rely on what we might call the 'facts', which can be drawn from many sources: documents, written texts, and also various artefacts. The historian's task is to make these mute objects speak again — but inevitably they do so with the historian's voice. This is ultimately because the big questions raised here do not have simple answers, and so studying the history of mathematics involves a certain amount of disagreement. Historians produce arguments, based on selections of the evidence; their conclusions are not so much facts as opinions. We can ask that their opinions are well-argued and well-informed, but opinions they remain, and other historians, perhaps bringing forth new facts, can disagree. This gives the history of mathematics a necessarily provisional character, but it opens the way to new and important work in the subject, and you will find much recent historical work reflected in these pages.

Because our intention is to bring original sources to life whenever appropriate, our approach raises the question of the prerequisites for reading this book. These are both more and less than a casual inspection might suggest. More, because the material on geometry has become increasingly remote. Less, because our interest is in the *history* of the mathematics we introduce, and it is often possible to see and appreciate the importance of a source without entirely grasping the mathematics it contains. We can often see what a text is about, how it differs from other comparable accounts, and why it was written even when the topic is technical and the exposition difficult to follow. And, depending on the historical question being investigated, that is often enough.

We do not assume knowledge of any specific piece of mathematics, but we assume some familiarity with the subject and a willingness to grapple with the details. The sources, and our commentaries on them, should be read pen in hand; drawing a diagram for yourself is the best way of seeing what is in it. We have also provided suggested exercises, many of which contain primary sources, and which — unlike the exercises in most textbooks on the subject — are firmly historical, rather than largely mathematical. They will be found at the end of the book, along with suggestions for how they can be approached.

This book is aimed at the general mathematically inclined reader. We hope that it will provide a rich introduction not only to the history of mathematics, but to mathematics, itself. The prerequisites gradually involve more mathematics, but we believe that each section of this book can be read as an introduction to the mathematics involved, as well as providing an absorbing account of history it describes.

No-one should have left school with too little mathematics to follow this book. But for those who were cut off from mathematics because they could not appreciate why it was done, and who could not connect to its excitements, we hope that this historical account offers a rewarding way in.

2

Early Mathematics

Introduction

In this chapter we look first at the very origins of mathematics, insofar as they can be discerned. We consider some of the archaeological evidence from pre-literate communities, and look at linguistic evidence drawn from languages spoken in the ancient Near East and in Europe.

We then turn to the two major literate civilisations of ancient Egypt and Mesopotamia (see Figure 2.1). There is written evidence from Egypt in the form of inscriptions and

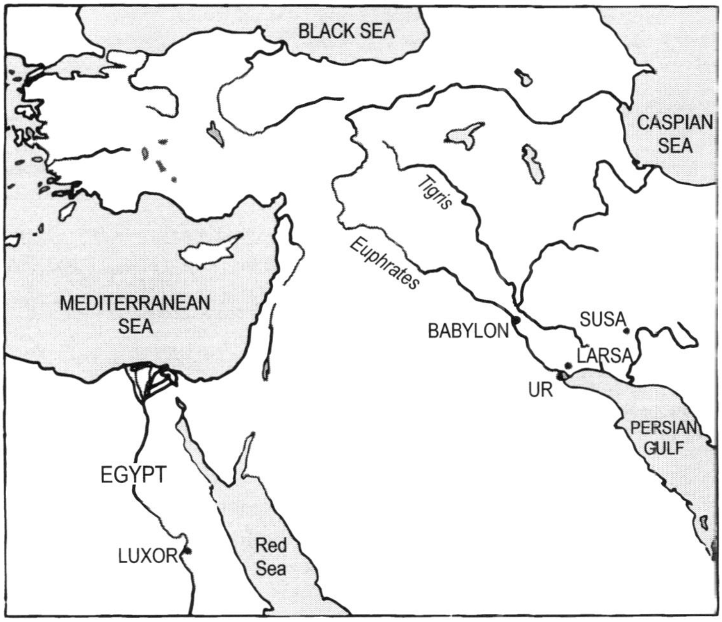

Figure 2.1. Map of ancient Egypt and Mesopotamia

papyri. In Mesopotamia people wrote on clay tablets, many thousands of which have survived. These sources, most of which date back nearly 4000 years and the oldest of which date back over 5000 years, enable us to form a good, if partial, impression of the range and role of mathematics in these long-lasting societies.

2.1 Early counting

Perhaps the first question that we can ask in the history of mathematics is: When did mathematics begin? Unfortunately, that is about the most difficult question that can be asked, since we cannot come to an answer — but understanding why we cannot, and why it is a difficult question, provides a helpful introduction to the study of the history of mathematics.

We can see that many different assumptions may be built into the question, and these can in turn influence the answer. A simple answer like '500 BC' or '5000 BC' invites us to leave it at that, whereas 'AD 1854' would surely surprise us. Yet this last date was the one given by Bertrand Russell,[1] who claimed that

> Pure mathematics was discovered by Boole, in a work which he called *The Laws of Thought* (1854).[2]

Let us not be distracted by precisely what Russell was getting at here. Clearly he was being provocative, but he was also saying something about the sense in which he was using the term *pure mathematics*. So this term too is problematical. The question then seems to beg a further one: What is mathematics? This may suggest that we are moving away from any easy answers at all. In fact, much of this book is devoted to unfolding the layers of meaning, activity, and perception represented by the single word *mathematics*.

Let us next turn to the word *history*. This is also ambiguous: there are at least two meanings that we need to disentangle. 'History' can mean 'the past' in a simple temporal sense, or it can mean 'organised knowledge of the past', something acquired and developed by historians. In this book, we use both meanings simultaneously: we study the mathematics of the past, and we shall also come to see how our knowledge of that past is gained — the two are inseparable. There is unfortunately no simple entity called 'the past' that is accessible to us independently of the process of organising our knowledge about it — that is, the knowledge that we have gained on the basis of the evidence that is currently available to us.

What can we say about the notion that mathematics started at some particular time in the past? This is still a problem, and we shall shortly look at various views. To prepare for this, we first survey some of the kinds of evidence that we can draw on.

The Ishango bone. The drawings in Figure 2.2 are of a bone dug up in the 1950s, at an African village called Ishango on the north shore of Lake Edward (one of the headwaters of the Nile, it straddles the border between the Democratic Republic of the Congo and Uganda).[3] Carbon dating has shown the bone to be about 20,000 years old. As you can see, there are marks engraved on it. What are we to make of them?

The bone's discoverer, Jean de Heinzelin,[4] gave an account in which he proposed that the markings may represent an arithmetical game, and that the patterns strongly suggest a

[1] See (Russell 1918, 59).

[2] George Boole was a prominent mathematician and logician of the 19th century.

[3] The bone is preserved in the Royal Belgian Institute of Natural Science, Brussels, Belgium.

[4] See (de Heinzelin 1962, 109–111).

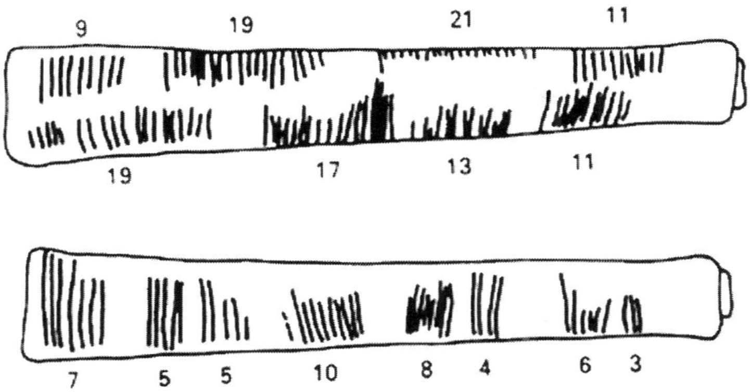

Figure 2.2. The arrangement of notches on the Ishango bone

counting system based on 10 and a knowledge of multiplication by 2 and of prime numbers.[5]

How plausible is this view? There is a big jump between noticing patterns that seem to us to exhibit various arithmetical properties, and claiming that these patterns formed part of the conscious knowledge of the bone's maker. Even though de Heinzelin expressed himself cautiously, the structure of his argument is close to a dichotomy: either the patterning is random and fortuitous, or it shows a knowledge of duplication and prime numbers — and then, assuming it is not the former, it must be the latter.

But he did not explore other possibilities. To bring in prime numbers, a more sophisticated concept than any others he mentioned, forces us to wonder what else we might need to know about the people of the Congo some 20,000 years ago. In other words, we can draw some conclusions about the object only by making some assumptions in the light of the rest of our knowledge of, or beliefs about, the period. The Ishango bone does not speak to us or carry its own interpretation: it is simply a bit of old bone, until we incorporate it into our organised knowledge of the past.

Alexander Marshack[6] criticised de Heinzelin's view and gave a different interpretation of the markings on the bone in terms of time-keeping and lunar count. He took exception to de Heinzelin's suggestion of 'an arithmetical game', on the grounds that this does not properly distinguish adult activities or meanings from childhood ones. So Marshack assumed that the people of that time had a division between childhood and adulthood that grants some cultural significance to the adult uses of arithmetic over and above the intrinsic properties of numbers as such. Marshack seems to be trying to combat the assumption, prevalent in some Western writings of the last century, that 'primitive' peoples were in some sense developmentally akin to children in our culture. Clearly, de Heinzelin's remarks about prime numbers fall by the wayside if Marshack is successful in showing that the bone's markings are better explained in terms of time.

Another view that some people subscribe to is that the Ishango bone has nothing to do with either counting or time-keeping. They believe that the marks on the bone are simply there as an aid for gripping the bone.

Similar problems arise with other bones. One such bone, discovered at Veěstonice in 1937 in what was then Czechoslovakia, is even older than the Ishango bone, dating from

[5]A prime number is a whole number, greater than 1, that is divisible only by itself and 1: the first few primes are 2, 3, 5, 7, 11, and 13.

[6]See (Marshack 1972, 25–27, 31).

around 30,000 BC.[7] It has 57 notches, 55 of which have the same length and are possibly grouped in fives, divided into groups of 25 and 30 separated by two much longer notches. The suggestion that whoever carved it was counting in fives may seem strong, but even if this turns out to be true we have no idea why. Another bone, found in the Lebembo mountains near Swaziland, has 29 notches and dates from around 35,000 BC. But neither separately nor together do these bones speak to us unambiguously.

Although we cannot be certain what significance to attach to such bones, we have learnt something useful and important about historical sources — namely, they do not speak without interpretation. The Ishango example was chosen for this purpose, but the same caution is needed for much later sources, even written ones. To understand past mathematics and past mathematicians from any period, we must assume that there are some similarities recognisable to us, without compromising the sense that people in history lived in different cultural environments and held different beliefs. So we must carefully steer our course between two opposites: the extreme relativism of holding that since the past was so different it is literally incomprehensible to us, and believing that mathematicians in the past were trying to do just what we are trying to do today, but not succeeding so well.

Counting systems. The evidence of archaeological artefacts is not the only source for our attempts to build knowledge about the early period. Another is the early development of our spoken counting system, which is accessible through the study of historical linguistics.[8]

The English language uses distinct counting words for numbers up to ten (or twelve, actually — we return to this later), while for larger numbers it uses words that are built up from multiples of ten: 'eighty-seven', for example, seems to be derived from a contraction of something like 'eight-tens-seven'. There is essentially a consistent decimal counting system, in which 'ten' is the key building block for higher number words; 'hundred' is then a fresh word, and so on. The main anomaly is that the words 'eleven' and 'twelve' do not seem to fit the rest of the pattern: we should perhaps use something like 'one-teen' and 'two-teen', but the actual words do not look or sound close to these. So something strange has happened.

In French, some higher numbers are formed differently. The word for 20 ('vingt') seems independent of the words for 2 ('deux') and 10 ('dix'), and is used in some higher number words such as 'quatre-vingts-sept' (four-twenties-seven) for 87. So French counting words have traces of a twenty-base, as well as a ten-base: an example of the latter is 'dix-huit' (ten-eight) for 18. Indeed, English also has traces of a twenty-base: we still understand the meaning of 'three score and ten', even though such forms are seldom used nowadays.

So our counting system has a structure present in the words we use, and comparing the systems in nearby languages can point up surprising differences. We also see that the counting system is quite a complex construction, with different historical structures overlaying one another. By analysing present languages and tracing their genealogies back into the past, we can make quite strong inferences about how and when our counting system developed. To see this, consider the number words of various languages in the next Table. Of these languages, only English, Welsh, and Basque are in current everyday use. Gothic is an old East Germanic language, and was used in Bishop Wulfila's translation of the Bible

[7] See (Barrow 1992, 31).

[8] The phrase 'our spoken counting system' refers to the words we use to count (one, two, three, ...), as contrasted with the written numerals (1, 2, 3, ...) that represent them. Historical linguistics is the study of how languages have changed over time.

around AD 350. Sanskrit is the dominant classical language of ancient India. It is also the liturgical language of Hinduism, Buddhism and Jainism.

English	Gothic	Latin	Ancient Greek	Welsh	Sanskrit	Basque
one	ains	unus	heis	un	eka	bat
two	twai	duo	dyo	dau	dva	biga
three	threis	tres	treis	tn	tri	hirur
four	fidwor	quattuor	tettares	pedwar	catur	laur
five	fimf	quinque	pente	pump	panca	bortz
six	saihs	sex	hex	chwech	sad	sei
seven	sibun	septem	hepta	saith	sapta	zazpi
eight	ahtau	octo	okto	wyth	asta	zortzi
nine	niun	novem	ennea	aw	nava	bederatzi
ten	taihun	decem	deka	deg	dasa	hamar
eleven	ainlif	undecim	hendeka	un ar ddeg	ekaadasha	hamaika
twelve	twalif	duodecim	dodeka	deuddeg	dvaadashan	hamabi
twenty	twaitigjus	viginti	eikosi	ugain	vimsatih	hogoi

We can gather a number of clues from this table. For instance, our words ‘eleven’ and ‘twelve’, which puzzled us earlier, seem quite close to their Gothic equivalents, but not to anything else here: these Gothic words mean something like ‘one left’ and ‘two left’, in the sense of ‘left over’. Since this construction occurs in most other old northern European languages, we may infer that at an early stage there were northern European tribes whose counting words went up only to ‘ten’. In order to count to ‘eleven’ and ‘twelve’, when needed, they took away ten and counted what was left.

We can even form a hypothesis about the age of our counting system, as seen in its number words up to ‘nine’ or ‘ten’. There are several similarities between the words that become more apparent when you say them aloud. The most striking example, because it is so unexpected, is how close to Latin, or even to Gothic and English, the Sanskrit number words can be made to sound. It thus seems plausible that Sanskrit and the Western languages (except for Basque, which seems rather different) may have had a common root, which from the given dating was before or during the second millennium BC.[9] Since Latin, Greek, Welsh, and Sanskrit seem to have much the same word for ‘ten’, we can reasonably infer that the counting system went up at least to ten by this time. To justify this hypothesis we clearly need to examine much more evidence of languages old and new and other historical information relating to the movement of tribes and peoples and cultural influences.

We shall not survey all the evidence here, but the result is that historians of language generally agree that English, Russian, Persian, Afghan, and most European languages are descended from an original Indo-European language. There is less agreement about when and where it was spoken: a commonly held view asserts that it was spoken around 4000 BC, perhaps somewhere in the area north of the Black Sea. The way that this language was diffused is even more hotly disputed, but military conquest is one method that cannot be ruled out.

A rival, and closely argued, view is that of the archeologist Colin Renfrew who concludes that

[9]A millennium (from the Latin word *mille*, meaning ‘thousand’) is 1000 years, so the second millennium BC ran from 2000 BC to 1001 BC. Similarly, the second century BC (from the Latin word *centum*, meaning hundred) ran from 200 BC to 101 BC.

> the first Indo-European languages came to Europe from Anatolia [Turkey] around 6000 BC ... and they were in fact spoken by the first farmers of Europe.[10]

Similarities in the words for 100 (Latin 'centum', Lithuanian 'simtas', Sanskrit 'satem') suggest that there was a counting word for 'hundred' in the original Indo-European languages; thus our spoken decimal counting system, with counting at least up to one hundred, is perhaps 5000 years old. Counting by twenties, which can be seen residually in French and Welsh, seems to have been the practice in languages such as Basque, which were spoken in Europe before the spread of Indo-European.

Our brief analysis shows that we can begin to build up some knowledge about the past by piecing together several different sorts of evidence.[11] It seems that our spoken counting system may be from 5000 to 8000 years old, a considerable age but yet only a recent part of human history.

What about the origins of counting as an activity in itself? This is not an easy question to answer, and it may indeed have none. One reason for this is that we may never find the evidence to throw sufficient light on the matter. A deeper reason is that the question of when counting started may turn out, on closer analysis, to be meaningless. This would be the case if whatever happened did so over such a long time period, and with so gradual a change of significance, that any attribution that historians could make would be more revealing of their own conceptions of counting than of anything else.

These possibilities are explored in two accounts of the origins of counting by Sir John Leslie and Kenneth Lovell.[12] These contrasting accounts, written over a century apart, bring out clearly the way that historians' perceptions of the past draw upon the interests and presuppositions of their own times. The authors have placed the origin of counting in different phases of human development: Leslie favoured the bloody conflicts of hunter-gatherers, while Lovell preferred peaceful shepherds. More tellingly, Leslie saw counting as having developed along the lines of the classification procedures of natural history, much in vogue during the late 18th century, while Lovell appealed to the more sophisticated notion of a *one-to-one correspondence*, which became prominent in mathematics only at the end of the 19th century.

The moral is not that these are egregiously subjective beliefs about the past, but that building up organised knowledge about the past is a process of dialogue in which present views and concerns are inescapably involved. This is not to say that all historical hypotheses and conclusions are equally valid; some historical judgements are sounder than others, but we must also acquire the skills needed to help us to decide which these are.

2.2 Egyptian mathematics

Ancient Egypt is perhaps best known for its famous pyramids. The largest of these, the Great Pyramid at Giza, erected around 2575 BC, is one of the seven wonders of the ancient world and is an impressive testimony to the mathematical and design skills of its builders. It was built for Khufu, the second King of the Fourth Dynasty, and originally stood 147 metres (481 feet) high, making it the tallest building in the world until the 19th century AD. Its sides are about 230 metres (756 feet) and point accurately in the four cardinal directions of the compass (the base is a perfect square to within one-hundredth of a percent). The burial chamber at its centre is reached by a series of passages, and even carving and laying

[10]See (Renfrew 1987, 288.)

[11]For a more thorough examination of these issues from a number of viewpoints, see (Menninger 1969) and (Renfrew 1987).

[12]See (Leslie 1820, 1–3) and (Lovell 1961, 26–27).

the more than two million stones that make it up with the necessary precision calls for our admiration. The Greek historian Herodotus wrote that he visited it in 453 BC and was told that it took 100,000 men twenty years to build it.[13] These figures, like so much of what Herodotus tells us, are now matters of dispute, but they cannot be far wrong.

The pyramids themselves tell us that there must have been a sophisticated grasp of a number of issues in Egyptian times. Indeed, for many centuries ancient Egypt was seen as the source of wisdom and knowledge, about mathematics as well as other things. There was a long classical Greek tradition to this effect, and in later centuries the indecipherability of the hieroglyphs did nothing to dispel this belief. But since the early 19th century, when the deciphering of the Rosetta Stone by Thomas Young and Jean-François Champollion enabled rapid progress to be made in translating extant Egyptian texts, the picture has changed to reveal a civilisation whose mathematics was more pragmatic and down to earth. In this section, we investigate what we now know of Egyptian mathematics, and how we know it.

Egyptian scripts. How do we know anything about Egyptian mathematics? The following passage, adapted from the British Museum website, has this to say about the Rosetta Stone and its significance:

> Soldiers in Napoleon's army discovered the Rosetta Stone in 1799 while digging the foundations of an addition to a fort near the town of el-Rashid (Rosetta). On Napoleon's defeat, the stone became the property of the English under the terms of the Treaty of Alexandria (1801), along with other antiquities that the French had found.
>
> The decree on it affirms the royal cult of the 13-year-old Ptolemy V on the first anniversary of his coronation (around 196 BC), and is inscribed on the stone three times, in hieroglyphic (suitable for a priestly decree), demotic (the native script used for daily purposes), and Greek (the language of the administration). The importance of this to Egyptology is immense. Soon after the end of the 4th century AD, when hieroglyphs had gone out of use, the knowledge of how to read and write them disappeared.
>
> In the early years of the 19th century, some 1400 years later, scholars were able to use the Greek inscription on this stone as the key to decipher them. Young, an English physicist, was the first to show that some of the hieroglyphs on the Rosetta Stone wrote the sounds of a royal name, that of Ptolemy. The French scholar Champollion then realized that hieroglyphs recorded the sound of the Egyptian language and laid the foundations of our knowledge of ancient Egyptian language and culture.

Thus equipped, Egyptologists can now read ancient Egyptian writings. In their daily lives Egyptian scribes wrote on papyrus, a paper-like medium made from reeds, which has survived only because of the remarkably dry conditions in Egypt, and their writings offer us glimpses of what was going on several thousand years ago. Unfortunately for us, only a dozen or so of the surviving Egyptian papyri are concerned with mathematical calculations, of which the earliest dates from about 1850 BC and the most recent from AD 750. The two major ones are the *Rhind Papyrus*,[14] an impressive document of about five-and-a-half metres (18 feet) long and 30 cm (1 foot) wide which dates from around 1650 BC and is now in the British Museum, and the *Golenischev* (or *Moscow*) *Papyrus*, which

[13] See (Herodotus 1996, 132).

[14] The Rhind Papyrus is named after Henry Rhind, the man who bought it on his holidays in Luxor in 1858.

dates from around 1850 BC and is in Moscow. These two papyri are authentic primary sources, examples of the foundational artefacts on which our knowledge of the history of mathematics is constructed.

The earliest Egyptian script is *hieroglyphic*, used from before 3000 BC until the early centuries AD. Initially an all-purpose script, it was eventually used for monumental stone carving and formal inscriptions. By about 2400 BC it had been superseded by the more fluid *hieratic* script which was used for more rapid writing on papyri. Most of the handful of extant mathematical papyri, including the Rhind Papyrus, are written in hieratic, and many metrological units originally had only a hieratic expression. Later still, there developed a cursive everyday script called *demotic* that looks like a mad doctor's handwriting at the end of a bad day. Because of the great variability of hieratic and demotic writing in the hands of different scribes, Egyptologists habitually transcribe all texts into the more legible and standardised hieroglyphic form. Note that hieroglyphic and hieratic scripts usually read from right to left, so the ordering has to be reversed when we translate it into English.

Hieroglyphic numerals were formed on a straightforward decimal repetitive principle, with each symbol for a power of ten being repeated as often as necessary. These symbols are shown in Figure 2.3 — the symbol for 100 often occurs mirror-reversed.

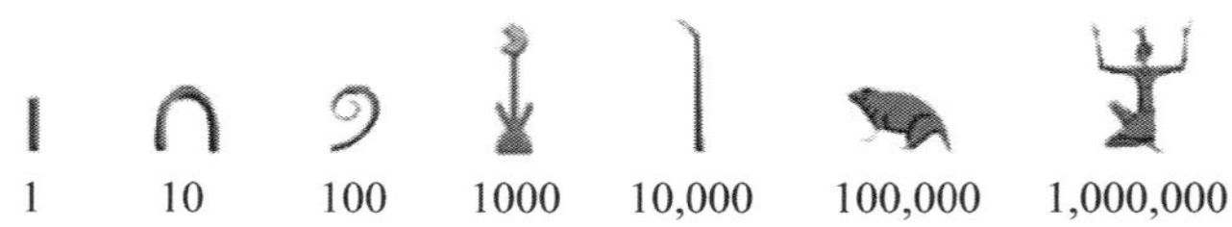

Figure 2.3. Egyptian numerals

In this script the number 1953 would appear as shown in Figure 2.4.

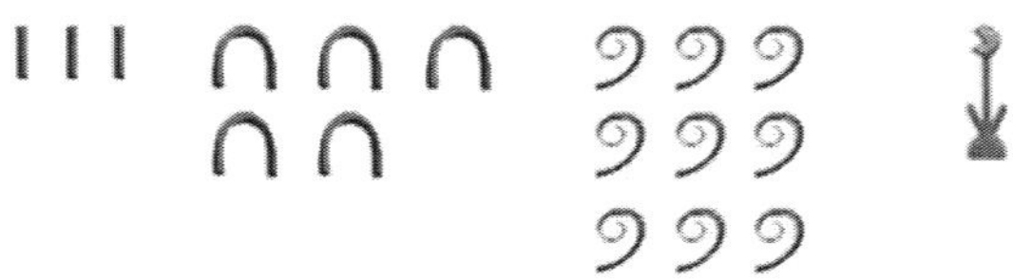

Figure 2.4. A representation of 1953 in hieroglyphic numerals

The representation of numerals in these scripts had an interesting development. In hieratic script, and even more so in demotic script, further new symbols were devised for each of the numbers 2 to 9, 20 to 90, and so on, so there was a great increase in the number of symbols used for different numbers. Some historians see this as a significant step in the development of numeration.

The Rhind Papyrus. We now take our first detailed look at the Rhind Papyrus, in order to understand how whole numbers were written, how multiplication and division were carried out, and how fractions were denoted. As we shall see it contains tables expressing the fractions $\frac{2}{5}, \frac{2}{7}, \frac{2}{9}, \ldots, \frac{2}{101}$, and then presents eighty-four problems of various kinds. However, for most of us there is an obstacle: the Rhind Papyrus is not meaningful until it has been translated. It is instructive to think about what we can do about this, because it is not generally practicable for us to learn afresh each new language or script of those cultures whose mathematics interests us. Here we must rely on the knowledge of Egyptologists for the material from which we can start to build our own understanding. Even when we have a translation we must still go through a process of interpretation.

In A.B. Chace's edition of the Rhind Papyrus, from which we have taken our extracts, each page of text has three sections. The top section is a handwritten copy of the papyrus text, written in hieratic. The middle section is a transcription of the text into hieroglyphic. The bottom section contains a literal translation into English and our modern numerals. At this stage the translation is not clear: not only is the calculation of an unfamiliar kind, but even what the problem is has yet to be put in an accessible form.

Let us start with Problem 40 (see Figure 2.5) and see whether we can work out what it means.[15]

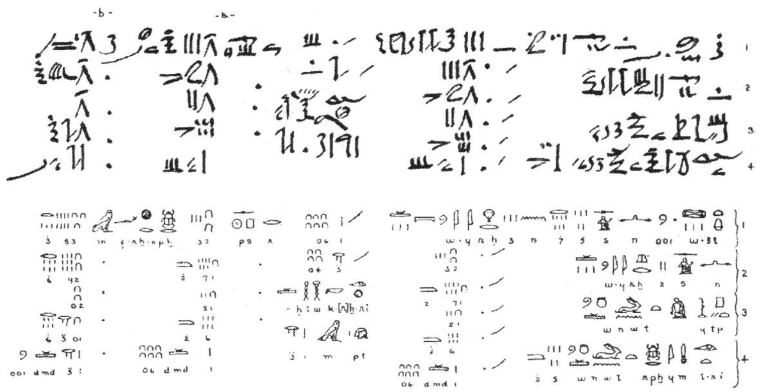

t;·w 100 n s 5 7̇ n 3 ḥry·w n s 2 ẖry·w pty twnw
Loaves 100 for man 5, $\frac{1}{7}$ of the 3 above to man 2 those below. What is the difference of share?

ir·t my ḫpr twnw 5 2̇ ir·ḫ[r]·k w;ḥ–tp m 1 3̈
The doing as it occurs. The difference of share being 5 $\frac{1}{2}$, Make thou the multiplication: 1 $\frac{2}{3}$

			r up to	sp times	ḫpr·ḫr·f m becomes it:	
\1	23			23		38 3̇
\1	17 2̇		"	17 2̇	"	29 6̇
\1	12		"	12	"	20
\1	6 2̇		"	6 2̇	"	10 3̈ 6̇
\1	1		"	1	"	1 3̈
dmd Total	60		dmd Total	60	dmd Total	100.

\1	60
\3̈	40

Figure 2.5. Rhind Papyrus, Problem 40

Rhind Papyrus, Problem 40.

> Loaves 100 for man 5, $\frac{1}{7}$ of the 3 above to man 2 those below.
> What is the difference of share?

This seems to be saying that a hundred loaves are to be divided between five men in a particular way. A careful look at the wording shows that each man is to receive a different number of loaves so that the two men who receive the least end up with a seventh of what the other three men get between them. Further, it requires that each man receives a fixed

[15]The numbering of the problems is that of a 19th-century German editor of the Rhind Papyrus; in the original, the problems are unnumbered.

number of loaves more than the next man. The scribe's way of putting the problem had at least the virtue of succinctness!

We shall leave the actual calculation for now, and return to it when we have looked at some more general principles governing the ways that Egyptians handled numbers, as inferred from the evidence of the Rhind and other papyri. As already mentioned, evidence from such papyri extends over a long period, although it is rather scanty and is reinforced by evidence from only a few other artefacts, such as bits of pot, tiles, and stone inscriptions. From the limited evidence we have, however, the fundamental spirit and mathematical approach of Egyptian mathematics seem to have changed very little over at least two thousand years.

Egyptian arithmetic. The most frequently used basic operations in any Egyptian calculation were doubling and halving; increasing and decreasing by a factor of ten was also common.

Here is an example of a calculation from the Rhind Papyrus (Problem 69, see Figure 2.6). To understand it, you will need to look back at Figure 2.3 to see what numbers the Egyptian numerals stand for. You will find it transcribed into our numerals below, but see if you can work out what is going on.

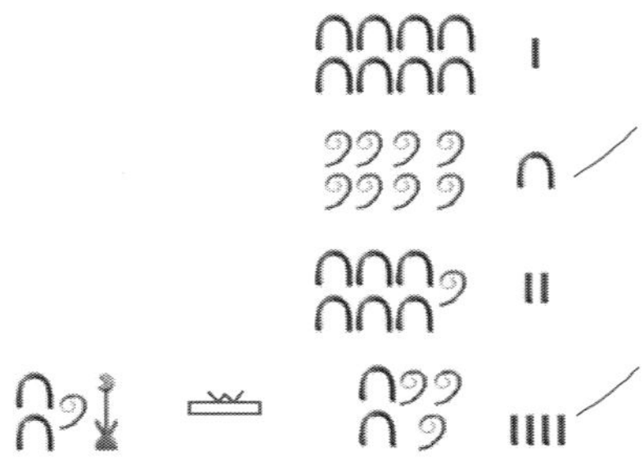

Figure 2.6. Rhind Papyrus, Problem 69

The transcription looks like this:

	80	1	
	800	10	/
	160	2	
1120 makes	320	4	/

The sum looks like a multiplication of 80 by 14: the scribe has considered the latter as 10 and 4, multiplied 80 by each (in the lines marked by /) and added the results to obtain 1120. Notice that to multiply 80 by 4, he first doubled 80 and then doubled it again. Notice also that multiplying by 10 is very easy in hieroglyphs; simply change each symbol into the next one in the ordered sequence of symbols for powers of 10 (although it would have been harder in hieratic!). The method is quite simple, and you can use it to compute with today. Here the basic idea is that of proportionality or scaling up. In this case, 80 has been scaled up by 10, 2, 4, and finally 14.

Division is treated in a similar way. The problem is seen in terms of discovering by how much one number must be multiplied to make another. So, for instance, the above calculation could equally have stood for the division of 1120 by 80: we scan down selected multiples of 80 to see which of them add up to 1120. This yields the answer 10 and 4 — that is, 14.

Of course, the answer to such a division sum may not always be exact if the calculation is restricted to whole numbers. Exploring what happens when a division does not yield a whole-number answer leads us to one of the most striking and influential features of Egyptian mathematics — fractions.

Nearly a third of one side of the Rhind Papyrus is taken up with calculating the doubles of the odd fractions $\frac{1}{3}, \frac{1}{5}, \frac{1}{7}, \frac{1}{9}, \ldots, \frac{1}{101}$, expressed in the equivalent form of dividing the number 2 by the odd numbers from 3 to 101. We can write these as 2:3, 2:5, and so on; for instance, in modern notation, some early results read:[16]

$$
\begin{array}{llll}
2:11 & \frac{1}{6} & \frac{1}{66} & \\
2:13 & \frac{1}{8} & \frac{1}{52} & \frac{1}{104} \\
2:15 & \frac{1}{10} & \frac{1}{30} & \\
2:17 & \frac{1}{12} & \frac{1}{51} & \frac{1}{68}
\end{array}
$$

We shall not pursue the details here of how these were calculated, but present another circumstance in which fractions can be expected, and we shall see that they wrote the fractions in the above form. Consider the following problem, this time going straight to the English translation.

> *Problem 24.* A quantity, $\frac{1}{7}$ of it added to it, becomes it: 19.

This problem can be seen to mean something like: A quantity and the seventh part of it is 19: what is the quantity? (In modern terms, $1\frac{1}{7}$ of the quantity is 19 — the solution is $16\frac{5}{8}$.)

As part of his solution of this problem (we complete our account of this problem later) the scribe computed 19 divided by 8, proceeding as follows.

$$
\begin{array}{ll}
8 & 1 \\
16 & 2/ \\
4 & \frac{1}{2} \\
2 & \frac{1}{4}\ / \\
1 & \frac{1}{8}\ /
\end{array}
$$

The scribe doubled 8 to give 16, which is 3 short of 19, so he had to find what multiplies 8 to produce the remaining 3. He halved the 8 (obtaining 4, which is still too much), then halved that and halved it again. Now the quarter of 8 and the eighth of 8 together make up the 3 that he wants. So his answer to this part of the problem, as indicated by the / marks, is the sum of 2 and $\frac{1}{4}$ and $\frac{1}{8}$; for the Egyptian scribe, this is $2\frac{1}{4}\frac{1}{8}$.

Perhaps the most striking thing about this procedure is that the scribe did not add the fractions together to produce $2\frac{3}{8}$, as we might do. In Egyptian mathematics, only *unit fractions* or *reciprocals* are used, such as $\frac{1}{2}, \frac{1}{3}$, and $\frac{1}{8}$, together with the fraction that we write as $\frac{2}{3}$. Why there should have been this seemingly strange restriction of Egyptian mathematics is an interesting question to which we return shortly.

The importance of the principle of doubling, combined with the use of unit fractions, made it necessary for scribes to know the result of doubling fractions. This is one reason why the table of results for 2:3 to 2:101 was important enough to write down.

[16]These should be interpreted, in modern notation, as $\frac{2}{11} = \frac{1}{6} + \frac{1}{66}$, $\frac{2}{13} = \frac{1}{8} + \frac{1}{52} + \frac{1}{104}$, etc.

Unit fractions. One thing to notice about the above results is that what might seem to us the 'obvious' answer — for instance that 2:17 is $\frac{1}{17}$ $\frac{1}{17}$ — was never one that the Egyptians presented. The possibility of repeating a fraction in order to double it did not count as an answer, it seems, for the Egyptian scribe. This observation gives us a clue concerning the question raised earlier, as to why we find mainly unit fractions in Egyptian mathematics. The answer to this question is important and revealing, because thinking it through informs us not only about Egyptian mathematics, but also more generally about what studying history can involve.

The question arose because it suggested itself from the source material and from our efforts to understand it. It seems that the Egyptians used fractions, yet only in a seemingly highly restricted way that drove them into subtle and complicated contrivances. Why did they not use them as we do? It would have saved them a lot of trouble! But notice that the more we spell out the problem, the more peculiar it begins to look. Imagine yourself asking the scribe, 'Why do you use only *unit* fractions?'. His look of bewilderment is explained if you go on to reflect that this question presupposes him to be making the deliberate choice of not using fractions in the way we do. Expressed like this, it is surely an unreasonable presupposition. So how have we got into this problem, and how do we get out of it?

The problem seems to have arisen partly conceptually and partly notationally — conceptually, because the very use of the word *fraction* carries our arithmetical expectations with it, whereas if we had used another word such as *reciprocal* then no-one would have expected us to try to combine them, and notationally, because the translation of an Egyptian symbol into 'our' fractions may set up unhelpful expectations. If you look back at the statement of Rhind Papyrus Problem 40 (Figure 2.5) you will see that our fraction $\frac{1}{7}$ is translated from a hieroglyphic symbol, which in turn is equivalent to a hieratic squiggle with a dot on top. The Egyptian form does not readily lend itself to writing other than unit fractions.

The upshot of this argument is that our initial question, of why Egyptian fractional usage was 'restricted' to unit fractions, was badly posed. It need not arise if we reconsider the question as: What was the Egyptian concept for which we have hitherto used the word 'fraction'?

The Egyptian symbols for what we have been thinking of as $\frac{1}{3}$, $\frac{1}{4}$, ... are better translated by the word 'part', as in 'the third part', 'the fourth part', and so on. The reason that these are better is because to say 'one-seventh' or use the symbol $\frac{1}{7}$ invites the idea familiar to us of there being 'two-sevenths' ($\frac{2}{7}$), 'three-sevenths' ($\frac{3}{7}$), and so on, whereas 'the seventh part' is unique. So this would explain why the result of doubling the eleventh part is not the eleventh and the eleventh; for there can be only one eleventh part. To put it another way, to write 2:11 would simply be a restatement of the problem, not an answer to it.

Thus equipped, we can more easily see how a calculational practice involving parts could form a coherent framework in itself, not needing the overtones of our concept of fractions. That it was coherent is attested by its extraordinary longevity. Two millennia after the Rhind Papyrus, for instance, we find many of the results in Ptolemy's *Almagest* (c. AD 150), the greatest of ancient astronomical texts, presented in Egyptian fashion, as in this passage:[17]

> One sighting we made in the year 18 of Hadrian, Egyptianwise Pharmouthi 2–3, according to which the morning Venus was at its greatest elongation from the Sun; and, sighted with the star called Antares, it was $11 + \frac{1}{2} + \frac{1}{3} + \frac{1}{12}^{\circ}$ within the

[17]Ptolemy, *Almagest* (1952), Book X, Chap. 3.

Goat, while the mean Sun was then 25 and $\frac{1}{2}^{\circ}$ within the Water Bearer. And so the greatest morning elongation was $43 + \frac{1}{2} + \frac{1}{12}^{\circ}$.

This discussion of unit fractions suggests that past mathematical procedures and concepts may be easier to understand if we do our best to meet them on their own terms. In that spirit, let us look again at Egyptian calculation methods. We return to Problem 24 from the Rhind Papyrus.

> *Problem 24* (again). A quantity, $\frac{1}{7}$ of it added to it, becomes it: 19.

The scribe solved this by producing the number 7 as a trial guess having the useful property that its seventh is easy to determine:

1	7
$\frac{1}{7}$	1

So had the quantity been 7, it and its seventh would make 8; but what we want is for it and its seventh to make 19; the scribe has now to scale up the numbers. If we knew how many times 8 goes into 19, we would know how many times 7 goes into the quantity we want. This part of the calculation, dividing 19 by 8, is what we did before, with answer $2\frac{1}{4}\frac{1}{8}$. So all that now remains is to multiply this by 7 to get the answer, which is done by doubling.

1	2	$\frac{1}{4}$	$\frac{1}{8}$
2	4	$\frac{1}{2}$	$\frac{1}{4}$
4	9	$\frac{1}{2}$	

The sum of these is $16\frac{1}{2}\frac{1}{8}$, which is therefore the desired 'The doing as it occurs'. The scribe added to the quantity its seventh ($2\frac{1}{4}\ \frac{1}{8}$) to show that their sum is indeed 19, as required.

The strategy followed in this problem — choosing any convenient number and working out the solution with respect to it — then scaling up by the ratio of your answer to the answer you want — is a common one in Egyptian mathematics. Indeed, this systematic method seems characteristic of the Egyptian style of arithmetic. For example, if you look again at the way division was described you will notice that it can be seen in terms of getting close to the desired result and then seeing what needs to be done to make up the remainder. This technique has become known as 'the method of false position'.

We conclude this brief account of Egyptian mathematics with one final problem from the Rhind Papyrus, number 48.

> *Problem 48.* Compare the area of a circle and of its circumscribing square.

The circle of diameter 9.		The square of side 9.	
1	8 setat	\1	9 setat
2	16 setat	2	18 setat
4	32 setat	4	36 setat
\8	64 setat	\8	72 setat
		Total	81 setat

On the right the scribe has explained how to calculate the area of a square of side 9, and finds that the answer is 81 *setat* — a setat is a unit of area equal to about 2576.5 square meters. What proportion of the square is occupied by the inscribed circle of diameter 9? The answer is that the circle occupies 64 *setat*. This very much suggests that the rule for giving the area of a circle of known diameter (such as 9) was to reduce the diameter by a factor of $8/9$ (giving 8) and then square the resulting amount — which is not a bad estimate in numbers that are easy to use.

Other geometry problems from the Rhind Papyrus involved rectangles, triangles, and pyramids (see Figure 2.7).

Figure 2.7. A geometry problem from the Rhind Papyrus

Who used the Rhind Papyrus? Now that we have some idea of what the Rhind Papyrus contains, we can raise the question of who, or what, it was for. What was the scribe's intention? His opening words were, in Chace's translation (which preserves the word order of the original):[18]

> Accurate reckoning of entering into things, knowledge of existing things, all mysteries …secrets all. Now was copied book this in year 33 month four of the inundation season [under the majesty of the] King of [Upper and] Lower Egypt, A-user-Rê, endowed with life, in likeness of old made in the time of the King of Upper [and Lower] Egypt [Ne-ma]et [Rê]. Lo the scribe A'h-mose writes copy this.[19]

This information has allowed scholars to date the papyrus to the 33rd year of the reign of Apophis, the last king but one of the Fifteenth Hyksos Dynasty (1650–1550 BC) and to say that it is a copy of a Twelfth Dynasty original from about 1985–1795 BC. The words about knowledge and secrets are interesting, if only because they tell us how important this material was, but they do not really clarify for us whether this kind of thing was taught to schoolchildren, or whether the Rhind Papyrus represented an advanced theoretical work by Egyptian standards.

---

[18]Chace (1927, 1).

[19]Chace added that the scribe's name means A'h [the Moon = god] is born.

When trying to understand this ancient mathematical activity, it is natural for us to ask how it compares with later mathematics. It would be interesting to know whether it was simply a collection of empirical computational techniques, brought together for applying to everyday practical problems, or whether there are visible traces of, say, the more abstract theoretical tone that you will meet in Greek mathematics. Historians have differed in the judgements they have reached on this question. For instance, G. J. Toomer has claimed that:[20]

> The Rhind and Moscow papyri are handbooks for the scribe, giving model examples of how to do things which were a part of his everyday tasks ... The sheer difficulties of calculation with such a crude numeral system and primitive methods effectively prevented any advance or interest in developing the science for its own sake. It served the needs of everyday life ... and that was enough.

while Chace has remarked that:[21]

> A careful study of the Rhind Papyrus convinced me several years ago that this work is not a mere selection of practical problems especially useful to determine land values, and that the Egyptians were not a nation of shopkeepers, interested only in that which they could use. Rather I believe that they studied mathematics and other subjects for their own sakes.

Which of these arguments do you find more convincing? On the limited evidence of Problems 24 and 40, there is perhaps more to be said for Toomer's view: the problem of dividing loaves among a group of men could certainly have been an everyday task for the overseers of large building works, and your experience of Egyptian calculation may well have inclined you to the view that 'crude' and 'primitive' are understandable epithets. But there are features of the problems that are harder to fit with everyday life in such a straightforward way. For instance, the circumstances under which real loaves are to be divided among a group of people as in Problem 40 are hard to imagine. Again, the fact that Problem 24 is posed in terms of 'a quantity'[22] does imply a certain level of abstraction, a recognition that the same techniques and rules apply to any of a range of real-world objects. So there is some recognisable mathematical activity 'for its own sake', which was Chace's claim. Indeed, the Rhind Papyrus is organised in a way that supports this claim: first we have the $2 : n$ table, which is used for everything, then problems about numbers, then problems in geometry, and finally problems about weights and measures.

Let us reflect on the situation where historians reach apparently contradictory views. It is immediately helpful to check out the judgements against any relevant source material you are aware of. In this case we did not reach any very firm conclusion, though, either because we have not yet looked at enough evidence, or perhaps because moving from the evidence to a judgement upon that evidence is more difficult than appeared at first sight. So we should go on to investigate more fully what evidence, and what arguments from that evidence, these historians used to reach their conclusions.[23] Note that what we are doing has more general import: our interest is not only in the pros and cons of this particular question, but also in trying to work towards a method that can apply in other contexts in order to evaluate historical judgements.

It seems that neither historian can be endorsed wholeheartedly. Chace's argument seems flawed by the point that unrealistic problems and answers do not in themselves

[20]See (Toomer 1971) in F&G 1.D7.

[21]See (Chace 1927) in F&G 1.D6.

[22]The Egyptian word can also be translated as 'heap' — something rather unspecific.

[23]See (Chace 1927, 42–43) in F&G 1.D6 and (Toomer 1971, 37–40, 45) in F&G 1.D7.

imply a study of mathematics 'for its own sake'. Toomer's argument is difficult because he made his judgement of Egyptian mathematics partly by means of a comparison with the Babylonians. Indeed, he seems engaged in a broader question than Chace was, one more to do with assessing how much the Egyptians contributed to the development or advance of mathematics.

What this exercise has also shown is that the endeavour to understand past mathematics naturally leads to our forming judgements about it, in much the same way as art history and art criticism are cognate activities. Some historians are disappointed that a civilisation favoured by the gods in so many ways did not contribute more advanced mathematics. There was perhaps no need for more sophisticated mathematical investigation, nor any perception of such a possibility. As the historian Otto Neugebauer has agreeably remarked:[24]

> Of all the civilisations of antiquity, the Egyptian seems to me to have been the most pleasant. The excellent protection which desert and sea provide for the Nile valley prevented the excessive development of the spirit of heroism which must often have made life in Greece hell on earth.

2.3 Mesopotamian mathematics

Mesopotamia — the valley of the two rivers, Euphrates and Tigris, which lead into the Persian Gulf — nourished cultures that were contemporary with those of ancient Egypt and maintained large cities and a sophisticated agriculture.[25] As in Egypt, these societies were run by royal dynasties; religions were sustained by priests, armies were maintained, and people were taxed to provide revenue for it all.

To keep this going, systems for keeping records evolved, and fortunately for us the preferred medium was wet clay upon which impressions could be made with sharpened sticks. The clay was then allowed to dry, and a record was thus created that was not only adequate for the purposes for which it was invented, but which was capable of surviving to the present day — unlike papyri, which perish easily. Indeed, Mesopotamia provides us with our most ancient evidence for writing, from perhaps as early as 4000 BC as some scholars claim. This system evolved over the centuries to a fully written script, but curiously the records we have suggest that it began with a system of about 1000 signs, as if it were a deliberate construction. As a result, we have a whole profusion of clay tablets about economic and administrative matters — 85% of the early writings are about economic matters, which has suggested to some that the system was created to cope with economic demands. Other tablets form literary, religious, and scientific works — including the flood myth that later became part of the biblical book of Genesis — there are word-lists, and there are many thousands of mathematical problems and tables.

By the middle of the third millennium BC, the writing style had evolved into a highly abstract *cuneiform* ('wedge-shaped') script. This script, used initially for writing down words in the Sumerian language, was later also adopted by neighbouring peoples. All of the Mesopotamian tablets are written in Akkadian, a Semitic language quite different from Sumerian, although some tablets do use Sumerian words.

Most of the texts that give us our fullest understanding of Mesopotamian mathematics — indeed, of any mathematics before the Greeks — date from about 1800–1600 BC. During this period, King Hammurabi unified Mesopotamia out of a rabble of small city-states

[24] See (Neugebauer 1969, 71).

[25] The word 'Mesopotamia' means 'between the rivers' in Greek.

(the remains of a previous empire) into a new empire whose capital was Babylon, which was on the Euphrates sixty miles south of present-day Baghdad.

Box 2.1. 'Mesopotamian mathematics' and 'Babylonian mathematics'. The term 'Mesopotamian mathematics' is used here to refer to the mathematics of the Mesopotamian region in antiquity, which covers the period from about 3200 BC to the end of the Seleucid empire in 125 BC. Between about 1850 BC and 1600 BC the region was dominated by the Babylonian kingdom, and many cuneiform mathematical tablets have survived from that period. To distinguish this kingdom from what is called the neo-Babylonian kingdom of more than a millennium later the first kingdom is often called the *Old Babylonian kingdom*, and tablets from that period will be labelled *Old Babylonian*.

An example. Before seeing how our knowledge has been acquired, let us see what a mathematical problem looks like when a modern cuneiform scholar has translated a tablet.[26]

The following example is taken from the clay tablet YBC 4652, now at Yale University. It was translated by Otto Neugebauer and Abraham Sachs, two scholars who helped bring about the first vigorous move to translate Mesopotamian mathematical tablets in the years between 1930 and 1945. Words in square brackets are their suggested reconstructions of what the tablet says (where it is damaged), and words in parentheses are the translator's additions so that the English is relatively more understandable.[27]

> I found a stone, (but) did not weigh it; (after) I subtracted one-seventh, added one-eleventh, (and) subtracted one-thir[teenth], I weighed (it): 1 ma-na.
>
> What was the origin(al weight) of the stone? [The origin(al weight)] of the stone was 1 ma-na, $9\frac{1}{2}$ gin (and) $2\frac{1}{2}$ she.

This tablet contained 22 such problems and answers, none indicating how the answer was reached, and each involving a stone of 1 ma-na (see Figure 2.8).[28]

We can make little progress without knowing how the units of weight are related — there are in fact 60 gin to 1 ma-na, and 180 she to 1 gin — but we can reach some conclusions. For example, could the text be describing a practical problem? Clearly not: if he were really interested in its weight, the scribe would have done better to have weighed the stone when he found it!

This being the case, what might the tablet have been for? It seems rather similar to the Rhind Papyrus Problem 24, with a stone in the place of a quantity, and it is formulated in terms of unit fractions. So, on the evidence of this tablet at least, similar things seem to have been taking place mathematically in Egypt and Mesopotamia, at much the same time.

The observations that there are so many similar problems on the tablet and that no working is shown suggest that it may have been for teaching purposes, with the method explained verbally. Just what was being taught is unclear, however, and various possibilities spring to mind: it could have been the method of solving problems like this, or it could

[26] We return to the issue of translations later.

[27] For the tablet see (Neugebauer and Sachs 1945, 100–101); see also F&G 1.E1 for more from the same tablet and a commentary. YBC stands for Yale Babylonian Collection.

[28] 1 ma-na is about 500 grams.

Figure 2.8. The tablet YBC 4652 with an indication of its size

have been the learning of units of weight, for the solution emerges only if these are understood correctly. It could also have been a question of how to handle these unit fractions — they are all awkward, in that they do not divide into the weights: one-seventh of a ma-na is not a whole number of gin or of she.

There is one further point that we should mention, in case you tackled the problem and did not obtain the scribe's answer. The fractions in the question are not all parts of the original stone, but are parts of whatever the previous step has been. So it is the stone less its seventh, plus the eleventh of all that, and so on. This makes for a slightly more complicated calculation than most of the otherwise similar problems from the Rhind Papyrus.

Historical scholarship and Mesopotamian scripts. So how did historical study reach the stage where Neugebauer and Sachs could pick up a tablet in a library and translate it with a fair degree of understanding?

As with Egyptian hieroglyphs, cuneiform studies date from the 19th century. Their equivalent of the Rosetta Stone was a sheer rock-face at Behistun in south-western Iran into which a text was carved in three languages (Old Persian, Elamite, and Babylonian), proclaiming the victories of Darius the Great (520 BC). The British Consul in Baghdad, Henry Rawlinson, rediscovered this inscription and between 1835 and 1851 copied it (at the risk to his life that any amateur mountaineer faces 300 feet up a precipice) and began to decipher both the script and the languages.

Shortly thereafter, the burgeoning science of archaeology resulted in excavations of cuneiform tablets from ancient sites in Mesopotamia. These have sometimes been unearthed in vast quantities, with the result that many more tablets are available in museums and universities throughout the world than have been translated or even catalogued. By 1900 a considerable re-evaluation of the Mesopotamians was under way, as they moved out of the domain of biblical studies (as in the Tower of Babel), and came to be seen as a series of cultures in their own right.

Only a small proportion of these cuneiform tablets have been shown to have mathematical content, perhaps a few thousand of the several hundred thousand extant tablets. The study of these from the 1920s to the 1950s led to the recognition that the mathematical attainments of Mesopotamia put those of the Greeks of 1200 years later into a fresh perspective. For a time it was thought that there were clear signs that Mesopotamian mathematics had influenced the early years of Greek mathematics, around 600 BC, but most historians now believe that there was probably no such influence.

Lessons from Larsa. The earliest understanding to emerge was that of the remarkable numeration system of Mesopotamia. This discovery was also due to Henry Rawlinson, who in 1855 studied a tablet from the ancient city of Larsa. Some of its main features are identifiable, as you can see from the drawing in Figure 2.9.

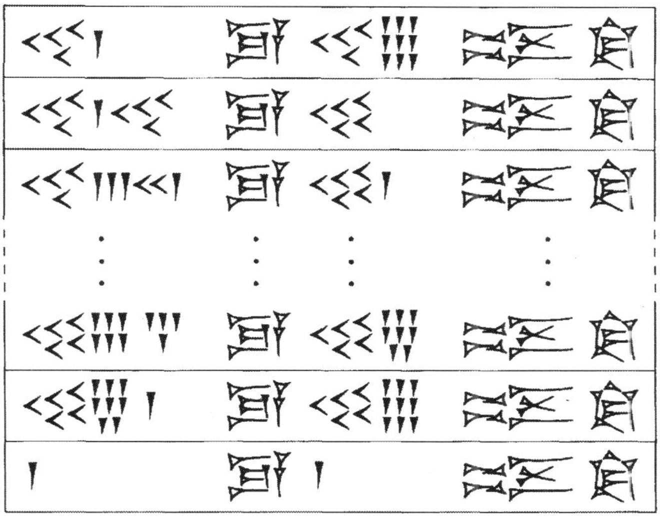

Figure 2.9. Drawing of part of a clay tablet from Larsa

This tablet seems to consist of four columns, of which the second and fourth do not change, but the first and third do. The third column, especially, changes so regularly that we may infer that it presents successive numbers, and is constructed on a principle like that of the Egyptian hieroglyphic numbers. If 𒁹 represents 1, and 𒌋 is 10, then the third column gives the numbers 49, 50, 51, ..., 58, 59, and then 1, for a reason that is not yet clear. The first column is less regular and has the curious feature that like symbols are not all collected together: the third line, for instance, has four 10s, then three 1s, then two 10s, then one 1. Rawlinson realised that all could be consistently explained if he dropped the assumption that a number sign can represent only one number value. So he suggested that the 1-symbol at the foot of the third column should be understood as 60, and that the third line's first-column number was forty-three 60s and twelve 1s; this seemed to be, then, a place-value system (see below), in which the value of each component number symbol depends on its position in the numeral as a whole.

With this in mind we can try, as Rawlinson did, to transcribe the Larsa tablet and to suggest what the cuneiform symbols in columns two and four might mean. To do this, we should form some initial hypothesis about the relationships among the numbers and see whether this is borne out elsewhere: for example, in the second line, what relation does '41 60s and 40' have to 50? Then we should see whether the third line confirms this, and so on through the whole tablet.

Since 41 60s and 40 come to 2500 in our numerals, which is the square of 50, we might hope the next line is the square of 51 (that is, 2601), which indeed is 43 60s and 21.

It looks as though our hypothesis is on the right lines. Now we can go back to the first line and note that if the first number is to be 49 squared (2401), then the cuneiform symbols cannot be 41 'anythings' (even though that is what they look like), but must instead be 40 60s and 1 unit.

It seems safe, then, to infer that column two carries the meaning 'is' or 'equals', and column four means 'squared', so the first line reads '2401 equals 49 squared', and the last line reads '3600 equals 60 squared'.

There is one further thing to remark about our interpretation of the tablet. Suppose that all the numbers but one, say, had fitted our conjectured pattern: how should we respond to the inconsistent entry? It is possible that some other much more complicated interpretation could be found to cover every number without exception — but historians generally adopt the simpler view that the scribe must have made a mistake. Primary sources are not necessarily 'correct' merely by virtue of being old!

The sexagesimal place-value system. We write our numerals in the *decimal place-value system*, based on 10 — each place has value 10 times the next. For example, our number '35' means $(3\times 10)+(5\times 1)$, while '305' means $(3\times 100)+(0\times 10)+(5\times 1)$. Also, '88' means $(8 \times 10) + (8 \times 1)$, where the value of the first 8 is 10 times that of the second 8, so the same symbol can represent different numbers, depending on its position. Note that we have enough distinct symbols 1, 2, ..., 9, 0 to put each one in each place without repetition; for example, we can write 3 rather than writing 1 three times.

For their mathematics Mesopotamian scribes used a 'sexagesimal place-value system', based on 60 — each place has value 60 times the next. But, as we have seen, it differs from our system in that it uses a 'floating sexagesimal point', giving no indication of the exact value of the number; for example, <<< could be

$$30\ (= 30 \times 1) \text{ or } 1800\ (= 30 \times 60) \text{ or } \tfrac{1}{2}(= 30 \times \tfrac{1}{60}) \text{ or } \tfrac{1}{120}(= 30 \times \tfrac{1}{3600}), \text{ etc.}$$

In any particular situation where the exact value of the number was significant, this would be clear to the scribe from the context.[29] Although this initially makes it harder for us to read the tablets, it seems to have given the Mesopotamians unprecedented flexibility in their calculations. Among other things, there is no symbolic distinction between 'whole numbers' and 'fractions'.

It follows that there is a translation problem in the task of finding our equivalents for what the scribe wrote down. A helpful notation was devised by Neugebauer, who represented the value within each sexagesimal place in our numerals, separating the places by commas; for example, the entries in the first column of the Larsa tablet would be transcribed as

$$40,01 \quad 41,40 \quad 43,21 \quad \text{and so on.}$$

This leaves unspecified the exact value of each number, just as the scribe did. However, if we have reason to believe where the 'integer part' of the number ends and the 'fractional part' begins, then we use a semi-colon to separate them; for example,

$$1,10;30 \text{ represents } 60 + 10 + \tfrac{30}{60}, \text{ which is } 70\tfrac{1}{2} \text{ in our decimal system.}$$

In spite of its disadvantages, the Mesopotamian notation completely sidesteps the relatively cumbersome Egyptian technique of handling fractional parts and, together with the use of multiplication tables, leads to computations that arguably can be even smoother than

[29]We similarly use context in this way — for example, if we say 'six-fifty' we might refer to time (10 minutes before 7 o'clock) or to cost (a taxi ride costing £6.50 or a long plane flight costing £650), depending on the context.

our own. This system, which appears to have been developed during the highly bureaucratic centralisation of 2100–2000 BC, was used consistently only within mathematics. In dating, weights and measures, economic records, and the like, there was a wide mixture of units with many local variations. In fact, the place-value system was used only for intermediate calculations and in mathematical texts proper; in any real-life application of mathematics (and sometimes, when given numbers or results were stated in mathematical texts) other, unambiguous, notations are used.

Another table. Now let us explain this notation, and see more of the flexibility of Mesopotamian calculations. Below we display an array of numbers from a different tablet, transcribed in Neugebauer's notation.[30]

2	30	16	3, 45	45	1, 20
3	20	18	3, 20	48	1, 15
4	15	20	3	50	1, 12
5	12	24	2, 30	54	1, 06, 40
6	10	25	2, 24	1	1
8	7, 30	27	2, 13, 20	1, 4	56, 15
9	6, 40	30	2	1, 12	50
10	6	32	1, 52, 30	1, 15	48
12	5	36	1, 40	1, 20	45
15	4	40	1, 30	1, 21	44, 26, 40

To see what this table is about, try multiplying the corresponding numbers in the first pair of columns and see whether a pattern emerges; for example, in line 6, multiply 8 by 7, 30. Now, 7, 30 may mean 7; 30 ($= 7\frac{1}{2}$), or 7, 30 ($= 450$), or 0;7, 30 ($= \frac{7}{60} + \frac{30}{3600} = \frac{450}{3600} = \frac{1}{8}$), and when we multiply by 8 we get 60 or 3600 or 1, respectively. But these are all represented by the same cuneiform symbol, and so it does not matter which one we choose.

You can reach an identical conclusion by looking at any other pair. In short, the table shows numbers that multiply together to give 60 (say), so we could call it a 'reciprocal table': the numbers in the right-hand columns are, in effect, 60 (or 1) divided by the left-hand numbers.

Next, we observe that in the left-hand columns, certain numbers such as 7, 11, 13, etc., do not appear. When this happens, the historian naturally looks for an explanation, and one that must make sense for the scribes and their audience at the time.

If we try to divide 7, 11, or 13 into 60, and to express the result in sexagesimal fractions, we will find that it does not work exactly; the process is never-ending. Once we have noticed this, we can go back to the table and see that the only numbers appearing are those that are the results of multiplying together 2s, 3s, and 5s in various combinations; so 50 ($= 2 \times 5 \times 5$) is in the table, but 21 ($= 3 \times 7$) is not. Since 2, 3, and 5 are the only prime factors of 60 ($= 2 \times 2 \times 3 \times 5$), we deduce that only those numbers whose prime factors are 2, 3, or 5 have reciprocals that can be expressed as terminating sexagesimal fractions. Following Neugebauer, we call such numbers 'regular numbers'.[31]

---

[30]This tablet is labelled BM 106444, meaning that it is number 106,444 in the collection in the British Museum.

[31]The sexagesimal place-value system has more regular numbers than our decimal system, for in our decimal system any number with prime factors other than 2 and 5 has an unending decimal fraction; for example, $\frac{1}{3} = 0.333\ldots$

There are a couple of further points to make at this stage. Historians discuss whether this is to be interpreted as a table of reciprocals, as suggested above, or whether its function was as a conversion table of fractional parts into their sexagesimal equivalents; so 2 would stand for 'the second part', 3 for 'the third part', and so on, down to 1,21 for 'the eighty-first part', in something like the Egyptian mode. In this interpretation, the columns would not be of numbers related reciprocally, but of the same number expressed in two different ways, as a unit fraction and as a sexagesimal fraction. This is an attractive idea, illustrating once again that it is sometimes hard to identify even simple-looking tables. Thus, the Mesopotamian scholar Eleanor Robson presents it as

Sixty: its 2nd part is	30
its 3rd part is	20

and so on.[32] Fortunately, for our purposes, we do not need to resolve this point, so we shall continue to refer to the table as a reciprocal table: indeed, scholars often call this one, which lists the reciprocals of every regular number from 1 to 81 (or 1^2 to 9^2) and which exists in many copies, the *standard reciprocal table*, because of its ubiquity.

What would happen if a Mesopotamian scribe had to record the result of dividing by a number that is not regular? In practice, this requirement arose rarely, and the few tablets that discuss it may have been exercises. But they show that scribes were able to do this, approximating their results to three or four sexagesimal places, and there are reciprocal tables from the period for sequences of numbers (containing both regular and non-regular ones). Any division problem could thus be converted into an equivalent multiplication problem with the use of such tables: in order to divide by a number, you multiply by its reciprocal. If the number you are trying to divide by is regular, then the answer is exact; otherwise, it is approximate.

Tables and problems. The extant mathematical tablets from the Old Babylonian period fall broadly into two categories, *tables* and *problems*. There is also a type of tablet that Robson has characterised as *rough work*, showing a student's attempt to solve a problem of some sort.

The weighing-the-stone problem with which we started is from a table of problems, while the others we have looked at — the table of squares and the reciprocal table — consist solely of tables of numbers. Several thousand tables have been found, and many types of calculations appear to have been carried out by means of them: as well as squares and reciprocals, there are multiplication tables, tables of square roots, tables of cubes and cube roots, tables of weights and measures, and so on. Numerical tables seem to have been a staple constituent of Mesopotamian life, as ubiquitous for them as is the desktop computer for us today. Or rather, as both Robson and the historian Jens Høyrup believe, the existence of so many surviving tablets shows that all this material had to be learned, over and over, by rote.

Problem tablets, by contrast, are much rarer — only 500 or so have been found — and they seem to relate to an educational context of advanced scribal training. Early Mesopotamian culture had seen the development of specialised occupations, as a part of the newly developing and highly complex urban structuring of the community, and the profession of scribe was central to the running of economic, bureaucratic, and other aspects of the state.

32 In (Katz 2007, 79).

It is in this educational context that both tables and problems seem to have been written and used. Some of the latter merely give the problem and the answer, while others are more forthcoming on what to do to reach the answer. Let us look at one of these now, Problem 2 on the tablet BM 13901, in the translation by Neugebauer:[33]

> I have subtracted the side of my square from the area: 14, 30. You write down 1, the coefficient. You break off half of 1. You multiply 0; 30 and 0; 30. You add 0; 15 to 14, 30, result 14, 30; 15. This is the square of 29; 30. You add 0; 30, which you multiplied, to 29; 30. Result 30, the side of the square.

This is not immediately comprehensible. The problem is given rather cryptically, in the first sentence, where some information is given from which the side of the square is to be found. The rest of the text gives the stages involved in finding the solution. After doing various things to the numbers initially given, the result 30 is reached. This indeed solves the original problem, for a square of side 30 has area $30^2 = 900$ (= 15,0 in sexagesimal notation), and subtracting 30 from 900 gives 870 (= 14,30).

Whenever one is dealing with a translated text, as we must here, it is advisable to check that the difficulties are not with the translation itself, rather than the original. The current generation of historians of Mesopotamian mathematics (Friberg, Høyrup, Ritter, Robson, and others) have thought more deeply than earlier scholars about how the language of the original tablets could (and should) be translated. They have noticed that, even today, we do not always use standard mathematical terms when doing arithmetic: we might *add* some numbers, or *sum* them, or *combine* them, or *take them together*, and so on. They also pay more attention to the actual words used on the tablets themselves, and do not jump to a modern term that makes arithmetical sense of the numbers on the tablet, in the hope that a more faithful translation might shed more light on what the scribe was actually saying. Robson, for example, translates the above problem as follows, using a slightly different convention for transcribing the numbers:[34]

> I took away my square-side from inside the area and it was 14 30. You put down 1, the projection. You break off half of 1. You combine 0;30 and 0;30. You add 0;15 to 14 30. 14 30;15 squares 29;30. You add 0;30 which you combined to 29;30 so that the square-side is 30.

This may not make the scribe's work any easier to understand, but the differences from Neugebauer's translation are interesting. The language is more dynamic: one projects, breaks off, combines, squares. It vividly suggests a process that was more geometrical, and less a matter of simply crunching numbers.

Squares and sides. To see whether we can make better sense of the above problem, we now try to follow the process using letters for numbers. Since this was not something the scribe did, or could have done, we also need to consider the extent to which this part of our analysis distorts the picture that we are trying to form.

Let us denote by x the unknown (the side), by b the coefficient (which here is 1), and by c the number in the statement of the problem (here, 14 30). In the present example it is not clear why we can legitimately replace the number 1 by the arbitrary constant b, but other tablets, where multiples of the square-side are used, make this clear. Then the problem is, in modern terms:

$$\text{to find } x, \text{ given that } x^2 - bx = c\,;$$

[33] In F&G 1.E1(f).

[34] In (Katz 2007, 104).

this quadratic equation represents what the scribe wrote as the result of taking the square-side from the area.

The solution consists of the following steps — on the right are the numbers from Robson's translation of what the scribe wrote:

put down b, the projection	1
break off half of the projection, to give $\frac{1}{2}b$	0; 30
combine it with itself, to give $(\frac{1}{2}b)^2$	0;15
add this to the original number c, to give $c + (\frac{1}{2}b)^2$	14 30; 15
find what it is the square of, to give $\sqrt{c + (\frac{1}{2}b)^2}$	29; 30
add this to the halved coefficient $\frac{1}{2}b$ to give $\frac{1}{2}b + \sqrt{c + (\frac{1}{2}b)^2}$	30
This is the answer.	

It can be verified that this algebraic formula is just the same as we obtain using modern algebra. This is most satisfactory — or perhaps, on another consideration, somewhat alarming: if our method of understanding what the Mesopotamian scribe might have been doing is to turn him into a 21st-century algebraist, it is possible that there has been some misunderstanding. So we must ask ourselves: how does the symbolic description given above compare with what the scribe did? What similarities are there, and what are the differences?

As for similarities, we find that the sequence of instructions given by the scribe seems to follow quite closely the procedure we now call *completing the square*,[35] in terms of actions on the particular numbers specified at the outset. Moreover, as with the weighing-the-stone problem, the scribe seems to be labelling the unknown in an abstract, symbolic, way.

To see this, consider the alternative possibility that this is a realistic geometrical problem, as the words 'square-side' and 'area' seem to imply. Surely the process of taking a side away from an area does not make sense to our way of thinking about geometry: they are different kinds of things — taking away a line does not alter the numerical measure of the area that is enclosed. But if the projection put down is thought of as a rectangle of a certain length and width 1, then it becomes more likely that the scribe was thinking about actual areas, and actually completing a square. We shall come back to this point later on.

As for differences, there are major notational ones, such as our use of x for the unknown and of b and c for fixed numbers, where we do not want to be specific about what they are. Most of the other symbols here ($+$, $\sqrt{}$, etc.) seem fairly harmless translations of Mesopotamian words and operations, as long as we remember that the scribe used words for them.

There are two significant distortions, however. One arises from our use of the equals sign, and our ability, even eagerness, to recognise this problem as one about 'quadratic equations'. The scribe had a problem about squares and their sides; it is we who have turned it into a piece of algebra that we now call a quadratic equation. Second, there is nothing in the Mesopotamian text that parallels our formula for the answer, a structure in which all the contributions of the original coefficients are still evident. But note that our 'formula' makes sense only when we understand the conventions about the order in which we perform the operations. Given a formula like $\sqrt{c + (\frac{1}{2}b)^2}$ we are taught to interpret it as 'square $\frac{1}{2}b$, *then* add c, and *then* take the square root', which is beginning to sound like the Mesopotamian instructions again. Indeed, the parallel is all the more marked in the

[35] This is the method usually taught at school for solving quadratic equations.

computational techniques developed over recent decades. Solving a quadratic equation on a pocket calculator or computer involves carrying out a sequenced program of operations that closely mirror the instructions on the Mesopotamian tablet, even down to pushing the 'square-root' button at just the stage that the scribe would have leant over to consult some square-root tables.

All the tablets featuring problems with solutions are of this sort, apparently instructing about a general approach through particular instances. In some cases, all the answers on a particular tablet turn out to be the same, indicating that it was the journey rather than the destination that mattered; this is confirmed by details within the calculations. For instance, a number may be explicitly multiplied by 1, which seems pointless until we realise that, *in general*, it might be some number other than 1 at that stage. This serves as a reminder that some multiplication is to be done there.

A general lesson. The previous exercise not only informs us about the Old Babylonian mathematical style, but on another level gives us more experience in the endeavour of trying to understand past mathematics. Our model can be characterised thus:

> Using any means, such as any symbolism or notation that occurs to you, find your way into the problem; then check rigorously to see how much of your new understanding is a projection backwards from your own time and techniques.

First, try to understand what the original writer might have been doing. Then address the harder questions of how and why. As this process becomes more familiar, you will find it increasingly easy to respond to past mathematics on its own terms, and to understand and evaluate historical questions and concerns.

The Old Babylonian 'quadratic' problem above is fairly characteristic of problem tablets. Some problems have two unknowns ('length' and 'width') and two conditions connecting them, and a few even have three unknowns and three conditions, while others involve finding cubes or higher powers of unknown numbers.

To understand further how Mesopotamian mathematicians may have thought about such problems, and how one particular historian has analysed their work, we now turn to Høyrup's detailed discussion of the opening problem of the above tablet BM 13901.[36]

We follow his translation line by line:

(1) The surfa[ce] and my confrontation I have accu[mulated]: 45′ is it. 1, the projection,

(2) you posit. The moiety of 1 you break, [3]0′ and 30′ you make hold.

(3) 15′ to 45′ you append: [by] 1, 1 is equalside. 30′ which you have made hold.

(4) in the inside of 1 you tear out: 30′ the confrontation.

The strangeness of his English translation is deliberate: Høyrup wishes the reader to stay as close as possible to the original, so as not to slip into anachronism or miss any detail. The word 'moiety' that he uses is obsolete English for 'half'. Words and fragments in square brackets are missing from the tablet, and he writes numbers differently from, say, Neugebauer, and more like the way we write minutes and seconds of arc.

Høyrup reads this, and many other tablets, in the following way.

Line 1: The area (the surface) and the side (the confrontation) of the square are accumulated. To add a surface and a line together, the line is thickened by positing the 'projection' 1 (it protrudes to the left in Figure 2.10). In this way the side of length s becomes an area

[36] See (Høyrup 2002, 50–51).

of size $s \times 1$. Added to the area of the square the result is an area of $45'$. (The term for projection, he tells us, was used on other tablets to mean 'something protruding or projecting, e.g., from a building', and in mathematical texts he finds that its value is always 1.)

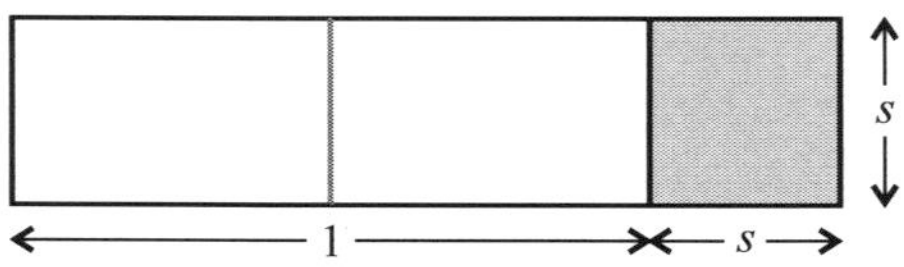

Figure 2.10. Adding a side to a square

Line 2: The outer moiety (or half) is broken off and made into the sides of a rectangle (in this case a square) with area $30' \times 30' = 15'$. For line 3 to make sense, the new arrangement of the original square and its moieties must be as shown in Figure 2.11, as an L-shaped figure nowadays called a *gnomon*.

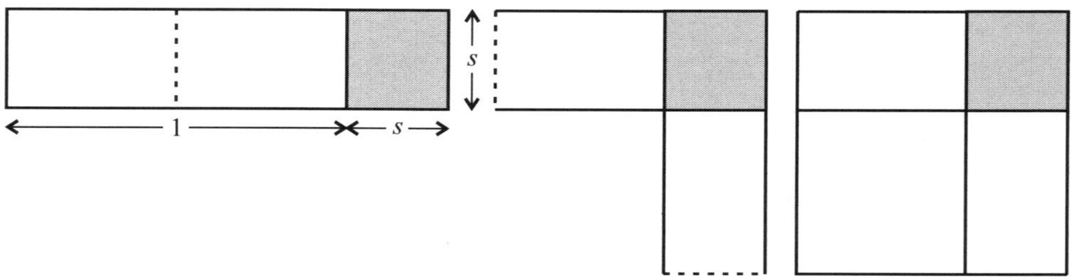

Figure 2.11. Making a rectangle and a gnomon

Line 3: The square $15'$ is now appended, and Høyrup urges us to take this quite literally, thus completing the gnomon to a larger square of area $45' + 15' = 1$, and therefore of side 1 (called by the scribe the 'equalside').

Line 4: The original moiety that was made into one side of the completing square is now torn out. What remains of the equalside is the answer: $30'$.

To help us to see what is going on, we invent a more general problem and, guided by our earlier discussion, we follow through this invented problem:

Line 1: The surface and b copies of my confrontation I have accumulated: c is it. 1, the projection,

Line 2: you posit. The moiety of b you break, $\frac{1}{2}b$ and $\frac{1}{2}b$ you make hold.

Line 3. $(\frac{1}{2}b) \times (\frac{1}{2}b)$ to c you append: of $c + (\frac{1}{2}b)^2$, d is equalside. b which you have made hold.

Line 4: in the inside of $c + (\frac{1}{2}b)^2$ you tear out $\frac{1}{2}b$: s the confrontation.

So in modern notation, if the unknown side of the square is s, then we have to solve the equation $s^2 + bs = c$.

As before, we make a rectangle of side s and length $b \times 1$ and area bs, which sticks out from the side of the original square (see Figure 2.11): this figure has area $s^2 + bs = c$. We break this rectangle in half and rearrange it into the gnomon shape, which therefore also has area $s^2 + bs = c$. The sides of the rectangle now enclose a square of area $(\frac{1}{2}b)^2$, and the gnomon and this square together make an area of $c+(\frac{1}{2}b)^2 = d^2$. But this larger square has sides $s + \frac{1}{2}b$, so we have $d = s + \frac{1}{2}b$, and the side we seek is found to be $s = d - \frac{1}{2}b$. This agrees with our modern formula, which says in the present case that $s = -\frac{1}{2}b + d$.

Problems with reciprocals. Two other types of mathematical problems from Old Babylonian times are worth selecting from the many that were tackled. These are the so-called *igûm-igibûm* problems and the *gate and ladder problems*.

In an igûm-igibûm problem, information is given about a number (the igûm) and its reciprocal the (igibûm), and both are to be found, as in this example, from the tablet YBC 6967.[37]

The tablet YBC 6967.

1. [The igib] ūm exceeded the igūm by 7.
2. What are [the igūm and] the igibūm?

3–5. As for you — halve 7, by which the igibūm exceeded the igūm, and (the result is) 3;30.

6–7. Multiply together 3;30 with 3;30, and (the result is) 12;15.

8. To 12; 15, which resulted for you,
9. add [1, 0, the produ]ct, and (the result is) 1,12;15.
10. What is [the square root of 1],12;15? (Answer:) 8;30.
11. Lay down [8;30 and] 8;30, its equal, and then

(Reverse)

1–2. subtract 3; 0, the takīltum, from the one,

3. add (it) to the other.
4. One is 12, the other 5.
5. 12 is the igibûm, 5 the igûm.

In this problem the igibūm exceeds the igūm by 7. Since the scribe understood powers of 60 from the context, we can accept his answer: the number is 5 and its reciprocal is 12 (since $5 \times 12 = 60$ and $12 - 5 = 7$).

Like the above problem, solutions to these problems were found by completing the square. In this case the tablet tells us to break in half the 7 by which the number exceeds its reciprocal, giving 3;30. Multiply 3;30 by 3;30 to get a square of area 12;15. This fits into a gnomon made by breaking the rectangle of length $x - 1/x$ into two, and makes a large square of area 1,00 + 12;15, which is the square of 8;30. From this 8;30 one subtracts the 3;30 to obtain $1/x$, which is 5, and adds the 'other' 3:30 to give its reciprocal, 12.

In a gate problem, two of the width, height, and diagonal of a rectangular gate are given, and the third is to be found. In a ladder problem the foot of a ladder of known length has slid a known distance away from a vertical wall, and the problem is to calculate how far up the wall the ladder now reaches. There are also variants where related numbers are given, such as the amount by which the head of the ladder has dropped and the distance from the foot of the ladder to the wall is to be found. This suggests that teachers then, like teachers today, rang the changes on a problem to be sure that the students understood all the details. These problems are found in some quantity, and it is clear that Mesopotamian scribes knew the relationship between the sides of a right-angled triangle that we would describe as given by the Pythagorean theorem.

[37] In (Neugebauer and Sachs 1945, 129–130); see also F&G 1.E1 for a commentary.

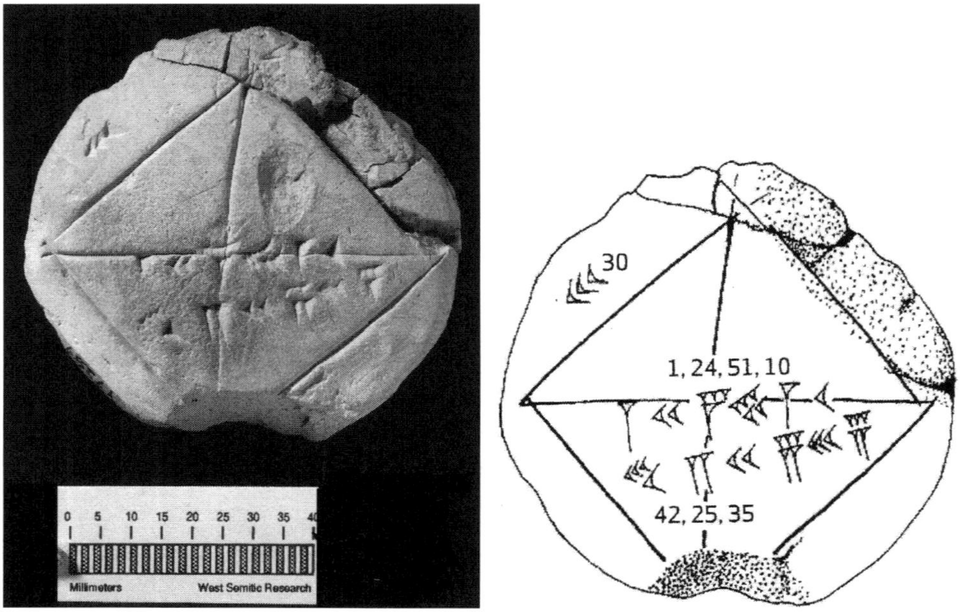

Figure 2.12. YBC 7289: photograph and drawing

The tablet YBC 7289 in Figure 2.12 provides further insight into this topic.[38] It displays the length of the diagonal of a square of side 30 as the excellent approximation $42, 25, 35$.

The number 30 indicates the side a of the square, and $1; 24, 51, 10$ means

$$1; 24, 51, 10 \approx \sqrt{2}.$$

The value for $\sqrt{2}$ is extremely good, as can be seen from its square

$$(1; 24, 51, 10)^2 = 1; 59, 59, 59, 38, 1, 40,$$

which is correct to within five parts in a million. This is an impressive tribute to their computational skills. We therefore find that the length of the diagonal is

$$d = a\sqrt{2} = 42; 32, 25.$$

The scribe probably knew that the number written down for the diagonal is not exact but only an excellent approximation — or is that too confident a judgement? Robson has argued persuasively that the tablet itself is typical of the work of trainee scribes; the person who wrote this tablet would have taken the excellent approximation we are interested in from a list, and may not have understood it deeply at all. But, as she recognises, this merely shunts the question backwards to whoever discovered this approximation and put it into circulation. Square roots were calculated by an iterative process, and can of course be checked by squaring them. So it is likely that the experts of the time knew that the number is indeed an approximation. But did any scribe know that no sexagesimal number, however far it is continued, can ever give an exact measure of the diagonal in relation to the side? This seems unlikely, because we have no evidence that they could prove an impossibility of this kind.

[38] See (Neugebauer 1969, 37 and Plate 6a) and Robson in (Katz 2007, 143) for the drawings; both authors give commentaries; see also F&G 1.E1.

An overview. As we have seen, Mesopotamian mathematical problems can be expressed fairly abstractly, although often in terms of concrete imagery that indicates training for practical problems that a scribe might be called upon to solve professionally. Yet even apparently practical scenarios can discuss flamboyantly unrealistic problems: on one tablet, the problem is to discover how large a field can be irrigated by a particular volume of water, and the answer turns out to be a field some $\frac{1}{2}$ kilometres square, covered with water to a uniform depth of one finger's breadth — a task that would tax even the most diligent of Mesopotamian farmers![39]

But we should not lose sight of the fact that many of these problems were conspicuously rooted in the world of surveyors: civil, religious, and military. Many of the tablets speak of walls, ditches, trenches, cisterns, water clocks, cylindrical granaries, and fields divided by rivers, with work being done at such-and-such a rate: How long will the job take? How much will it cost? Conversions must be made between different systems of units, just as we now convert pounds sterling into dollars or euros, and pounds weight into kilograms. Brick making and wall making figure prominently, and there are problems about dividing wages unequally, some of which hint at legal divisions for inheritance purposes.

2.4 A historical case study

We have now seen how the sexagesimal place-value system works, and you have met two examples of tables, one of squares and one of reciprocals, that give an idea of the level, spirit, and flexibility of Mesopotamian computations. We did not try to ascertain how the results on the tables were arrived at, as that seemed either obvious or not very interesting. But thus equipped, we are now ready to tackle a tablet whose method of construction does turn out to be rather interesting, and which has provoked considerable discussion among historians of ancient mathematics.

Plimpton 322. The tablet is known as Plimpton 322, and was described by Neugebauer as 'one of the most remarkable documents of Old-Babylonian mathematics' (see Figure 2.13).[40] The name arises from the fact that the tablet has catalogue number 322 in the George A. Plimpton collection at Columbia University, New York. Plimpton bought it in about 1923 from a Mr. Banks of Florida; it is not certain where Banks obtained it, but it may have been dug up at Larsa in southern Iraq. The left-hand side of the original tablet has been broken off, and traces of modern glue suggest that this has happened since its excavation. All told, it is a sorry tale of how a tablet can be robbed of its archaeological context.

Our task is to make sense of it. Is it a problem or a table, or something else? What might its relation be to other Old Babylonian mathematics? Why was it constructed? How could it have been used? We shall see that historians of mathematics have disagreed about this, as have mathematicians drawn to the challenges posed by this famous text.

Look at the tablet and notice the main features of what remains: there are four columns of numbers, with words at the head of each column. Now look at the transcription (we have labelled the columns *A, B, C, D*, for ease of reference), and see whether any pattern is evident to you.

---

[39] See, for example, the collection of problems of Robson in (Katz 2007, 120–122) and for an earlier translation of some of them F&G 1.E1(e). These concern a number of workers digging a ditch for so many days at a certain expected productivity. The problem of irrigating the field is in F&G 1.E1(b).

[40] See (Neugebauer 1969, 40).

Figure 2.13. Plimpton 322

The tablet Plimpton 322.

A	B	C	
[1;59,0,]15	1,59	2,49	1
[1;56,56,]58,14,50,6,15	56,7	3,12,1	2
[1;55,7,]41,15,33,45	1,16,41	1,50,49	3
[1;]5[3,1]0,29,32,52,16	3,31,49	5,9,1	4
[1;]48,54,1,40	1,5	1,37	5
[1;]47,6,41,40	5,19	8,1	6
[1;]43,1 1,56,28,26,40	38,1 1	59,1	7
[1;]41,33,45,14,3,45	13,19	20,49	8
[1;]38,33,36,36	9,1	12,49	9
1;35,10,2,28,27,24,26,40	1,22,41	2,16,1	10
1;33,45	45	1,15	11
1;29,21,54,2,15	27,59	48,49	12
[1;]27,0,3,45	7,12,1	4,49	13
1;25,48,51,35,6,40	29,31	53,49	14
[1;]23,13,46,40	56	53	15

At first sight, this is not very promising! Something seems to be being listed, as the lines are numbered in the final column; and the numbers in column *A* diminish fairly regularly, from just under 2 to just over $1\frac{1}{3}$ — but if you noticed this, it was partly the effect of the editorially informed semi-colons and Neugebauer's reconstructions in square brackets. It is not yet clear why so many entries in column *A* are prefixed by an added [1], for example. Otherwise, the numbers look fairly random, and there is little to tell that this is not the equivalent of a supermarket till receipt (indeed, historians had originally classified it as an accountancy document). But as Neugebauer discovered — presumably after a considerable amount of conjecture and refutation — the numbers in each line can be related by

$$A = C^2/(C^2 - B^2)$$

and this was the basis for his reconstruction of the illegible entries.

Let us check this out on the simplest line, line 11:

$$\text{(working in decimals) } B^2 = 45^2 = 2025\,; \quad C^2 = 75^2 = 5625;$$

$$\frac{C^2}{(C^2 - B^2)} = \frac{5625}{3600} = 1\frac{2025}{3600} = 1\frac{33}{60}\frac{45}{3600} = 1;33,45 = A,$$

as claimed. So it is possible to find a relation, albeit a somewhat devious one, between the columns of the tablet. In fact, for this to hold consistently throughout, the underlined numbers have to be considered as mistakes by the scribe, a point to which we shall return.

Neugebauer's analysis. Let us investigate these numbers further. In his book *The Exact Sciences in Antiquity*, Neugebauer calculated $C^2 - B^2$ for each entry, and then took its square root, D. What he found may strike you as surprising. We give those values in decimal form below: the scribe's 'errors' have been corrected.

B	C	D (decimal)	D (sexagesimal)
119	169	120	$2,0$
3367	4825	3456	$57,36$
4601	6649	4800	$1,20,0$
12709	18541	13500	$3,45,0$
65	97	72	$1,12$
319	481	360	$6,0$
2291	3541	2700	$45,0$
799	1249	960	$16,0$
481	769	600	$10,0$
4961	8161	6480	$1,48,0$
45	75	60	$1,0$
1679	2929	2400	$40,0$
161	289	240	$4,0$
1771	3229	2700	$45,0$
56	106	90	$1,30$

Notice that, compared with columns B and C, the values computed in D are remarkably simple looking numbers: this is particularly noticeable in their sexagesimal representation. You may indeed have recognised some as old friends from the reciprocal table. In fact, all of them are regular numbers, whereas B and C are all non-regular (except for line 11), which explains why their squares divide into C^2, yielding exact sexagesimal expressions.

How should we feel about this speculative calculation of Neugebauer's? On the one hand, nothing on the tablet looks like column D. On the other hand, it seems too good to be true: surely the original scribe must have known of these numbers, and possibly recorded them on another tablet, now lost. So this might be a clue; perhaps Neugebauer has discovered something that we should analyse further.

In fact, there is information on the tablet about the significance of these numbers that Neugebauer scarcely took into account — namely, the text of the column headings themselves (omitted here). The text heading of column A is partly destroyed, but the heading for B says something like 'Solving number of the width', and C is 'Solving number of the diagonal'. The term 'Solving number', Neugebauer remarked, 'is a rather unsatisfactory rendering for a term that is used in connection with square roots and similar operations and has no exact equivalent in our modern terminology'.[41]

However, the word 'diagonal' seems to have suggested to Neugebauer that the numbers on the tablet should be viewed as relating to right-angled triangles. Indeed, as the numbers in the above table have the property that $B^2 + D^2 = C^2$ (which follows from the way that D was defined), it appears that columns B and C of Plimpton 322 list the shortest side and diagonal of right-angled triangles whose third side is a simple regular number. The numbers B, C, and D are what are called *Pythagorean triples* and, said Neugebauer,[42]

> We know that all Pythagorean triples are obtainable in the form $D = 2pq, B = p^2 - q^2, C = p^2 + q^2$, where p and q are arbitrary integers subject only to the condition that they are relatively prime and not simultaneously odd, and $p > q$.

This gave him the following table.

B	C	D (decimal)	D (sexagesimal)	p	q
119	169	120	2,0	12	5
3367	4825	3456	57,36	64	27
4601	6649	4800	1,20,0	75	32
12709	18541	13500	3,45,0	125	54
65	97	72	1,12	9	4
319	481	360	6,0	20	9
2291	3541	2700	45,0	54	25
799	1249	960	16,0	32	15
481	769	600	10,0	25	12
4961	8161	6480	1,48,0	81	40
45	75	60	1,0		
1679	2929	2400	40,0	48	25
161	289	240	4,0	15	8
1771	3229	2700	45,0	50	27
56	106	90	1,30	9	5

He remarked that the ratio C/D can therefore be written as $\frac{1}{2}(p/q + q/p)$, and noted that this ratio is expressible as a finite sexagesimal fraction if and only if both p and q are regular numbers (as is the case for Plimpton 322). He then checked the values of p and q, and found that they are the regular numbers that appear in what he called the 'standard table' of reciprocals, except for $p = 125$, which he said was well known in other tables. He concluded that: 'This seems to me a strong indication that the fundamental formula for the construction of Pythagorean numbers was known'.

Buck's analysis. In 1980, some years after Neugebauer published his analysis, the mathematician R. Creighton Buck took up the study of Plimpton 322.[43] Buck followed Neugebauer in seeing Pythagorean triples in the tablet, and wrote out the appropriate values of p and q. He then turned to the errors in the table, and explained them mostly as transcription errors. He argued that the scribe would have done a series of calculations, starting from p and q, that involve numbers and their squares, copying results from other tablets to this one, and that all these mistakes can be seen to be likely to occur if indeed the scribe had started with these values of p and q.

[41] See (Neugebauer 1969, 37).
[42] See (Neugebauer 1969, 39).
[43] See (Buck 1980).

Only now did Buck turn to the possible geometrical interpretation of the tablet. He looked at the value of $(B/D)^2$ and noted that the angles of the B, C, D triangle can be seen to decrease from about $45°$ to $30°$ in steps of about $1°$.

Buck then looked at other interpretations of the numbers on the tablet from the small community of historians of ancient mathematics. He selected an observation of E. M. Bruins that all the numbers can be calculated from the single parameter p/q, and so they can all be obtained from a specially constructed reciprocal table. He also noted that D. L. Voils, another historian, had observed that the numbers $(B/D)^2$ also occur in another context: the igûm–igibûm problems discussed above.

On this interpretation, the tablet has nothing to do with triangles or Pythagorean triples, but would be a tool for constructing lots of examples of igûm-igibûm problems that are suitable in a teaching context because they have answers in regular numbers.

Buck therefore concluded that all these reconstructions had something to recommend them, and that no final resolution was possible.

Robson's analysis. The third interpretation we consider is more recent. Robson looked at Plimpton 322 in the light of scholars' improved grasp of the ancient written language, their further study of the known collection of Mesopotamian tablets, and archaeological evidence.[44] She noted that Plimpton 322 is typical of tablets written by temple administrators in Larsa in the short period 1822–1784 BC. In particular, the rows run parallel to the long side of the tablet, and there are traces of words of Sumerian as well as Akkadian. So, she suggested, this fixes the tablet as one written in that period by someone familiar with temple administration in Larsa. Next she remarked that no other Mesopotamian tablets from this period, or for many years later, display any concept of angle, so she ruled out any suggestion that the tablet was a table of trigonometrical values: this does not, of course, rule out the idea that it is about right-angled triangles with various slopes.

She then noted that there are many surviving sets of the standard table of reciprocals, and that while some of the reciprocals underlying Plimpton 322 are not in the standard table, they are well attested in other tablets. She also noted that the numbers in column A appear in descending order — typical of the Larsa temple tablets mentioned earlier — but the corresponding values of p and q do not. She thus rejected the idea that the missing column on the right would have given the p and q numbers.

What about the words at the head of the columns? She agreed with Neugebauer and Sachs that B and C are called the 'square of the short side' and the 'square of the diagonal'. But, unlike them, she found that the heading of column A is more-or-less legible, and indeed that it reads (with one difficult word that we shall come back to): 'the takiltum-square of the diagonal from which 1 is torn out, so that the short side comes up'.

Robson therefore proposed the following process whereby Plimpton 322 was written. A table of reciprocal pairs was drawn up, and its values were listed in decreasing order and written on the part of the tablet that is now missing. From those numbers the short sides B and the diagonals C of right-angled triangles with third side of length 1 were calculated by the igûm-igibûm method:

Start with a rectangle with sides x and $1/x$ and area 1. Rearrange it as a gnomon enclosing a square of side $s = \frac{1}{2}(x - 1/x)$ and forming part of a larger square of side d. (In this context, the word 'takiltum' refers to the larger square.) Then 1, s, and d are the sides of a right-angled triangle, because, by construction, $1 + s^2 = d^2$.

[44] See (Robson 2002).

Since, moreover, it is almost certain that the igûm–igibûm tablet we looked at, YBC 6967, also came from Larsa around the same time, Robson concluded that Plimpton 322 was produced from the reciprocal pairs in this way, with the B and C columns preceded by an intermediate column A. The reciprocal pairs are made up of regular numbers for convenience, but it is likely that the scribe began with a list of reciprocal pairs, either the standard one or another well-known one.

Robson went further. Although there were female temple scribes, they were known only in more northern and central parts of what is now Iraq, so the scribe was male. He must have been highly numerate, employed for these skills rather than a student. In the absence of any religious use for the tablet, she asserted that he was probably a teacher, concerned to produce a convenient set of numbers for use in the sorts of educational problems they were preoccupied with in ancient Mesopotamia — for example, as an aid to posing and solving problems about diagonals of rectangles, and the gate and ladder problems.

If you look back at the interpretations that Robson rejected, you can see more clearly some reasons for disquiet. When Neugebauer introduced the numbers p and q he noted[45] that 'all Pythagorean triples are obtainable in the form $D = 2pq \quad B = p^2 - q^2 \quad C = p^2 + q^2$' and indeed this is a well-known result today. But that does not tell us whether ancient Babylonian scribes knew it, and there seems to be no evidence that they did; Neugebauer inferred this from the tablet itself but from no other source. It is the same with talk about angles, as opposed to slopes, which also seems to have been a later insight. All in all, Robson's deflationary interpretation suggests that the high level of attention paid to this one tablet may say more about the mathematics that has been read into it by mathematical historians than about the mathematics it actually contains.

A further analysis. A still more recent analysis of Plimpton 322 offers cautious agreement that the tablet was written in Larsa, but suggests a new interpretation for it.[46] The authors read the tablet as being about rectangles whose sides and diagonals are all finite sexagesimal numbers, and, as they put it,

> We argue that P 322 presents a coherent *mathematical* project, which is to generate a series of finite sexagesimal rectangles from a basic algorithm, or, in other terms, that P 322 is a problem text including a complete solution. To a large extent, this approach revives Neugebauer's original interpretation as modified by Bruins' introduction of reciprocals and other subsequent improvements.

They compared the tablet with some that they suggest dealt with related problems, and may even have been written by someone aware of the content of Plimpton 322, and proposed that the task facing the scribe was to

> Make the list of all the rectangles with length (us) equal to 1 and width (sag) and diagonal (s.iliptum) finite sexagesimal numbers, (columns -1 and 0) and represent the dimensions, in reduced form, without common sexagesimal factor (columns IIB and IIIB).

So, rather than seeing it as a teacher's aid, they see the tablet as a challenging mathematical problem and its solution, and they conclude (p. 561)

> Whatever the case, P 322 is *mathematics* as Neugebauer recognised, not exactly as Neugebauer understood it, but real mathematics focused on a discovery which significantly expanded the capabilities of the art. As such it appears far more deeply rooted than simply a utilitarian exercise in pedagogical convenience.

---

[45] See (Neugebauer 1969, 39).

[46] See (Britton, Proust, Shnider 2011).

It would seem that we still have much to understand about some of our earliest written mathematical sources and the cultures that produced them.

Conclusion. We note that Robson's approach is not rooted in a modern mathematical insight, but in a wide reading of clay tablets. She advocates bringing to bear as broad an awareness of contemporary sources as possible — contemporary, that is, with the ancient texts themselves.

In keeping with that approach, and as the conclusion to this chapter, we can ask: What was Old Babylonian mathematics about? Although it is not easy to answer this question precisely, because of difficulties of interpretation (such as you saw with Plimpton 322), the overwhelming impression is of the study and use of numbers, and various techniques for solving problems involving numbers. Where the numbers arise from — whether from land measurement, economic questions, idealised geometrical objects (cubes, triangles, and so on), or just fairly abstractly — seems a relatively secondary matter. As Neugebauer has put it:[47]

> The central problem in the early development of mathematics lies in the numerical determination of the solution which satisfies certain conditions. At this level there is no essential difference between the division of a sum of money according to certain rules and the division of a field of given size into, say, parts of equal area.

Old Babylonian techniques are sometimes described as algebraic, although most historians would now reject the applicability of this term. In our view, discussing the style of mathematics used in some culture or period can be a better way of trying to understand and empathise with past mathematical activity than is the desire to fit topics into pre-ordained, modern, slots. For, as we have already seen, the ostensible subject matter, the language used, and the techniques applied do not always fit together in past cultures in the way that they do in ours — it is some combination of all these things that is characterised by the word *style*.

This idea is, we admit, a slippery one. It will become clearer when we have investigated the mathematics of a culture that contrasts rather strongly with the ones we have so far examined — the mathematical activity and traditions of classical Greece.

2.5 Further reading

Barrow, J.D. 1992. *Pi in the Sky: Counting, Thinking and Being*, Clarendon Press, Oxford. This is an interesting and accessible account of counting, both ancient and modern.

Corry, Leo. 2015. *A Brief History of Numbers*, Oxford University Press, Oxford. The book takes the development of the number concept from the early history of written numbers, through their use in geometrical problems in ancient Greece and medieval Islam, and their uses in algebra, to their many ramifications at the start of the 20th century.

Doblhofer, E. 1973. *Voices in Stone* (1st edn., 1957), Paladin. A good account of the decipherment of hieroglyphs and cuneiform texts, without being especially mathematical.

Imhausen, A. 2015. *Mathematics in Ancient Egypt: A Contextual History*, Princeton University Press. The author has drawn on texts from a variety of sources to provide a fresh and readable account of the use of mathematics in Egyptian society over a period of three thousand years.

[47] See (Neugebauer 1969, 44–45).

Katz, V.J. (ed.) 2007. *The Mathematics of Egypt, Mesopotamia, China, India, and Islam: A Sourcebook*, Princeton University Press. This book should be the first port of call for anyone interested in early source material. It is a carefully edited accessible account by five leading historians: Annette Imhausen on Egypt, Eleanor Robson on Mesopotamia, Kim Plofker on India, Joseph Dauben on China, and J. Lennart Berggren on Islam.

Menninger, K. 1969. *Number Words and Number Symbols* (1st edn., 1958), MIT Press. A book that gives more detail on the history of numbers, and is particularly attentive to linguistic evidence.

Neugebauer, O. 1969. *The Exact Sciences in Antiquity* (1st edn., 1949), Dover. A classic: a highly readable account by one of the founders of the study of mathematics in Mesopotamia, although some of its findings are now disputed.

Renfrew, C. 1987. *Archaeology and Language. The Puzzle of Indo-European Origins*, Pimlico (Penguin, 1998). Renfrew, a leading archaeologist, gives an unorthodox account, arguing for a new model of cultural change.

Robson, E. 2001. Neither Sherlock Holmes nor Babylon: a reassessment of Plimpton 322, *Historia Mathematica* 28, 167–206. This and the essay mentioned below contain a discussion of Plimpton 322 which is more detailed than ours and is part of the author's thorough re-examination of what we can learn about how and why the Mesopotamians used mathematics.

———. 2002. Words and pictures: new light on Plimpton 322, *American Mathematical Monthly* 109, 105–120, reprinted in *Sherlock Holmes in Babylon* (ed. M. Anderson, V.J. Katz, and R.J. Wilson), Mathematical Association of America, 2004, pp. 14–26.

Robson, E. 2008. *Mathematics in Ancient Iraq*, Princeton University Press. This book is an intellectual history of ancient Iraq, integrating archeological studies with the history of mathematics.

3

Greek Mathematics: An Introduction

Introduction

In the next four chapters we survey the historical development of Greek mathematical activities from our earliest records to the transmission of classical Greek texts in the Byzantine Empire. This chapter provides a chronological overview up to the time of Euclid, and in Chapters 4 to 6 we fill out and analyse those aspects of the period during which the characteristic Greek approach to mathematics was formed and developed.

We have divided the time-scale of Greek mathematics into five periods:

- up to 400 BC: the time before Plato, when an independent style of mathematics emerged in Greece, but for which our sources are poor
- 400–300 BC: the period from Plato to Euclid[1]
- 300–200 BC: the time of Euclid, Archimedes, and Apollonius, the most important and influential writers whose works have come down to us in any quantity
- 200 BC–AD 450: the later Alexandrian period, which includes Ptolemy the astronomer, Diophantus, and Pappus
- after AD 450: the time of the commentating tradition, which takes us beyond the end of the Roman Empire and its continuation in Byzantium.

In turning to Greek mathematics we have moved forward about a thousand years from the Mesopotamian texts. By the 6th century BC, Greek-speaking peoples had settled around the eastern Mediterranean, from Sicily to Asia Minor and from northern Africa to the Black Sea, living in relatively independent 'city-states' such as Athens, Sparta, Corinth, and Syracuse.

Despite political differences and varying forms of government, there were strong cultural bonds across the Greek-speaking world. This makes it possible for us to speak of

[1] A note on names: we have conformed to general English usage and chosen Plato, Euclid, Pappus, and Diophantus over Platon, Euclides, Pappos, and Diophantos.

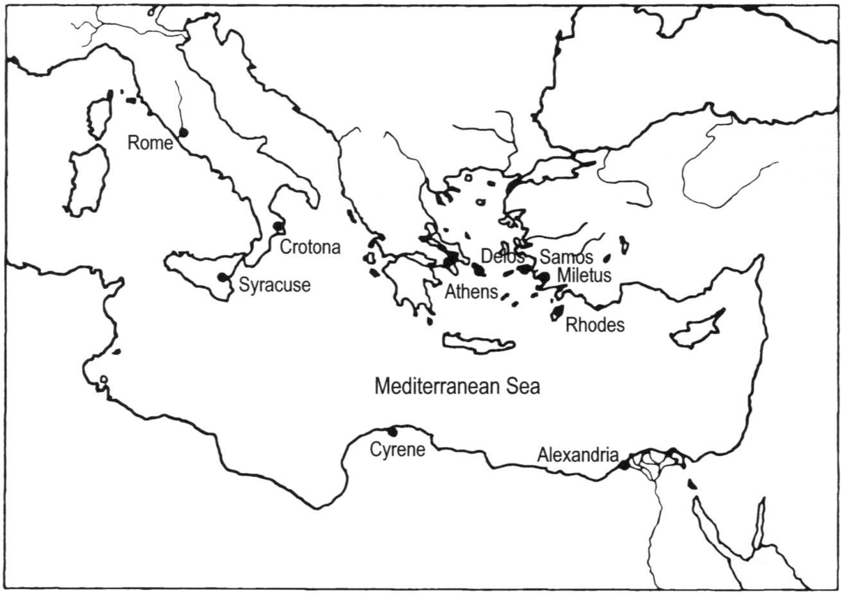

Figure 3.1. Some towns of the Greek-speaking world

'Greek mathematics', in that the people with whom we are concerned spoke and wrote in the Greek language. They did not necessarily have much to do with the territory of modern Greece, however: indeed, the earliest Greek scientific thought is associated with people of Ionia (the western coast of what is now Turkey), and the best-known of Greek mathematical works, Euclid's *Elements*, was produced in Alexandria, in Egypt (see Figure 3.1). In addition, although we are primarily concerned with mathematical developments from about 450–200 BC, the Greek culture of which we are speaking lasted until around the 6th century AD and was consciously imitated for several centuries thereafter. So our 'Greek mathematics' is widely spread in both space and time.

In this chapter our chronological overview starts with one of the most important mathematical passages in all of Greek philosophy: this celebrated dialogue, taken from Plato's *Meno*, was written about 385 BC. Since Euclid's *Elements* is often said to have been written towards the end of the same century, around 300 BC (but see our discussion in Section 3.4), the two works give us a convenient way to frame the growth of mathematics in ancient Greece — from a setting in which mathematics was already important to the writing of one of its most enduring documents.

We shall range back two centuries before *Meno* was written to learn about the origins of Greek mathematics. We shall also find it helpful to stratify Greek mathematics into different levels, in order to consider what was to be transmitted to later times.

3.1 A dialogue from Plato's *Meno*

Greek mathematical activity differs in fundamental ways from Egyptian and Mesopotamian mathematics. We can begin to see this by studying one of the earliest extended pieces of reliable evidence about Greek mathematics, and working outwards from there. This evidence occurs in the course of Plato's dialogue *Meno*, and derives particular importance because Plato's writings are among the most important and influential from the ancient world.[2]

The dialogue has three characters: the philosopher Socrates (Plato's teacher), Meno (a young aristocrat), and a slave boy. Socrates and Meno are discussing the nature of virtue, and whether it can be taught. In so doing, they move on to the more general question of whether a knowledge of anything can be taught, and if so how. In order to show that true knowledge can be elicited from someone who initially holds false beliefs, Socrates holds a mathematical discussion with the slave boy. The lengthy extract that follows is well worth reading for the light it sheds on many different aspects of early Greek mathematics — and indeed because it is one of the earliest texts in any language showing how mathematics can be discussed. We follow it with a commentary on several of these points.

Socrates and the slave boy.

MENO: Yes, Socrates; but what do you mean by saying that we do not learn, and that what we call learning is only a process of recollection?

Can you teach me how this is?

SOCRATES: I told you, Meno, just now that you were a rogue, and now you ask whether I can teach you, when I am saying that there is no teaching, but only recollection; and thus you imagine that you will involve me in a contradiction.

MENO: Indeed, Socrates, I protest that I had no such intention. I only asked the question from habit; but if you can prove to me that what you say is true, I wish that you would.

SOCRATES: It will be no easy matter, but I will try to please you to the utmost of my power. Suppose that you call one of your numerous attendants, that I may demonstrate on him.

MENO: Certainly. [*To a slave boy*] Come hither, boy.

SOCRATES: He is Greek, and speaks Greek, does he not?

MENO: Yes, indeed; he was born in the house.

SOCRATES: Attend now to the questions which I ask him, and observe whether he learns of me or only remembers.

MENO: I will.

SOCRATES: Tell me, boy, do you know that a figure like this is a square? *Socrates begins to draw figures in the sand at his feet. He points to the square $ABCD$* [See Figure 3.2].

BOY: I do.

SOCRATES: And you know that a square figure has these four lines equal?

BOY: Certainly.

SOCRATES: And these lines which I have drawn through the middle of the square are also equal? [*The lines EF, GH.*]

BOY: Yes.

SOCRATES: A square may be of any size?

[2]We look at Plato's writings in more detail in Section 3.3. The extract comes from (Plato 1892, 41–47). Another translation can be found in F&G 2.E1.

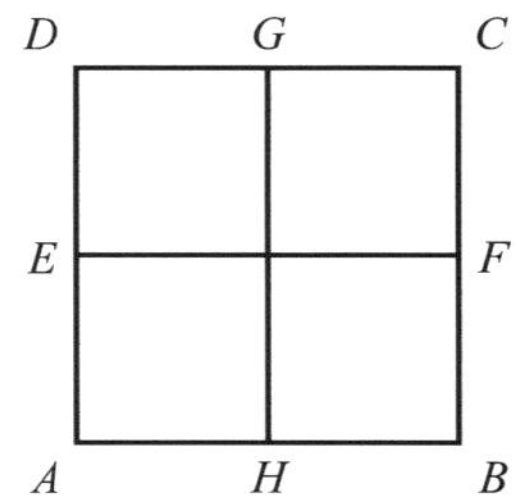

Figure 3.2. Four squares

BOY: Certainly.

SOCRATES: And if one side of the figure be of two feet [*points to* AB], and the other side be of two feet [*points to* AD], how much will the whole be? Let me explain: if in one direction the space was of two feet, and in the other direction of one foot, the whole would be of two feet taken once?

BOY: Yes.

SOCRATES: But since this side is also of two feet, there are twice two feet?

BOY: There are.

SOCRATES: Then the square is of twice two feet?

BOY: Yes.

SOCRATES: And how many are twice two feet? Count and tell me.

BOY: Four, Socrates.

SOCRATES: And might there not be another square twice as large as this, and having like this the lines equal?

BOY: Yes.

SOCRATES: And of how many feet will that be?

BOY: Of eight feet.

SOCRATES: And now try and tell me the length of the line which forms the side of that double square: this is two feet — what will that be?

BOY: Clearly, Socrates, it will be double.

SOCRATES: Do you observe, Meno, that I am not teaching the boy anything, but only asking him questions; and now he fancies that he knows how long a line is necessary in order to produce a figure of eight square feet; does he not?

MENO: Yes.

SOCRATES: And does he really know?

MENO: Certainly not.

SOCRATES: He only guesses that because the square is double, the line is double.

MENO: True.

SOCRATES: Observe him while he recalls the steps in regular order. [*To the Boy*] Tell me, boy, do you assert that a double space comes from a double line? Remember that I am not speaking of an oblong, but of a figure equal every way, and twice the size of this — that is to say of eight feet; and I want to know whether you still say that a double square comes from double line?

BOY: Yes.

SOCRATES: But does not this line AB become doubled if we add another such line here BJ?

BOY: Certainly.

SOCRATES: And four such lines will make a space containing eight feet?

BOY: Yes.

SOCRATES: Let us describe such a figure [*that is AJ, and adding JK, KL and making LA complete by drawing in its second half LD.* (see Figure 3.3)] Would you not say that this is the figure of eight feet?

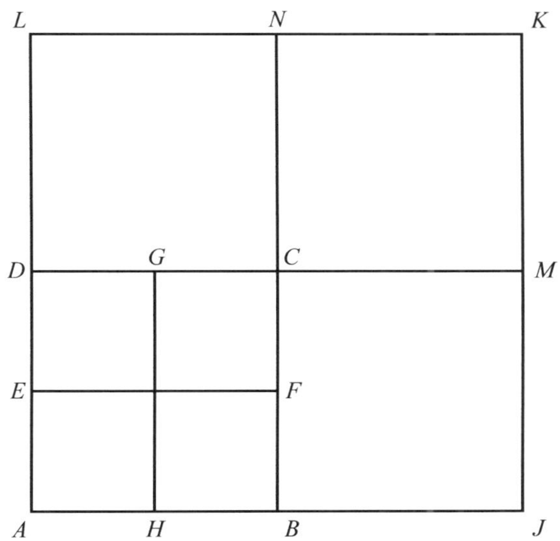

Figure 3.3. An incorrect doubling refuted

BOY: Yes.

SOCRATES: And are there not these four divisions in the figure, each of which is equal to the figure of four feet? [*Socrates has drawn in the lines CM, CN to complete the squares he wishes to point out.*]

BOY: True.

SOCRATES: And is not that four times four?

BOY: Certainly.

SOCRATES: And four times is not double?

BOY: No, indeed.

SOCRATES: But how much?

BOY: Four times as much.

SOCRATES: Therefore the double line, boy, has given a space, not twice, but four times as much.

BOY: True.

SOCRATES: Four times four are sixteen — are they not?

BOY: Yes.

SOCRATES: What line would give you a space of eight feet, as this gives one of sixteen feet; — do you see?

BOY: Yes.

SOCRATES: And the space of four feet is made from this half line?

BOY: Yes.

SOCRATES: Good; and is not a space of eight feet twice the size of this [$ABCD$], and half the size of the other [$AJKL$]?

BOY: Certainly.

SOCRATES: Such a space, then, will be made out of a line greater than this one, and less than that one?

BOY: Yes; I think so.

SOCRATES: Very good; I like to hear you say what you think. And now tell me, is not this a line of two feet and that of four?

BOY: Yes.

SOCRATES: Then the line which forms the side of eight feet ought to be more than this line of two feet, and less than the other of four feet?

BOY: It ought.

SOCRATES: Try and see if you can tell me how much it will be.

BOY: Three feet.

SOCRATES: Then if we add a half to this line of two [BO, *half of* BJ], that will be the line of three. Here are two and there is one; and on the other side, here are two also and there is one: and that makes the figure of which you speak? [*Socrates completes the square* $AOPQ$ (see Figure 3.4).]

BOY: Yes.

SOCRATES: But if there are three feet this way and three feet that way, the whole space will be three times three feet?

BOY: That is evident.

SOCRATES: And how much are three times three feet?

BOY: Nine.

SOCRATES: And how much is the double of four?

BOY: Eight.

SOCRATES: Then the figure of eight is not made out of a line of three?

BOY: No.

SOCRATES: But from what line? — tell me exactly; and if you would rather not reckon, try and show me the line.

BOY: Indeed, Socrates, I do not know.

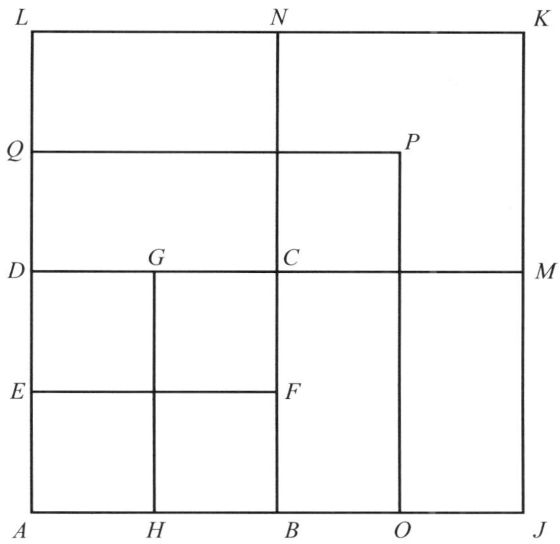

Figure 3.4. Another incorrect doubling refuted

SOCRATES: Do you see, Meno, what advances he has made in his power of recollection? He did not know at first, and he does not know now, what is the side of a figure of eight feet: but then he thought that he knew, and answered confidently as if he knew, and had no difficulty; now he has a difficulty, and neither knows nor fancies that he knows.

MENO: True.

SOCRATES: Is he not better off in knowing his ignorance?

MENO: I think that he is.

SOCRATES: If we have made him doubt, and given him the 'torpedo's shock,' have we done him any harm?

MENO: I think not.

SOCRATES: We have certainly, as would seem, assisted him in some degree to the discovery of the truth; and now he will wish to remedy his ignorance, but then he would have been ready to tell all the world again and again that the double space should have a double side.

MENO: True.

SOCRATES: But do you suppose that he would ever have enquired into or learned what he fancied that he knew, though he was really ignorant of it, until he had fallen into perplexity under the idea that he did not know, and had desired to know?

MENO: I think not, Socrates.

SOCRATES: Then he was the better for the torpedo's touch?

MENO: I think so.

SOCRATES: Mark now the farther development. I shall only ask him, and not teach him, and he shall share the enquiry with me: and do you watch and see if you find me telling or explaining anything to him, instead of eliciting his opinion.

[*Socrates here rubs out the previous figures and starts again.*]

Tell me, boy, is not this a square of four feet [$ABCD$] which I have drawn?

BOY: Yes.

SOCRATES: And now I add another square equal to the former one? [$BCEF$]

BOY: Yes.

SOCRATES: And a third, which is equal to either of them? [$CEGH$]

BOY: Yes.

SOCRATES: Suppose that we fill up the vacant corner? [$DCHJ$]

BOY: Very good.

SOCRATES: Here, then, there are four equal spaces?

BOY: Yes.

SOCRATES: And how many times larger is this space than this other?

BOY: Four times.

SOCRATES: But it ought to have been twice only, as you will remember.

BOY: True.

SOCRATES: And does not this line, reaching from corner to corner, bisect each of these spaces? [*Points to* BD]

BOY: Yes.

SOCRATES: And are there not here four equal lines which contain this space? [$BEHD$, (*see Figure* 3.5).]

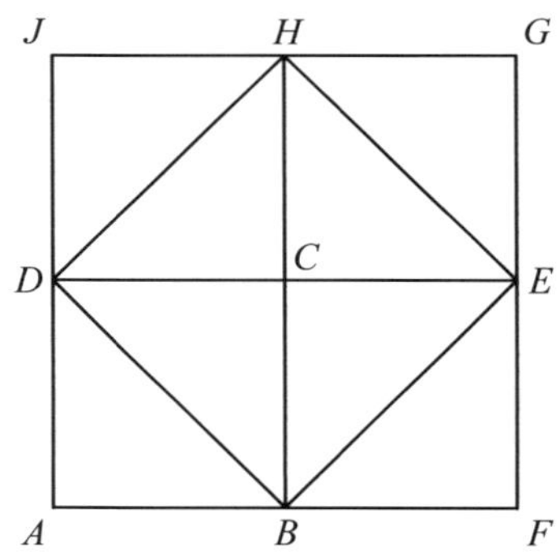

Figure 3.5. A doubling of the square

BOY: There are.

SOCRATES: Look and see how much this space is.

BOY: I do not understand.

SOCRATES: Has not each interior line cut off half of the four spaces?

BOY: Yes.

SOCRATES: And how many spaces are there in this section? [$BEHD$]

BOY: Four.

SOCRATES: And how many in this? [$ABCD$]

BOY: Two.

SOCRATES: And four is how many times two?

BOY: Twice.

SOCRATES: And this space is of how many feet?

BOY: Of eight feet.

SOCRATES: And from what line do you get this figure?

BOY: From this. [*Points to* BD]

SOCRATES: That is, from the line which extends from corner to corner of the figure of four feet?

BOY: Yes.

SOCRATES: And that is the line which the learned call the diagonal. And if this is the proper name, then you, Meno's slave, are prepared to affirm that the double space is the square of the diagonal?

BOY: Certainly, Socrates.

SOCRATES: What do you say of him, Meno? Were not all these answers given out of his own head?

MENO: Yes, they were all his own.

SOCRATES: And yet, as we were just now saying, he did not know?

MENO: True.

SOCRATES: But still he had in him those notions of his–had he not?

MENO: Yes.

SOCRATES: Then he who does not know may still have true notions of that which he does not know?

MENO: He has.

SOCRATES: And at present these notions have just been stirred up in him, as in a dream; but if he were frequently asked the same questions, in different forms, he would know as well as any one at last?

MENO: I dare say.

SOCRATES: Without any one teaching him he will recover his knowledge for himself, if he is only asked questions?

MENO: Yes.

SOCRATES: And this spontaneous recovery of knowledge in him is recollection?

MENO: True.

SOCRATES: And this knowledge which he now has must he not either have acquired or always possessed?

MENO: Yes.

SOCRATES: But if he always possessed this knowledge he would always have known; or if he has acquired the knowledge he could not have acquired it in this life, unless he has been taught geometry; for he may be made to do the same with all geometry and every other branch of knowledge. Now, has any one ever taught him all this? You must know about him, if, as you say, he was born and bred in your house.

MENO: And I am certain that no one ever did teach him.

SOCRATES: And yet he has the knowledge?

MENO: The fact, Socrates, is undeniable.

SOCRATES: But if he did not acquire the knowledge in this life, then he must have had and learned it at some other time?

MENO: Clearly he must.

SOCRATES: Which must have been the time when he was not a man?

MENO: Yes.

SOCRATES: And if there have been always true thoughts in him, both at the time when he was and was not a man, which only need to be awakened into knowledge by putting questions to him, his soul must have always possessed this knowledge, for he always either was or was not a man?

MENO: Obviously.

SOCRATES: And if the truth of all things always existed in the soul, then the soul is immortal. Wherefore be of good cheer, and try to recollect what you do not know, or rather what you do not remember.

MENO: I feel, somehow, that I like what you are saying.

SOCRATES: And I, Meno, like what I am saying. Some things I have said of which I am not altogether confident. But that we shall be better and braver and less helpless if we think that we ought to enquire, than we should have been if we indulged in the idle fancy that there was no knowing and no use in seeking to know what we do not know; — that is a theme upon which I am ready to fight, in word and deed, to the utmost of my power.

MENO: There again, Socrates, your words seem to me excellent.

Further remarks on the dialogue in *Meno*. The purpose of the mathematical discussion was to elicit from the slave boy the solution of the following problem:

Find a square whose area is twice that of a given square.

The slave boy first established that the given square, whose side is 2 feet, has an area of 4 square feet. It follows that the square he seeks must have an area of 8 square feet. The problem is to find the length of its side.

The slave boy first suggests a length of 4 feet: to double the area, just double the sides. But Socrates leads him to realise that this gives a square with area 16, rather than 8, and so the correct length must lie between 2 and 4. The slave boy then proposes a length of 3 feet, but this gives an area of 9, which is still too large. The boy is perplexed by this until Socrates draws another figure in the sand that enables him to deduce that the diagonal of the original square is the side that is required.

This dialogue illustrates a number of ways in which Greek mathematical activity appears to differ from what we have learned of Egyptian and Mesopotamian mathematics. There are three main features to notice.

First, note the character of the source. This is an articulate literary composition, in marked contrast to the staccato instructions, cryptic calculations, and tables that we saw earlier. Although the contrast can partly be attributed to the particular source, a philosophical conversation written with more than purely mathematical intentions, the impression remains of a much fuller explanation of what is going on, and why.

Second, the subject matter is different, being more explicitly geometrical: it is about the relationships of shapes, and such numbers as are introduced seem fairly incidental. The slave boy's guesses at the length of the side prove to be an unsatisfactory approach to the problem, and the solution comes about only when he stops guessing numbers and considers a geometrical construction.

Third, a major concern is to show *why* the result is true. Throughout the argument, great care is taken to show which statements follow from other statements and what are the implications of various conjectures: there is a logical air to the proceedings that we have not met before. The result comes about not so much by a series of instructions — 'to double a square, do this, this, and this' — but by a process whose aim is to enable the boy to feel convinced that the construction yields the right answer: having seen how the square is constructed on the diagonal, it becomes the boy's 'personal opinion' that this is indeed the solution, because he has seen why. Moreover, as Socrates observes, when the boy has done this a few times, 'he would know as well as anyone at last'. In other words, he will finally know the subject as well as anyone.

The structure of part of the mathematical argument is interesting. Let us take the boy's first belief, that to double the area we double the side. Socrates knows that this is wrong, but in order to show why this is the case he goes along with it in order to pursue its implications: 'doubling the side has given us not a double but a four-fold figure'. Since this is not what was wanted, the initial belief must have been false.

Socrates also makes a significant remark half-way through, after persuading the boy that his side of length three feet does not work either:

> But from what line? — tell me exactly; and if you would rather not reckon, try and show me the line.

The casual conversational tone here is deceptive. Socrates is giving a strong hint that the required length cannot be told exactly, in numbers like 2, 3, or 4, but can be shown only on a diagram: the length can be pointed to, but not 'reckoned' or counted up. So even the apparently straightforward procedure of attaching numbers to lengths of lines seems inadequate for elementary geometrical investigations.

Finally, note the clear distinction made here between having opinions and having knowledge: in this early source, mathematics appears as exemplifying knowledge of an especially secure kind. We receive no hint of this conception in pre-Greek sources.

This completes our survey of this passage. We have raised a number of questions, and there are puzzling features of which earlier non-Greek sources gave no warning. Explaining these will take up much of the next few chapters. Without further evidence we cannot assume that this source is typical of the mathematical level or concerns of the period (early 4th century BC), but we can use it to focus on features in the Greek mathematical landscape that we need to explore further.

3.2 Geometry before Plato

In this section we look at some of the earliest names associated with mathematics in Greek times, starting with Thales, Pythagoras, and Hippocrates of Chios. While doing so, we shall need to confront the difficult question of our sources for early Greek mathematics, which are generally written by people other than mathematicians and which have reached us only by a complicated process of transmission.

We take as our two 'fixed points' the two dates already mentioned: the composition of Plato's *Meno* (385 BC) and of Euclid's *Elements* (300 BC).[3] These give us the first three periods of our chronological division of Greek mathematics: before Plato (6th and 5th centuries BC); between Plato and Euclid (4th century BC); and the time of Euclid (3rd century BC). This division is not entirely arbitrary — in their different ways Plato and Euclid were important figures in the development of Greek mathematics, and so there are good reasons for framing our analysis in this way.

Although Greek writings can be more chatty and informative than earlier ones, we rarely have the original sources. Whereas the Egyptian Rhind Papyrus and Mesopotamian clay tablets are the physical objects that left the scribes' hands, there are no Greek equivalents of any significance. Like the Egyptians, the Greeks generally wrote on papyrus, which did not survive well in the moist climate of the Greek Empire. Instead, what have come down to us are copies of copies of copies, and it is not unusual for the earliest extant manuscript to be closer in time to us than to the original source. Furthermore, the copyists sometimes made mistakes, introduced further material, reorganised and 'improved' the texts (by numbering the propositions, for example), or omitted parts of it. So establishing or reconstructing the original text can be very difficult.

One implication of this is that the works that have come down to us are primarily those that each successive generation has thought worthy of preserving, copying, and handing on, for one reason or another. On the other hand, many works whose names we know and which we greatly wish to study have not survived. For example, Euclid's *Elements* seems to have superseded earlier works of the same name that are no longer extant: presumably they were discarded, just as we now discard outdated textbooks.

Most distressingly for historians, an account of the development of geometry written in the 4th century BC by Eudemus of Rhodes, a pupil of Aristotle, has been lost. It was accessible to our next author Proclus, who quotes him by name in his commentary on Euclid's *Elements*, so it survived at least as far as the start of the 5th century AD. Today we can only speculate on what it contained, guided by the few fragmentary quotations from it that appear in other sources.

Proclus. Our evidence for the existence of these early *Elements* comes from a commentary on Euclid written by the scholar Proclus, who was born in Byzantium (now Istanbul) around AD 410 and was educated in Alexandria and in Athens, where he eventually became head of a philosophical academy. In the long extract that follows, which is taken from Proclus's summary of the history of geometry up to the time of Euclid, it is interesting to balance his attention to his sources of information with his own identification with one particular school of thought.[4]

Proclus's historical summary.

> Thales, who had travelled to Egypt, was the first to introduce this science into Greece. He made many discoveries himself and taught the principles for many others to his successors, attacking some problems in a general way and others more empirically. Next after him Mamercus, brother of the poet Stesichorus, is remembered as having applied himself to the study of geometry; and Hippias of Elis records that he acquired a reputation in it.

[3] Both of these dates are approximate: they are our best estimates in the light of the evidence available and, as we discuss below in Section 3.4, the date of the *Elements* is particularly uncertain.

[4] From (Morrow 1970, 52–57) in F&G 2.A1. (Morrow 1970) is the source of all references to Proclus's writings in this chapter.

Following upon these men, Pythagoras transformed mathematical philosophy into a scheme of liberal education, surveying its principles from the highest downwards and investigating its theorems in an immaterial and intellectual manner. He it was who discovered the doctrine of proportionals and the structure of the cosmic figures. After him Anaxagoras of Clazomenae applied himself to many questions in geometry, and so did Oenopides of Chios, who was a little younger than Anaxagoras.

Both these men are mentioned by Plato in the *Erastae* as having got a reputation in mathematics. Following them Hippocrates of Chios, who invented the method of squaring lunules, and Theodorus of Cyrene became eminent in geometry. For Hippocrates wrote a book on elements, the first of whom we have any record who did so.

Plato, who appeared after them, greatly advanced mathematics in general and geometry in particular because of his zeal for these studies. It is well known that his writings are thickly sprinkled with mathematical terms and that he everywhere tries to arouse admiration for mathematics among students of philosophy.

At this time also lived Leodamas of Thasos, Archytas of Tarentum, and Theaetetus of Athens, by whom the theorems were increased in number and bought into a more scientific arrangement. Younger than Leodamas were Neoclides and his pupil Leon, who added many discoveries to those of their predecessors, so that Leon was able to compile a book of elements more carefully designed to take account of the number of propositions that had been proved and of their utility. He also discovered *diorismi*, whose purpose is to determine when a problem under investigation is capable of solution and when it is not. Eudoxus of Cnidus, a little later than Leon and a member of Plato's group, was the first to increase the number of the so-called general theorems; to the three proportionals already known he added three more and multiplied the number of propositions concerning the 'section' which had their origin in Plato, employing the method of analysis for their solution. Amyclas of Heracleia, one of Plato's followers, Menaechmus, a student of Eudoxus who also was associated with Plato, and his brother Dinostratus made the whole of geometry still more perfect. Theudius of Magnesia had a reputation for excellence in mathematics as in the rest of philosophy, for he produced an admirable arrangement of the elements and made many partial theorems more general.

There was also Athenaeus of Cyzicus, who lived about this time and became eminent in other branches of mathematics and most of all in geometry. These men lived together in the Academy, making their inquiries in common. Hermotimus and Colophon pursued further the investigations already begun by Eudoxus and Theaetetus, discovered many propositions in the *Elements*, and wrote some things about locus-theorems. Philippus of Mende, a pupil whom Plato had encouraged to study mathematics, also carried on his investigations according to Plato's instructions and set himself to study all the problems that he thought would contribute to Plato's philosophy.

> All those who have written histories bring to this point their account of the development of this science. Not long after these men came Euclid, who brought together the *Elements*, systematising many of the theorems of Eudoxus, perfecting many of those of Theaetetus, and putting in irrefutable demonstrable form propositions that had been rather loosely established by his predecessors. He lived in the time of Ptolemy the First, for Archimedes, who lived after the time of the first Ptolemy, mentions Euclid. It is also reported that Ptolemy once asked Euclid if there was not a shorter road to geometry than through the *Elements*, and Euclid replied that there was no royal road to geometry. He was therefore later than Plato's group but earlier than Eratosthenes and Archimedes, for these two men were contemporaries, as Eratosthenes somewhere says. Euclid belonged to the persuasion of Plato and was at home in this philosophy; and this is why he thought the goal of the *Elements* as a whole to be the construction of the so-called Platonic figures.

Proclus's historical survey of Greek mathematics is the fullest we have that was written relatively near to the time, but while reading his account of Euclid we should note that he was writing some 750 years after Euclid flourished. This is roughly our distance from the signing of Magna Carta, so where Proclus acquired his information is an interesting question. His survey pays serious attention to what he can assert in the light of the sources at his disposal. His weighing of the evidence for when Euclid lived is a tribute to his endeavour to be truthful and accurate, and also reminds us that Euclid was already an inhabitant of the dim past when Proclus wrote.

We might also be suspicious of his remarks about Plato's writings. Proclus was the head of a neo-Platonic Academy,[5] which could explain his emphasis on Plato, and it would not be surprising if he exaggerated a little in a spirit of retrospective pride.

However, 800 years is a long time in the life of any system of belief, and it could easily have evolved into a philosophy quite different from what its founder intended. On the other hand, there are good reasons for thinking that Proclus's source for this period was Eudemus, who must have been writing within a few years of Plato's death. Since Proclus documents his sources carefully when writing about Euclid, his account of Plato and his contemporaries seems likely to be fairly reliable.

Proclus asserted that Greek geometry was learned from the Egyptians. Greek sources are remarkably unanimous about this, even while disagreeing on why it occurred. This is a problem for historians, because the Egyptian geometry that we know about bears little resemblance to Greek geometry (as exemplified in the *Meno* passage or in Euclid's *Elements*), being more in the nature of arithmetic applied to the measurement of shapes. So it is surprising that ancient authors saw a continuity of tradition between Egypt and Greece, when the qualitative leap made by the Greeks is more apparent to us.

Thales and Pythagoras. The earliest Greek mathematical investigations that we know of took place during the 6th century BC and are associated with the name of Thales, who lived at Miletus (in modern-day Turkey), the wealthiest town of Ionia at the centre of major trade routes by land and sea. A little later in the same century we hear of Pythagoras, who came from Samos, an island north-west of Miletus. But what these men achieved is now difficult for us to ascertain.

[5]The word 'neo-Platonic' means following the ideas of Plato. We discuss this in more detail in Chapter 6.

Thales of Miletus is one of the most shadowy figures in the history of Greek mathematics. We have tantalising pieces of information, about which scholars reach conflicting opinions, but little actual knowledge.

Proclus's *Commentary* is very clear: Thales travelled to Egypt and was the first to introduce mathematics into Greece. This sounds straightforward, but you already know enough to be suspicious — what could Thales have brought back from Egypt (or Mesopotamia, for that matter) that would have started the very different enterprise of Greek geometry? Later, Proclus says that Thales was the first to demonstrate that *the circle is bisected by the diameter*, but one's confidence in this assertion is somewhat dented when Proclus continues 'the cause of this bisection is the undeviating course of the straight line through the centre'.[6] This does not sound like the language of later Greek geometry, and one wonders about the implications of the fact that Proclus lived almost a thousand years after Thales. On the other hand, there seems to be no reason for Proclus to invent this information. Once again, we are reminded of how much we have lost.

As for Pythagoras, almost nothing is known for certain about the man himself. There seems to have been a mystically or philosophically inclined group called the Pythagoreans, and we can suppose that he was influential in it, but nothing ties him explicitly to the mathematical aspects of their philosophy. One of our few sources is Aristotle, who wrote that the 'so-called Pythagoreans' were the first to advance the study of mathematics. Indeed, according to Aristotle in his *Metaphysics*:[7]

> they thought its principles were the principles of all things . . . such and such a modification of numbers being justice, another being soul and reason, another being opportunity . . .

This does not sound much like mathematics to us. Indeed, Serafina Cuomo, in line with most historians today, goes further:[8]

> Most scholars will agree that there was a Pythagorean school of philosophy from the sixth until probably the fourth century BC, that they were involved in politics and that they had certain beliefs about life and the universe, including perhaps the tenet that 'everything is number', or that number holds the key to understanding reality. But most scholars today also think, for instance, that Pythagoras never discovered the theorem that bears his name.

The Pythagorean theorem states that the area C of the square on the hypotenuse (the longest side) of a right-angled triangle is equal to the sum $A+B$ of the areas of the squares on the other two sides (see Figure 3.6). We have seen that some such result was known much earlier to mathematicians in Mesopotamia, but this does not mean that the theorem could not have been rediscovered independently in Greece, or supplied with a proof (or a new or better one). The historical record does not permit us to make any such claim, because there is no good evidence to support it. It seems that an ancient result has been ascribed to an old group of philosophers with some interest in mathematics and in the absence of any compelling evidence to the contrary.

It is probably safe to concede that Thales and Pythagoras existed, but the specific discoveries attributed to them in later sources seem improbable at best, and palpably legendary in some instances. However, even legends can be informative, as they tell us what people of later generations wanted to hear; the existence of a legend (whether true or not) is a historical fact.

[6]See (Proclus 1970, 77) and F&G 3.B1(b).

[7]See pages 985b–986a, quoted in (Cuomo 2001, 34).

[8]See (Cuomo 2001, 30–31).

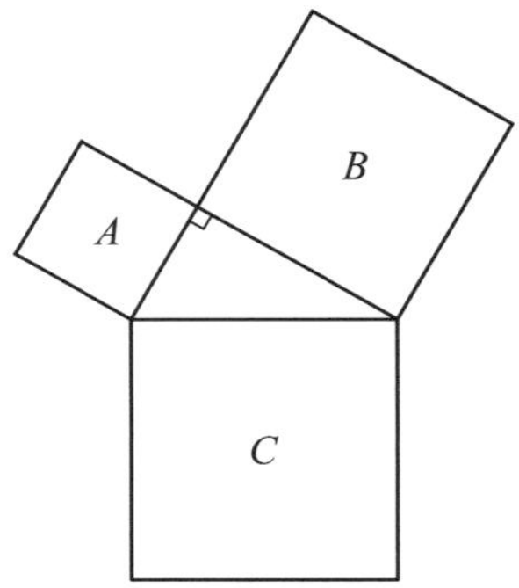

Figure 3.6. $A + B = C$

Returning to Proclus's survey, there are two aspects to note. Proclus writes of a transformation of geometrical study into 'a scheme of liberal education', and mentions that Thales disclosed 'principles ... to his successors'.[9]

The first of these is significant because one of the most notable new aspects of Greek mathematics was the extent to which it was consciously different from that of calculation for practical purposes. The doubts that arose in our evaluation of Egyptian mathematics, over the extent to which it was anything more than practical computation, cannot arise with respect to Greek mathematics. This is because we learn quite explicitly what mathematics was (for educated Greeks), and what it was for: it was linked, at least in the influential work of Plato, with a high educational ideal quite divorced from problems of everyday calculation and measuring.[10]

Second, from Proclus's mention of 'successors' and the names of several of them, we learn of something else of which we have no record in earlier cultures. This is a *research tradition*, whereby people saw themselves as studying mathematics and developing it, solving certain problems arising in the work of their teachers and predecessors, and perhaps creating fresh ones. However, to establish this historically we must go beyond the writings of Proclus.

Hippocrates of Chios. We can analyse several aspects of the geometrical research tradition as it developed in the later 5th century BC, just before Plato's time. Any such tradition comprises:

- a set of problems that change as some are solved or give rise to others
- particular styles of method for approaching these problems
- various means of justifying or explaining the solutions of these problems

We now look at how these aspects are exemplified in the work of Hippocrates,[11] who came from the island of Chios (between Samos and Lesbos, off the Ionian coast) and lived in Athens in the latter half of the 5th century, perhaps from 450–430 BC.

Proclus refers to Hippocrates as having discovered the quadrature of the lune. A *lune*, as its name suggests, is a moon-shaped crescent (see Figure 3.7). Finding a *quadrature* of a lune means finding a square with the same area, preferably in a general way that

[9] In Chapter 4 we discuss what these principles might have been.

[10] We discuss Plato's views in the next section.

[11] Hippocrates of Chios is not the same person as Hippocrates of Cos, who initiated the study of medicine and after whom the 'Hippocratic oath' is named.

applies to lunes of various shapes and sizes. Unless we have reason to suppose that lunes were especially significant, we can probably infer that this is but one example of a class of problems that involved trying to find the quadratures of all kinds of shapes.

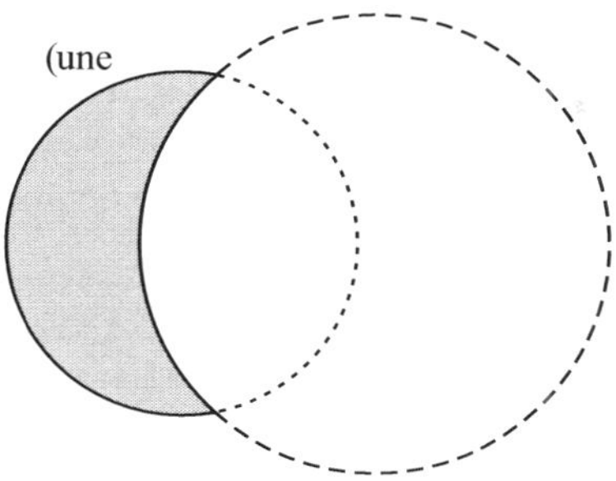

Figure 3.7. A lune

Fortunately, we have a full description of Hippocrates's work on lunes — indeed, we have no account as detailed as this of any other mathematics from the later 5th century BC. It appears in a copy made by Simplicius in the 6th century AD of an extract from the now-lost *History of Geometry* by Eudemus. However, we cannot know what alterations were made, if any, in this process of transmission.

The account yields a surprising amount of information, even before we attempt a full mastery line by line. In the following extract Hippocrates did not deal with lunes in general, but took a particular case of a lune whose outer edge is a semicircle and whose inner edge is 'similar' to the outer arcs formed in the course of his construction.

The key word that describes his method is *construction*. Hippocrates *did* something ('circumscribing') to produce a figure, in much the same spirit as Socrates drew squares in the sand for the slave boy, and then supplied an argument to show how this led to what was required.

The justification for what Hippocrates did is a logical one, and is more sophisticated than that in the *Meno* passage. The truth of Hippocrates's claim (see our analysis of Hippocrates's account in Box 3.1, which follows the extract) about the quadrature of the lune depends upon several geometrical results that were presumably known to him. The construction was then devised so as to ensure that these proved or known truths can be applied to the constructed situation.

The following extract from Eudemus's *History of Geometry*, handed down by Simplicius, has been called 'one of the most precious sources for the history of Greek geometry before Euclid', in the judgement of Sir Thomas Heath,[12] and is especially interesting as it records in detail such an early attempt at finding the area of figures with *curved* sides, from which the theorems that must have been known in order to establish this rigorously can be inferred. Simplicius said that he was copying out Eudemus 'word for word, adding only for the sake of clearness a few things taken from Euclid's *Elements*'. In the passage as given here, these additions (as far as they can be determined) have been removed, so what remains is believed to be close to the original Eudemus. How accurate an account Eudemus, writing a century after Hippocrates, gave of this work is hard to say. It runs as follows.[13]

[12] See (Heath 1921, Vol. I, 182).

[13] See (Thomas 1939, Vol. I, 239–249) in F&G 2.B.

Eudemus on Hippocrates's quadrature of lunes.

> The quadratures of lunes, which seemed to belong to an uncommon class of propositions by reason of the close relationship to the circle, were first investigated by Hippocrates, and seemed to be set out in correct form; therefore we shall deal with them at length and go through them. He made his starting-point, and set out as the first of the theorems useful to his purpose, that similar segments of circles have the same ratios as the squares on their bases. And this he proved by showing that the squares on the diameters have the same ratios as the circles.
>
> Having first shown this he described in what way it was possible to square a lune whose outer circumference was a semicircle. He did this by circumscribing about a right-angled isosceles triangle a semicircle and about the base a segment of a circle similar to those cut off by the sides. Since the segment about the base is equal to the sum of those about the sides, it follows that when the part of the triangle above the segment about the base is added to both the lune will be equal to the triangle. Therefore the lune, having been proved equal to the triangle, can be squared. In this way, taking a semicircle as the outer circumference of the lune, Hippocrates readily squared the lune.

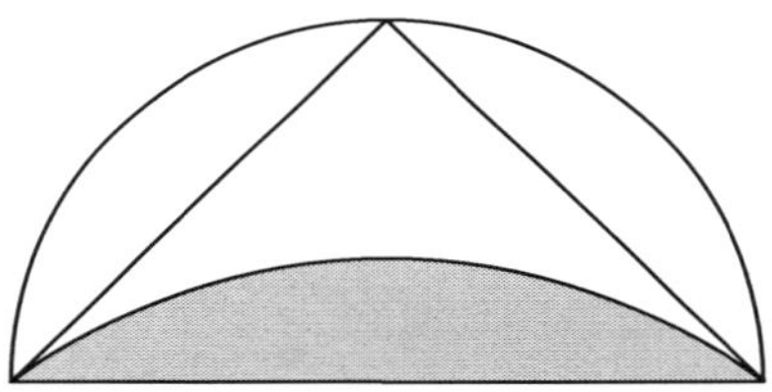

Figure 3.8. The first squaring of a lune

Hippocrates then squared some other lunes, but a discussion of what he did would take us too far afield.[14] Eudemus then concluded:

> Thus Hippocrates squared every lune, seeing that he squared not only the lune which has for its outer circumference a semicircle, but also the lune in which the outer circumference is greater, and that in which it is less, than a semicircle.

From this extract we have learned something about the level, style, and problems of geometrical research during the period. More generally, we seem to have found ourselves in the middle of a research tradition concerned with properties of figures and using methods of construction and logical justification. Assuming our sources to be reliable, this is interesting and rather remarkable.[15]

---

[14] See F&G 2.B.

[15] In Chapters 4–6 we develop these concerns more fully.

Box 3.1. Hippocrates's quadrature of a lune.

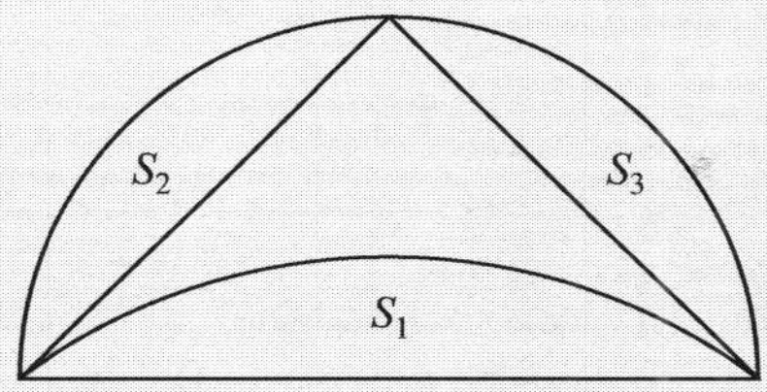

Figure 3.9. Three areas

Hippocrates's construction. Starting with a right-angled triangle whose two shortest sides are equal, he treated the hypotenuse as the diameter of a circle and drew a semicircle through the three corners of the triangle.[a]

On this hypotenuse (diameter) he then constructed the inner edge of the lune, a circular arc similar to (but bigger than) each of the two circular arcs forming the outer edge of the lune[b] (see Figure 3.9). This completes the construction of the lune.

Next Hippocrates proved that the area of the large segment S_1 is the sum of the areas of the two smaller segments S_2 and S_3: it then follows that his lune is equal in area to his original triangle.[c] This accomplishes the quadrature.

Since all three segments have the same shape, the ratio of their areas is equal to the ratio of the squares on their bases: this follows from the Pythagorean theorem. It follows from the equality of ratios that the area of the large segment is the sum of the areas of the two smaller segments.

[a] This required him knowing that such a semicircle does indeed pass through these three corners.
[b] This required him knowing that such a circular arc can be constructed.
[c] This required him knowing that any triangle can be changed into a square of equal area.

3.3 Plato and Aristotle

Plato was born in Athens in about 427 BC and is of great significance to our story, for several reasons.

First, as Proclus noted, he 'greatly advanced mathematics in general, and geometry in particular, because of his zeal for these studies'. Plato was a philosopher and not primarily a mathematician, and the precise nature of his contribution is still not fully agreed upon, but he seems to have been an important influence upon the mathematicians of his time, by inspiration and direction.

Second, the works of Plato are our fullest and best source of information about the mathematical developments at that time. As in the *Meno* extract he generally introduced mathematics into his discussion illustratively or metaphorically, rather than writing mathematical treatises as such, and the meaning and significance of some of his remarks still give rise to lively controversy.

Third, the historical significance of Plato's thought on succeeding generations is almost unparalleled in our culture. It is barely an exaggeration to say that Plato 'defined the complex of general ideas forming the imperishable origin of Western thought'.[16]

One of the dominant conceptions of what mathematics is about, from Greek times until now, is Plato's. Much of Plato's influence upon his contemporaries was through founding a school in about 387 BC, in a part of Athens called *Academy* (from which all subsequent educational uses of this word are derived). Here he lectured,[17] wrote, and directed studies for most of the remaining forty years of his life, dying around 347 BC.

Nearly all the 4th-century mathematicians we know of were connected either as students or as teachers with the Academy. Later we shall learn about the mathematicians associated with Plato. For now, it is important to retain the sense of the research tradition in geometry being stimulated and developed especially by people working in Athens, with Plato's Academy becoming latterly a focal point for such studies. In 367 BC, some twenty years after its founding, the 17-year-old Aristotle arrived at the Academy as a student.

The influence of Aristotle, our other main contemporary source for 4th-century mathematics, has been scarcely less than Plato's. Aristotle, who lived from 384 to 322 BC, was in many respects more pragmatic and down to earth than Plato, and while he lacked Plato's literary elegance he wrote on a much wider range of subjects, including biology, physics, cosmology, and many topics in philosophy. He often disagreed with Plato, which can be helpful for our understanding of what Plato was taken to be saying.

Among Aristotle's philosophical interests was logic. His analysis of logical arguments essentially defined and constituted the science of logic for over 2000 years, and his view of the universe's structure and operating principles lasted (in Western thought) nearly as long. Like Plato, he wrote no explicitly mathematical work as such, but he referred to contemporary mathematics in the course of illustrating points of discussion, especially its logical aspects.

Aristotle was associated with the Academy until Plato's death in 347 BC, and then left Athens for a while. In 342 BC he became tutor to the young prince Alexander of Macedon, who became king after his father Philip's assassination in 336 BC, and who, by the time of his own death only thirteen years later, had conquered and assembled an empire from Egypt to the Black Sea and from Italy to the Indus. Alexander the Great's conquest of the Persian empire brought with it the ancient Mesopotamian cities of Babylon and Susa, as well as the Persian capital Persepolis, and he pushed beyond the old Persian frontiers to the edge of the Indian subcontinent, the site of the equally ancient Indus valley culture (see Figure 3.10). The possibility thus existed for cultural contacts and the spread of knowledge between the Mediterranean and India, reinforcing earlier trade routes.

Aristotle, meanwhile, had returned to Athens and founded a school, the *Lyceum*. More empirically scientific than the Academy, the Lyceum was supplied by Alexander (according to later sources) with both money and scientific specimens from his expeditions. For, somewhat like Napoleon two millennia later, Alexander's conquests were attended by a scholarly entourage of engineers, geographers, historians, philosophers, and naturalists, and this spirit of intellectual inquiry lasted longer than his empire. In Egypt, Alexander had founded the city of Alexandria. After his death from malaria in Babylon in 323 BC his

[16]See (Whitehead 1933, 132).

[17]Strictly, we know of only one lecture by Plato, on 'The good': see Aristoxenus in (Thomas 1980, Vol. 1, 389–391) and F&G 2.E8.

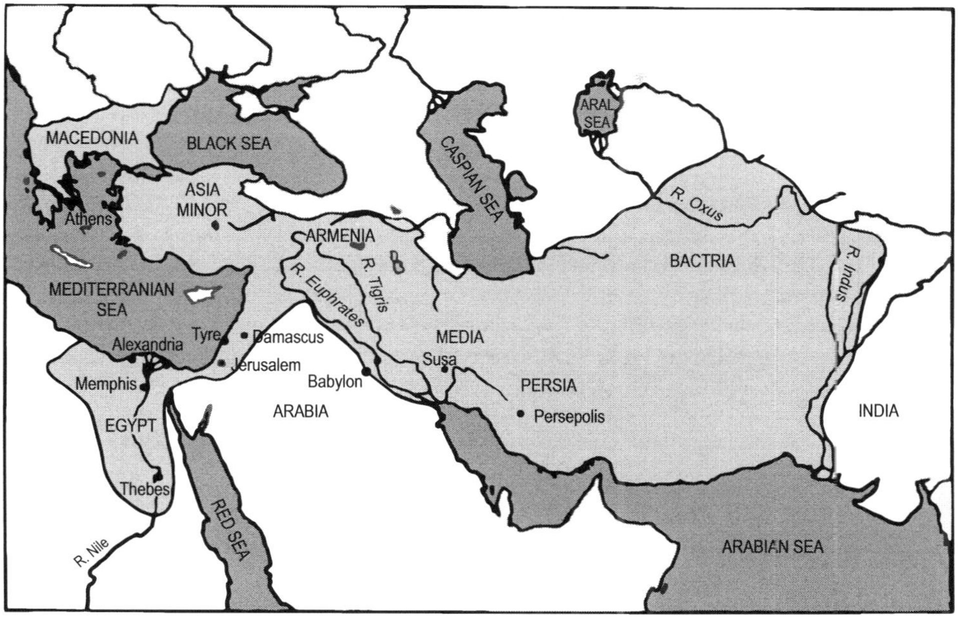

Figure 3.10. Alexander's empire

empire became partitioned among three of his generals, one of whom, Ptolemy, became ruler of Egypt.[18]

Although Egypt was the smallest third of Alexander's empire, it was also the richest and most easily governed, and Ptolemy's shrewdness may be further gauged by his securing possession of Alexander's body and bringing it to Alexandria, where he set about planning and constructing the new capital city of Egypt. He did this with such success, furthered by his son Ptolemy II who succeeded him in 285 BC, that Alexandria became the cultural centre of the Greek-speaking world for several centuries.

After the 3rd century BC, the state of our sources and the kind of mathematics done (where we know about it) move into a different phase, so this seems a good opportunity to pause and consider the nature of the subject matter. In this section, the words 'mathematics' and 'geometry' have been used somewhat interchangeably. Historically this is not inappropriate for the period between Plato and Euclid, but you can assess that better once we have looked more generally at the place of mathematics in Greek culture.

Plato's writings. The writings of Plato and Aristotle enable us to see what mathematics meant to educated Greeks of the 4th century BC. The very word *mathematics* is of Greek origin, and meant 'that which can be known', or 'any subject of study': it did not consistently take on something like its present meaning, as a particular kind of high-quality knowledge, until about the time of Aristotle. This shift of meaning itself tells us about how our subject was regarded — namely, as the most important. Most of our evidence upon the matter comes from Plato, who was not necessarily typical of all Greeks or even of all Athenians. Bearing this reservation in mind, we shall look at his views in more detail as representing an informed and articulate contemporary judgement.

[18]Do not confuse this Ptolemy, as many Renaissance writers did, with the astronomer Claudius Ptolemy (whom we meet in Chapter 6).

We begin with Plato's late dialogue, *Laws*, which alerts us to two things: Plato's ideas of what the various branches of mathematics are (which vary from one dialogue to another), and his firm view that some parts of education were only for an elite and were different in kind from what should form a general education.

A look at Plato's *Philebus* will help us to spell out what this division of the educational task entailed for mathematics, and we can juxtapose it with what we learn from other sources about the practical mathematics of Plato's day.

In *Timaeus*, Plato discussed the implications of geometry for cosmology and theology.

In Plato's most famous dialogue, *Republic*, we get an even closer look at Plato's opinions about mathematics, its nature, and purpose. Plato made a sharp distinction between 'pure' mathematics and the 'merely useful'. This makes for an interesting reading of what may be the earliest parts of Euclid's *Elements*, Books VII–IX on number theory: indeed, the content of some of these Books may well have been known in Plato's day.

More information can be gained by reading Aristotle. Among the 'elite' or 'advanced' topics in Plato's opinion is *incommensurability*,[19] and one of the earliest arguments we have for the existence of incommensurability is to be found in Aristotle's writings. Inevitably, this brings us to a consideration of proof in Greek mathematics, with which this section concludes, and which is studied in more detail in the next chapter.

Plato's *Laws*. In Plato's *Laws* a character called the 'Athenian Stranger' — usually taken to represent Plato himself — discusses mathematical education.[20] It becomes clear that Plato was concerned only with what freeborn boys should study, which makes the performance of the slave boy in *Meno* all the more striking.

While mathematics is useful, he argued, and useful mathematics is enough for some people, there is also a higher purpose. Basic arithmetic and calculations are useful for military purposes and in household management, and also to make boys 'more awake'. But the purpose of studying the measurement of geometrical objects is to free people from 'ridiculous and shameful ignorance', the nature of which he then went on to explain. Plato claimed that, for the masses, a mathematical education on a par with what was taught in Egypt is sufficient, but to study it more thoroughly and accurately is only 'for a select few'. It does seem, though, that he had more than elementary calculation in mind for the masses, whose studies should advance at least to the point where they could redeem the shameful disgrace of not understanding incommensurability.

The language that Plato used in relation to this last point is quite strong. It is shameful, he says in Book VII, and makes us more like pigs than men, not to know about this geometrical fact. His agitation should give us pause, not to dismiss it as the petulant irritation of a weary teacher, but as evidence of how, for Plato, the concept of mathematical knowledge and truth was very close to a defining characteristic of what distinguishes human beings from animals. For better or worse, Plato's exalted vision has not always been shared by subsequent generations.

The mathematical point at issue here is one implicit in *Meno*: not all lines are *commensurable* – that is, you cannot always find a small line that fits a whole number of times into two others that you are trying to compare. For example, two lines of lengths 2 and 5 are commensurable because each can be measured by a ruler of length 1, while the diagonal and the side of a square are incommensurable,[21] as Socrates hinted to the slave boy.

---

[19] We explain incommensurability at the end of this section.

[20] Plato, *Laws*, 817e–820d, in F&G 2.E6.

[21] We prove this later in this section.

Likewise, the lines of length 2π and 5π are commensurable with each other, but not with a line of length 1.

In *Laws*, Plato divided mathematical subjects into three parts: arithmetic, geometry, and astronomy. Elsewhere, he gave a four-part or five-part classification, which we consider shortly.[22] Cutting across this is a separate division into knowledge appropriate for the masses and that to be studied by the select few. Just how this latter division was made is not yet clear; Plato gave an image of old men in Athenian coffee-shops setting each other problems concerning incommensurability as readily as they played draughts, but this is perhaps an optimistic one.

The Athenian Stranger seems to suggest that the difference between mass studies and select ones is just a difference of thoroughness and accuracy. But, by the end, he proposes that the distinction must go further, into method and justification and problems — in short, into all aspects of the research tradition, as described in the previous section.

Plato's *Philebus*. It is an important historical fact that the first explicit differentiation between two kinds (or levels) of mathematical studies (ordinary and higher) is found in Plato's dialogue *Philebus*, where the classification is explained and made more explicit.

In this dialogue, Philebus advocates that life should be lived for the pursuit of pleasure, and he remains unconvinced by Socrates' arguments that the distinction between two arithmetics, and between two geometries, is made in order to investigate whether one kind of knowledge is purer than another. Protarchus, however, is gradually persuaded, concluding that the arithmetic and geometry of the philosophers are 'immensely superior in point of exactness and truth'. So the calculation and measurement involved in everyday practical activities was sharply distinguished from that of the philosopher, studied by the 'select few' in *Laws*. In making this distinction Plato used words such as *precision, exactness, purity,* and *truth*.

These considerations remind us that Plato was a philosopher of mathematics who also saw the subject as a stepping stone to higher things. Moreover, Plato did not repeat the doctrine of recollection that we saw in *Meno* — the slave boy must have recollected his knowledge of geometry from a previous existence because he could not have been taught it — although he could have done so; this suggests that by this late dialogue he had abandoned it.

As we shall see in the next chapter, mathematical truth and how it can be established were of the greatest concern to Greek mathematicians from the mid-5th century BC. So although our source for the explicit description of the distinction between everyday and 'higher' mathematics is in Plato's works, he was analysing a real and profound difference that started in Greek times and has persevered ever since.

Plato's *Timaeus*. Another use that Plato had for geometry was related to cosmology or theology, and concerns the five regular solids (see Figure 3.11) in which the faces are identical regular polygons:

- the *tetrahedron*, whose four faces are equilateral triangles
- the *cube* or *hexahedron*, whose six faces are squares
- the *octahedron*, whose eight faces are equilateral triangles
- the *dodecahedron*, whose twelve faces are regular pentagons
- the *icosahedron*, whose twenty faces are equilateral triangles

Because of Plato's interest in them, they are now known as the *Platonic solids*.

[22] His four-part classification appeared in five sections, because he split geometry into plane geometry and solid geometry.

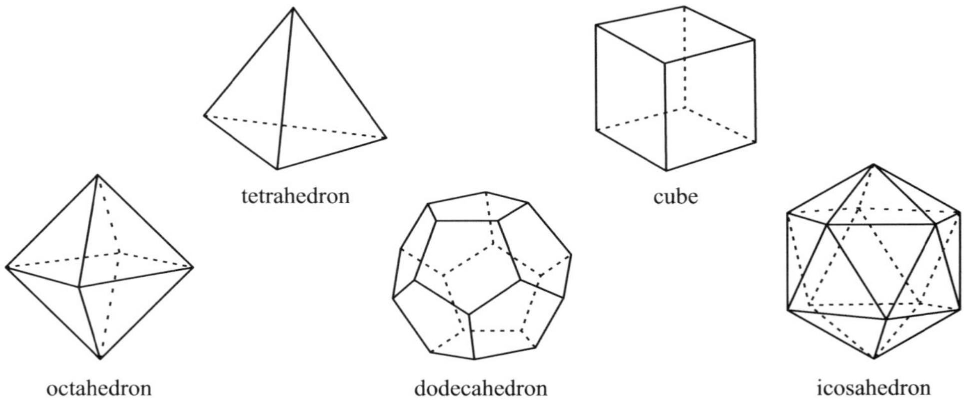

Figure 3.11. The five Platonic solids

Plato's friend Theaetetus, who died in 369 BC, may have been the first to recognise that there are only five regular solids, and only a few years later Plato incorporated a discussion of his work into *Timaeus*, a work of great importance in the later developments of neo-Platonism, and thus of Western thought in general. Here, Plato expressed most fully his conception of the creator-god as a mathematician who designed the world on mathematical principles. This explains how the physical world of change, transience, and decay is nonetheless rationally understandable, because it partakes in a pure mathematical design through the craftsman-like activity of its creator. *Timaeus* can thus be read as offering a theological view of how mathematics applies to the world — it is because the world is designed mathematically to begin with. This perspective was to inspire many generations, right up to our own time.

In *Timaeus* Plato described the relationship between the four elements *earth, fire, air,* and *water*.[23] He asserted that *air* and *water* are the two 'mean proportionals' between *fire* and *earth* — that is,

$$\textit{fire} : \textit{air} = \textit{air} : \textit{water} = \textit{water} : \textit{earth}.$$

We do not need to probe this formulation too closely — the significant point is that Plato explained the cohesion of the world through the language of proportion. Plato then introduced the regular solids and assigned one to each of the four elements (cube = *earth*, tetrahedron = *fire*, octahedron = *air*, icosahedron = *water*), leaving the dodecahedron to represent the *cosmos*.[24]

Plato went on to explain what he meant by the construction of the world-soul; it is a difficult passage, but it gives genuine insights into Platonic conceptions of the cosmos. In particular, Plato's division of mathematics into high and low, and his insistence that even high mathematics is to be studied as a stepping-stone to more fundamental issues, makes even more sense when we see how Plato intertwined the arithmetic of musical harmony with astronomical geometry in a mysterious vision of cosmic mathematics.

Everyday mathematics. Alongside the high geometrical research tradition ran an undercurrent of everyday computational mathematics. What did it look like? Ironically, we know less about mathematics at this level, practised by most of the citizens, than we

[23] *Timaeus* 32–33 and F&G 2.E5(a).

[24] *Timaeus* 53–56 and F&G 2.E5(a).

know about the high mathematics of the 'select few', because the former did not write books about it. Some things are clear, however.

Greek numerals are of interest because letters of the alphabet were used to denote numbers. To be precise 24 letters that were in daily use and 3 ancient ones brought back for the purpose (see Figure 3.12). For a long time historians believed that these Greek numerals were unsuited to calculation and must have been used just to record the result of calculations done on an abacus or a pebble counting-board (which the Greeks also used). However, it is quite possible to carry out multiplication (in a way not dissimilar to ours) with these alphabetic numerals.

1	2	3	4	5	6	7	8	9
α	β	γ	δ	ε	ς	ζ	η	θ
10	20	30	40	50	60	70	80	90
ι	κ	λ	μ	ν	ξ	o	π	ϙ
100	200	300	400	500	600	700	800	900
ρ	σ	τ	υ	ϕ	χ	ψ	ω	ϡ

Figure 3.12. The Greek numerals and their modern equivalents

To assess the uses to which calculations were put would amount to a survey of practical mathematics within ancient civilisations, and we cannot do justice to that here. But you have seen several pointers already: in *Laws*, Plato mentioned household management, and we could add that in *Philebus* he mentioned navigation, military science, building, agriculture (cows), and carpentry. Two other aspects of the latter passage are worthy of note: his mention of Egypt as an exemplar of arithmetic teaching, and the problems he gave as illustrations.

The spread of knowledge among ancient cultures, especially among Egypt, Mesopotamia, and Greece, is a question that has long taxed historians. While the evidence is not as full as we might wish, it seems plausible that knowledge of different computational practices spread along trade routes, for example. A later Hellenistic commentator spoke of 'the so-called Greek and Egyptian methods in multiplications and divisions, and the additions and subtractions of fractions',[25] showing that the Egyptian computational techniques we studied in Chapter 2 were known in later Greek culture. Again, the Mesopotamian sexagesimal system was used (with Greek numerals) in later Greek astronomical calculations, although how much other Mesopotamian mathematical lore filtered through to the Greeks is unclear. It would have been easy for sexagesimal numbers to spread after the conquests of Alexander, and harder (and more interesting) if they had spread before, but nothing is known for certain about this.

The particular teaching problems that Plato mentioned in *Laws* (the distribution of apples, and something to do with bowls and different metals) are significant in that they sound similar to some early Egyptian problems, such as the distribution of loaves in Problem 40 of the Rhind Papyrus (see p.12). Later sources contain similar problems: for instance, in

[25]Quoted in (Heath 1921), Vol. 1, 14.

the 6th century AD the scholar Metrodorus collected some arithmetical epigrams that concern problems such as dividing apples or nuts among a group of people in a particular way, and about finding the weights of bowls. There were also problems of the form, 'if a pipe fills a vessel in one day, another pipe takes two days to fill it and a third takes three days, how long will all three running together take to fill the vessel?'. So we can see a long tradition, probably extending much further back than the sophisticated geometrical research tradition of the select few, of elementary problems of an educational or recreational nature with some practical everyday overtones — a kind of 'mathematics of the people'.

Plato's *Republic*. We now return to Plato's classification of the mathematical sciences. Its fullest description and explanation appear in *Republic*, his great work about the ideal state, written around 380 BC. There Plato, in a dialogue between his brother Glaucon and Socrates, described the mathematical disciplines appropriate for the education of those who are to rule the state: *arithmetic*, *plane and solid* (three-dimensional) *geometry*, *astronomy*, and *musical harmony*. It is a long extract, but important for what it says about some relatively early opinions about the nature and purpose of mathematics in Greek times and also for the echoes it has created down to the present day.[26] It gains an extra layer of significance by being, at least in Plato's opinion, the right education for leaders, and this is captured by the distinctions that Plato draws between his aims and those of his much more down-to-earth brother. Each section starts off with Glaucon taking a robust, utilitarian view — one that reminds us, incidentally, of just how ready the Greeks were to go to war — and ends with Socrates transporting him to a philosophical view about the nature of knowledge and how it can be acquired. What it says about mathematics it says indirectly and in passing, which makes it an interesting challenge to discover.

Mathematical studies for the philosopher ruler.

> [*Socrates to Glaucon*] What sort of knowledge is there which would draw the soul from becoming to being? And another consideration has just occurred to me: You will remember that our young men are to be warrior athletes?
>
> Yes, that was said.
>
> Then this new kind of knowledge must have an additional quality?
>
> What quality?
>
> Usefulness in war.
>
> Yes, if possible.
>
> There were two parts in our former scheme of education, were there not?
>
> Just so.
>
> There was gymnastic which presided over the growth and decay of the body, and may therefore be regarded as having to do with generation and corruption?
>
> True.
>
> Then that is not the knowledge which we are seeking to discover?
>
> No.

[26]Becuase of its length, we have split this extract into six individual extracts. See (Plato 1898, 222–235). Another translation can be found in F&G 2.E2.

But what do you say of music, which also entered to a certain extent into our former scheme?

Music, he said, as you will remember, was the counterpart of gymnastic, and trained the guardians by the influences of habit, by harmony making them harmonious, by rhythm rhythmical, but not giving them science; and the words, whether fabulous or possibly true, had kindred elements of rhythm and harmony in them. But in music there was nothing which tended to that good which you are now seeking.

You are most accurate, I said, in your recollection; in music there certainly was nothing of the kind. But what branch of knowledge is there, my dear Glaucon, which is of the desired nature; since all the useful arts were reckoned mean by us?

Undoubtedly; and yet if music and gymnastic are excluded, and the arts are also excluded, what remains?

Well, I said, there may be nothing left of our special subjects; and then we shall have to take something which is not special, but of universal application.

What may that be?

A something which all arts and sciences and intelligences use in common, and which everyone first has to learn among the elements of education.

What is that?

The little matter of distinguishing one, two, and three — in a word, number and calculation:— do not all arts and sciences necessarily partake of them?

Yes.

Then the art of war partakes of them?

To be sure.

Then Palamedes, whenever he appears in tragedy, proves Agamemnon ridiculously unfit to be a general. Did you never remark how he declares that he had invented number, and had numbered the ships and set in array the ranks of the army at Troy; which implies that they had never been numbered before, and Agamemnon must be supposed literally to have been incapable of counting his own feet — how could he if he was ignorant of number? And if that is true, what sort of general must he have been?

I should say a very strange one, if this was as you say.

Can we deny that a warrior should have a knowledge of arithmetic?

Certainly he should, if he is to have the smallest understanding of military tactics, or indeed, I should rather say, if he is to be a man at all.

I should like to know whether you have the same notion which I have of this study?

What is your notion?

It appears to me to be a study of the kind which we are seeking, and which leads naturally to reflection, but never to have been rightly used; for the true use of it is simply to draw the soul towards being.

Plato has led his readers to the point where they can perceive the importance of mathematics, and now heads off the strictly utilitarian arguments that Glaucon (and, one might suppose, many of Plato's readers) would advocate.

Plato on arithmetic.

> Then this is a kind of knowledge which legislation may fitly prescribe; and we must endeavour to persuade those who are to be the principal men of our State to go and learn arithmetic, not as amateurs, but they must carry on the study until they see the nature of numbers with the mind only; nor again, like merchants or retail-traders, with a view to buying or selling, but for the sake of their military use, and of the soul herself; and because this will be the easiest way for her to pass from becoming to truth and being.
>
> That is excellent, he said.
>
> Yes, I said, and now having spoken of it, I must add how charming the science is! and in how many ways it conduces to our desired end, if pursued in the spirit of a philosopher, and not of a shopkeeper!
>
> How do you mean?
>
> I mean, as I was saying, that arithmetic has a very great and elevating effect, compelling the soul to reason about abstract number, and rebelling against the introduction of visible or tangible objects into the argument. You know how steadily the masters of the art repel and ridicule anyone who attempts to divide absolute unity when he is calculating, and if you divide, they multiply, taking care that one shall continue one and not become lost in fractions.
>
> That is very true.
>
> Now, suppose a person were to say to them: O my friends, what are these wonderful numbers about which you are reasoning, in which, as you say, there is a unity such as you demand, and each unit is equal, invariable, indivisible, — what would they answer?
>
> They would answer, as I should conceive, that they were speaking of those numbers which can only be realized in thought.
>
> Then you see that this knowledge may be truly called necessary, necessitating as it clearly does the use of the pure intelligence in the attainment of pure truth?
>
> Yes; that is a marked characteristic of it.
>
> And have you further observed, that those who have a natural talent for calculation are generally quick at every other kind of knowledge; and even the dull, if they have had an arithmetical training, although they may derive no other advantage from it, always become much quicker than they would otherwise have been.
>
> Very true, he said.
>
> And indeed, you will not easily find a more difficult study, and not many as difficult.
>
> You will not.
>
> And, for all these reasons, arithmetic is a kind of knowledge in which the best natures should be trained, and which must not be given up.
>
> I agree.

This defence of arithmetic continues to advocate that the more important aspect of mathematics is not its usefulness 'with a view to buying and selling', but that in learning to appreciate the nature of number the mind learns how to think. This has been the argument of educators on and off ever since. By insisting on it here, Plato may have denied historians a chance to find out about the practical arithmetic of his day, but he may also be shedding light on the remarkable insistence of Greek mathematicians not only on proof but on geometry, to which, with seeming inevitability, he next turned.

Plato on plane geometry.

> Let this then be made one of our subjects of education. And next, shall we enquire whether the kindred science also concerns us?
>
> You mean geometry?
>
> Exactly so.
>
> Clearly, he said, we are concerned with that part of geometry which relates to war; for in pitching a camp, or taking up a position, or closing or extending the lines of an army, or any other military manoeuvre, whether in actual battle or on a march, it will make all the difference whether a general is or is not a geometrician.
>
> Yes, I said, but for that purpose a very little of either geometry or calculation will be enough; the question relates rather to the greater and more advanced part of geometry — whether that tends in any degree to make more easy the vision of the idea of good; and thither, as I was saying, all things tend which compel the soul to turn her gaze towards that place, where is the full perfection of being, which she ought, by all means, to behold.
>
> True, he said.
>
> Then if geometry compels us to view being, it concerns us; if becoming only, it does not concern us?
>
> Yes, that is what we assert.
>
> Yet anybody who has the least acquaintance with geometry will not deny that such a conception of the science is in flat contradiction to the ordinary language of geometricians.
>
> How so?
>
> They have in view practice only, and are always speaking, in a narrow and ridiculous manner, of squaring and extending and applying and the like — they confuse the necessities of geometry with those of daily life; whereas knowledge is the real object of the whole science.
>
> Certainly, he said.
>
> Then must not a further admission be made?
>
> What admission?
>
> That the knowledge at which geometry aims is knowledge of the eternal, and not of aught perishing and transient.
>
> That, he replied, may be readily allowed, and is true.
>
> Then, my noble friend, geometry will draw the soul towards truth, and create the spirit of philosophy, and raise up that which is now unhappily allowed to fall down.

> Nothing will be more likely to have such an effect.
>
> Then nothing should be more sternly laid down than that the inhabitants of your fair city should by all means learn geometry. Moreover the science has indirect effects, which are not small.
>
> Of what kind? he said.
>
> There are the military advantages of which you spoke, I said; and in all departments of knowledge, as experience proves, anyone who has studied geometry is infinitely quicker of apprehension than one who has not.
>
> Yes indeed, he said, there is an infinite difference between them.
>
> Then shall we propose this as a second branch of knowledge which our youth will study?
>
> Let us do so, he replied.

With this section of the dialogue it becomes clear that the discussion is not between the advocates of pure and applied mathematics, to use our modern terms with deliberate anachronism, but between those who advocate the applications of mathematics and those who find in studying mathematics some higher purpose altogether, one that is not to do with mathematics but the ability to see 'the idea of good'. We are reminded that the *Republic* is Plato's attempt to set out the right education for an elite of rulers. He is an advocate of (a certain kind of) mathematics only insofar as it serves another purpose, and we must be careful to read him as the advocate he was, not as a dispassionate historian writing for posterity. That said, he next deliberately stayed with the division of geometry into the topics of his time and did not rush on to his principal concern.

Plato on solid geometry.

> And suppose we make astronomy the third — what do you say?
>
> I am strongly inclined to it, he said; the observation of the seasons and of months and years is as essential to the general as it is to the farmer or sailor.
>
> I am amused, I said, at your fear of the world, which makes you guard against the appearance of insisting upon useless studies; and I quite admit the difficulty of believing that in every man there is an eye of the soul which, when by other pursuits lost and dimmed, is by these purified and re-illumined; and is more precious far than ten thousand bodily eyes, for by it alone is truth seen. Now there are two classes of persons: one class of those who will agree with you and will take your words as a revelation; another class to whom they will be utterly unmeaning, and who will naturally deem them to be idle tales, for they see no sort of profit which is to be obtained from them. And therefore you had better decide at once with which of the two you are proposing to argue. You will very likely say with neither, and that your chief aim in carrying on the argument is your own improvement; at the same time you do not grudge to others any benefit which they may receive.
>
> I think that I should prefer to carry on the argument mainly on my own behalf.

Figure 3.13. A Roman mosaic, c. 1st century AD, from Pompeii, showing Plato at his Academy

Then take a step backward, for we have gone wrong in the order of the sciences.

What was the mistake? he said.

After plane geometry, I said, we proceeded at once to solids in revolution, instead of taking solids in themselves; whereas after the second dimension the third, which is concerned with cubes and dimensions of depth, ought to have followed.

That is true, Socrates; but so little seems to be known as yet about these subjects.

Why, yes, I said, and for two reasons:— in the first place, no government patronises them; this leads to a want of energy in the pursuit of them,

> and they are difficult; in the second place, students cannot learn them unless they have a director. But then a director can hardly be found, and even if he could, as matters now stand, the students, who are very conceited, would not attend to him. That, however, would be otherwise if the whole State became the director of these studies and gave honour to them; then disciples would want to come, and there would be continuous and earnest search, and discoveries would be made; since even now, disregarded as they are by the world, and maimed of their fair proportions, and although none of their votaries can tell the use of them, still these studies force their way by their natural charm, and very likely, if they had the help of the State, they would someday emerge into light.
>
> Yes, he said, there is a remarkable charm in them. But I do not clearly understand the change in the order. First you began with a geometry of plane surfaces?
>
> Yes, I said.
>
> And you placed astronomy next, and then you made a step backward?
>
> Yes, and I have delayed you by my hurry; the ludicrous state of solid geometry, which, in natural order, should have followed, made me pass over this branch and go on to astronomy, or motion of solids.
>
> True, he said.
>
> Then assuming that the science now omitted would come into existence if encouraged by the State, let us go on to astronomy, which will be fourth.

The 'omission' of solid geometry and its difficulty are apparent in education today.

Plato on astronomy.

> The right order, he replied. And now, Socrates, as you rebuked the vulgar manner in which I praised astronomy before, my praise shall be given in your own spirit. For every one, as I think, must see that astronomy compels the soul to look upwards and leads us from this world to another.
>
> Everyone but myself, I said; to everyone else this may be clear, but not to me.
>
> And what then would you say?
>
> I should rather say that those who elevate astronomy into philosophy appear to me to make us look downwards and not upwards.
>
> What do you mean? he asked.
>
> You, I replied, have in your mind a truly sublime conception of our knowledge of the things above. And I dare say that if a person were to throw his head back and study the fretted ceiling, you would still think that his mind was the percipient, and not his eyes. And you are very likely right, and I may be a simpleton: but, in my opinion, that knowledge only which is of being and of the unseen can make the soul look upwards, and whether a man gapes at the heavens or blinks on the ground, seeking to learn some particular of sense, I would deny that he can learn, for nothing of that sort is matter of science; his soul is looking downwards, not upwards,

whether his way to knowledge is by water or by land, whether he floats, or only lies on his back.

I acknowledge, he said, the justice of your rebuke. Still, I should like to ascertain how astronomy can be learned in any manner more conducive to that knowledge of which we are speaking?

I will tell you, I said: The starry heaven which we behold is wrought upon a visible ground, and therefore, although the fairest and most perfect of visible things, must necessarily be deemed inferior far to the true motions of absolute swiftness and absolute slowness, which are relative to each other, and carry with them that which is contained in them, in the true number and in every true figure. Now, these are to be apprehended by reason and intelligence, but not by sight.

True, he replied.

The spangled heavens should be used as a pattern and with a view to that higher knowledge; their beauty is like the beauty of figures or pictures excellently wrought by the hand of Daedalus, or some other great artist, which we may chance to behold; any geometrician who saw them would appreciate the exquisiteness of their workmanship, but he would never dream of thinking that in them he could find the true equal or the true double, or the truth of any other proportion.

No, he replied, such an idea would be ridiculous.

And will not a true astronomer have the same feeling when he looks at the movements of the stars? Will he not think that heaven and the things in heaven are framed by the Creator of them in the most perfect manner? But he will never imagine that the proportions of night and day, or of both to the month, or of the month to the year, or of the stars to these and to one another, and any other things that are material and visible can also be eternal and subject to no deviation — that would be absurd; and it is equally absurd to take so much pains in investigating their exact truth.

I quite agree, though I never thought of this before.

Then, I said, in astronomy, as in geometry, we should employ problems, and let the heavens alone if we would approach the subject in the right way and so make the natural gift of reason to be of any real use.

That, he said, is a work infinitely beyond our present astronomers.

Yes, I said; and there are many other things which must also have a similar extension given to them, if our legislation is to be of any value. But can you tell me of any other suitable study?

No, he said, not without thinking.

Astronomers were not to be persuaded to 'let the heavens alone ' in order to approach the subject in the right way — but this 'conclusion' should not be rejected out of hand; Plato's arguments are worth contesting even at their most paradoxical. Nor have the paradoxes quite ended, as Plato now turned to give Socrates' reasons for the study of musical harmony.

Plato on harmonics.

Motion, I said, has many forms, and not one only; two of them are obvious enough even to wits no better than ours; and there are others, as I imagine, which may be left to wiser persons.

But where are the two?

There is a second, I said, which is the counterpart of the one already named.

And what may that be?

The second, I said, would seem relatively to the ears to be what the first is to the eyes; for I conceive that as the eyes are designed to look up at the stars, so are the ears to hear harmonious motions; and these are sister sciences — as the Pythagoreans say, and we, Glaucon, agree with them?

Yes, he replied.

But this, I said, is a laborious study, and therefore we had better go and learn of them; and they will tell us whether there are any other applications of these sciences. At the same time, we must not lose sight of our own higher object.

What is that?

There is a perfection which all knowledge ought to reach, and which our pupils ought also to attain, and not to fall short of, as I was saying that they did in astronomy. For in the science of harmony, as you probably know, the same thing happens. The teachers of harmony compare the sounds and consonances which are heard only, and their labour, like that of the astronomers, is in vain.

Yes, by heaven! he said; and 'tis as good as a play to hear them talking about their condensed notes, as they call them; they put their ears close alongside of the strings like persons catching a sound from their neighbour's wall — one set of them declaring that they distinguish an intermediate note and have found the least interval which should be the unit of measurement; the others insisting that the two sounds have passed into the same — either party setting their ears before their understanding.

You mean, I said, those gentlemen who tease and torture the strings and rack them on the pegs of the instrument: I might carry on the metaphor and speak after their manner of the blows which the plectrum gives, and make accusations against the strings, both of backwardness and forwardness to sound; but this would be tedious, and therefore I will only say that these are not the men, and that I am referring to the Pythagoreans, of whom I was just now proposing to enquire about harmony. For they too are in error, like the astronomers; they investigate the numbers of the harmonies which are heard, but they never attain to problems — that is to say, they never reach the natural harmonies of number, or reflect why some numbers are harmonious and others not.

That, he said, is a thing of more than mortal knowledge.

A thing, I replied, which I would rather call useful; that is, if sought after with a view to the beautiful and good; but if pursued in any other spirit, useless.

Very true, he said.

Now, when all these studies reach the point of inter-communion and connection with one another, and come to be considered in their mutual affinities, then, I think, but not till then, will the pursuit of them have a value for our objects; otherwise there is no profit in them.

I suspect so; but you are speaking, Socrates, of a vast work.

Commentary. It is clear from Plato's description that these subjects are to be treated in their highest and purest form: to 'draw the soul from becoming to being', or as another translation puts it 'minds are to be drawn from the world of change to reality' and 'conversion of the soul from the world of becoming to that of reality and truth'.[27] In this conception, even the apparently empirical sciences of astronomy and harmony are quite separate from the physical world: they are a matter for the mind, not the senses.

Glaucon always ends up by agreeing with Socrates, but he noticeably had a different conception of the 'use' of mathematics. As each new topic arises, Glaucon cites its everyday practical uses, which are irrelevant to the high educational needs that Socrates had in mind — indeed, it is not even the same subject. In the case of astronomy, for instance, Glaucon takes it to be something to do with telling the seasons, whereas Socrates explains that it has to do with pure numbers and perfect figures, 'to be apprehended by reason and intelligence, but not by sight'. This dramatic device enabled Plato to make his distinction both clearly and effectively.

One historical significance of this passage is that from it stemmed the *quadrivium*, the advanced part of medieval higher education from about AD 1000 onwards. This consisted of four subjects known as the *mathematical arts* — arithmetic, geometry, astronomy, and music — and was preceded by the *trivium*, the three subjects of grammar, logic, and rhetoric. The actual content of the medieval studies sometimes had little to do with Plato's conception, being probably rather closer to Glaucon's in places; however, the categorisation and the placing of mathematics as the culmination of the liberal arts course derive from Plato through various later Greek and Latin writers and educationalists.

It is probable that this division of mathematical studies did not originate with Plato, but developed during the previous century. For there is one other contemporary reference to it, in a fragment by Plato's friend Archytas. Archytas lived in Tarentum, in Southern Italy, and was referred to by Proclus as one of those 'by whom the theorems were increased in number and brought into a more scientific arrangement'. Archytas wrote:[28]

> I think that those concerned with the sciences [*mathemata*] are men of discernment, and it is not strange that they should think correctly about the nature of particular things ... And so they have handed down to us clear knowledge of the speed of the heavenly bodies and their risings and settings, of geometry, numbers and, not least, of the science of music. For these sciences seem to be related: they are concerned with the first two kinds of what is, which are related.

Archytas's explanation of why these subjects are related is perhaps not very clear, but he states plainly that the four quadrivium subjects have been 'handed down to us'. According to later sources, this was by the Pythagoreans; indeed, Archytas is always spoken of as a member of the Pythagorean brotherhood, so this may have been the case.

---

[27] See (F&G 2.E2, 67)

[28] See (Bowen 1982, 82) in F&G 2.D1.

At the end of the *Republic* extract, Plato made a similar point to that of Archytas, that all these studies are related and have 'connection with one another'. How they do so is rather clearer in the way that Plato explained the nature of the subjects.

Plato on arithmetic. We know what Plato meant by geometry, and we have looked at the geometrical research tradition. But what of arithmetic, the first topic in Plato's classification? The *Republic* discussion was clearer about what this subject should achieve than what it actually is; perhaps Plato told us, but his meaning is too unfamiliar for a 21st-century reader to recognise, or perhaps he gave what we would regard as a philosopher's answer rather than a mathematician's.

At all events, it becomes clear that arithmetic seems to be about the nature of numbers — a contrast between 'unity' and 'plurality'. This is still hard to pin down, but we are given a substantial clue when Socrates described an argument involving 'the experts'. We learn that all units are the same and cannot be divided into parts. We also learn that if you did try to divide the unit, then the experts would reply by multiplying it: for example, if you try to divide the unit into halves (say), the expert redefines the unit so that your half is the new unit, and the old unit is double what it was before. However this may be, it seems that units and their multiples (that is, plurality) lie at the centre of Plato's concept of arithmetic.[29]

On the other hand, Plato was aware that arithmetic cannot capture every discussion of quantity. As you saw in *Meno*, the slave boy's initial attempts to double the square by arithmetic led nowhere, so Socrates diverted the boy's attention towards a geometrical solution.

Aristotle on arithmetic. The inapplicability of unit-arithmetic to some geometrical figures seems to have become known towards the end of the 5th century BC, perhaps during Plato's youth. Plato discussed it in several places — his diatribe against 'shameful ignorance' in the *Laws*[30] was related to this, and it was one of Aristotle's favourite examples:[31]

> For all who argue *per impossibile* infer by syllogism a false conclusion, and prove the original conclusion hypothetically when something impossible follows from a contradictory assumption, as, for example, that the diagonal [of a square] is incommensurable [with the side] because odd numbers are equal to even if it is assumed to be commensurate. It is inferred by syllogism that odd numbers are equal to even, and proved hypothetically that the diagonal is incommensurate, since a false conclusion follows from the contradictory assumption.

As is usual with Aristotle, he here uses a mathematical example to illustrate a more general point; presumably the mathematics would have been familiar to his original hearers, as his treatment was rather allusive.

You may have recognised his general logical point — he is discussing an argument structure in which we assume the opposite of what we wish to prove and then derive a contradiction. This is known as an indirect or *reductio* argument (short for *reductio ad absurdum*, or reduction to an absurdity) or as a proof by contradiction.[32] Historians have given various reconstructions of the details of this proof; we give the most time-hallowed one in Box 3.2, expressed in modern algebraic notation.

---

[29] *Republic*, 521–531 in F&G 2.E2.

[30] *Laws*, 817e–820d and F&G 2.E6.

[31] Aristotle, *Prior Analytics*, 41a 23–27, and F&G 2.H6(b).

[32] We discuss proofs by contradiction in greater detail in Chapter 4.

Box 3.2. The diagonal and the side of a square are incommensurable.

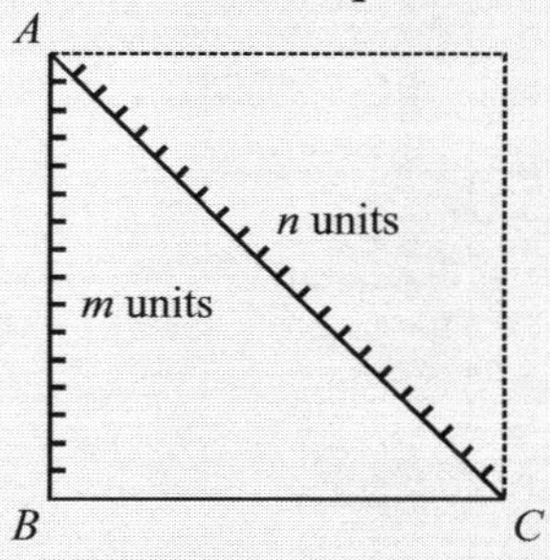

Figure 3.14

Suppose that the diagonal AC and the side AB are commensurable, as in Figure 3.14. Then there exists a unit length in terms of which AC is n units and AB is m units, and n and m are as small as possible. This means that the numbers n and m can have no factors in common, other than the unit.

Since the square on the diagonal AC is twice the square on the side AB,

$$n^2 = 2m^2.$$

It follows that n^2 is twice some number, so is an even number, and so n is also even.

So we can write $n = 2k$ (say). We deduce that

$$n^2 = (2k)^2 = 4k^2, \text{ giving } 2m^2 = 4k^2, \text{ so } m^2 = 2k^2.$$

Thus m^2 is an even number, and so m is also even.

The above discussion shows that n and m are both even, so each has a factor of 2.

This contradicts our assumption that n and m have no factors in common. The contradiction arose from our initial assumption that the diagonal and the side of a square are commensurable. It follows that they are incommensurable.

The overall structure of this mathematical proof is as follows. We assume that the diagonal of a square is commensurable with its side. An absurdity then arises from the fact that the numbers n and m are both even, yet have no factors in common. Thus our initial assumption must be wrong: the diagonal is not commensurable with the side, and unit-arithmetic is inapplicable.

The importance of proof. We do not know whether the case of the square was the first in which the phenomenon of incommensurability was recognised; some historians argue that this recognition originated with the regular pentagon, whose side and diagonal are also incommensurable (see Box 3.3). Nor do we know whether the above argument was the original proof of this result. But notice one significant aspect of the above discussion. Assuming that you found it convincing and believe the result, you do so because of the proof: the result has little plausibility without the proof that accompanies it.

Box 3.3. The side and diagonal of a pentagon are incommensurable
In Figure 3.15 triangle ACH is isosceles, so $AH = AC$, and triangles ABC and BCH are similar. So if $AC = AH = 1$ and $HB = a$ then $1/a = (1 + a)/1$ or $a^2 + a = 1$, and $a = \frac{1}{2}(\sqrt{5} - 1)$. (Compare Euclid's *Elements*, Book 2, Prop. 10 and Book IV, Props. 10, 11.)

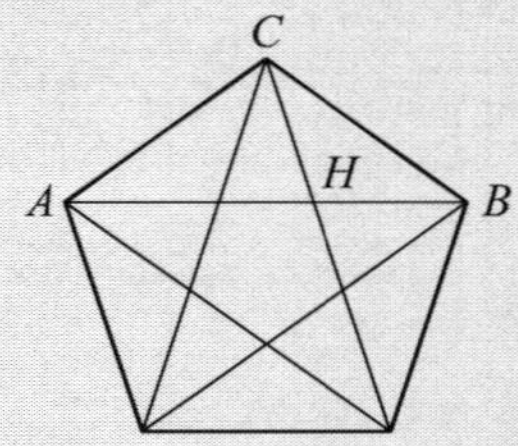

Figure 3.15

This is an entirely new situation. The proofs of other results, such as Hippocrates's quadrature of a lune, act more to corroborate what might have seemed quite likely beforehand. But the discovery and proof that two lines are incommensurable must have been more-or-less simultaneous. Indeed, we might go further and speculate that its first proof may have constituted its discovery, even though no details of this event are known.

It follows that the phenomenon of incommensurability could have arisen only in a culture that employed some notion of logical proof. This is why it seems unlikely that the Mesopotamian scribe who wrote the tablet YBC 7289 that we discussed earlier (see Figure 2.12), with its impressive approximation to the length of the diagonal of a square, knew that this length cannot be written correctly as a rational multiple of the length of the side.

This example illustrates the large gap between the two levels of mathematics emphasised by Plato. Perhaps Plato and Aristotle paid so much attention to incommensurability arguments — results whose very credibility is inconceivable without their proofs — because they symbolised, better than any other mathematical results of the time, the power of the newly emerging techniques of logical justification.

3.4 Euclid's *Elements*

Euclid's *Elements* is one of the great written works of the Western World, and one that has done much to shape our intellectual, scientific, and practical lives. So before we go any further it would be best to give a brief account of what we know about the author of this famous work. Unfortunately, we know almost nothing! Although much of what Euclid wrote has survived — more than just the *Elements* – he never thought to provide any prefaces from which we can glean information about him.

The first reliable reference to Euclid that we have in the work of others comes from the geometer Apollonius. He observed in his *Conics*, which we shall look at below in Section 5.3, that he was writing about a topic that Euclid had treated only imperfectly. However, from the way that Apollonius expressed himself we cannot infer that when he wrote Euclid was already dead — he could have been still alive and working. Now, Apollonius's *Conics* can be reliably dated to about 185 BC. We can set this alongside three much less secure claims that have been inferred from other ancient evidence, namely that

Euclid was born in 365 BC or even as late as 325, perhaps in Athens, where as a young man he was a student in Plato's Academy; that Ptolemy I of Egypt summoned Euclid to Alexandria about 300 BC; and that Euclid was one of the scholars employed in the famous Museum, until he died around 275 BC.

Our problem here is that our sources for these claims are authors writing several centuries after Euclid. When the authors cannot be shown to be plain wrong — as on occasion they can — they are nonetheless open to the charge that they add to and embroider the sources they had. On the other hand, many of those sources are now lost — perhaps the authors were simply quoting from a reliable text no longer available to us. Our conclusion must be that we know nothing about Euclid the person, and even Euclid's dates could move by up to a century, so the usual ascription that Euclid flourished around 300 BC must be regarded as uncertain.

Euclid's *Elements* has been one of the best-selling books of all time: this indicates its remarkable importance in the medieval and modern West, primarily, but it was also a major work in the Islamic world. It comes in thirteen parts, called Books, whose contents are outlined as follows:

The Books of Euclid's *Elements*

Book I	Foundations of plane geometry
Book II	The geometry of rectangles
Book III	The geometry of circles
Book IV	Regular polygons in circles
Book V	Magnitudes in proportion
Book VI	Geometry of similar figures
Book VII	Basic arithmetic
Book VIII	Numbers in continued proportion
Book IX	Prime numbers
Book X	Incommensurable line segments
Book XI	Foundations of solid geometry
Book XII	Areas and volumes
Book XIII	The Platonic solids

We must be careful with our descriptions of the *Elements*, for various reasons. We know little about how it was regarded in Greek times, although there is evidence that it became the standard work in its subject and evidence from Hellenistic and Roman writers that attests to its value. We also learn from Proclus that Euclid was not the first to write a work on *Elements*: Hippocrates and others had done so previously, but their writings have not survived. So it seems that Euclid's success was such as to overshadow his predecessors. Even so, we know little about how it was used, and we should probably not call it a textbook because this suggests that it was regularly used in a school setting, which seems rather too modern: it is likely that the later parts of the *Elements* were regarded in their day as rather advanced work.

Elements. Just as in modern English, where the word 'element' (as in 'chemical element') may have a more specialised meaning than the adjective 'elementary' (in the sense of simple), so too in classical Greece there was a range of meaning for the equivalent word *stoicheia*. Initially it meant any constituent of a line or row of things (so a soldier might be an element of a line of soldiers). Later, Menaechmus used 'element' for any proposition used in the proof of another one, so in this sense 'the squares on the diameters have the same ratios as the circles' is an element of Hippocrates's quadrature of lunes.

By the time of Euclid, an 'element' in mathematics meant a starting point for other theorems. Proclus compared an element in this sense with a letter of the alphabet: from these letters (also called *stoicheia* in Greek) all words are constructed; and similarly the elements of geometry are the foundational propositions from which other geometrical truths can be derived. In a passage which gives a strong indication of why Euclid was so valued by later mathematicians, Proclus commented that:[33]

> It is a difficult task in any science to select and arrange properly the elements out of which all other matters are produced and into which they can be resolved ... in general many ways of constructing elementary expositions have been individually invented.
>
> Such a treatise ought to be free of everything superfluous, for that is a hindrance to learning; the selections chosen must all be coherent and conducive to the end proposed, in order to be of the greatest usefulness for knowledge; it must devote great attention both to clarity and to conciseness, for what lacks these qualities confuses our understanding; it ought to aim at the comprehension of its theorems in a general form, for dividing one's subject too minutely and teaching it by bits make knowledge of it difficult to attain.
>
> Judged by all these criteria, you will find Euclid's introduction superior to others.

This brief summary of changing meanings of the word 'element' shows that we can pick up significant evidence about Greek development of proof through this kind of linguistic analysis, and also that the name of Euclid's work is not an arbitrary label but provides a deep clue to the importance of its contents.

Euclid's importance, then, lay in the choice and arrangement of 'the elements out of which all other matters are produced'. We have seen examples of deductive proof, results shown to be true by logical inference from other results or assumptions. But Euclid's *Elements* is the earliest example we have of a full-blown axiomatic system in mathematics, in which all the results are proved by reference back to foundational starting points. It is evident from the works of Aristotle, and to a lesser extent of Plato, that such concerns were becoming more developed throughout the 4th century BC, at the end of which Euclid compiled what became the definitive *Elements*.

An overview of Euclid's *Elements*. We now look at Euclid's *Elements* in more detail to see what it can tell us about the topics studied and the nature of proof in Greek mathematics.

Historians of mathematics sometimes speak loosely of Euclid's *Elements* as a 'book', even though it was originally written on papyrus rolls. Some small fragments dating from the 1st century AD exist (see Figure 3.16). The earliest complete text that survives is a handsome volume in Byzantine Greek from AD 888, a date almost exactly halfway between Euclid's times and ours (see Figure 3.17).[34] There are also later texts from which we can deduce that the work was thought well worth preserving. Since these texts differ in various respects, it is no easy matter to determine what Euclid actually wrote.

From these texts derived many editions of the *Elements*, and some of these differ greatly. Apart from the different sources, there is an Arabic textual transmission that is

[33] See (Morrow 1970, 58–61) and F&G 3.A1.

[34] This copy is housed in the Bodleian Library, Oxford University, and is available online at http://www.claymath.org/library/historical/euclid/.

Figure 3.16. A papyrus fragment showing a theorem from Euclid's *Elements*, Book II, Prop. 6

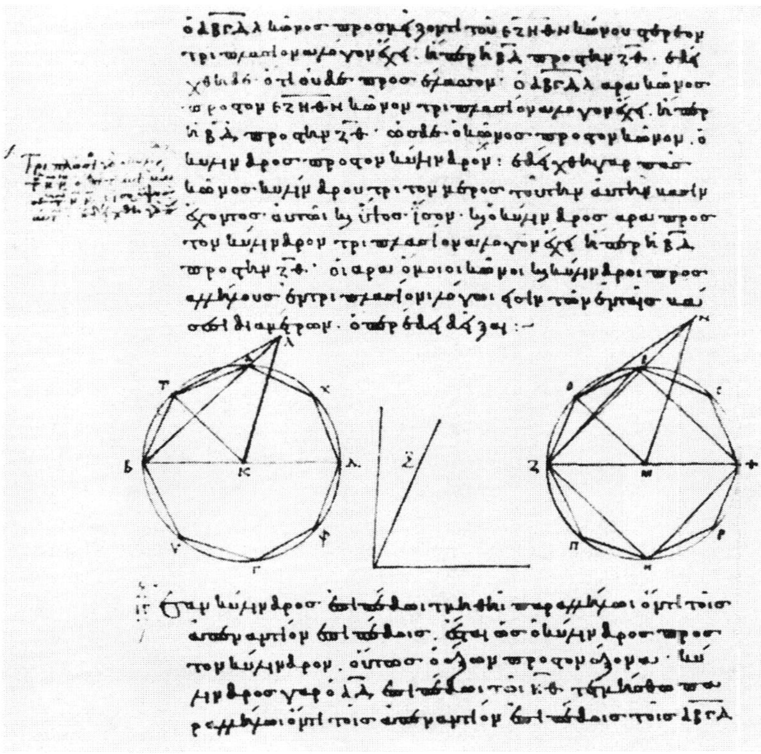

Figure 3.17. Euclid's *Elements*, Book XII, Prop. 12, AD 888.

only now being properly studied, two separate Western traditions (one in Greek and one in Latin) that differ over many of the technical terms, and innumerable later Western versions. Many of these are incomplete: as we shall see, there were sound educational reasons for publishing only the first four or six of the thirteen Books. On the other hand, some editions overshot: during the 19th century, scholars removed two spurious Books numbered XIV and XV.

With these cautions noted, we now summarise the Books of Euclid's *Elements* in the version that modern scholarship puts on our desks.

Books I–IV. These four Books deal with the geometry of plane figures made from straight lines and circles.

Book I introduces the main geometrical *definitions*: point, line, surface, various types of angles (acute, right, and obtuse), triangles and quadrilaterals, and parallel lines. There are five explicit assumptions (called *postulates*) concerning the properties of the objects so defined, of which the first three specify that there is a straight line segment joining any two points, that a straight line segment can always be extended in a straight line, and that a circle can always be drawn with any centre and radius. There are also five explicit statements (called *common notions*) about how to reason about quantities in general, such as 'If equals be added to equals, the wholes are equal' and 'The whole is greater than the part'. Thereafter, definitions of other figures and technical terms are given as needed, and more definitions are given in later books, but no further postulates or common notions are introduced.

Book I also establishes what is meant by the *congruence* of two figures — that is, when they are the same figure but in different positions. Euclid described such figures as *equal*, and then gave a number of crucial results, some of which we return to below: the isosceles triangle theorem, properties of triangles and parallelograms, and the Pythagorean theorem.

Book II deals with dissections of plane figures, and the implications that these have for the areas of such figures.

Book III introduces circles in the fashion of Book I, with explicit definitions, and discusses properties of chords and angles in circles. One celebrated example is the proposition that any triangle drawn in a semicircle with one side as a diameter is a right-angled triangle, as shown in Figure 3.18.[35]

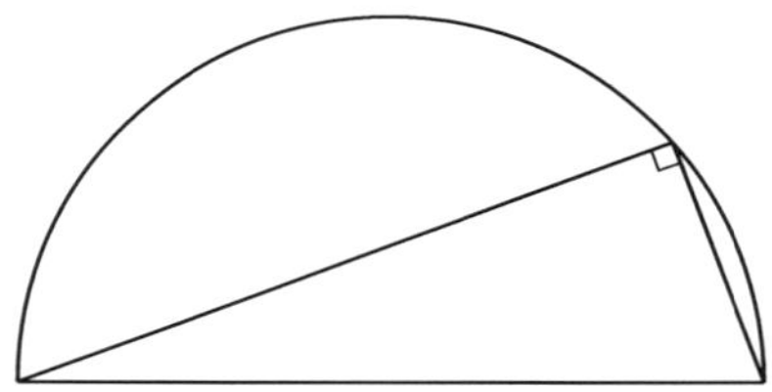

Figure 3.18. The angle in a semicircle is a right angle

Book IV deals with constructions involving circles and lines, such as those for drawing certain regular figures (such as a pentagon) inscribed in a circle.

Books V–VI. These Books differ greatly from the first four. Book V takes nineteen definitions to introduce the delicate topic of ratio in geometry. It then shows how ratio and proportion appear in connection with incommensurable quantities; this theory of proportion was probably due to Plato's pupil Eudoxus. The Book also develops a theory of similar figures. While Book V is rather conceptual, Book VI is more concerned with constructions of the type: 'given three magnitudes, find a fourth magnitude that satisfies a certain condition'.

[35]We met this result in connection with Hippocrates's quadrature of a lune.

Books VII–IX. These three Books are different again, and may consist of earlier material that was rewritten by Euclid. The topic is arithmetic and number theory, and includes discussions of the divisibility of integers and the theory of prime numbers.

Book VII begins with what is now called the *Euclidean algorithm* for finding the highest common factor of two numbers.

Book IX contains Euclid's famous proof that there are infinitely many prime numbers (see Chapter 4).

Book X. This is the longest of the thirteen Books, amounting to one-quarter of the whole work, and is also the most daunting. It is devoted to working out results on incommensurable magnitudes along the lines that seem to be hinted at in Plato's *Laws*. These magnitudes are needed in the construction of regular solids in Book XIII.

Books XI–XIII. The last three Books are on solid geometry — the geometry of three dimensions.[36]

Books XI and XII extend much of the material in the first four Books to three dimensions, while Book XIII introduces the five regular solids and describes how they can be constructed.

Book XIII also contains a proof that there are only five regular solids. This result is of a kind that we have not met before, and is one of the earliest examples of a classification theorem in mathematics: it is not about a single object, but about a class of objects, and it specifies completely all those objects with a particular property. The significant point here is not the knowledge of any particular solid such as the cube, many of whose properties had doubtless been studied earlier, but that a defining property can be singled out that is possessed only by these five solids. It has been argued that:[37]

> The real history of the regular solids therefore begins at the point when men realised there was such a subject. The discovery of this or that particular body was secondary; the crucial discovery was the very concept of a regular solid. Long familiarity has made the idea seem almost obvious, but it is not.

Even this brief survey of Euclid's *Elements* brings up several features. The *Elements* is often said to be mainly about plane geometry, and the geometry of figures bounded by straight lines and circles at that, but this is the theme of only the first few Books. The first four are more elementary, which is why they were often presented separately. The next two introduce the geometrical concept of similarity, but charge a high price in rigour and difficulty, so it makes more sense to include them in demanding contexts than in elementary ones. The next three Books are on a topic that we no longer regard as geometrical, and the concerns of Book X are somewhat subtle. So the *Elements* are not held together by any strong thematic unity. Rather, as we argue later, they exhibit a strong methodological unity based around the idea of axiomatic proof.

Some propositions from Book I. We now discuss some specific concerns that arise in Book I, starting with Euclid's lists of postulates, common notions, and definitions, and continuing with a selection of his forty-eight propositions.[38]

The postulates and common notions at the start of Book I lie at the heart of the axiomatic structure of the *Elements*. Essentially, a postulate is a geometrical axiom, whereas

---

[36]This is the subject that Plato had claimed, in his *Republic*, 521–531 (F&G 2.E2), to have been neglected by mathematicians and should be investigated further.

[37]See (Waterhouse 1972, 214).

[38]For the postulates, common notions, and definitions, see Heath's edition of Euclid's *Elements* and F&G 3.B1(a). All references to Euclid's *Elements* will be to Heath's edition.

a common notion is a logical axiom, a general truth, or an assumption that applies to wider inquiries. It is upon these, and on the twenty-three definitions in Book I (together with the definitions in later books), that the structure of the *Elements* is built.[39]

In the first proposition of the *Elements* Euclid shows how to construct an equilateral triangle with any given straight line segment as base. Let us briefly review his construction before discussing what is so interesting about it. Let the segment be AB, and draw the circle with centre A and radius AB and the circle with centre B and radius BA (see Figure 3.19). Here as always in elementary geometry you will find it very helpful to follow the instructions line by line — this will also help make sense of Euclid's conventions in labelling figures. These circles meet at two points, either of which is the third vertex of the required equilateral triangle. Euclid wrote:[40]

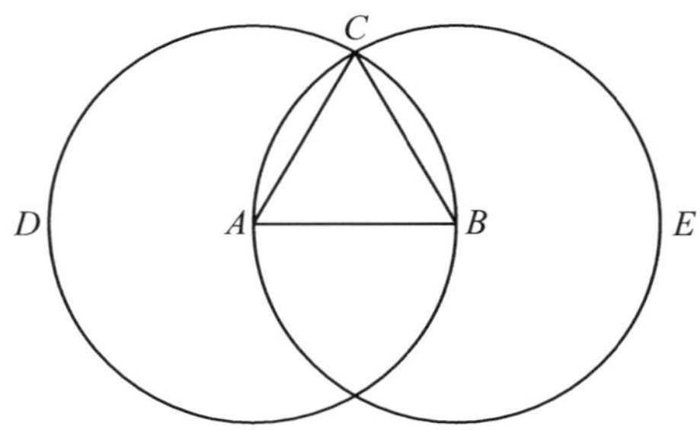

Figure 3.19. The construction of an equilateral triangle

Euclid, *Elements* Book I, Proposition 1.

Proposition 1:

On a given finite straight line to construct an equilateral triangle.

Let AB be the given finite straight line.

Thus it is required to construct an equilateral triangle on the straight line AB.

With centre A and distance AB let the circle BCD be described; [Post.3]

Again with centre B and distance BA let the circle ACE be described; [Post.3]

And from the point C, in which the circles cut one another, to the points A and B let the straight lines CA and CB be joined. [Post.1]

Now, since the point A is the centre of the circle CDB, AC is equal to AB. Again, since the point B is the centre of the circle CAE, BC is equal to AB. [Def.15]

But CA was proved equal to AB; therefore each of the straight lines CA and CB is equal to AB.

And things which are equal to the same thing are also equal to one another, [C.N.1] therefore CA also equals CB.

---

[39]We return to this topic in Chapter 4.

[40]Euclid, *Elements*, Book I, Prop. 1. In Heath's additions, 'Post.' stands for Postulate and 'C.N.' for Common Notion.

Therefore the three straight lines CA, AB, and BC are equal to one another.

Therefore the triangle ABC is equilateral, and it has been constructed on the given finite straight line AB.

(Being) what it was required to do.

You might think there is nothing very remarkable in this, and in a way you would be right: any designer or architect with a fixed compass has surely carried out this construction many times. It is remarkable that a book on the elements of geometry does not take it for granted, and instead supplies an explicit construction.

Euclid also *proved* that this construction does indeed produce the required triangle. This tells us straightaway that Euclid had high standards, and that we can expect similar attention to detail throughout the rest of the work. The clear intention is to argue from the definitions alone by means of explicit rules of deduction, and this is what singles Euclid's *Elements* out from many other accounts of geometry.

We emphasise that Euclid offered both a construction and a proof. This is a constant theme of his: there are theorems in the *Elements*, and there are constructions, as here, but these are regarded as two different mathematical activities. Indeed, the results in the *Elements* are generally of two kinds: deductions of stated results, and constructions. The Latin commentating tradition highlighted this important distinction by marking the end of a proof with the letters QED (*quod erat demonstrandum*: that which was to be shown) and of a construction with QEF (*quod erat faciendum*: that which was to be done).

We may add that commentators have had a field day with Proposition 1. The philosopher Zeno of Elea, who is best known for his paradoxes of motion, remarked that Euclid had assumed that two straight lines cannot have a segment in common, but Euclid's defenders (Proclus among them) claimed that this is implied by the formulation of Postulate 2 (which allows one to produce a finite straight line continuously in a straight line).

However, even Proclus barely dealt with the next objection, which was that one must also show that two circles cannot have an *arc* in common without coinciding completely. Either of these objections, if valid, would mean that the construction yields a curve of some kind, rather than a point. And modern commentators have called attention to an even bigger problem: one must either show that the constructed circles do indeed have two points in common, or assume it as a postulate. Euclid could not have proved it, but nor did he make it an explicit assumption: by modern standards it is a gap in his work

Proposition 5 is a celebrated result that shows that in an isosceles triangle (one in which two sides are equal) two of the angles are equal,[41] while Proposition 6 establishes the converse result, that if a triangle has two equal angles then the opposite sides are equal.[42]

There is an interesting difference in the proofs of these two results. The proof of Proposition 5 is a direct argument: if you draw this, and this, and consider these lengths, and use this previously established result, you can deduce that these angles are equal. The proof of Proposition 6 is a *reductio* argument, or proof by contradiction: it says that if the desired conclusion were false, then something would be the case, and then something else, until a contradiction is reached (see Box 3.4).

[41] In F&G 3.B2(a).

[42] In F&G 3.B3(a).

Box 3.4. Euclid's *Elements*, Book I, Prop. 6.
In a triangle with two equal angles, the sides opposite those angles are equal.
To prove this, Euclid said that either the two sides in question are equal, or they are unequal. He then took the second alternative and deduced a contradiction from this, showing that, on this assumption, the two triangles ABC and DBC are equal (see Figure 3.20). But the existence of the triangular sliver ACD shows that DBC is actually smaller than ABC by that amount.

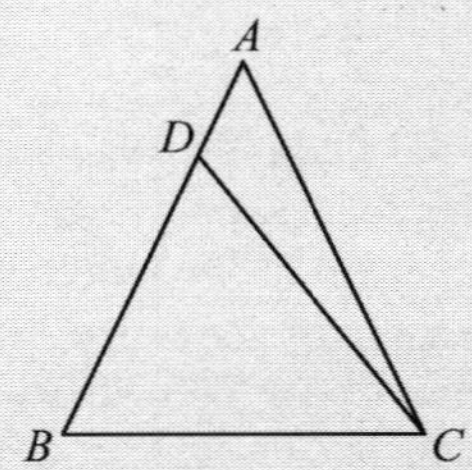

Figure 3.20. The isosceles triangle theorem

So one triangle is both equal to, and smaller than, another one: this is the required logical absurdity or contradiction. It follows that the two sides are equal.

The fifth or parallel postulate. In the *Elements*, parallel lines are defined as 'straight lines that do not meet'. The fifth postulate of Book I states:

> If a straight line falling on two straight lines makes the interior angles on the same side less than two right angles, then the straight lines, if produced indefinitely, meet on the side containing these angles.

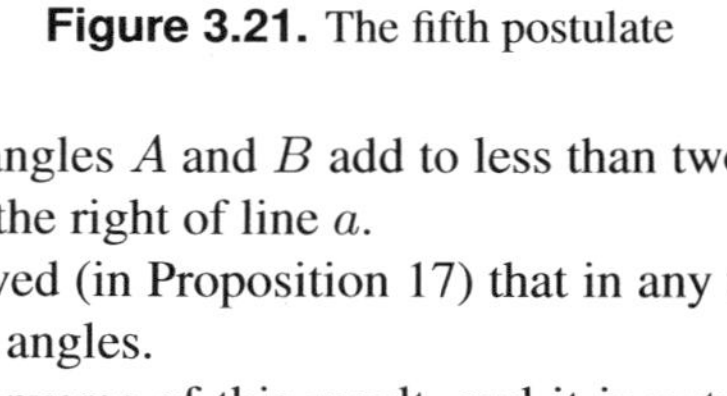

Figure 3.21. The fifth postulate

Thus, in Figure 3.21 if the angles A and B add to less than two right angles then the lines b and c meet somewhere to the right of line a.

Euclid had already proved (in Proposition 17) that in any triangle the sum of any two angles is less than two right angles.
The fifth postulate is the converse of this result, and it is noteworthy that Euclid felt the need to assume it.

The fifth postulate is also often called the *parallel postulate*, even though it makes no specific mention of parallel lines; this is because of the way that it is used in the *Elements*. Its use is in fact delayed for some time: Book I, Prop. 27 marks the first appearance of

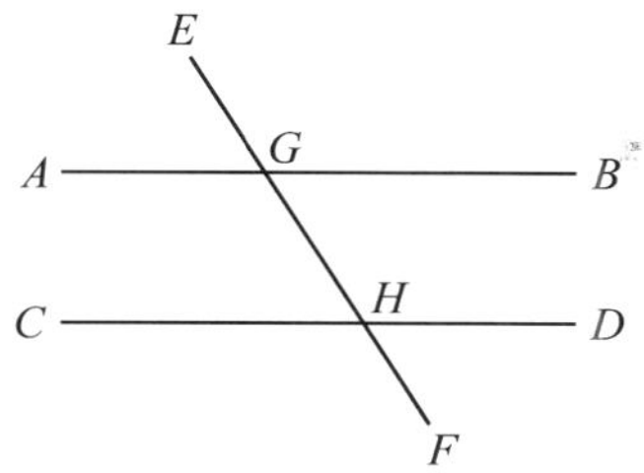

Figure 3.22. Euclid's *Elements*, Book I, Prop. 29

parallel lines, and Book I, Prop. 29 presents the first use of the fifth postulate, asserting that parallel lines make equal angles with any line that crosses them (see Figure 3.22).

The fifth postulate is not needed to show that parallels exist. What it does show, given all Euclid's other assumptions, is the uniqueness part of what is sometimes called *Playfair's axiom* (see Figure 3.23):[43]

Given any line ℓ and any point P not on the line ℓ there is in the plane of P and ℓ a unique line m through P that does not meet ℓ.

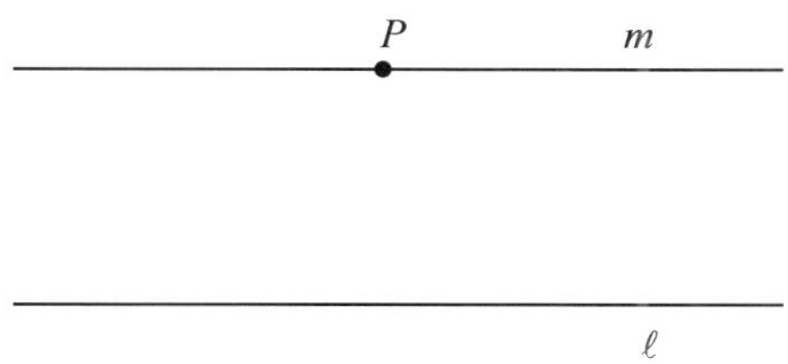

Figure 3.23. Playfair's axiom

Playfair's axiom is easier to understand than the fifth postulate, so it is helpful to know that Playfair's axiom and the fifth postulate are equivalent (given all Euclid's other assumptions): each implies the other. But Playfair's axiom makes it easier to see the uniqueness of the line through P parallel to ℓ, and it is this uniqueness of parallels that makes it possible to establish Proposition 29 and the important propositions that follow. Among these is Proposition 32 which states that:[44]

the angles of a triangle add up to two right angles.

Results on areas also follow from the parallel postulate, and are crucial to the proof of the Pythagorean theorem.

The delayed appearance of the parallel postulate in Euclid's *Elements* has attracted much attention, because it suggests that Euclid was aware that it is a strange assumption to make. How can we be sure that there is only one parallel to a given line through a given point? The point where two lines cross can surely be arbitrarily far away, so how can experience suggest that there cannot be many lines through the given point that do not meet the given line? On the other hand, without the parallel postulate none of the results we have just mentioned can be proved, and the whole geometrical enterprise grinds to a halt. Perhaps the parallel postulate was assumed as a matter of sheer expediency.

[43] Playfair's axiom is named after John Playfair, an 18th-century British editor of the *Elements*.

[44] F&G 3.B4(a).

Certainly there is evidence that it was much discussed; we have room here for just one example. Proclus[45] objected that it

> ought to be struck from the postulates altogether. For it is a theorem ...

This would be a satisfactory way around the lack of evidence for the parallel postulate, for if the uniqueness of parallels follows from the other assumptions in Euclid's *Elements*, just as their existence does, then there is no need to assume it — and yet all the theorems of the *Elements* remain. Unfortunately, Proclus's attempt at a proof was flawed. So were other attempts, and the parallel postulate began to draw a constant stream of attention as one mathematician after another tried to repair it, only to fail in one way or another.[46]

By the 17th century the Oxford scholar Sir Henry Savile openly called the parallel postulate a 'blot on Euclid', and so it remained until its dramatic resolution in the 19th century.[47]

The Pythagorean theorem.

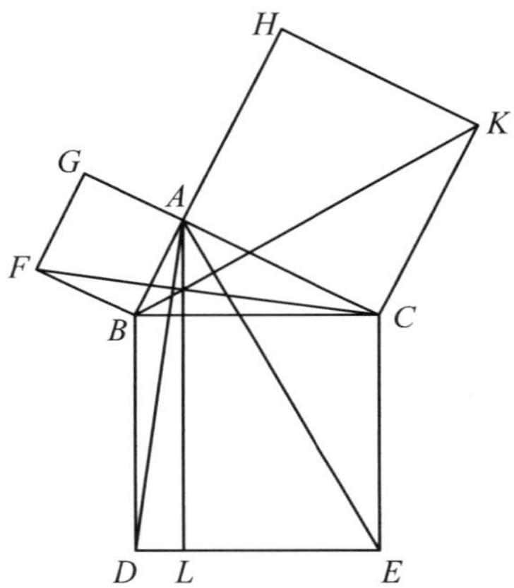

Figure 3.24. Euclid's *Elements*, Book I, Prop. 47

In Proposition 46, there is another geometrical construction, that of the square. This is followed by the most famous result in the *Elements*: Book I, Prop. 47, the Pythagorean theorem (see Figure 3.24):

> In right-angled triangles the square on the side subtending the right angle is equal to the squares on the sides containing the right angle.

It is clear from the existence of various problems in Egyptian and Mesopotamian texts that this result was known to earlier cultures, although we have little idea how they proved it, if indeed they did so: it may have struck them as more like a fact than a theorem. Interestingly, Mesopotamian scribes seem to have formulated this result as one about the sides of a right-angled triangle or the diagonal of a rectangle, whereas in Euclid's *Elements* it is firmly expressed as a result about areas. The Mesopotamian scribes were oddly closer to our way of thinking in this respect at least.

---

[45] See (Morrow 1970, 58–59) and F&G 3.B1(b).

[46] We take up this story in Chapter 8, where we look at mathematics in the Islamic world.

[47] See (Savile 1621, 140), quoted in (Gray 1989, 60); the reference there should be to Heath, Euclid's *Elements*, Book II, 190.

Euclid, *Elements*, Book I, Prop. 47.

In right-angled triangles the square on the side subtending the right angle is equal to the squares on the sides containing the right angle.

Let ABC be a right-angled triangle having the angle BAC right; I say that the square on BC is equal to the squares on BA, AC.

For let there be described on BC the square $BDEC$, and on BA, AC the squares GB, HC [I. 46]; through A let AL be drawn parallel to either BD or CE, and let AD, FC be joined.

Then, since each of the angles BAC, BAG is right, it follows that with a straight line BA, and at the point A on it, the two straight lines AC, AG not lying on the same side make the adjacent angles equal to two right angles; therefore CA is in a straight line with AG [I. 14].

For the same reason BA is also in a straight line with AH.

And, since the angle DBC is equal to the angle FBA: for each is right: let the angle ABC be added to each; therefore the whole angle DBA is equal to the whole angle FBC [C.N. 2].

And, since DB is equal to BC, and FB to BA, the two sides AB, BD are equal to the two sides FB, BC respectively; and the angle ABD is equal to the angle FBC; therefore the base AD is equal to the base FC, and the triangle ABD is equal to the triangle FBC [I. 4].

Now the parallelogram BL is double of the triangle ABD, for they have the same base BD and are in the same parallels BD, AL [I. 41].

And the square GB is double of the triangle FBC, for they again have the same base FB and are in the same parallels FB, GC [I. 41].

[But the doubles of equals are equal to one another.]

Therefore the parallelogram BL is also equal to the square GB.

Similarly, if AE, BK be joined, the parallelogram CL can also be proved equal to the square HC; therefore the whole square $BDEC$ is equal to the two squares GB, HC [C.N. 2].

And the square $BDEC$ is described on BC, and the squares GB, HC on BA, AC. Therefore the square on the side BC is equal to the squares on the sides BA, AC. Therefore etc. Q.E.D.

Euclid's proof is intimidating at first sight (see Box 3.5), and many mathematicians down the ages have amused themselves by offering simpler ones. As we shall see, the apparent complexity of this proof provides a rich source of information about Greek mathematics.

Box 3.5. Euclid's proof of the Pythagorean theorem.
Euclid took the small square $ABFG$ on the left-hand side, and divided it into two equal triangles by a diagonal.

He then slid the vertex A along the line GAC which is parallel to the base of the square FB,[a] obtaining from the triangle FAB a new triangle FAC. These triangles have the same base FB and lie between the same parallels, and so have the same area.

The triangle FBC is congruent to the triangle ABD (think of FBC being rotated about the vertex B until it coincides with the triangle ABD), and this triangle ABD has the same area as the triangle BDL since they have the same base BD and lie between the same parallels BD and AL.

Let X be the point where AL meets BC. The triangle BDL is one-half of the rectangle $BDLX$, and this rectangle is a piece of the square on the hypotenuse: we have moved half of the square on the side AB of the right-angled triangle ABC into half of the rectangle $BDLX$. The area of the other half of the square $ABFG$ fits exactly into the area of the other half of the rectangle $BDLX$. Thus, the square with side AB fits exactly into the rectangle $BDLX$, and so the square $ABFG$ and the rectangle $BDLX$ have the same area.

But exactly the same argument applies to the rectangle $ACKH$ and the square $CELX$ on the right-hand side of the figure. Since the square $BCED$ is made up of the rectangles $BDLX$ and $CELX$,[b] it follows that the square on the hypotenuse is equal in area to the sums of the squares on the other two sides.

---

[a]Note that a square is sometimes specified by two opposite corners, rather than by all four corners. It will help you if you draw the diagonal FA.

[b]You can think of this proof as 'pouring' the areas of the little squares into the big one, making sure that they fit exactly.

The commentary on Euclid's proof of the Pythagorean theorem in Box 3.5 shows us several things. First, it is recognisably an argument — indeed, a proof. It is not a mere assertion (a statement of a result with no special reason for believing it), and nor is it a special case in which the sides have given lengths (say, 3, 4, and 5), followed by an unsubstantiated claim to generality. We have not seen proofs of this complexity in earlier mathematical cultures; indeed, nothing we have seen in Egyptian and Mesopotamian mathematics was exhibited as a proof in this sense.

Second, as is clear from the many back references, the proof uses results established earlier. This is a major feature of the *Elements*; it is not a random collection of theorems, but a highly structured collection in which later results depend on earlier ones, and putting it together was a highly non-trivial task. Moreover, it cannot be fully understood without those earlier results: if you genuinely do not agree that the triangles ABD and BDL have the same area, then you are morally bound to stop here until insight comes to you; Euclid dealt with this difficulty by putting all the necessary information earlier.

Third, the concept of 'area' was evidently a primitive concept for Euclid, in that it cannot be reduced conceptually to a simpler concept, such as length. In our current mathematical culture, the Pythagorean theorem is often expressed as follows:

> In a right-angled triangle with sides of lengths a, b, and c, where side c is opposite the right angle,
> $$a^2 + b^2 = c^2.$$

There are many differences here from the Greek version, even though in some sense they are about the same thing. The key difference is that the modern version deals with the sides of the squares and 'squares' them to get the areas, which it then adds. Euclid worked with the areas directly.

Fourth, if you know any of the simpler proofs of the Pythagorean theorem you may wonder why Euclid did not present one of these. Even clever people can miss obvious arguments, and there are occasions in the history of mathematics where a later argument simplifies an earlier one, but Euclid had his reasons here. As we saw in the *Meno* example, it is not easy to specify the length of the diagonal of a square. If the Pythagorean theorem is to apply to the simple case of an isosceles right-angled triangle, then we must either give an account of length that allows us to talk about the length of the diagonal of a square, or we must find a route that avoids doing so. Euclid took the second approach and also gave an account of these mysterious lengths — but only much later, and very complicated it is too. The same considerations govern a well-known proof by similar triangles: one first needs a theory of similarity, and Euclid chose to give that later as well, for essentially the same reason.

The axiomatic approach — Hobbes meets the *Elements*. If anything gives the *Elements* its coherence and the feeling that we are reading a single work, rather than a compilation, it is the use of the axiomatic method. Definitions and assumptions are explicitly given, and theorems are then presented and proved in a strict and generally cumulative fashion. We can understand the nature of this axiomatic structure better and illustrate what Euclid achieved, both in logical terms and in its capacity to astonish and delight, by approaching it through the experience of the English philosopher Thomas Hobbes around 1628, as later described by the English antiquary and writer John Aubrey.[48]

> He was 40 years old before he looked on Geometry; which happened accidentally. Being in a Gentleman's Library, Euclid's *Elements* lay open, and 'twas the 47 *El. libri* I. He read the Proposition. By G–, sayd he (he would now and then sweare an emphaticall Oath by way of emphasis) *this is impossible*! So he reads the Demonstration of it, which referred him back to such a Proposition; which proposition he read. That referred him back to another, which he also read. *Et sic deinceps* [and thus in succession] that at last he was demonstratively convinced of that trueth. This made him in love with Geometry.

The proposition in question, Prop. 47 of Book I, is the Pythagorean theorem, and to retrace Hobbes's steps is a good way of investigating the structure of Book I. The demonstration refers back to other propositions on which steps in the proof depend. These propositions in turn depend on other propositions, and we can construct a diagram summarising these dependencies (see Figure 3.25). Note that not all of the propositions in Book I appear in the diagram as not all of them are used in the proof of Prop. 47.

It is clear from this that the logical structure is not a simple linear progression, as implied by Aubrey's account, but is a dense and complicated interaction of chains of reasoning. Aubrey has conveyed the spirit of Hobbes's journey of discovery while somewhat oversimplifying it.

Aubrey also took it for granted that his readers would appreciate the force of the impact that Euclid's *Elements* had on Hobbes. Hobbes had developed a mechanistic or materialist

---

[48]Aubrey lived from 1626 to 1697, but his collection of short biographies of his contemporaries was first published in 1949. For this extract, see F&G 3.F2(a).

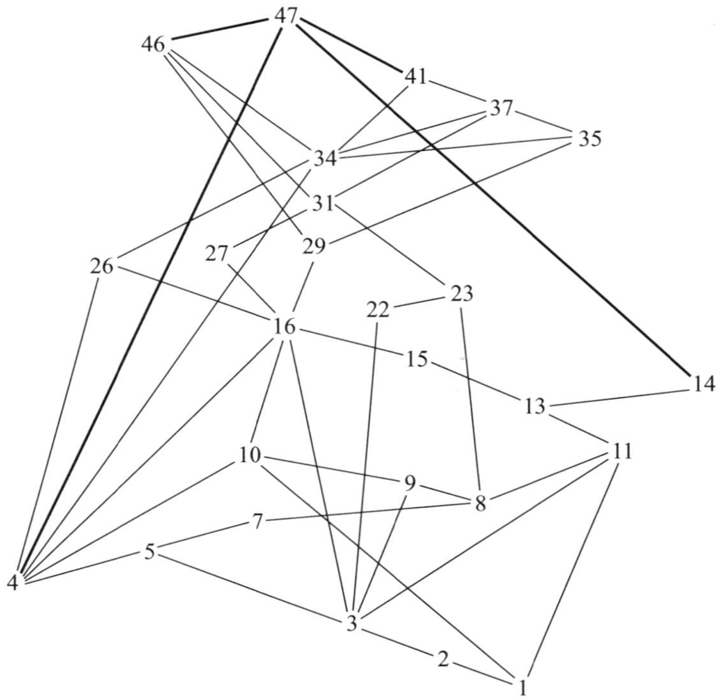

Figure 3.25. A diagram showing the propositions of Book I that enter into the proof of Prop. 47

view of the world, according to which the universe is ultimately explained by the movements of physical objects, and these motions and their consequences are therefore to be accounted for:[49]

> Seeing then that truth consisteth in the right ordering of names in our affirmations, a man that seeketh precise truth had need to remember what every name he uses stands for, and to place it accordingly; or else he will find himself entangled in words, as a bird in lime twigs; the more he struggles, the more belimed. And therefore in geometry (which is the only science that it hath pleased God hitherto to bestow on mankind), men begin at settling the significations of their words; which settling of significations, they call definitions, and place them in the beginning of their reckoning.

Another thing to notice from our Hobbesian exercise of working backwards through Book I is that the downward progression has to stop at propositions that are proved, not from others, but from the postulates and common notions at the start of the Book, the heart of the axiomatic structure of the *Elements*. Although much of the content of the *Elements* seems to have been discovered earlier, it is generally agreed that Euclid himself either formulated the postulates, or selected them from those that may have been under discussion during the previous century. Euclid showed remarkable insight and intuition in making

[49] See Hobbes, *Leviathan* (1651, 19).

his choice of postulates — in particular, the parallel postulate — so that the entire superstructure was deducible from them, and yet they were 'free of everything superfluous', in Proclus's words.[50]

It is this extended demonstration of the power of abstract reasoning that has been part of the appeal of Euclid's *Elements* down the ages, quite independently of the importance of its contents. Indeed, some writers and educators have found the content tedious, but have appreciated the logical rigour and organisation of the work.

The later discovery of imperfections in it was a matter for comment, and the whole topic of the definition and properties of parallel lines became an adventure in itself, as we shall see. But these 'blots', as they were once called, were never held to damage the fundamental achievement of the work as a whole. Rather, it is an intriguing fact about the presentations of geometry that were promoted at the end of the 19th century, when it was finally felt that Euclid's *Elements* was no longer sufficiently rigorous, that some mathematicians revived the axiomatic style of the *Elements*. This had not been a prominent feature of mathematical writing until then, and it is a deep compliment to the *Elements* that its architecture was imitated in order to surpass it.

3.5 Further reading

Suggestions for further reading are given at the end of Chapter 5.

[50]See (Morrow 1970, 58–59), in F&G 3.B1(b).

4

Greek Mathematics: Proofs and Problems

Introduction

In this chapter, we look at two complementary aspects of Greek geometry: the Greeks' emphasis on proofs, as exemplified by Euclid's *Elements*, and their involvement with specific problems.

We start by discussing the nature and historical development of proof, and consider a number of specific types of proof that were used by the Greeks. One early type of argument involves appealing directly to what one can see in a given geometrical diagram. Another approach involves the indirect arguments, or proofs by contradiction, that we met briefly in Chapter 3: here we assume the opposite of what we wish to prove and show that such an assumption must lead to an absurdity. A third type of proof involves carrying out a specific geometrical construction and then justifying our belief that this construction does indeed lead to the object we require. Using these methods we can build a hierarchical structure, step by step and proposition by proposition, from a small number of basic assumptions or postulates. In particular, the geometry in Euclid's *Elements* emerges as a remarkable structure that arises from little more than a basic understanding of what we mean by a circle and a straight line.

For the second aspect of Greek geometry, there is evidence that Greek mathematicians responded to challenges by tackling problems that seemed to defy the means they had to solve them. In so doing they extended their technical mastery, but also had to decide which new arguments counted as genuine proofs and were not merely plausible statements. So the problems led both to new discoveries and to increasingly sophisticated ideas about what could be proved from what.

Three of these problems became particularly famous, and our account focuses on them (see Sections 4.3 and 4.4); they are known as the 'three classical problems': doubling the cube, trisecting an angle, and squaring the circle. As the Greeks rightly suspected, none of them can be solved by using lines and circles and the construction methods of Euclid's *Elements*, but the reasons for this were not to be discovered for over two thousand years.

4.1 The development of proof

If there is one aspect of Greek mathematics, more than any other, that distinguishes it from the mathematical activity of earlier cultures, it is the notion of *proving* results. In Chapter 3 we saw examples of proofs from different stages of Greek mathematical development, including Euclid's proof of the Pythagorean theorem and a proof that the diagonal and side of a square are incommensurable. This aspect of mathematics did not spring up overnight, and nor was it divorced from other social, political, and ideological developments in classical Greece.

In this section we discover what we can of the Greek development of mathematical proof, in what seems to have been the critical period: from Thales to Euclid. This period ends with the imposing and influential axiomatic and deductive structure of proved results that make up the thirteen Books of Euclid's *Elements*, but starts more mistily with what are probably legendary attributions to Thales. Let us look at one of these now, as it will lead us usefully into the subject.

Writing in the 5th century AD, Proclus reported (as we observed in Section 3.2) that:

> The famous Thales is said to have been the first to demonstrate that the circle is bisected by the diameter. The cause of this bisection is the undeviating course of the straight line through the centre.

This is an intriguing claim. Proclus probably took it from the now-lost *History of Geometry* by Aristotle's pupil Eudemus, written 200 years after Thales and 800 years before Proclus. It seems best to take it as a 4th-century BC perception of what might have (or should have) happened in the time of Thales.

But it is intriguing, because it encapsulates so neatly the unprecedented Greek approach to mathematics. At an earlier stage it may not have occurred to anyone that such a statement — *a diameter of a circle bisects the circle* — needs proving: it might simply have seemed obvious. Nor would they necessarily have the concept of proof in order to formulate that thought. And even when one has the general idea of proving things, it is far from obvious how to go about justifying this particular claim. What other property might we start from so as to *deduce* the 'obvious' one that we are interested in?

There are other problems. What do we actually mean by 'bisecting the circle'? Are we dividing the circle into two equal areas, two equal shapes, two bits of circumference that are equal in length, or what? In this instance these all amount to the same thing, but how such possibilities are related in general raises a fresh set of problems that open up when we try to decide even what the meaning of our property is.

One experiment we could try is to draw a circle on paper, cut it out, and fold it along a diameter. Then the two halves coinciding would show all that three properties hold: equal areas, equal shapes, and equal edge lengths. But this procedure raises the question of the relation between our physical pieces of paper and the more abstract concept of a general 'circle'. It may lead us to believe the result, but have we proved it?

Proclus himself offered the following argument that his modern editor conjectures may have been due to Thales:[1]

> If you wish to demonstrate this mathematically, imagine the diameter drawn and one part of the circle fitted upon the other. If it is not equal to the other, it will fall either inside or outside it, and in either case it will follow that a shorter line is equal to a longer. For all the lines from the center to the circumference are equal, and hence the line that extends beyond will be equal to the line that falls

[1] See (Morrow 1970, 58–59) in F&G 3.B1(b).

> short, which is impossible. The one part, then, fits the other, so that they are equal. Consequently the diameter bisects the circle.

Such musings suggest that the notion of proving things involves quite a complicated set of considerations, and unravelling and clarifying these historically will occupy us for the remainder of this section. However, before leaving this property of the circle, let us observe an ironic twist to any attempt to unravel its origins.

If we turn to Euclid's *Elements* to discover how he proved it, we find that he side-stepped the problem entirely by incorporating it as part of the definition of a circle's diameter. Indeed, Book I, Definition 17 runs as follows:[2]

> A diameter of the circle is any straight line drawn through the centre and terminated in both directions by the circumference of the circle, and such a straight line also bisects the circle.

So whatever Thales may or may not have done, Euclid did not do it at all! Euclid's approach is unsatisfactory, because his definition is mixed up with one of its consequences. His confusion may have been more editorial than conceptual, however, because the result can be derived as a consequence of propositions proved later in his Book III.

Parmenides on knowledge and belief. One reason for scepticism that Thales 'demonstrated' any mathematical results in the 6th century BC is that there is no reliable supporting evidence. Furthermore, such a claim does not fit in with our understanding of general philosophical developments in the subsequent century. The earliest example of sustained deductive argument that we know of occurs in a poem by Parmenides written in the early 5th century BC called *The Way of Truth*. Parmenides lived in Elea, in southern Italy, and we infer when he lived because Plato relates that Parmenides, when quite old, once visited Athens with his pupil Zeno of Elea, where he met the young Socrates; this would have been about 450 BC. In his poem Parmenides discussed how knowledge is possible, and gave an important role to logical principles, as instructed by a goddess:[3]

> And the goddess greeted me kindly, and took my right hand in hers, and addressed me with these words: 'Young man, you who come to my house in the company of immortal charioteers with the mares which bear you, greetings. No ill fate has sent you to travel this road — far indeed does it lie from the steps of men — but right and justice. It is proper that you should learn all things, both the unshaken heart of well-rounded truth, and the opinions of mortals, in which there is no true reliance.'

Parmenides — or his goddess — sought to establish a fundamental distinction between two sorts of knowledge: that acquired through the senses ('the opinions of mortals, in which there is no true reliance') and that attained through reason alone ('the unshaken heart of well-rounded truth'). Whereas the former sort of knowledge would be better described as 'belief' or 'opinion', the latter is the only sort that can be trusted, the only true knowledge: this is what lies behind Parmenides saying, later in the poem: 'judge by reason (*logos*) the strife-encompassed refutation spoken by me'. Knowledge, as opposed to opinion, has a 'force' or 'compulsion' about it — Parmenides went on to speak of 'the force of conviction'. That true knowledge is to be contrasted with beliefs acquired from the everyday world is a startling claim that has underpinned much Western thought ever since.

---

[2] See F&G 3.B1(a).

[3] See (Kirk, Raven, and Schofield 1983, 243, 245, 248–250) in F&G 2.C1.

During the 5th and 4th centuries BC, other philosophers clarified this distinction and its consequences for modes of argument, as well as using it in a taken-for-granted way. Thus the *Meno* passage in Chapter 3 — written within a century of Parmenides's poem — includes a lengthy discussion about 'opinions' and 'knowledge', how to get from one to the other, and Plato put into Socrates' mouth logical structures that are akin to the use that Parmenides made of the phrase 'it is or it is not'. Recall that the *Meno* extract concluded:

> If then there are going to exist in him, both while he is and while he is not a man, true opinions which can be aroused by questioning and turned into knowledge, may we say that his soul has been for ever in a state of knowledge? Clearly he always either is or is not a man.

So it was during the 5th century BC that startling and profound developments took place in two related aspects of thought: the status of different kinds of knowledge, and the techniques for acquiring (or persuading others of) true knowledge. The latter concerns the question of proof, while the former concerns what the proof relates to, or achieves.

The nature of mathematical knowledge. Where does mathematics fit into this picture? We must bear in mind that what we now identify as the mathematics of the period was not a separate field of study to the extent that it would be today, but was part of the general realm of intellectual discourse that had political, moral, cosmological, and other aspects to it.

There were, however, traditions of teachers and learners that might emphasise one aspect or another, or have particular preoccupations. Both the *Eleatics* and the *Pythagoreans* flourished in southern Italy in the 5th century BC. The Eleatics — Parmenides and his followers, such as Zeno — seem to have emphasised logical concerns in particular, while the Pythagoreans developed thought of a more religious and ethical cast, with mathematical overtones whose precise nature is still disputed. In Athens during this same century were Sophists, teachers of argument and rhetoric, against whose style and approach Socrates and Plato argued eloquently. However, there seems to have been a wide measure of agreement that mathematical knowledge was special, and of a higher kind of certainty than the everyday knowledge received through the senses. Mathematical knowledge was therefore seen to be inextricably connected with the way that it was acquired.[4]

4.2 Methods of proof

Our examination of Euclid's *Elements* showed us at least three distinct styles of proof: *direct*, *indirect*, and *axiomatic*. In order to study the development of specific styles of mathematical proof, as opposed to the general idea that proving mathematical results is a desirable and necessary aspect of doing mathematics, we need to undertake some historical reconstruction and inference.

***Deiknume* proofs.** The earliest fully recorded proof, Hippocrates's quadrature of the lune, is in quite a sophisticated and logically complicated style, and it is reasonable to infer that there were simpler methods of justification earlier in the 5th century. However, the discussion in *Meno* between Socrates and the slave boy culminates in what seems to be a different style of proof. Recall that Socrates drew the final diagram in the sand, from which the boy could essentially see (with the aid of a few leading questions) what the answer is

[4]There are two aspects to this: finding a mathematical result and demonstrating that it must be true. Here our main concern is with the latter, but we return to both aspects later.

and why it must be so. If we re-read the proof, in order to ascertain how rigorous it is, we notice that Socrates gains the boy's ready assent to the assertion that the diagonal of the square bisects it ('Now does this line going from corner to corner cut each of these squares in half?'), a proposition analogous to what Thales is said to have proved for the circle: it is, indeed, fairly 'obvious' that this is true. History does not record how Socrates would have responded had the boy replied: 'No — prove it to me!'.

In one type of proof, things are presented so that one can immediately 'see' that the result is true: this style of proof the Greeks called *deiknume*.[5] We saw two geometrical examples of *deiknume* in the *Meno* passage: the overall perception of the final diagram, and the case of the diagonal bisecting the square, although the latter slipped past so quickly that it may be stretching a point to claim it as proved at all.

Some early results about odd and even numbers were probably first proved in *deiknume* style. For example, let us consider the claim that *the sum of any two odd numbers is an even number*. It can easily be checked for individual cases ($3+5 = 8, 15+13 = 28$, and so on), but this does not amount to proving the general statement. The *deiknume* proof arises if we represent the numbers as dots or pebbles:[6] if we add two odd numbers by joining up the patterns, we can see that the resultant number must be even (see Figure 4.1).

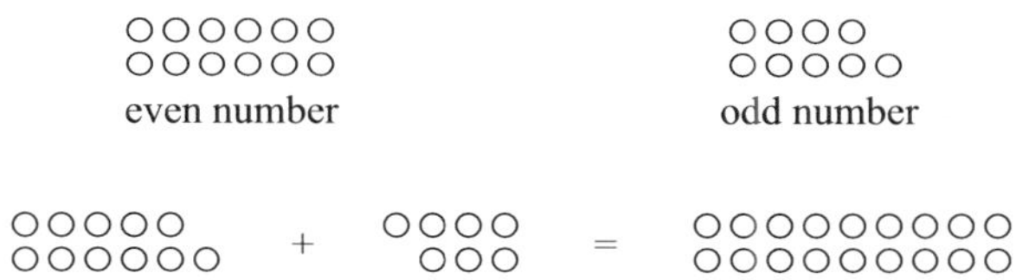

Figure 4.1. A *deiknume* proof

Evidence for the antiquity of these concerns can be inferred from examining the treatment of odd and even numbers in Euclid's *Elements* Book IX, Props. 21 and 22.[7]

Euclid, *Elements,* Book IX, Propositions 21 and 22.

Proposition 21

If as many even numbers as we please be added together, the whole is even.

For let as many even numbers as we please, AB, BC, CD, DE, be added together; I say that the whole AE is even.

For, since each of the numbers AB, BC, CD, DE is even, it has a half part [VII. Def. 6]; so that the whole AE also has a half part.

But an even number is that which is divisible into two equal parts [id.]; therefore AE is even. Q.E.D.

---

[5]The Greek word *deiknume* (pronounced 'dayk-noomy') means 'to show something to be so' or 'to make visible or evident'.

[6]A 5th-century BC work by Epicharmus provides evidence that numbers were represented as pebbles, so that the oddness or evenness was apparent. The Latin word for pebble is 'calculus', from which our word 'calculate' derives.

[7]F&G 3.D2(b) and D2(c).

Proposition 22

If as many odd numbers as we please be added together, and their multitude be even, the whole will be even.

For let as many odd numbers as we please, AB, BC, CD, DE, even in multitude, be added together; I say that the whole AE is even.

For, since each of the numbers AB, BC, CD, DE is odd, if an unit be subtracted from each, each of the remainders will be even [VII. Def. 7]; so that the sum of them will be even [IX. 21].

But the multitude of the units is also even.

Therefore the whole AE is also even [IX. 21]. Q.E.D.

Despite the fact that Euclid represented numbers as lines, and not as dots or pebbles, we can see the proofs of these propositions as translations into a more overtly logical framework of arguments in the *deiknume* style. If the proof in Proposition 22 arose in this way, something more elaborate was done with the pebbles than in the case of two odd numbers, but the translation is still clear. It can thus be argued that Euclid incorporated into the *Elements* some 5th-century BC results on 'odd and even', with something plausibly close to their original demonstration.

There are two major problems with *deiknume* proofs. First, however convincing a proof may be in itself, each new result has to be considered afresh on its merits. Because of the directness and immediacy of each proof, it is difficult to build up a structure in which results depend on other results. Second, they appear to depend on our sense impressions — on literally seeing — in a way that lays itself open to criticism that the proof-style is too closely tied to the transient everyday world of human opinions, rather than to that of mathematical truth. This need not be a fatal objection — it can be argued that the pebbles or the sand-diagram are simply aids to enable inner reason to attain true knowledge — but the style is clearly different from the path to knowledge advocated by Parmenides and the Eleatics.

***Reductio* proofs.** Can we prove things with absolute certainty, by pure reason alone and making no concessions to knowledge aroused by the senses? The Eleatics appear to have thought so, and the proof style they introduced, known as *indirect proof, proof by contradiction*, or by its Latin name *reductio ad absurdum*,[8] turned out to be enormously influential in Greek and later mathematics.

The historical question of whether mathematicians adopted this proof style as a direct consequence of the philosophical arguments of Parmenides and his followers is unresolved. The best we can say is that such *reductio* proofs were employed in mathematical argument some time after their philosophical advocacy by the Eleatics. What they rely on are purely logical assertions whose truth no-one can deny, regardless of the state of the world or of one's sense impressions.

You saw that Parmenides argued from the statement 'it is or it is not', and that in *Meno*, Socrates said 'Clearly he always either is or is not a man'. Statements of the form 'Either something is the case, or it is not' must receive universal acceptance, regardless of which alternative is actually the case, and indeed even when we do not know which one is the case. They are truths of pure reason, and are thus a suitable starting point for attaining true knowledge by enabling us to deduce further truths by the use of reason alone.

[8]Recall that 'reductio ad absurdum' means reducing something to an absurdity or contradiction.

The way that this deduction works, in *reductio* proofs, is to take one alternative and disprove it; from this it follows that the other alternative must be true. Disproving the alternative is a matter of pure logic: you deduce something that contradicts what you already know or have laid down, which is therefore a logical absurdity. A good example is the first *reductio* proof in Book I, Prop. 6 of Euclid's *Elements*, the isosceles triangle theorem, which we discussed in Box 3.4. There, we observed that 'if AB is unequal to AC, one of them is greater', and deduced a contradiction — the triangle ABC is equal to a strictly smaller triangle inside it.

There is an important general point to notice about *reductio* proofs. The initial statement that generates the proof structure, of the 'it is or it is not' kind, must consist of incompatible alternatives that exhaust all the possibilities; otherwise, disproving one alternative would say nothing of consequence about the other. For example, in the case above it would be unavailing to set out from the statement 'either the two sides are equal or the two angles are equal', since these are not incompatible. Similarly, to set out from 'either AB equals AC, or AB is less than AC' would be equally fruitless, since it does not take account of the possibility that AB is greater than AC.

Another of Euclid's *reductio* proofs (Book IX, Prop. 20) shows that there are infinitely many prime numbers, or 'Prime numbers are more than any assigned multitude of prime numbers'.[9] This proof has long been considered one of the most beautiful and elegant in all mathematics: the distinguished English mathematician G. H. Hardy said that it is:[10]

> as fresh and significant as when it was discovered — two thousand years have not written a wrinkle on [it.]

The overall strategy of Euclid's proof is revealed in the first two lines, following directly from the statement of what is to be proved. He assigns a multitude of primes (A, B, and C) and shows that there are more prime numbers than this (Figure 4.2). To do this, he constructs another number (EF), which is effectively $(A \times B \times C) + 1$, and then argues as in Box 4.1. We present both Euclid's proof and a modern one for comparison. (In Euclid's proof, the term 'measured by' can be understood as 'divisible by'.)

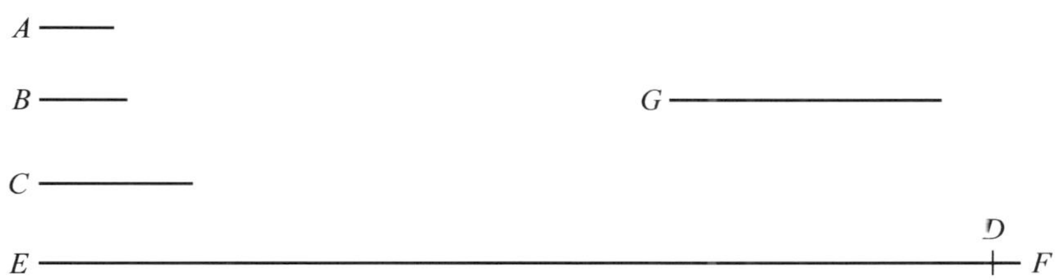

Figure 4.2

Notice what happens in Box 4.1: a statement is proved by showing that its opposite leads to an absurdity or contradiction (here, that the unit can be divided). Note also that Euclid's proof seems to show only that if there are three prime numbers, then there must be a fourth. There seems to be an assumption, which he does not spell out, that for any other assigned multitude of primes the proof is exactly analogous — or, to put it another way, that the case of three primes adequately represents the general case. If we grant this assumption, the proof is then perfectly general.

[9]F&G 3.D2(a).

[10]See (Hardy 1967, 92).

Box 4.1. Prime numbers are more than any assigned multitude of prime numbers.

Euclid's proof

Let A, B, C be the assigned prime numbers; I say that there are more prime numbers than A, B, C. For let the least number measured by A, B, C be taken, and let it be DE; let the unit DF be added to DE. Then EF is either prime or not.

First, let it be prime; then the prime numbers A, B, C, EF have been found which are more than A, B, C.

Next, let EF not be prime; therefore it is measured by some prime number [VII, 31]. Let it be measured by the prime number G. I say that G is not the same with any of the numbers A, B, C.

For, if possible, let it be so.

Now A, B, C measure DE; therefore G will also measure DE. But it also measures EF.

Therefore, G being a number, will measure the remainder, the unit DF: which is absurd.

Therefore G is not the same with any one of the numbers A, B, C. And by hypothesis it is prime.

Therefore the prime numbers A, B, C, G have been found which are more than the assigned multitude of A, B, C. Q.E.D.

Modern proof

Suppose that the only prime numbers are $p_1, p_2, \ldots, p_k$.

Let $N = (p_1 \times p_2 \times \cdots \times p_k) + 1$. Then either N is prime or not.

If N is prime, it is different from $p_1, p_2, \ldots, p_k$ and so is a new prime.

If N is not prime, then it is divisible by some prime p different from $p_1, p_2, \ldots, p_k$. But if $p = p_k$, for some k, then p divides both

$$p_1 \times p_2 \times \cdots \times p_k \quad \text{and} \quad (p_1 \times p_2 \times \cdots \times p_k) + 1,$$

which is impossible. Thus, again we have discovered a new prime.

Therefore, in each case there is a prime different from $p_1, p_2, \ldots, p_k$. Q.E.D.

In analysing the nature of *reductio* proofs we are running ahead of the history. We cannot be sure that mathematicians in the 5th century BC thought about the nature of proof in the explicit way that we have done, an analysis that owes more to the work of Aristotle in the middle of the following century. It was not until Aristotle that we find a full and careful analysis of deductive argument and modes of proof, so earlier mathematicians may have operated on a more intuitive and less self-conscious plane. A similar difficulty arises in considering what 5th-century conceptions of mathematical knowledge may have been — here the later influence of Plato is so strong as to colour inescapably our perceptions of his predecessors.

Although *reductio* proofs are powerful and were greatly used, they have an undesirable consequence when used to attain mathematical knowledge, compared (for example) with the state of mind achieved through a *deiknume* proof. With the latter you can see clearly and convincingly *why* the result is true, whereas *reductio* proofs demonstrate unarguably *that* the result is true, but do not necessarily provide any understanding of *why* or *how*. We prove that things cannot be other than they are, and that other possibilities are absurd, but this is not the same as knowing why a result *must* be true.

Construction proofs. The Greeks did indeed develop other proof styles, more logically convincing than *deiknume* and leading to deeper understanding than the indirect *reductio* style. Logical proofs involving longer chains of direct deduction were in use by the latter part of the 5th century BC, as you saw in Hippocrates's quadrature of the lune. There, the proof involved a construction of the lune whose quadrature was sought. But even when the object we wanted to prove something about was already given, the proofs often involved further constructions on which a deductive argument would be based, as illustrated by Euclid's construction of an equilateral triangle in *Elements*, Book I, Prop. 1 (see Section 3.4).

Here we take another example, from Book I, Prop. 32 and related extracts from Aristotle and Proclus.[11] Euclid and Proclus both gave proofs, using different constructions, that the sum of the internal angles of a triangle equals two right angles. In Euclid's proof, a side is extended in one direction, and at that corner a line is drawn parallel to the opposite side (see Figure 4.3). The other construction, ascribed by Eudemus to 'the Pythagoreans', consists simply of drawing a line through one corner of the triangle, parallel to the opposite side (see Figure 4.4).

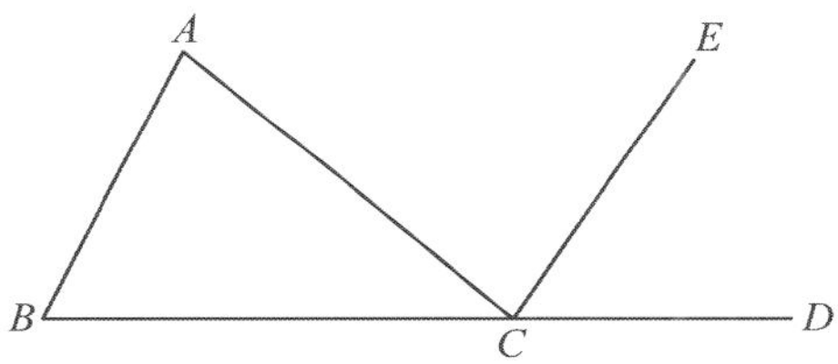

Figure 4.3. Euclid's diagram for *Elements*, Book I, Prop. 32

The angle sum of a triangle.

The sum of the angles of a triangle equals two right angles.

Euclid's Proposition 32

In any triangle, if one of the sides be produced, the exterior angle is equal to the two interior and opposite angles, and the three interior angles of the triangle are equal to two right angles.

Let ABC be a triangle, and let one side of it BC be produced to D; I say that the exterior angle ACD is equal to the two interior and opposite angles CAB, ABC, and the three interior angles of the triangle ABC, BCA, CAB are equal to two right angles.

For let CE be drawn through the point C parallel to the straight line AB [I. 31]. Then, since AB is parallel to CE, and AC is fallen upon them, the alternate angles BAC, ACE are equal to one another [I. 29].

Again, since AB is parallel to CE, and the straight line BD has fallen upon them, the exterior angle ECD is equal to the interior and opposite angle ABC [I. 29].

But the angle ACE was also proved equal to the angle BAC; therefore the whole angle ACD is equal to the two interior and opposite angles BAC, ABC.

[11] F&G 3.B4.

Let the angle ACB be added to each; therefore the angles ACD, ACB are equal to the three angles ABC, BCA, CAB.

But the angles ACD, ACB are equal to two right angles [I. 13]; therefore the angles ABC, BCA, CAB are also equal to two right angles.

Therefore etc. Q.E.D.

Aristotle[12]

Propositions too in mathematics are discovered by an activity; for it is by a process of dividing-up that we discover them... Why does the triangle make up two right angles? Because the angles about one point are equal to two right angles. If then the parallel to the side had been drawn up, the reason why would at once have been clear from merely looking at the figure.

Proclus[13]

Eudemus the Peripatetic attributes to the Pythagoreans the discovery of this theorem, that every triangle has internal angles equal to two right angles, and says they demonstrated it as follows. Let ABC be a triangle, and through A draw a line DE parallel to BC [*see Figure* 4.4]. Then since BC and DE are parallel, the alternate angles are equal, and angle DAB is therefore equal to ABC and EAC to ACB. Add the common angle BAC. Then angles DAB, BAC, CAE — that is, angles DAB and BAE, which are two right angles — are equal to the three angles of the triangle ABC. Therefore the three angles of a triangle are equal to two right angles. Such is the proof of the Pythagoreans.

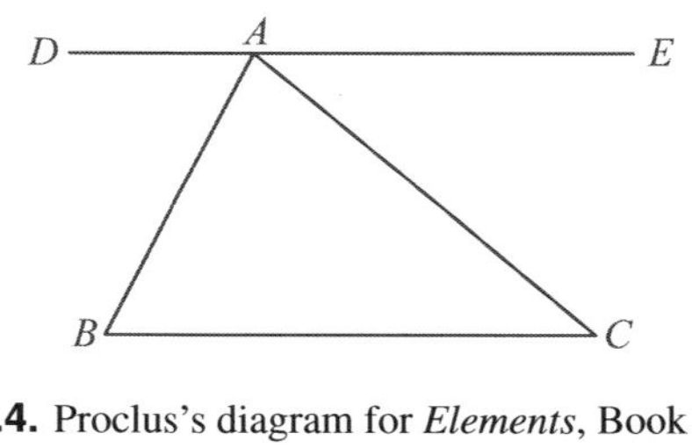

Figure 4.4. Proclus's diagram for *Elements*, Book I, Prop. 32

In both Euclid's and Proclus's arguments the rest of the proof consists of direct reasoning about the figure thus constructed. Notice that it cannot literally be seen in the figure; we cannot measure angles precisely by eye. The proof enables us to see the stated equality.

Aristotle's comments (while illustrating a more general point) are interesting in suggesting that after the construction has been done, the result follows by the process we have described as *deiknume*.

What Aristotle's discussion brings to light is something that we skated over in our earlier account of *deiknume* proofs; seeing something to be obviously true depends on the state of mind and knowledge we bring to bear on the perception. For example, as you saw in the *Meno* proof, Socrates needed to ask several questions in order to awaken in the slave boy his own 'personal opinion' that the result is true.

[12] *Metaphysics*, 1051a, 21–24, from (Heath 1949), in F&G 3.B4(b).

[13] See (Morrow 1970, 298), in F&G 3.B4(c).

In the case of the internal angle-sum of the triangle, the result and the reason for it are certainly 'clear from merely looking at the figure', provided that several other results are so well known as not to need thinking about: knowledge of parallel lines, of the relation between particular angles, of adding angles just by looking at them, and so on. In this sense, Aristotle's claim is right.

On the other hand, the detailed argument given by Euclid or Proclus is necessary if we wish to establish a logically convincing proof that can be shared with others more readily than can an inner conviction. The proofs spell out in more detail what it is about the constructed situation that leads to the *deiknume* knowledge that the result is true. Without a detailed logical argument, we cannot be sure that our feelings of clear and certain knowledge are not mistaken.

So proofs involving constructions can be described as sophisticated versions of *deiknume* proofs, in which the initial mathematical object is changed or added to by the construction into one enabling the truth of the proposition to be seen more readily, or argued towards more easily. For example, in Book I, Prop. 1, Euclid began with a construction and the proved that it is correct. We have not yet specified clearly what 'a construction' is, however.

Line-and-circle constructions. The constructions in the above proofs involve drawing and extending lines, drawing lines parallel to other lines, and drawing circles. The construction in *Meno* consists of drawing more squares and drawing a line diagonally across each of them, while Hippocrates constructed a lune by drawing arcs of circles in a particular relationship to a certain triangle, and Euclid drew lines and circles in order to construct an equilateral triangle with a given side as base. What all of these line-and-circle constructions have in common is that they can be effected by just two instruments, an *unmarked ruler* (often called a *straight-edge*) for drawing straight lines, and a *pair of compasses* for drawing circles — indeed, they are often called *ruler and compass constructions*.[14]

Such constructions have a somewhat practical-sounding air. We seem to be in the realm of everyday physical activity, far from the disembodied world of pure reason that was beginning to be taken as the mathematical ideal. Yet we can abstract from the physical instruments and actions, and consider their effects (straight lines and circles) as pure geometrical objects.

We do not know how fully this point was debated among mathematicians of the 5th century BC, but it seems to have been in the latter part of that century that the idea emerged of seeing what could be constructed by using only an unmarked ruler and a pair of compasses. This idea proved very fertile, both in the remarkable number of things that can be done using these apparently restricted means, and (as we see later) for stimulating the investigation of problems that proved recalcitrant by these means. Indeed, the emergence of such line-and-circle constructions signalled the beginning of the deeper interest in geometry that characterised the following two centuries.

Direct arguments about a constructed object are intuitively more appealing than *reductio* proofs, since we emerge with an object that can be constructed, rather than an assurance that things could not be otherwise. But notice that they did not start from undeniable logical truths: some things have to be taken for granted. Here lines and circles are the building blocks from which everything else is constructed.

This is an interesting departure. It seems that some 5th-century BC Greeks took the straight line as the basic object from which all other rectilinear geometrical objects (such as squares and triangles) are made, and the circle as the basic curved object. From these

[14] See Chapter 3.

two building blocks they could then construct further objects and describe their properties, thereby deducing new results from previous ones. This ensured knowledge that was both logically secure and also grounded in objects convincing to the senses. Using lines and circles alone, for example, we can give rigorous constructions for bisecting and trisecting any line segment and for bisecting any given angle (see Box 4.2).

***Neusis* constructions.** Line-and-circle constructions were those most frequently used, and in Euclid's *Elements* almost no others are found. But other kinds of constructions are found elsewhere in Greek mathematics, involving instruments other than a pair of compasses and an unmarked ruler. Most notable was the *neusis* construction, in which we insert a line segment of pre-determined length between two other (straight or curved) lines so that it points in a given direction towards a fixed point. In physical terms we can think of *neusis* as moving around a marked ruler on which the desired length has been marked until it passes through the fixed point and the marks coincide with the two lines.[15]

The earliest indication of this construction is in Hippocrates's quadrature of lunes.[16] Hippocrates was dealing with a lune whose outer curved edge is less than a semicircle. The *neusis* occurs in the 'preliminary construction', prior to the construction of the lune itself:

> Let the straight line EF be placed between this and the circumference verging towards B so that the square on it is one-and-a-half times the square on one of the radii.

We start with a semicircle with diameter AB and centre K, and let C be the midpoint of BK. Let CD be a line perpendicular to AB. We now imagine a ruler with the distance EF marked on it, where $EF^2 = 1\frac{1}{2}AK^2$. This is then moved about, always passing through B, until E lies on the circumference and F lies on the vertical line CD (see Figure 4.5).

Hippocrates's *neusis* construction was only part of his argument. He also produced a line-and-circle construction, and the result involves quite a complicated diagram. The early date of 450–430 BC for this work is interesting, because it shows that sophisticated construction proofs were around early in the period we are discussing.

The purpose of a construction proof is to build up a geometrical situation on which a rigorous argument can be based, in order to show the truth of the result required. In the course of the argument, previous geometrical knowledge is introduced, both to make

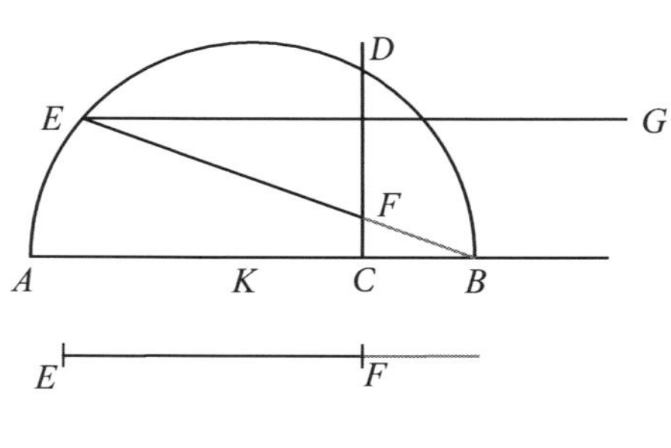

Figure 4.5. A *neusis* construction by Hippocrates

[15]'*Neusis*' is sometimes translated as 'verging'.

[16]See Section 3.2 and F&G 2.B.

Box 4.2. Three line-and-circle constructions.

Bisecting a line segment

Let the line segment be AB (see Figure 4.6). Draw the circle with centre A and radius AB, and the circle with centre B and radius BA; these circles intersect at two points C and D. Draw the line CD, and let E be the point where CD meets AB. Then $AE = EB$, and the line segment AB has been bisected, as required. (This is the argument in Euclid's *Elements*, Book I, Prop. 10.)

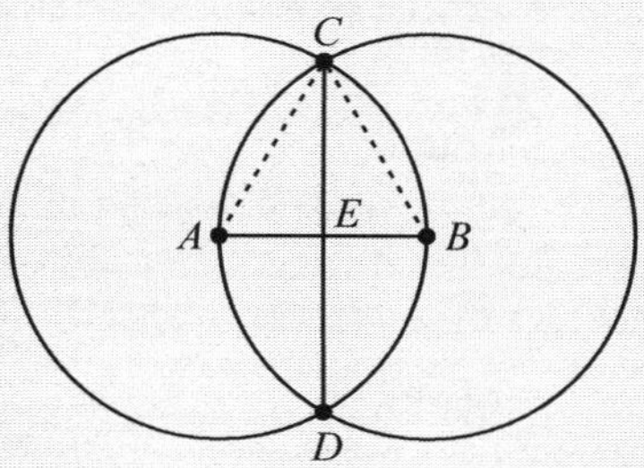

Figure 4.6. The segment AB is bisected at E

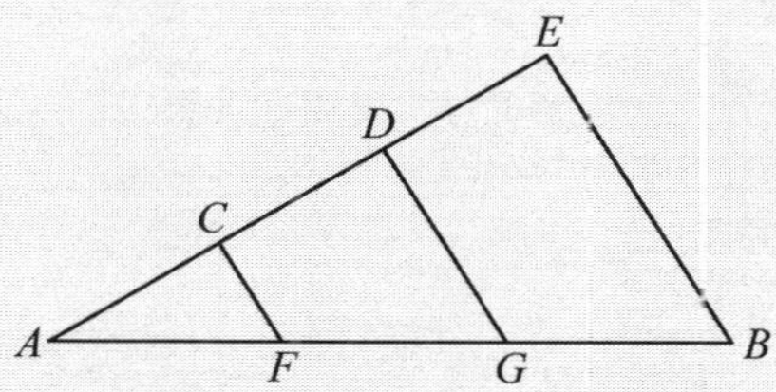

Figure 4.7. The segment AB is trisected at F and G

Trisecting a line segment

Let the line segment be AB (see Figure 4.7). Draw any line segment AC such that C does not lie on the line AB. Extend the line AC and by means of compasses mark off three equal segments $AC = CD = DE$ on the line. Draw the line EB and draw lines through C and D parallel to EB, meeting AB at F and G, respectively. Then $AF = FG = GB$, and the line segment AB has been trisected, as required.

Bisecting an angle

Let the angle be BAC (see Figure 4.8). Take any point D on AB, and let E be the point on AC such that $AD = AE$. Construct the equilateral triangle DEF with base DE. The line segment joining A and F bisects the angle BAC. (This is the argument in Euclid's *Elements*, Book I, Prop. 9.)

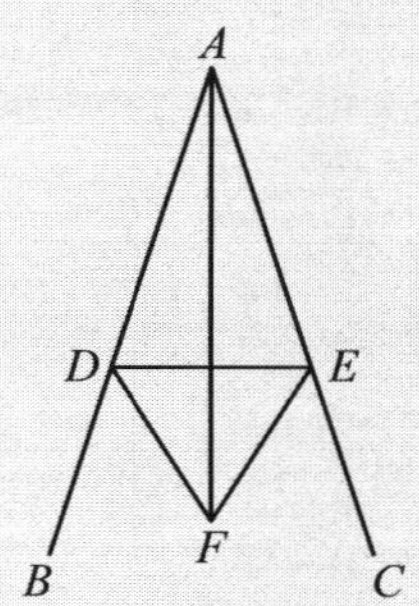

Figure 4.8. Bisecting the angle at A

deductions about the geometrical situation and also to ensure that the constructions are valid and have the required properties.

In this connection, the status of *neusis* constructions is puzzling. On the one hand, they seem less fundamental than line-and-circle constructions, and when physically realised by juggling marked rulers around they are less intuitively appealing. On the other hand, some proofs are possible with *neusis* constructions that cannot be done by line-and-circle methods, although the Hippocrates example is not such a case: there is a line-and-circle construction that draws a line of the required length in the right place, using techniques later set down in Book II of Euclid's *Elements*. It is notable, however, that neither Hippocrates nor his commentators Eudemus and Simplicius felt it necessary or desirable to 'reduce' the *neusis* construction in this way.

There is also a passage in Aristotle that implies that *neusis* constructions were accepted as a fundamental part of geometry. Talking about how the first principles of mathematics must be assumed, Aristotle went on to claim:[17]

> thus arithmetic assumes the answer to the question what is meant by 'odd' or 'even', 'a square' or 'a cube', and geometry to the question what is meant by 'the irrational' or 'deflection' or 'verging' (*neuein*); but that there are such things is proved by means of the common principles and of what has already been demonstrated.

In this context it is interesting to consider Euclid's postulates, since they define the scope of the *Elements*. Postulates 1–3 amount to setting line-and-circle constructions as the axiomatic foundation of the *Elements*. But there is no postulate allowing a *neusis* construction. Since the *Elements* consists only of results that can be deduced from the initial axiomatic foundations, we should expect there to be large and important areas of Greek geometry that are not dealt with in Euclid's *Elements* — this is the case, as we shall see, with the study of conics.

Discovering and proving: Analysis and synthesis. We conclude this section by looking at a distinction that we raised earlier, the difference between *discovering* and *proving* mathematical knowledge. When considering Hippocrates's construction for lunes, you may have felt that the condition on the *neusis* (the square on one line is $1\frac{1}{2}$ times the square on some other line) was strangely unmotivated, and that the point of it became evident only at the end. Clearly Hippocrates did not just kneel in the sand one day and start sketching out the construction as we have it. He must have been 'working backwards' in order to discover that it was precisely this condition that he would need for his logical argument later.

This method of 'working backwards' was formalised by Greek mathematicians under the name *analysis*, whereas the usual Greek method of 'working forwards' through a proof is known as *synthesis*. The fullest account we have of these terms was given by Pappus many centuries later, around AD 320, but his explanations are hard to follow because of his rather technical language.[18]

Basically, in *synthesis* 'we finally arrive at the construction of what was sought', whereas in *analysis* 'we take that which is sought as though it were admitted'. So the proofs that you have been studying are synthetic, working *from* generally agreed starting points ('Let there be a circle ...', etc.) *towards* the construction and proof of what is required. But even though most Greek proofs are synthetic, we may suppose that the presentation of a synthetic proof was preceded by some 'analysis', in which mathematicians analysed the conclusion they wanted to reach, breaking it down in stages until they got

---

[17] Aristotle, *Posterior Analytics*, 76a31–77a4, and F&G 2.H1(a).

[18] See F&G 5.B3.

to something known to be true or allowable, such as a *neusis* or line-and-circle construction.[19] Only by some such process, however informal, could Hippocrates have known in what direction to take his subsequent synthetic proof.

However, if the analytic method was regularly used in conjunction with the synthetic one, most traces of this have been lost. Indeed, the synthetic proofs that have survived (particularly those by Archimedes and Apollonius[20]) later presented such an unmotivated appearance to mathematicians of the 16th and 17th centuries as to suggest that 'the ancients' had a hidden method of discovery — for otherwise, how could their originators have known that their synthetic proofs were heading in the right direction?

4.3 Doubling the cube and trisecting an angle

In this section and the next we investigate three of the most important challenges that Greek mathematicians set themselves, provoking some of their most significant and ingenious discoveries. Collectively they are known as the *three classical problems*: we look at two of them here, *doubling the cube* and *trisecting an angle*, and at the third one, *squaring the circle*, in Section 4.4.

Doubling the cube: Given a cube, construct another cube with twice its volume (Figure 4.9).

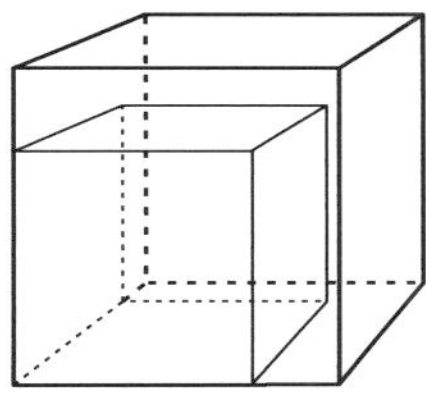

Figure 4.9. Doubling the cube

Trisecting an angle: Given an angle between two straight lines, construct two lines that divide the angle into three equal parts (Figure 4.10).
Squaring the circle: Given a circle, construct a square with the same area (Figure 4.11).

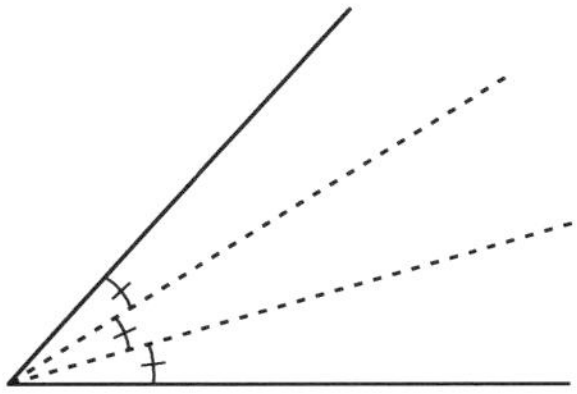

Figure 4.10. Trisecting an angle

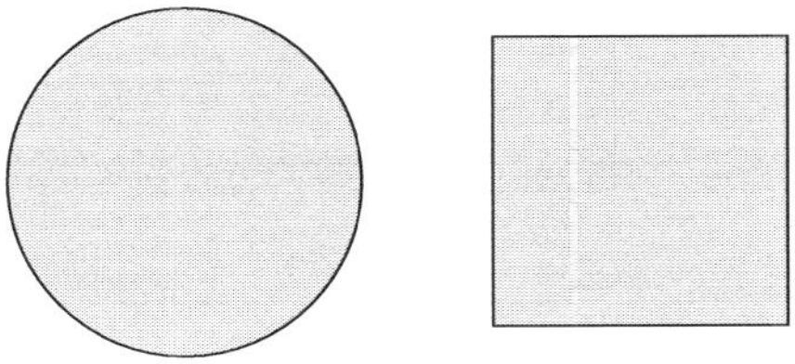

Figure 4.11. Squaring the circle

As we have seen, construction techniques were increasingly used towards the end of the 5th century BC for obtaining and justifying mathematical results: this went hand in hand with an increasing concentration on *geometrical* aspects of mathematics. We have spoken

[19] We return to analysis later, when we see that modern uses of the word do not carry its Greek connotations.
[20] We consider these in Chapter 5.

of a research tradition as comprising *problems, methods* of approach, and means of *justification*, and you are now seeing how interrelated these are. The mathematical problems that the Greeks tackled, and indeed the whole geometrical bias of so much Greek mathematics, both stimulated and was stimulated by construction methods, and the question naturally arose as to which geometrical problems can be solved by the basic line-and-circle constructions, and which ones cannot.

It was soon discovered that the three easy-to-describe problems above do not yield readily to such constructive methods. This realisation made these problems enormously influential and led to profound and fruitful mathematical advances as people attempted to solve them. It must have been frustrating for mathematicians of the 5th century BC to find that although doubling a square is straightforward (as you saw in the *Meno* extract), *doubling a cube* is not. Similarly, bisecting an angle and trisecting a line segment by straight-edge and compasses are quite simple, as we saw in Box 4.2, whereas *trisecting an angle* seems to be impossible, except in a few specific cases (such as a right angle). And by the end of the 5th century BC, the problem of squaring the circle was sufficiently well known in Athens for the popular comic dramatist Aristophanes to make a joke out of it in his play *The Birds* of 414 BC:[21]

> METON: These are my special rods for measuring the air. You see, the air is shaped — how shall I put it — like a sort of extinguisher; so all I have to do is to attach this flexible rod at the upper extremity, take the compasses, insert the point...
>
> PEISTHETAERUS: No.
>
> METON: Well, I now apply the straight rod — so — thus squaring the circle; and there you are ...
>
> PEISTHETAERUS: Brilliant — the man's a Thales.

The early Greek mathematicians seem to have quickly realised that these three problems are impossible to solve by line-and-circle constructions, and turned their attentions to devising constructions that would produce solutions, albeit not by straight-edge and compasses alone. As a result of this activity, striking and important geometrical developments took place — notably, the early development of *conic sections*. It was not until the 19th century that these three classical problems were *proved* to be insoluble by line-and-circle constructions.

So this enterprise marks a new kind of problem, one created by the methods of approach used within the developing research tradition. It was the failure of the line-and-circle construction method to solve these apparently elementary problems that led to much further enrichment of the mathematical methods, tools of analysis and proof methods of Greek geometry. We now consider some of the issues that arise from the ways in which solutions to the three classical problems were sought.

Doubling the cube. It is not known for certain how the problem of doubling the cube arose, but there are a number of colourful and informative stories. Two are from the commentators Theon of Smyrna (2nd century AD) and Eutocius (early 6th century AD), each of whom claimed works by Eratosthenes, the librarian of Alexandria in the 3rd century BC, as his source.

[21] Aristophanes, *The Birds*, in Barrett (1978, 187–188). A fuller version of this extract appears in F&G 2.C2.

Theon and Eutocius on doubling the cube. Theon's account is as follows:[22]

> In his work entitled *Platonicus* Eratosthenes says that, when the god announced to the Delians by oracle that to get rid of a plague they must construct an altar double of the existing one, their craftsmen fell into great perplexity in trying to find how a solid could be made double of another solid, and they went to ask Plato about it. He told them that the god had given this oracle, not because he wanted an altar of double the size, but because he wished, in setting this task before them, to reproach the Greeks for their neglect of mathematics and their contempt for geometry.

Eutocius's account, in his *Commentary on Archimedes' Sphere and Cylinder*, is as follows:[23]

> The story goes that one of the ancient tragic poets represented Minos having a tomb built for Glaucus, and that when Minos found that the tomb measured a hundred feet on every side, he said: 'Too small is the tomb you have marked out as the royal resting place. Let it be twice as large. Without spoiling the form quickly double each side of the tomb.' This was clearly a mistake. For if the sides are doubled the surface is multiplied fourfold and the volume eightfold.
>
> Now geometers, too, sought a way to double the given solid without altering its form. This problem came to be known as the duplication of the cube, for, given a cube, they sought to double it.

In each story the problem originated in a practical way, through seeking to double the volume of an altar or a tomb, but the *motive* for this doubling is variously given. The explanation that Theon ascribed to Plato is consistent with our study of his *Laws* — that he was ready to cite any unsolved mathematical problem to berate his fellow Greeks for ignorance and apathy towards mathematics. At all events, it seems that a whole battery of high-powered mathematicians — Archytas, Eudoxus, and Menaechmus, all friends or associates of Plato — were soon at work devising solutions to a related problem, due to Hippocrates, of finding 'two mean proportionals' (a term we explain below).

So the problem was swept up into the highly theoretical geometrical research tradition. It may in fact have originated there, as we do not know what credence to give to the tomb and altar stories, and on chronological grounds Hippocrates's involvement is likely to have been earlier. But Eutocius also points out the practical advantages of a mechanical construction to solve the problem, in terms useful for engineers. Although we cannot infer from this that Greek engineers used mechanical solutions to the problem, it is a useful reminder to us that a practical mathematical tradition existed in Greek times. So although the logical status of the problem of doubling the cube is clear, when seen within the theoretical geometrical research tradition, practical knowledge or utility was later invoked in traditions where the high Platonic ideals held less sway, or had perhaps been modified.

Hippocrates's reduction to two mean proportionals. We have seen that Greek mathematical language makes greater use of ratio and proportion than we do. For example, in his quadrature of lunes, Hippocrates was said to have proved that 'similar segments of circles have the same ratios as the squares on their bases', by showing that 'the squares on the diameters have the same ratios as the circles'.[24] So where we would write

[22] Theon of Smyrna, in (Thomas 1939, Vol. 1, 257) and F&G 2.F1.

[23] Eutocius, in (Cohen and Drabkin 1948, 62–66) and F&G 2.F3.

[24] F&G 2.B.

d^2 *is proportional to* A, or $A = \frac{1}{4}\pi d^2$ (for a circle of diameter d and area A), the Greeks wrote something like $d_1^2 : d_2^2 = A_1 : A_2$. Even though these can seem to amount to the same thing, the differences in expression are important: for example, where we wrote d^2 Greek mathematicians would have understood the geometrical square on the line d, and not a number multiplied by itself. In particular, this language enabled geometrical entities to be treated as such, and their integrity preserved, rather than being swept up into the arithmetical language of numbers as we tend to do.

The most significant advance in the history of doubling the cube was made by Hippocrates: reducing the problem to that of finding two *mean proportionals*.

Eutocius and Proclus on two mean proportionals. Eutocius's account continues:[25]

> Now when all had sought in vain for a long time, Hippocrates of Chios first discovered that if a way can be found to construct two mean proportionals in continued proportion between two given straight lines, the greater of which is double the lesser, the cube will be doubled. So that his difficulty was resolved into another no less perplexing.

We have seen examples of the reduction of a problem to other problems whose solution enables the original one to be solved. In the present connection, Proclus commented:[26]

> 'Reduction' is a transition from a problem or a theorem to another which, if known or constructed, will make the original proposition evident. For example, to solve the problem of doubling the cube geometers shifted their inquiry to another on which this depends, namely, the finding of two mean proportionals; and thenceforth they devoted their efforts to discovering how to find two means in continuous proportion between two given straight lines.
>
> They say the first to effect reduction of difficult constructions was Hippocrates of Chios, who also squared the lune and made many other discoveries in geometry, being a man of genius when it came to constructions, if there ever was one.

So what are these 'mean proportionals'? Given two lines a and b, a *mean proportional* between them is a line segment x with the property that

$$a : x = x : b.$$

This is the equivalent, for line segments, of what we call the *geometric mean* of two numbers — the square root of their product. For, if we write these lines as numbers,[27] then the property becomes

$$a/x = x/b\ , \text{ or } x^2 = ab\ , \text{ or } x = \sqrt{ab}.$$

From the $x^2 = ab$ formulation, we can see that the problem is geometrically that of constructing a square equal (in area) to a rectangle whose sides are a and b; the side x of the square is then the mean proportional between a and b. In *Elements* Book VI, Prop. 13, Euclid showed how this construction can be carried out by straight-edge and compasses.

---

[25]Eutocius, *Commentary on Archimedes' Sphere and Cylinder* in (Cohen and Drabkin 1948, 62–66) and in F&G 2.F3.

[26]See (Morrow 1970, 167) in F&G 2.F2.

[27]Recall that the Greeks represented a number as a line.

The problem of *two mean proportionals* cannot be solved by straight-edge and compasses. Here, given two line segments a and b, the problem is to construct *two* line segments x and y such that

$$a : x = x : y = y : b.$$

It was Hippocrates who seems to have established that if this general problem can be solved, then the doubling of the cube arises as a special case: it is the case of *constructing two mean proportionals between the lines s and* $2s$, where s is the side of the cube to be doubled. In our algebraic language, this condition becomes

$$s/x = x/y = y/2s, \quad \text{so} \quad (s/x)^3 = (s/x) \times (x/y) \times (y/2s) = \tfrac{1}{2},$$

giving $x^3 = 2s^3$.

So, if two mean proportionals x and y can be constructed between s and $2s$, then the cube on x doubles the cube on s, as required. It appears that the Greeks thereafter always investigated the original problem in this reduced formulation.

Many different solutions, both mechanical and theoretical, to the problem of constructing two mean proportionals were put forward in Greek times: in particular, Eutocius described about a dozen solutions. The theoretical constructions ranged from ones involving a *neusis* construction to an astonishing construction of Archytas that determines the point of intersection of three surfaces in three-dimensional space.

We look at just one solution, simpler than that of Archytas. It is based on one that is probably due to Menaechmus, a younger associate of Plato, and is significant historically. From Menaechmus's work in tackling the problem of two mean proportionals may have arisen the Greek research tradition of investigating conic sections.

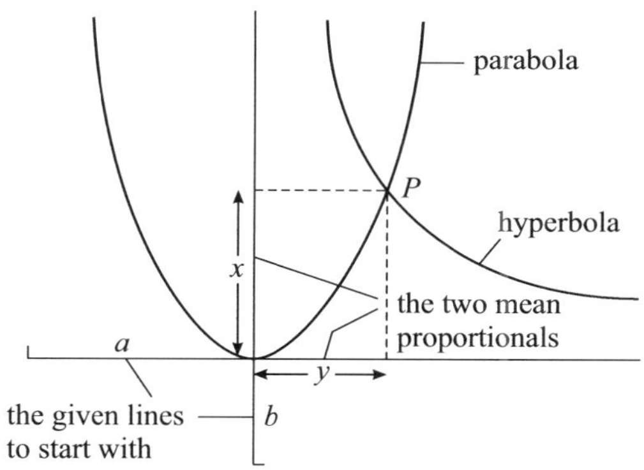

Figure 4.12. Menaechmus's construction of two mean proportionals

Menaechmus started his solution by using the method of *analysis*: he supposed the problem solved, and put down the two given line segments a and b and the two mean proportionals x and y, as in Figure 4.12. By rewriting the condition

$$a : x = x : y = y : b \quad \text{as} \quad \frac{a}{x} = \frac{x}{y} = \frac{y}{b},$$

we deduce that the line segments are related by the equation $bx = y^2$ (in modern notation), so the point $P = (x, y)$ lies on this curve (later called a *parabola*). The line segments are also related by the equation $ab = xy$, so P lies on this curve (later called a *hyperbola*). The mean proportionals x and y are where these curves intersect.[28]

[28]The lines are also related by the condition $ay = x^2$, so P lies on this parabola, too

A *synthetic* proof of this would go in the opposite direction: we would assume the parabola and hyperbola to be constructed (somehow), and we would then reverse the argument. The two mean proportionals would then be found, and in the particular case $a = s$ and $b = 2s$, the doubling of the cube would be achieved.[29]

Trisecting an angle. The earliest known method for trisecting an angle used a curve called the *quadratrix* (sometimes called the *trisectrix*), attributed to Hippias of Elis who flourished as a well-known teacher of the quadrivium subjects towards the end of the 5th century BC.[30]

A quadratrix is a curve formed by the intersection of two lines moving in the following way. Starting from the square $ABCD$, imagine the line BC to move downwards at a constant speed (so that it ends up coinciding with AD) in exactly the same time as the line AB sweeps out a quarter-circle at constant speed, hinged as it were at A, so that AB also ends up coinciding with AD (see Figure 4.13). The quadratrix is the path traced by the point where these two moving lines cross, and makes possible the trisection of any angle (whence its alternative name). In Box 4.3 we explain why this construction trisects the angle.

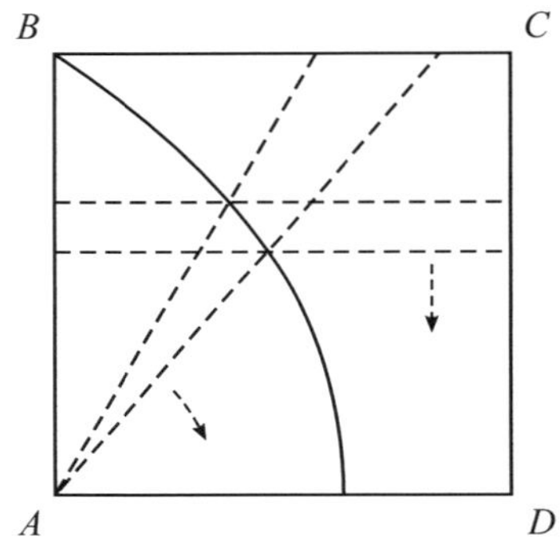

Figure 4.13. Forming a quadratrix

Our source for the above construction of the quadratrix is the commentator Pappus (4th century AD), and its ascription to Hippias has been inferred by historians from various other references by ancient writers. It is from Pappus, too, that we learn of a *neusis* construction for trisecting an angle, which we look at next.

This construction for trisecting an angle also exemplifies the method of analysis. Given any angle ABC, we suppose that it has been trisected by the line DB (Figure 4.14). This diagram is too sparse as it stands, so we introduce some more angles and triangles as follows. From A, drop a perpendicular down to BC and also draw a line from A parallel to BC, meeting BD extended at E. Finally join A to the mid-point G of DE. We can now deduce that the lengths DG, GE, and AG are all equal. Further, it can be shown that the length of DE is twice that of AB.

[29]We return to parabolas, hyperbolas, and related curves in Chapter 5.

[30]Plato's *Protagoras* portrayed Hippias as a teacher of the Sophist school in favour of compulsory education.

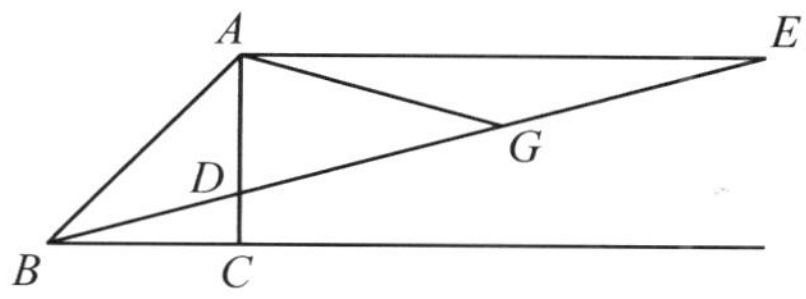

Figure 4.14. A *neusis* construction for trisecting an angle

Box 4.3. The quadratrix.
Suppose the angle to be EAD, where E is a point on the quadratrix. Then drop the perpendicular EF from E onto AD, and trisect it: this can be done by straight-edge and compasses, as we saw in Box 4.2. So G is one-third of the way along FE (Figure 4.15). G then determines a point H on the quadratrix, level with it. Draw the line from A to H. Then the angle HAD is one-third of the angle EAD.

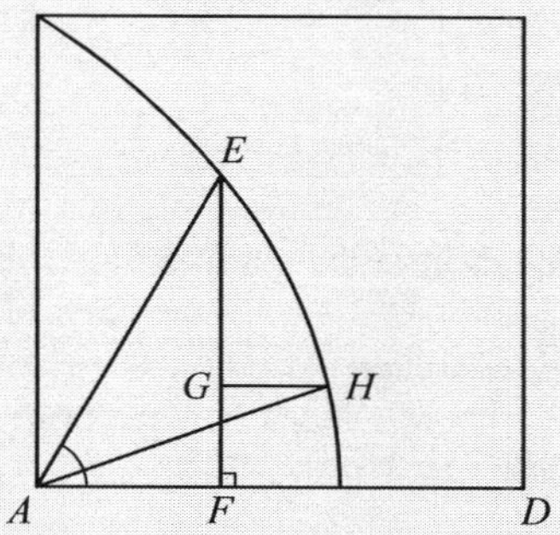

Figure 4.15. Trisecting an angle

So *if* the angle is trisected, *then* DE is twice AB: this is the result of the analysis. With this as a starting point, the geometrical argument would, we hope, work in reverse and we would end up by trisecting the angle: such would be the process of synthesis. The condition we have reached is a *neusis* condition: *if* we can draw a line segment of twice the length of AB in such a way that its ends lie on the two constructed lines *and* it points towards B, then we have trisected the angle.

You can imagine how this could be done with a *marked* ruler: mark the length $2AB$ on it and move the ruler about until the condition is fulfilled. Pappus explained how the *neusis* can be constructed, using the intersection of a circle and another curve (a hyperbola) to determine the point E. But this *neusis* (unlike the one in Hippocrates's quadrature of the lune) cannot be done with a straight-edge and compasses construction.

4.4 Squaring the circle

The problem of squaring the circle is intrinsically more difficult than the other two classical problems, involving the relationship between areas bounded by curved lines and straight lines, a relationship of paradox and perplexity in Greek (and later) times.

As you have seen, our earliest recorded result of Greek geometry was Hippocrates's quadrature of certain lunes — itself presumably part of an investigation of squaring the circle — which demonstrated that at least some things with curved sides can be 'squared'.

Ever since those early Greek times, 'squaring the circle' has been a by-word for the unattainable: the Italian poet Dante[31], for instance, spoke of a geometer failing to square the circle because he lacked the right principle, and the 18th-century English poet Matthew Prior[32] summarised it succinctly in the lines:

> Circles to square, and Cubes to double,
> Would give a Man excessive Trouble.

Origins of the problem. According to Proclus, the problem of squaring the circle arose by analogy from the early Greek success in determining exactly the area bounded by any figure with straight sides.[33] In his *Elements* (Book I, Prop. 45), Euclid asserts that any such polygon can be transformed into a parallelogram of any desired angle, and so can be squared. Proclus comments on this as follows:

> It is my opinion that this problem is what led the ancients to attempt the squaring of the circle. For if a parallelogram can be found equal to any rectilinear figure, it is worth inquiring whether it is not possible to prove that a rectilinear figure is equal to a circular area.

By *Elements* Book II, Prop. 14, Euclid has shown that any polygon can be transformed into a square of equal area, so the problem of squaring the circle would be solved if we could find any polygon of equal area. Proclus offered this account only as his opinion (writing many centuries later), and there is no historical tradition of how the problem came to prominence. But it seems that already by the later 5th century BC its challenging nature was already recognised. Not only was the problem notoriously difficult, but attempts to solve it exposed the kind of problems to be faced in any geometrical investigation of areas bounded by curves.

Squaring and rectifying. We first show the connection between squaring a circle (which involves its area) and 'rectifying' it (finding a line segment equal in length to its circumference): if you could solve either of these problems by straight-edge and compasses, then you could solve the other.[34]

If the circle of radius r and circumference $2\pi r$ can be rectified, we can produce a side of length $2\pi r$ and form the rectangle with sides πr and r. Then the construction in *Elements*, Book II, Prop. 14, produces a square with the same area as this rectangle. It follows that if a circle can be rectified by straight-edge and compasses, then it can be squared.

Conversely, if the circle of radius r has been squared, then the corresponding square has area πr^2. By a straight-edge and compass construction described in *Elements*, Book VI, Prop. 12, we can produce from the square a rectangle of equal area with one side of a given length: so we can construct a rectangle with sides r and πr. Thus, if a circle can be squared then its circumference can be rectified.

The quadratrix and the rectification of the circle. By the time of Archimedes, and probably earlier, the problem of squaring the circle had been reduced to that of rectifying its circumference. We know this because, as we see shortly, the problem of finding

---

[31] Dante, *Paradiso* XXXIII, 133–136.

[32] See (Prior 1718, 372).

[33] See (Morrow 1970, 335) in F&G 2.G1.

[34] Our analysis uses modern symbolism for explanatory purposes: it is not derived from any known Greek text.

such a reduction is one that Archimedes solved. This does not mean that the problem of squaring the circle had been solved, but it allows us to shift our attention to rectification. This can be done, it emerged, by using the quadratrix that served to trisect angles.

The way that the quadratrix is used to rectify the circumference may have been devised by Hippias, or by Dinostratus (a brother of Menaechmus), around 350 BC.[35]

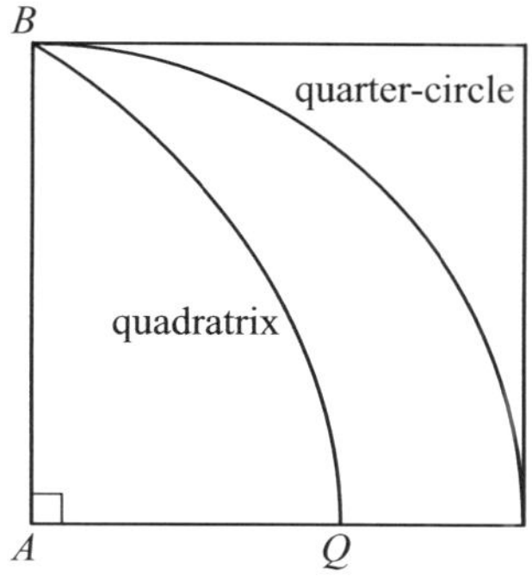

Figure 4.16. Rectifying a quarter-circle

It consists of proving (see Figure 4.16) that the following ratios are equal:

$$\text{quarter-circumference} : AB = AB : AQ.$$

This gives the quarter-circumference in terms of the radius of the circle (AB) and the point Q at which the quadratrix cuts the horizontal line. We shall not go through the proof of this result, which is a *reductio* argument.[36] If this proof is the original one and was indeed due to Hippias, then it would be one of the earliest *reductio* proofs of which we have a full record.

Given the above result, we can produce the rectification (a line segment equal in length to the quarter-circumference) by line-and-circle methods. For the result says that the circle's radius AB is the mean proportional between the arc length and the quadratrix distance AQ. Thus the problem becomes that of constructing a rectangle whose area is that of the square on AB, and with AQ as a side. Knowledge that this can be done is sufficient, and we shall not go into details.

But one aspect of this construction of the quadratrix is of interest, for it reminds us again of the high standards of rigour developed by the Greeks. Although it seems plausible, this construction is invalid for rectifying the circumference — or so Pappus claimed: his reasons for this were two-fold.

First, the end towards which the quadratrix is being devised (the quadrature of the circle) is assumed in its very construction. Remember that a quadratrix is defined by two simultaneous motions: the top line moves downwards in exactly the same time as AB pivots round on A so as to reach the bottom of the square together. So to construct this you need to know the ratio of their speeds, which is possible only if you already know the ratio of the side of the square to the quarter-circumference of the circle. But this is what you set up the construction to find out! So the logic seems to be going round in circles.

Second, the point Q, where the quadratrix cuts the bottom line, cannot be identified: this is unfortunate, since the length AQ is needed for the rectification. The difficulty again arises from the construction. Remember that the quadratrix is defined as the intersection

[35]The sources are ambiguous about this.

[36]For the proof, see F&G 2.G4.

of the two moving lines, but when they both reach the bottom of the square they no longer intersect in a unique point — they lie on top of one another, and so intersect everywhere. If you could look at the situation a moment earlier, it would be clear where the quadratrix is heading, but that is not good enough by the standards of Greek proof: the location of Q must be justified by a further logical argument, which would itself rely on something equivalent to the rectification of the circle.

Eudoxus of Cnidos. The earliest approaches we have to the problem of squaring the circle illustrate both the difficulties inherent in the problem and the high standards that the Greeks set anyone who attempted them. Indeed, our sources give only the names of two authors, Antiphon (later 5th century BC) and Bryson (probably a contemporary of Plato), and later writers' criticisms and refutations of their work. This is hardly an unbiased record of achievement.

The problem with these writers is that either they assumed that a circle is identical to a polygon with increasingly many sides (which is false), or they found polygons that successively approximate the area of the circle but they had inadequate control over the approximations. On one interpretation, they produced a sequence of *inscribed* polygons that get closer and closer to the circle, and another sequence of *circumscribed* polygons lying outside the circle and touching it at more and more points. Their hope was that these approximations would converge on the correct value for the area — but why should the inscribed polygons not all fall short of the area by a certain amount, while the circumscribed ones be always in excess of the area by another fixed amount? There was something to be proved here, and not merely asserted.

The mathematician credited with resolving these difficulties, and placing quadrature on a rigorous foundation, was Eudoxus. He was born in Cnidos (on the Ionian coast, just north of Rhodes) around 400 BC and studied geometry under Plato's friend Archytas at Tarentum. In his early twenties he came to study at Plato's Academy in Athens, and he returned there later, around 360 BC, at around the time that Aristotle came to the Academy as a young student. No work of Eudoxus survives, and his achievements can be inferred only from what Archimedes and others have said. But he seems to have been remarkably gifted, for besides his resolution of the quadrature problem, the proportion theory of Book V of the *Elements* reflects his ideas.[37]

The essence of Eudoxus's technique for overcoming the above difficulties was twofold. First, he changed the proof structure so as to make comparisons only between various polygonal areas: because there is no problem in finding such areas, this part of the proof became unproblematic. Second, to circumvent an awkward technical problem, he produced an explicit axiom to clarify the argument.

The first innovation worked as follows. Suppose that we seek the quadrature of a circle C, and a certain rectilinear area A seems a likely candidate for the area that some inner and outer polygons are approaching. Eudoxus would use a *reductio* argument, arguing that either C equals A, or it does not. If the latter assumption leads to a contradiction, then the former must hold and the quadrature of the circle has been found. Such an argument hinged on his being able to show that the second alternative leads to a contradiction: this contradiction was produced from a comparison of two polygonal areas.

[37]Eudoxus was also renowned for his work in mathematical astronomy.

To see how this can be done, we look at an example. Since nothing by Eudoxus has survived, we stay with the circle and outline a proof from Archimedes' *Measurement of a Circle*.[38] He was trying to prove the following result:[39]

> The area of any circle is equal to a right-angled triangle in which one of the sides about the right angle is equal to the radius, and the other to the circumference, of the circle.

His approach is now known as the *method of exhaustion* and was developed (we believe) by Eudoxus. Starting from the circle and a triangle K, Archimedes asserted that either the circle equals K (that is, their areas are equal), or it is greater than K (Case I), or it is less than K (Case II) — see Figure 4.17. He showed that the last two possibilities lead to contradictions and deduced that the first must hold. So the proof is a *reductio* argument.

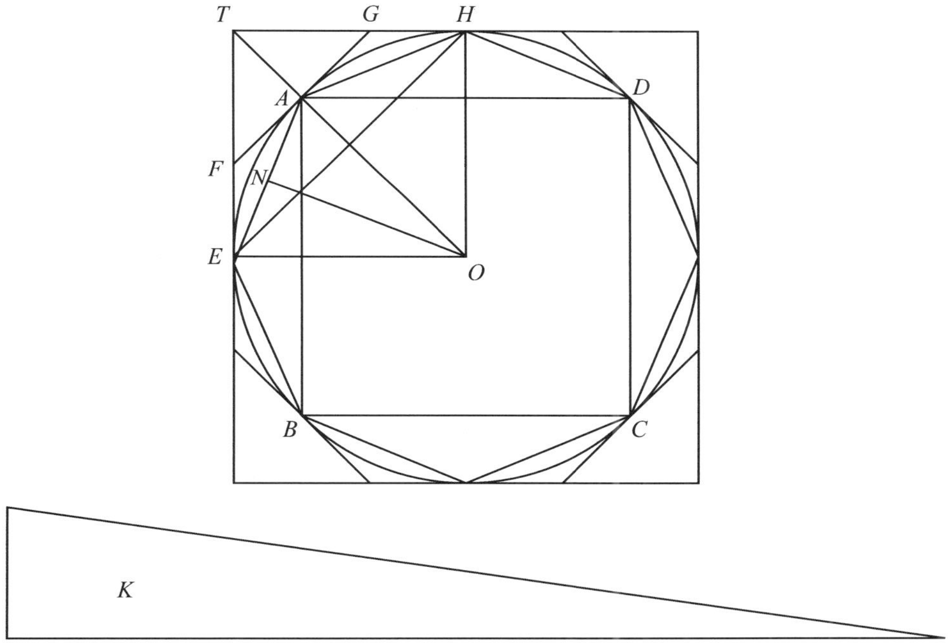

Figure 4.17. Archimedes on the circumference of the circle

Archimedes on the area of a circle.

Let $ABCD$ be the given circle, K the triangle described.

Then, if the circle is not equal to K, it must be either greater or less.

Case I. *If possible, let the circle be greater than K.*

Inscribe a square $ABCD$, bisect the arcs AB, BC, CD, DA, then bisect (if necessary) the halves, and so on, until the sides of the inscribed

---

[38] We discuss Archimedes in greater detail in Chapter 5.

[39] See Heath, *The Works of Archimedes*, 91–98; this extract appears in F&G 4.A1. All references to the works of Archimedes are to Heath's edition.

polygon whose angular points are the points of division subtend segments whose sum is less than the excess of the area of the circle over K.

Thus the area of the polygon is greater than K.

Let AE be any side of it, and ON the perpendicular on AE from the centre O.

Then ON is less than the radius of the circle and therefore less than one of the sides about the right angle in K. Also the perimeter of the polygon is less than the circumference of the circle, i.e. less than the other side about the right angle in K.

Therefore the area of the polygon is less than K; which is inconsistent with the hypothesis.

Thus the area of the circle is not greater than K.

Case II. *If possible, let the circle be less than K.*

Circumscribe a square, and let two adjacent sides, touching the circle in E, H, meet in T. Bisect the arcs between adjacent points of contact and draw the tangents at the points of bisection. Let A be the middle point of the arc EH, and FAG the tangent at A.

Then the angle TAG is a right angle.

Therefore $TG > GA > GH$.

It follows that the triangle FTG is greater than half the area $TEAH$.

Similarly, if the arc AH be bisected and the tangent at the point of bisection be drawn, it will cut off from the area GAH more than one-half.

Thus, by continuing the process, we shall ultimately arrive at a circumscribed polygon such that the spaces intercepted between it and the circle are together less than the excess of K over the area of the circle.

Thus the area of the polygon will be less than K.

Now, since the perpendicular from O on any side of the polygon is equal to the radius of the circle, while the perimeter of the polygon is greater than the circumference of the circle, it follows that the area of the polygon is greater than the triangle K; which is impossible.

Therefore the area of the circle is not less than K.

Since then the area of the circle is neither greater nor less than K, it is equal to it.

In Case I, Archimedes started from a square and built up polygons with more and more sides, progressively filling up the circle. Assuming that the area of the circle is greater than K, he claimed that he would ultimately obtain a polygon whose area is larger than K and smaller than the circle. But this polygon is made up of little triangles whose height is less than the radius of the circle (being strictly inside it) and whose total perimeter is less than the circumference of the circle (because a straight-line segment between any two points is shorter than a curved line). So, on either count, the area of the polygon must be less than K, contradicting our assumption that it is bigger than K. So the critical comparison is indeed between two polygonal areas. Case II is similar, and we omit it.

Where the discussion becomes difficult is Archimedes' claim that we can carry out this process and ultimately obtain a polygon whose area lies between that of K and the circle.

It is entirely reasonable to respond: 'How do you know?' Although he did not spell it out here, Archimedes knew because of an axiom of Eudoxus about how magnitudes behave. This is the second of Eudoxus's innovations.

We do not know just how Eudoxus phrased his axiom, but it was something along the following lines (in Aristotle's words):[40]

> If I add continually [the same magnitude] to a limited magnitude, I shall at length exceed any assigned magnitude whatever, and if I continually subtract from it, I shall similarly make it fall short of any assigned magnitude.

or in a more precise and slightly different form (in Archimedes' words):[41]

> Of unequal lines, unequal surfaces, and unequal solids, the greater exceeds the less by such a magnitude as, when added to itself, can be made to exceed any assigned magnitude among those which are comparable with it and with one another.

or in a different form again, as a result in Euclid's *Elements*, Book X, Prop. 1:

> Two unequal magnitudes being set out, if from the greater there be subtracted a magnitude greater than its half, and from that which is left a magnitude greater than its half, and if this process be repeated continually, there will be left some magnitude which will be less than the lesser magnitude set out.

These three formulations differ in various ways, which we do not explore here. But if the original idea is indeed due to Eudoxus, the endorsement of Aristotle, Archimedes, and Euclid is impressive. There seems to be little doubt that Eudoxus provided the logical lynchpin on which the determination of complicated quadratures could proceed. This is not unconnected with his theory of magnitudes presented in *Elements*, Book V, whose significance we now briefly consider.

The above three versions of Eudoxus's idea are elaborations of what appear as Definitions 3 and 4 of *Elements*, Book V:[42]

3. A ratio is a sort of relation in respect of size between two magnitudes of the same kind.
4. Magnitudes are said to have a ratio to one another which are capable, when multiplied, of exceeding one another.

At first sight these may look innocuous or banal; in fact, they are quite momentous. First, by requiring that magnitudes in a ratio be of the same kind they imply that magnitudes come in different kinds. This splits the rather broad concept of 'magnitudes' into categories (areas, lines, volumes, forces, motions, for example), and spells out that you can compare only comparable things: presumably, although it not made explicit here, one area with another area, or one line segment with another line segment; an area cannot be compared with a line segment, nor a volume to a motion, and so on. Second, things that do not behave like this are excluded altogether. In particular, the infinitely small and the infinitely large do not appear, and this goes a long way towards ruling out a worrying and paradoxical aspect of earlier mathematics.

Thus, what may at first have appeared as tiresome or long-winded — the language of ratio and proportion in which so much Greek mathematics was couched — makes good sense. It was generally used, in geometry at least, until comparatively recently. Through the work of Eudoxus it also made possible the strong logical development that fostered the triumphs of Greek geometry in the next century.[43]

---

[40] Aristotle, *Physics* VIII, in (Heath 1949, 153).

[41] Archimedes, *On the Sphere and Cylinder*, Assumption 5, in *The Works of Archimedes*, 1–90.

[42] F&G 3.C3.

[43] We study these in Chapter 5.

Archimedes and the rectification of the circle. We now return to Archimedes and to Proposition 1 of his *Measurement of a circle*.[44]

> The area of any circle is equal to a right-angled triangle in which one of the sides about the right angle is equal to the radius, and the other to the circumference, of the circle.

This result does not 'square the circle' in the terms of the traditional problem. Archimedes showed that the circle does indeed have a squarable area, but the problem was to obtain that area by a line-and-circle construction. In effect, he reduced the problem to that of constructing the circle's circumference as a straight line, but this is not possible by straight-edge and compass constructions.

The above proposition is remarkable, for the best previous result on the matter had been *Elements* Book XII, Prop. 2:[45]

> Circles are to one another as the squares on the diameters.

This is the result that Hippocrates had used in his quadrature of lunes a century or more earlier, although presumably he could not have proved it rigorously since Euclid's proof is by the method of exhaustion, using a *reductio* argument similar to that of Archimedes given above.

The significant development marked by Archimedes' proposition is that, for the first time that we know of, two questions about circles were shown to be reducible to each other. The first question concerns the ratio of the area of the circle to the square on the radius: given the radius r, we find the area by multiplying r^2 by a constant (now called π). It is by no means obvious that for all circles the area of a circle has the same constant ratio to the square of its radius, but this seems to have been recognised implicitly from quite early times, in both Egypt and Mesopotamia: it was first proved in *Elements*, Book XII, Prop. 2, in the form given above. The second question concerns the ratio of the circumference of the circle to its radius: given the radius r, the circumference is $2\pi r$. What Archimedes showed was that the area question and the circumference question involve the same constant.

We conclude with a striking and influential result of Archimedes on the ratio of the diameter of a circle to its circumference. In Prop. 3 of his *Measurement of a circle*, he established bounds for this ratio: it is greater than $3\frac{10}{71}$ and less than $3\frac{1}{7}$.

He did this by considering inscribed and circumscribed polygons of 96 sides, and basing his bounds on his calculations of their perimeters. The number 96 arises by starting with a regular hexagon (6 sides) and progressively doubling the number of sides so as to get regular polygons of 12, 24, 48, and 96 sides, becoming ever closer to the circle. Archimedes observed how the estimate of the ratio he was seeking improves each time that the number of sides is doubled, and apparently he doubled this number four times. By taking this doubling process further we can get bounds as close as we wish, but even in the case considered by Archimedes the computations demanded high number-handling skills.[46]

Unfortunately the text that has come down to us lacks the full details of his calculations, but it is clear that Archimedes was a skilled numerical calculator. This is something that we might not expect, to judge by the earlier products of the Greek geometrical tradition in which the very notion of using numbers to calculate with — still less to approximate with

[44] F&G 4.A1.

[45] In F&G 3.E3.

[46] Later mathematicians greatly extended this doubling process, thus conjuring up polygons with thousands, even millions, of sides, as we shall see in Chapter 7.

— was rather foreign. So these approximations provide evidence that Archimedes was not working wholly within a strict geometrical research tradition, and that the computational tradition and skills evident in some later Hellenistic mathematics date back at least to the time of Archimedes (notably in the work of Heron of Alexandria, and in the mathematical astronomy culminating with Ptolemy in the 2nd century AD).[47] Whether there was a significant Greek computational tradition even earlier, cast into historical shadow by the glare of Platonism, or whether this speaks of influences from Mesopotamia or elsewhere in the Middle East, is unclear.

4.5 Further reading

Suggestions for further reading are given at the end of Chapter 5.

[47]We consider Heron and Ptolemy in Chapter 6.

5

Greek Mathematics: Curves

Introduction

In this chapter we develop our discussions of the previous chapter, where we met several curves that were studied by Greek mathematicians. These curves varied widely, from the basic circle to the ingenious quadratrix.

We now look more systematically at the definitions and properties of curves in Greek times, leading to the works of the two great, but very different, 3rd century BC geometers, Archimedes and Apollonius. This study is interesting in its own right, and is important for us because of the critical role that these Greek achievements subsequently played. From the 17th century onwards, the Greek conception of curves and their properties, their methods of investigating them, and even the way that curves were thought of, were variously built upon or consciously rejected, but always with a deep debt to the Greek past, and to Archimedes and Apollonius in particular.

5.1 Problems with curves

What are curves? What were they taken to be in classical Greek times? How were they defined? These are clearly important questions, because the way that something is defined is bound to influence, if not determine, what can be proved about it. Let us first examine how to define curves, and then see what properties of curves were studied by the Greeks.

Definitions of curves. One interesting aspect of the various curves you have already seen — such as the *circle*, the *parabola*, and the *quadratrix* — is that their definitions were of quite different kinds. The circle can be defined in terms of the property that all straight-line segments from the centre to the curve have the same length; the parabola was to be defined as a certain plane section of a cone; and the quadratrix definition stands out by invoking motion, a concept that does not arise with the other two.

Note that the definitions of the quadratrix and parabola tell you what to do to produce the curve, in terms of intersecting two moving lines and of intersecting a cone and a plane, but the circle was defined in terms of a property that it has, rather than what you do to get it. We can describe these two ways of defining a curve as *definition by construction* and *definition by property*. Although most curves seem to have been defined by construction, there are hints that Plato, at least, considered definition by property the better approach; this is in keeping with his doctrine of eternal and immutable forms. You may recall that, in his discussion of geometry in the *Republic*,[1] Socrates described geometry as:

> a science quite the reverse of what is implied by the terms its practitioners use ... They talk about 'squaring' and 'applying' and 'adding' and so on, as if they were doing something and their reasoning had a practical end, and the subject were not, in fact, pursued for the sake of knowledge ... it must, I think, be admitted that the objects of that knowledge are eternal and not liable to change and decay.

But what mathematicians were told they ought to do, and what they actually did, are not necessarily the same. There is not even consistency on the matter in Euclid's *Elements* — for example, in Book I, Definition 15, a *circle* is defined as a curve with a certain property that distinguishes circles from other curves, while in Book XI, Def. 14 a sphere is defined as the figure obtained by rotating a semicircle around its diameter until it returns to its original position.[2] So a sphere arises as the result of a construction, and its properties must then be derived as theorems.

In these cases, definitions of either kind are equally easy to formulate, although for other curves an appropriate defining property might be harder to enunciate. But, as we shall see, Apollonius defined the conic sections by property. Many Greek curves came into existence only through being constructed for solving problems: the 'three classical problems' were especially fertile in this respect. Such curves had no other known properties until further investigated.

We shall shortly see what kinds of properties of curves were studied, but first we consider a particular classification of curves.

Pappus's classification of curves. In the 3rd century AD, Pappus classified problems according to the curves needed in their solutions. This in effect put curves into three classes:[3]

- *plane*: the circle and the straight line;
- *solid*: conic sections (the parabola, hyperbola, and ellipse), formed by taking plane sections of a cone;
- *line-like*: everything else, such as the quadratrix, spiral, conchoid, and cissoid (see Box 5.1).

Other curves were known too, beyond those mentioned by Pappus, as we learn from different classifications due to Geminus (c.1st century BC) and described by Proclus.[4] Proclus's text is rather complicated. There are composite or broken lines that form an angle, and incomposite ones. Incomposite ones might go on indefinitely (such as some sections of cones) or make figures (such as the circle, ellipse, and cissoid). The classification, due to Geminus, was into simple and mixed lines. Simple lines may make figures or, like the

[1] Plato, *Republic*, 521–531, in F&G 2.E2.

[2] F&G 3.E2.

[3] Pappus, *Collection* IV, in (Cohen and Drabkin 1948, 67–68), and in F&G 5.B4. This three-part classification was to be important in the 17th century, when Descartes developed it further.

[4] See (Morrow 1970, 90–92).

Box 5.1. The conchoid and the cissoid.

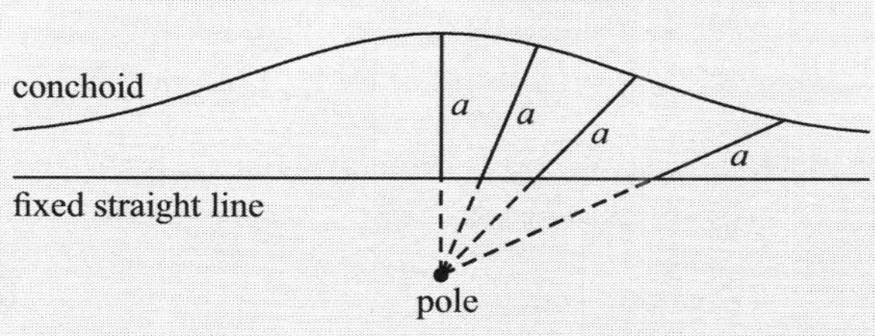

Figure 5.1. The conchoid

The *conchoid of Nicomedes* (3rd century BC) is the curve traced by the end of a line of constant length whose other end moves on a fixed straight line, while the moving line always points towards a fixed pole (see Figure 5.1). It serves for any problem that can be solved by means of a *neusis* construction, such as trisecting an angle. This curve was of interest because it approaches ever closer to a fixed straight line (an asymptote), but never reaches it.

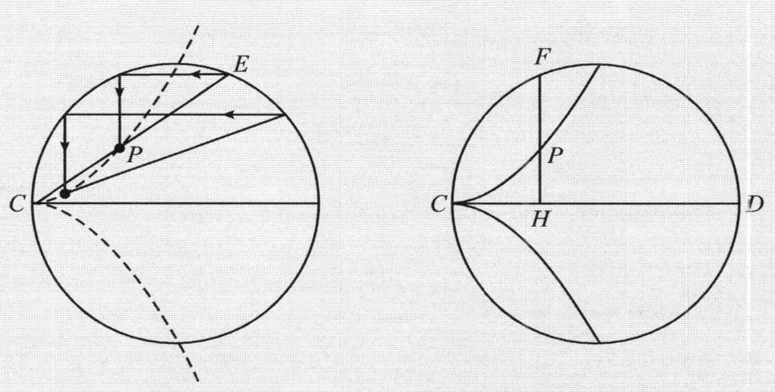

Figure 5.2. Constructing a cissoid and two mean proportionals

The *cissoid of Diocles* (early 2nd century BC) was devised to solve the problem of finding two mean proportionals, in order to double the cube. It is constructed pointwise, and consists of all points reached as follows (see Figure 5.2). Join any point E on a circle to the end of one fixed diameter C, and from E move horizontally across and vertically down until you hit the line CE at a point P; the cissoid consists of all such points P as you move around the circle.

The cissoid can be used to find two mean proportionals: in Figure 5.2

$$DH : HF = HF : HC = HC : HP,$$

so HF and HC are the two mean proportionals between DH and HP.

straight line, be unbounded, whereas mixed lines may lie in planes or on solids. If in planes they may return to themselves (like the cissoid), or not, whereas those in solids may be sections of them, or lie around them. Should we read this as an indication of the many and various kinds of curves that Greek mathematicians now prided themselves on having discovered, or as one of those exhaustive but ultimately vacuous exercises in classification favoured by the Gradgrinds of the world?[5] Proclus's answer was that he had given the list to 'encourage the able student to enquire into them',[6] but found it superfluous in the present work to enquire precisely into each line because Euclid had used only the simple and fundamental lines in his *Elements*.

Properties of curves. We now move from the definition and classification of curves to see what properties were studied. We consider a curve not as a solution to some problem, but as an object of geometrical interest in its own right, about which theorems can be proved. Broadly speaking, the properties investigated were analogous to those of figures made up of straight lines: angles, tangents, lengths, areas, etc. But these were more problematical for curves than the equivalent straight-line investigations, as we now see.

Angles and tangents. Both Hippocrates in the 5th century BC and Aristotle in the 4th century used the concept of the angle between a straight and a curved line. In the *Elements* Euclid explicitly defined the angle between two straight lines (Book I, Def. 9) as a subclass of plane angles (Def. 8), thus implying that the latter could be angles between curved lines.[7] Moreover, in Book III, Prop. 16, Euclid included a result concerning these two sorts of angle: the result says that if you choose a point P on a circle and draw the straight line at right angles to the diameter of the circle at P, then this straight line lies outside the circle,

> and into the space between the straight line and the circumference another straight line cannot be interposed; further the angle of the semicircle is greater, and the remaining angle less, than any acute rectilineal angle.

This is not the most luminous of statements. The claims are that the straight line you are to draw lies outside the circle ('the circumference'), that it is the only straight line to do so ('another straight line cannot be interposed'), and that the angle between the circle and the diameter is greater than any acute angle, while the angle between the circle and the straight line you draw is less than any acute angle (see Figure 5.3).

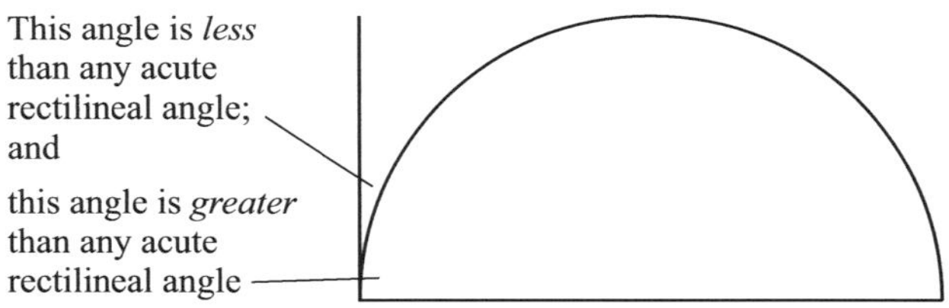

Figure 5.3. Euclid's *Elements*, Book III, Prop. 16

These last two claims show that Euclid was prepared to consider angles between straight lines and curves. They also illustrate a problem that can arise: the size of the angle between the circle and the line we are instructed to draw is presumably not 0°, as there is space between them, but is less than any rectilineal angle. But if rectilineal angles are not the correct things with which to measure these new sorts of angles, then it is not

[5] In Charles Dickens's novel *Hard Times*, Chapter 3, Gradgrind's children 'had only been introduced to a cow as a graminivorous ruminating quadruped with several stomachs'.

[6] See (Morrow 1970, 92).

[7] F&G 3.B1(a).

clear what can be used instead (see Figure 5.4). Much later, Proclus[8] recorded a classification of angles due to Geminus, every bit as elaborate and complicated as his classification of lines.

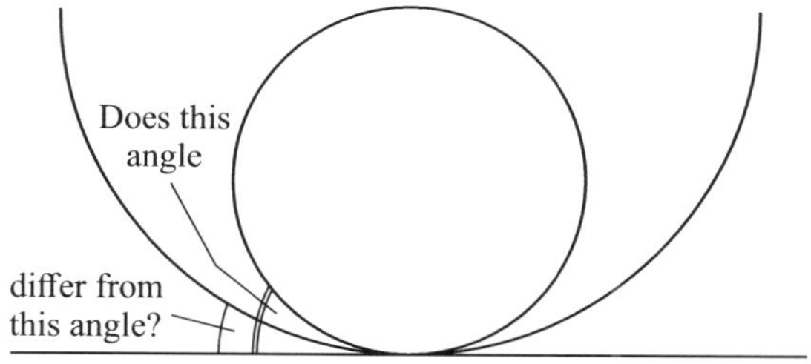

Figure 5.4. Tangents and angles

There is evidence that in the last few centuries BC there was much discussion of different definitions of angles, and of what kind of a mathematical object or property they are. However, nothing very satisfactory emerged. The difficulties and paradoxes that arose out of trying to consider angles between curved lines were perhaps too great: the question of how to compare two curvilinear angles seems to have made little progress.

Another question concerning the meeting of lines was developed more productively: where and how do curves *touch*?

In his *Elements*, Book III, Def. 3, Euclid calls a straight line that touches a curve a *tangent*. The above construction in Book III, Prop. 16 amounts to a construction of the tangent to a circle at any chosen point. Euclid showed that constructing a straight line at right angles to the diameter gives rise to a line with the tangent property, which he defined as 'meeting the circle but not cutting it'.

Two other properties of tangents to circles emerge from this proposition: any particular tangent is unique (at most one line can be tangent to the circle at the given point), and the tangent touches the curve at a single point. However, in the late 5th century BC the latter claim had been the subject of controversy, like so much else in Greek geometry. Aristotle, referring to the difference between geometrical lines and visible reality, remarked that:[9]

> a circle in fact touches a straight-edge, not at one point only, but in the way that Protagoras, in his refutation of the geometers, said that it did

— that is to say, they touch at many points. So there seem to have been discussions over whether a tangent touches a curve at one point, or at many.

Lengths and areas. Problems concerning the length of curved lines, areas enclosed by them, and volumes enclosed by curved surfaces, were also difficult, but in the hands of Archimedes, and others, they were solved with astonishing virtuosity, as we saw in Chapter 4. We have seen that such concerns were at the forefront of the geometrical research tradition from at least the time of Hippocrates in the 5th century BC. In the next section we look at developments in finding areas and volumes from then up to Archimedes, two centuries or so later.

By now you should have some feeling for the sort of enterprise that constituted the Greek study of curves. Later we return to how results were proved, which was one of the most impressive, or daunting, legacies of Greek mathematics to 16th-century and 17th-century Europe.

[8] See (Morrow 1970, 98–100), in F&G 3.B1(b).

[9] Aristotle, *Metaphysics*, 998a, in (Heath 1949, 204).

5.2 Archimedes

You may have noticed that whenever we needed an interesting or significant example, we were often drawn towards the work of Archimedes. The reason for this is easily explained: he was the most creative and prolific of those mathematicians whose works have in large measure survived. As with Euclid's *Elements*, this survival was no accident, but is testimony to his influence and the high regard in which he was held by succeeding generations. No-one we have met so far showed such astonishingly broad interests, or excelled in so many fields.

In this section, in order to broaden our perspective on the richness of Greek mathematical activities, we survey Archimedes' life, achievements, and death. We then discuss his contributions to the Greek study of curves.

The life of Archimedes. Born around 287 BC, Archimedes seems to have spent most of his life in his home town of Syracuse, in Sicily. He may also have spent some time in Alexandria, perhaps studying mathematics there a generation or two after Euclid: he certainly had friends there. Several of Archimedes' works come down to us with the covering letters he wrote, a revealing source for learning about the organisation of mathematical research in the 3rd century BC.

In these letters Archimedes had two correspondents. Most of his letters were to Dositheus, whom Archimedes does not seem to have known personally, but whom he adopted as a correspondent after the death of their mutual friend Conon:[10] one letter was to Eratosthenes, the librarian at Alexandria. Archimedes seems to have used these people as channels through whom he could communicate with 'persons engaged in mathematical studies',[11] setting them problems and subsequently supplying the proofs if the answers were not forthcoming, as was generally the case.

The impression we get, on reading between the lines, is of a few groups or individuals, dotted around the Mediterranean, who kept in touch by letter on what was happening at the forefront of geometrical research. The material under discussion was quite advanced, and Archimedes clearly saw himself in a historical tradition: he spoke of past mathematicians, including 'all the many able geometers who lived before Eudoxus',[12] and also of the advances that could be made by his contemporaries or by his successors. It could be that much of the mathematical activity at the time of Archimedes took place in Alexandria, or else that Alexandria was a clearing-house for communication to and from more scattered researchers, and there is no hard evidence that Archimedes spent any time there.

With the possible exception of music, Archimedes was interested in all the mathematical sciences: geometry, arithmetic, astronomy, mechanics, statics (levers and centres of gravity), and hydrostatics (bodies floating in water). What is especially noticeable, by comparison with many earlier mathematicians, is the way he combined the two streams of activity that Plato had been so keen to separate — pure geometrical analysis, and the mechanical or practical: his work on the lever is purely geometrical, with propositions derived logically from postulates in a Euclidean-looking way, whereas his astronomical book *On Sphere-making* (now lost) was about constructing a planetarium that modelled the motions of heavenly bodies. Indeed, his reputation in the centuries after his death was more as a maker of mechanical marvels than as a geometer. Two centuries later his planetarium was

[10] *The Works of Archimedes*, 233, in F&G 4.A3.

[11] *The Works of Archimedes*, 151, in F&G 4.A7.

[12] *The Works of Archimedes*, 2, in F&G 4.A5(a).

still extant, and was seen by the Roman statesman and 1st century BC author Cicero, who wrote:[13]

> When Gallus set the sphere in motion, one could, at every turn, see the Moon rise above the earth's horizon after the Sun, just as occurs in the sky every day; and then one saw how the Sun disappeared and how the Moon entered into the shadow-cone of the earth with the Sun on the opposite side ...

It would be interesting to know how Archimedes himself regarded his mechanical inventions. This would help us to understand whether he was essentially a 'pure' mathematician working within the geometrical research tradition but who happened to have mechanical interests as a sideline, or whether he was a more complex person who defied categorisation and was consciously working within a broader context. We have some evidence on this point in a most interesting account of Archimedes by the Greek historian Plutarch.

Plutarch's *Life of Archimedes*. Plutarch lived from about 46 to 120 AD in Boeotia, in mainland Greece (by then part of the Roman Empire). Having studied Platonic philosophy at Athens, he devoted himself mainly to a life of learning. The *Lives* he wrote of great Greeks and Romans is an important historical document, because so many of the sources he drew upon are now lost. However, Plutarch's interest in biography was primarily ethical, concerned with the moral character of his subjects, and so his attentiveness to historical detail and accuracy is not necessarily that of a historian. His account of Archimedes comes in his life of the Roman general Marcellus, head of the Roman troops whose siege of Syracuse (214–212 BC) ended in its capture and the death of Archimedes. So although Plutarch said of Archimedes[14] that he regarded

> the whole business of mechanics and the useful arts as base and vulgar, but placed his whole study and delight in those speculations in which absolute beauty and excellence appear unhampered by the necessities of life,

He offered almost no evidence for this assertion. In the next extract the observation that Archimedes wrote little about his mechanical discoveries is a telling supporting point; and his choice of tomb inscription (later in the extract) can reasonably be taken as showing what Archimedes considered to be his most important achievement. In the next extract, Plutarch is describing the siege of Syracuse, which started in 214 BC, led by the Roman general Marcellus (Figure 5.5).

Marcellus on the death of Archimedes.

> Marcellus directed a fleet of sixty quinqueremes full of armed men and missile weapons. He raised a vast engine upon a raft made by lashing eight ships together, and sailed with it to attack the wall, trusting to the numbers and excellence of his siege engines, and to his own personal prestige. But Archimedes and his machines cared nothing for this, though he did not speak of any of these engines as being constructed by serious labour, but as the mere holiday sports of a geometrician. He would not indeed have constructed them but at the earnest request of King Hiero, who entreated him to leave the abstract for the concrete, and to bring

[13] Cicero, *De Republica*, in (van der Waerden 1961, 211).

[14] See (Plutarch 1899, 45–51). Another translation can be found in F&G 4.B1.

Figure 5.5. Archimedes planning the defences of Syracuse, from A. Thévet, *Les Vrais Portraits et Vies des Hommes Illustres*, 1584

his ideas within the comprehension of the people by embodying them in tangible forms.

Eudoxus and Archytas were the first who began to treat of this renowned science of mechanics, cleverly illustrating it, and proving such problems as were hard to understand, by means of solid and actual instruments, as, for instance, both of them resorted to mechanical means to find a mean proportional, which is necessary for the solution of many other geometrical questions. This they did by the construction, from various curves and sections, of certain instruments called mesographs. Plato was much vexed at this, and inveighed against them for destroying the real excellence of geometry by making it leave the region of pure intellect and come within that of the senses, and become mixed up with bodies which require much base servile labour. So mechanics became separated from geometry, and, long regarded with contempt by philosophy, was reckoned among the military arts.

However Archimedes, who was a relative and friend of Hiero, wrote that with a given power he could move any given weight whatever, and, as it were rejoicing in the strength of his demonstration, he is said to have declared that if he were given another world to stand upon, he could move this upon which we live. Hiero wondered at this, and begged him to put this theory into practice, and show him something great moved by a small force. Archimedes took a three-masted ship, a transport in the king's navy, which had just been dragged up on land with great labour and many men; in this he placed her usual complement of men and cargo, and then sitting at some distance, without any trouble, by gently pulling with his hand the end of a system of pullies, he dragged it towards him with as smooth

> and even a motion as if it were passing over the sea. The king wondered greatly at this, and perceiving the value of his arts, prevailed upon Archimedes to construct for him a number of machines, some for the attack and some for the defence of a city, of which he himself did not make use, as he spent most of his life in unwarlike and literary leisure, but now these engines were ready for use in Syracuse, and also, the inventor was present to direct their working.

Plutarch then described how the Syracusans defended themselves with a variety of Archimedes' devices. Some fired missiles, there were cranes for keeping the Roman navy at bay: ships 'were suddenly seized by iron hooks, and by a counter-balancing weight were drawn up and then plunged to the bottom. Others they caught by irons like hands or claws suspended from cranes, and first pulled them up by their bows till they stood upright upon their sterns, and then cast down into the water, or by means of windlasses and tackles worked inside the city, dashed them against the cliffs and rocks at the base of the walls, with terrible destruction to their crews'.

Marcellus taunted his own engineers, saying 'Are we to give in to this Briareus of a geometrician, who sits at his ease by the seashore and plays at upsetting our ships, to our lasting disgrace, and surpasses the hundred-handed giant of fable by hurling so many weapons at us at once?'[15]

> Yet Archimedes had so great a mind and such immense philosophic speculations that although by inventing these engines he had acquired the glory of a more than human intellect, he would not condescend to leave behind him any writings upon the subject, regarding the whole business of mechanics and the useful arts as base and vulgar, but placed his whole study and delight in those speculations in which absolute beauty and excellence appear unhampered by the necessities of life, and argument is made to soar above its subject matter, since by the latter only bulk and outward appearance, but by the other accuracy of reasoning and wondrous power, can be attained: for it is impossible in the whole science of geometry to find more difficult hypotheses explained on clearer or more simple principles than in his works.
>
> Some attribute this to his natural genius, others say that his indefatigable industry made his work seem as though it had been done without labour, though it cost much. For no man by himself could find out the solution of his problems, but as he reads, he begins to think that he could have discovered it himself, by so smooth and easy a road does he lead one up to the point to be proved. One cannot therefore disbelieve the stories which are told of him: how he seemed ever bewitched by the song of some indwelling syren of his own so as to forget to eat his food, and to neglect his person, and how, when dragged forcibly to the baths and perfumers, he would draw geometrical figures with the ashes on the hearth, and when his body was anointed would trace lines on it with his finger, absolutely possessed and inspired by the joy he felt in his art. He discovered many beautiful problems, and is said to have begged his relatives and friends to place upon his tomb when he died a cylinder enclosing a sphere, and to write on it the proof of the ratio of the containing solid to the contained.

[15]Briaresus is mentioned in Book I of Homer's *Iliad* as one of three sons of Uranus and Gaia who were hundred-armed and fifty-headed; difficult opponents indeed.

> Such was Archimedes, who at this time rendered himself, and as far as lay in him, the city, invincible.
>
> [*Eventually the city was captured.*]
>
> Marcellus was especially grieved at the fate of Archimedes. He was studying something by himself upon a figure which he had drawn, to which he had so utterly given up his thoughts and his sight that he did not notice the assault of the Romans and the capture of the city, and when a soldier suddenly appeared before him and ordered him to follow him into the presence of Marcellus, he refused to do so before he had finished his problem and its solution. The man hereupon in a rage drew his sword and killed him. Others say that the Roman fell upon him at once with a sword to kill him, but he, seeing him, begged him to wait for a little while, that he might not leave his theorem imperfect, and that while he was reflecting upon it, he was slain. A third story is that as he was carrying into Marcellus's presence his mathematical instruments, sundials, spheres, and quadrants, by which the eye might measure the magnitude of the sun, some soldiers met with him, and supposing that there was gold in the boxes, slew him. But all agree that Marcellus was much grieved, that he turned away from his murderer as though he were an object of abhorrence to gods and men, and that he sought out his family and treated them well.

What are we to make of this major source? It is important to bear in mind that Plutarch was a Platonist. He attributed to Plato strong views about Eudoxus and Archytas having 'corrupted and destroyed the ideal purity of geometry' through invoking mechanical instruments in geometry. Consciously or otherwise, Plutarch would have wanted to see the great mathematician Archimedes as following the precepts of Plato. So Archimedes traditional enthusiasm for mechanical devices could be something that, for Plutarch, needed explaining away.

More generally, Plutarch's account of the role of Archimedes' inventions in the defence of Syracuse reads as though it had grown in the telling. Plutarch wrote two hundred years after the events he was describing, he offered three versions of the death of Archimedes, and all are subservient to what Plutarch is really interested in: the response to his death by Marcellus, the noble victor. So Plutarch gives us no great confidence that he had access to a reliable primary source, but rather that he is regaling us with the legends that grew up around the events. This leaves us in some doubt about how reliable Plutarch is on the subject of Archimedes. The implication is that it would not be safe to draw any conclusion from Plutarch about Archimedes's views on the question of the right balance between pure geometry and mechanical devices.

More generally, however, this account does offer first-hand evidence that the status of geometry versus mechanics was still a live issue in the 1st century AD, and that the views of Plato continued to be influential in the living academic tradition after nearly half a millennium.

Archimedes' work. There are a number of other interesting perceptions in what Plutarch wrote, to which we shall make reference when appropriate. We turn now, however, to Archimedes' mathematical work. We can highlight only a few significant aspects of this, described by Heath (perhaps with an element of hyperbole) as 'a sum of mathematical achievement unsurpassed by any one man in the world's history'.[16] When we discussed

[16]See (Heath 1981, 20).

Archimedes' *On the Measurement of a Circle* in Chapter 4, we saw, among other things, his labour-intensive but splendid estimations of the value of π (to use modern symbolism).

The Sand-reckoner. Archimedes seems to have enjoyed numbers, and wrote on ways of expressing very large ones. There was no ready means of doing this in everyday Greek numerals (see Section 3.3) — unlike our place-value system, which is capable of indefinite extension — but in a work called *The Sand-reckoner* Archimedes described how to form very large numbers.[17]

The basic idea is to count as far as one can — in Greek that would be a myriad myriad (= 100,000,000) — and then to use this as a unit for the next stage, and so on until a myriad myriad to the power myriad myriad is reached; then to use that as a fresh unit, and so on. In this way Archimedes reached a massive number that in our notation would be 1 followed by 80,000,000,000,000,000 zeros.

Archimedes then proceeded to show that this is sufficient to count the number of grains of sand that would fill the universe, even a universe as large as that required by the hypothesis of Aristarchus. So this work by Archimedes is of further historical significance as the place where we learn of the views of Aristarchus on a heliocentric universe, which Copernicus was aware of when devising his not-dissimilar system in the 16th century.

Aristarchus was about 25 years older than Archimedes, and was known to his contemporaries as Aristarchus the mathematician, because of his prowess. He worked on the theories of music, geometry, and astronomy, and was the first to propose that all the planets, the Earth included, go round the Sun. This bold counter-intuitive idea met with few supporters, however, and was displaced by the later astronomer Hipparchus's restoration of the theory of epicycles that had been used earlier by Apollonius. Aristarchus also did important work estimating the apparent and true sizes of the Sun and the Moon; he put the distance of the Sun from the Earth at between 18 and 20 times the distance of the Moon, a larger number than any previous estimate, even though it fell well short of the correct value (which is about 370 times). His argument was a clever mixture of eclipse data, novel geometry, and the estimation of various square roots that occurred in the course of the argument. The error came about because it required an accurate estimation of something that is almost impossible to determine: the precise point at which exactly half the Moon is in shadow.

There is a further point worth making about Archimedes' *The Sand-reckoner*. In much Greek cosmology, Aristotle's in particular, the heavens were constituted differently from the Earth, and different laws applied (albeit ones that could be described in terms of pure geometry); to break from this assumption was one of the great 17th-century developments, as we shall see in Chapter 11. So for Archimedes to conceive the very earthy image of the whole cosmos filled with sand, even in a spirit of exuberant fantasy, is further testimony to his unfettered free-thinking approach.

On spirals. We next look at several themes that come together in an interesting curve we have not met before, the *spiral of Archimedes*, defined in his *On Spirals*.

Archimedes' definition of a spiral[18] involves two simultaneous motions. In one, a straight line revolves at constant speed about a fixed point O, and on the other a point moves at constant speed out along the line. The spiral is the path of the moving point.[19] This is a definition by construction, because it tells you what to do to produce the curve.

---

[17] *The Works of Archimedes*, 221–232, and an extract in F&G 4.A2.

[18] *The Works of Archimedes*, 165, in F&G 4.A7(f), Definition 1.

[19] In polar coordinates, the spiral has the equation $r = a\theta$.

Any properties that the curve may have must be deduced from this description of this double motion.

Archimedes described his four major results in the penultimate paragraph of his introductory letter to Dositheus.[20] Three of these results are to do with areas, and one is concerned with line length in relation to the tangent to the curve. So he is interested in essentially those curve properties we mentioned earlier — tangents, line length, and area.

As Archimedes explained in the final paragraph of the letter, the other propositions in the book are mostly those needed to establish his major results. In part, their function is to mediate between the constructive definition of his curve and the eventual proofs of its significant properties, for in the latter proofs no concept of motion or time appears. We shall not study the structure of his proofs in detail, but his results on spirals are so remarkably beautiful and simple that we should outline what they are.[21]

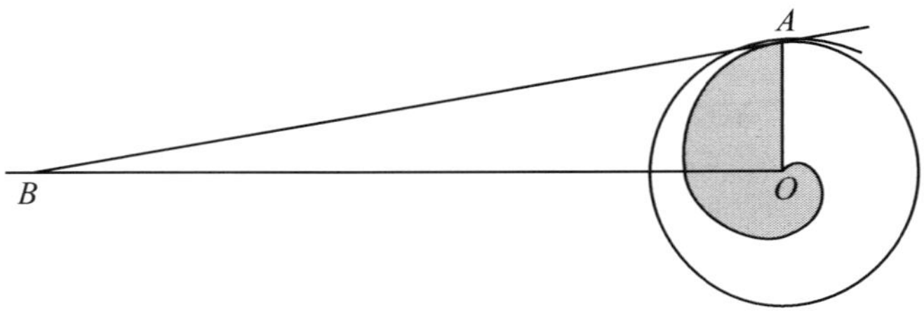

Figure 5.6. The area of an Archimedean spiral

In Figure 5.6 the point O is the centre of the spiral, A is the point on the spiral after the line has rotated exactly once, the lines OA and OB are at right angles, and the line BA is the tangent to the spiral at A.

In Proposition 18 Archimedes showed that the length of OB is equal to the circumference of the circle of radius OA. This is a striking result: a curved line (the circle) has been shown to have the same length as a straight line.

Proposition 24 is an important result on areas. The area of the spiral (the shaded area in Figure 5.6) is equal to one-third of the area of the circle of radius OA.

Archimedes also extended these results to cover any number of rotations of the rotating line.

Archimedes on areas and volumes. When finding areas and volumes Archimedes was within the geometrical tradition, and his work marks its Greek culmination: nothing so powerful was to be seen again for nearly two thousand years. It is difficult to convey this convincingly, however, without more detailed study. It seems paradoxical that what is generally held to be the arena of his greatest achievements is one where his contribution is not visibly different from the work of his predecessors. It is just the kind of thing that one finds in Euclid's *Elements*, Book XII (areas and volumes of this, that, and the other), only more so. He broke no radically new ground here, either in subject-matter or in approach, but a closer examination reveals that the problems are intrinsically very difficult and that Archimedes' handling of them is flexible and versatile. Plutarch must have been referring to books such as *On the Sphere and Cylinder* and *On conoids and spheroids*, when he said:[22]

---

[20] *The Works of Archimedes*, 153–155, in F&G 4.A7(a).

[21] *The Works of Archimedes*, 171, in F&G 4.A7(g).

[22] Plutarch, in F&G 4.B1, p. 175.

It is not possible to find in all geometry more difficult and intricate questions, or more simple and lucid explanations.

Examples of what Archimedes proved include:[23]

- *The surface of any sphere is equal to four times the greatest circle in it* (in modern terms, the surface area of a sphere of radius r is $4\pi r^2$).
- *Any sphere is equal to four times the cone which has its base equal to the greatest circle in the sphere and its height equal to the radius of the sphere* (the volume of a sphere of radius r is $\frac{4}{3}\pi r^3$).

A corollary of this last remark is the following great result said to have been engraved on his tombstone:

- *A sphere is two-thirds the size of the smallest cylinder that contains it* (this is because the volume of its circumscribing cylinder is $2\pi r^3$, which is $1\frac{1}{2}$ times $\frac{4}{3}\pi r^3$). See Figure 5.7.

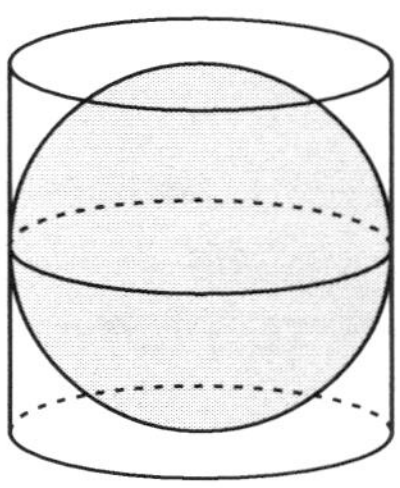

Figure 5.7. A sphere in its enclosing cylinder

We now look in more detail at one of the results in *On conoids and spheroids*, as our main example of what an Archimedean proof looks like. Conoids and spheroids are surfaces of revolution of the conic sections[24] — just as a sphere can be thought of as a circle revolved round its diameter, so too we obtain surfaces by revolving other conic sections (ellipse, hyperbola, and parabola) around their diameters or axes.

Proposition 21 of *On conoids and spheroids* shows that the volume of a right-angled conoid (a paraboloid of revolution, in modern terms) is $1\frac{1}{2}$ times that of the cone on the same base and axis.[25] Like the above results, the proof is by the method of exhaustion. After describing the construction and labelling of the paraboloid, Archimedes produced a cone with volume X and used a *reductio* proof to show that the volume of the paraboloid is $1\frac{1}{2}$ times X. In doing this, Archimedes fitted cylindrical discs, inside and outside the surface, so that the paraboloid is roughly approximated by two 'step pyramid' formations. The volume of each separate little cylinder is known, so the volume when they are all piled up is known too.

The third crucial aspect of the method of exhaustion was the axiom of Eudoxus in some form. This does not appear explicitly here, but we can infer that it was invoked in the previous supporting proposition. The revealing sentence is:[26]

[23] *The Works of Archimedes, On the Sphere and the Cylinder.* These appear, respectively, in F&G 4.A5(a), 4.A5(b), and 4.A5(b) again.

[24] We discuss the conic sections in Section 5.3 below.

[25] F&G 4.A8(b).

[26] F&G 4.A8(b).

We can then inscribe and circumscribe, as in the last proposition, figures made up of cylinders or frusta of cylinders with equal height and such that

$$(\text{circumscribed figure}) - (\text{inscribed figure}) < (\text{segment}) - X.$$

This says that, *whatever* the difference between the paraboloid and the cone X, we can construct cylindrical discs so that the inner and outer step pyramids are closer together than this difference. This tells us that the magnitudes are behaving as they ought to, and the rest of the proof is just spelling out the consequences.

Archimedes' path to his discoveries. The above proof uses the method of exhaustion. From the perspective of logical rigour, nothing could be more satisfying, nor a greater tribute to the sophistication of the Greek geometrical research tradition. But there is one problem outstanding. How did Archimedes know that this was the result he wanted to prove? How did he know, in advance, that the volume of the paraboloid is one-and-a-half times that of some cone? Why is it not twice, or 81 times, or even something that is not a commensurable ratio at all?

It is characteristic of a *reductio* proof that the result to be proved must be known in advance, so this problem occurs with all exhaustion proofs. Brilliant though Eudoxus's method was for enabling logically rigorous proofs to be formulated, and proving beyond doubt that some result is the case, it made no helpful contribution to finding the result in the first place.

Once mathematicians of the 17th century began to study the works of Archimedes with care, this struck them as puzzling. The suspicion grew that the method of analysing a problem so as to discover the answer before embarking upon its synthetic proof had been more developed in Greek times than the few scattered references suggested. In 1685 the English mathematician John Wallis[27] wrote that Archimedes seemed

> as it were of set purpose to have covered up the traces of his investigation, as if he had grudged posterity the secret of his method of inquiry, while he wished to extort from them assent to his results

and that

> not only Archimedes but nearly all the ancients so hid from posterity their method of Analysis (though it is clear that they had one) that more modern mathematicians found it easier to invent a new Analysis than to seek out the old.

There matters would have remained were it not for a remarkable find made early in the 20th century by J.L. Heiberg, a Danish scholar whose critical editions of Greek mathematical texts form the foundation of modern historical scholarship on the period, as we now see.

The Method. Until fairly recently, all existing Archimedean manuscripts were copies or translations of a single 9th-century manuscript, now lost. So the discovery in 1899 of a new text, the so-called *Archimedes palimpsest*,[28] caused real excitement when it was listed in a catalogue of the library of the Metochion of the Holy Sepulchre in Istanbul. Heiberg was able to examine the manuscript in 1906 and 1908, and in 1910–1915 he published the results of his study in the second edition of his critical text of Archimedes' works.

Most excitingly, this palimpsest gave alternative readings of four mathematical treatises, making it independent of the two lost manuscripts, and included the original Greek

[27]Wallis, cited in (Heath 1981, 21).

[28]A *palimpsest* is a text that has been scraped clean and written over.

text of *On Floating Bodies*, until then known only from the medieval Latin translation. Better yet, it contained the text of a treatise called *The Method of Mechanical Theorems*, in which Archimedes explained how he used mechanical means to discover the theorems for which he subsequently provided logical mathematical proofs. This provides exceptional insight into how Archimedes worked, and on the careful distinction he observed between discovery and subsequent proof in mathematics. It is unique among ancient scientific writings for its treatment of methodology.

The palimpsest itself carries the Archimedean text as it was copied in Constantinople in the mid-10th century on vellum leaves. These leaves had been washed clean in the 12th century, folded in half to make a smaller book, and covered with Greek religious texts; the lines of the second script run perpendicularly across those of the first. To read the Archimedean material one must rotate the pages, peer between the lines, and grapple with the fact that the 'Archimedean' order was scrambled in the process (see Figure 5.8).

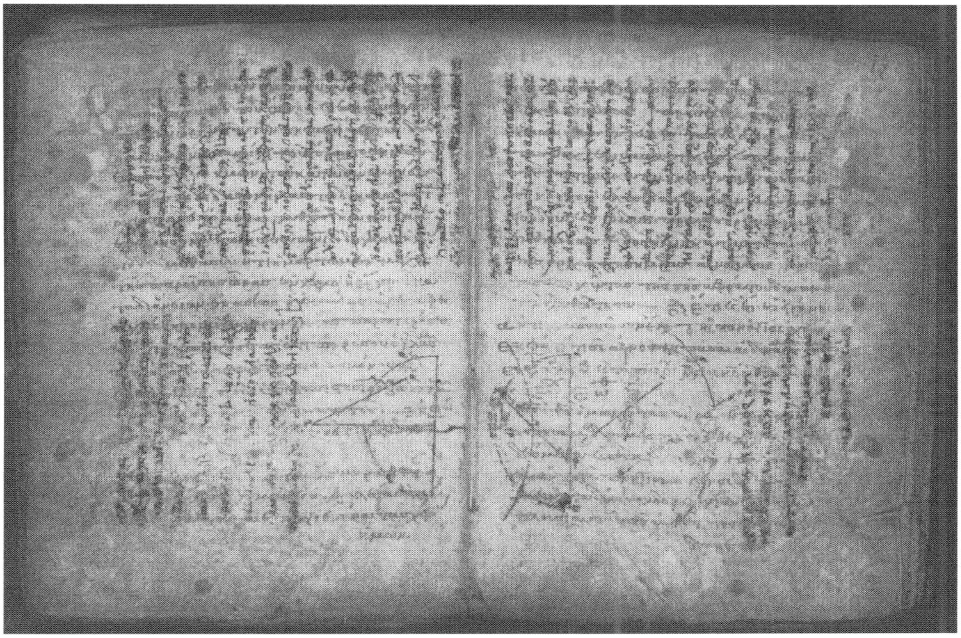

Figure 5.8. Part of the Archimedes palimpsest

There the matter might have rested, with a modern scholarly edition in place and the precious manuscript housed in Istanbul. But the terrible history of the 20th century intervened. After the First World War, a brutal struggle erupted between Greece and Turkey in which innocent civilians were massacred in great numbers. The survival of the library of the Metochion and the monks there could not be assured, and in 1922 the Greek Orthodox Patriarch of Jerusalem ordered that its 890 manuscripts be sent secretly to the National Library of Greece for safe keeping. In the event, 823 manuscripts were sent, and these remain in the library, the property of the Patriarchate.

But the Archimedes palimpsest was not one of them. Officially it was one of the 67 that were lost, and the manuscript was inaccessible to scholars from the 1920s until the 1980s, when rumours circulated that it was being privately offered for sale, but the doubtful legal situation made it extremely difficult for likely purchasers to proceed.

Then in 1998 the New York branch of the auction house Christie's informed the Greek Government that they intended to auction the palimpsest that August. Where had it been? Who was legally entitled to sell it? Christie's argued that the manuscript had been bought in the 1920s by an unnamed Frenchman, but admitted that the way he had come by the manuscript was entirely unclear. Perhaps it had been stolen, or perhaps one of the monks had sold it illegally to buy his way out of the raging war. Christie's agreed that proof of theft would make the sale impossible and if established they would withdraw it. But, they pointed out, it was well known that manuscripts from the Metochion had left the library apparently quite legally at various times. A page of the palimpsest had made its way to Cambridge University Library in 1876, another to the Bibliothèque Nationale de France around 1900, and two more pages migrated to America in the 1920s and 1930s.

The controversy surrounding the sale aroused world-wide interest, and doubtless did no harm to the final sale price, and matters were satisfactorily resolved. On 29 October 1998, the Archimedes palimpsest was sold to a still-as-yet unnamed buyer for $2.2 million. It has since been put on exhibition at the Walters Art Gallery, Baltimore, Maryland USA, who are charged with conserving it.

An example from *The Method*. Because the binding had been broken open, it was now possible to read the 10th-century Archimedean text that had been trapped in the spine when Heiberg saw it. Those, like the historian Reviel Netz, who were involved in attempting to read those texts with modern equipment — much of it state of the art — expected to find his reconstructions of such passages valid, as indeed they did. But one gap in the Heiberg's edition of the text stood out: the one in Proposition 14. A large section at the end of *The Method* is simply lost, and the text starts to break up because of the way the pages were shuffled around when the rebinding was done. As a consequence, Heiberg had little more than the start and the very end of Proposition 14 to read, while the rest was lost in the binding, and he could only write 'I shall not speculate as to what could have been written in such a large gap'.[29]

Now the binding could be opened. After a considerable amount of painstaking work, including a month when novel digitalisation techniques were applied to the palimpsest, Netz was able to conclude that in this proposition Archimedes had divided a certain solid figure into infinitely many slices and found them to be equal in multitude to another infinite collection. The sometimes legible proof revealed that Archimedes used this comparison four times. Nor had he done so in a naive way: he gave reasons why this equality was true and he did not say that all infinite sets have the same size.[30]

Netz summarised his findings this way:[31]

> ...we find that Archimedes calculated with actual infinities — in direct opposition to everything historians of mathematics have always believed about their discipline.

Such a remarkable claim must await the discussion of experts before it can pass into the body of accepted historical knowledge, but that it can be made on the basis of one new text reminds us to be cautious in our opinions about Greek mathematics in the face of all that we have lost.

Netz also noted that the diagrams in the palimpsest are interesting, not only because Heiberg did not copy them but asked a mathematical colleague to draw new ones, but also

[29]Quoted in (Netz and Noel 2007, 187).

[30]That, as mathematicians of the 17th century were to discover, is a sure route to error.

[31]See (Netz and Noel 2007, 199). There were other significant discoveries, but we have space only for this one.

because at least two of them differ significantly from those in the previously known text. When such a discrepancy happens, historians of ancient mathematics argue, as Netz did here, that because it was the scribe's job to copy what they were transmitting and not to re-work it (although mistakes can always occur), the diagrams in the palimpsest and those in the only other extant text must be derived from a common predecessor.

The Method is of considerable interest in helping us to resolve the problem of how Archimedes reached some of his results. It has an introductory letter that partly confirms and partly rebuts Wallis's conjecture quoted above. Archimedes did indeed have a method for discovering results, so the 17th-century suspicions were quite correct. On the other hand, Wallis's aspersions on Archimedes, that he 'grudged posterity the secret', is quite wrong, since Archimedes sent *The Method* to Eratosthenes hoping to make it available to other mathematicians. He was not responsible for its having been forgotten thereafter.

We can best see what Archimedes' method was by looking at an example: Proposition 4 on the paraboloid of revolution.[32]

Figure 5.9 shows a vertical slice through the middle of something with four components:[33] the paraboloid in question $(BOAPC)$, a cone BAC inside it, a cylinder $BEFC$ outside it, and a bar DAH going along the common axis, with A as its midpoint. Note that the paraboloid, cone and cylinder all have a circular base with diameter BC. We can think of the bar as that of a balance: since horizontal balance bars may be more familiar to you, we have rotated the diagram through $90°$.

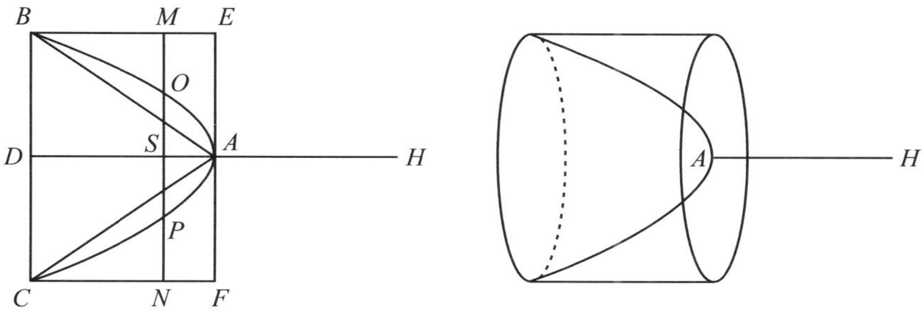

Figure 5.9. In a paraboloid of revolution the ratio $HA : SA$ = cross-section of cylinder : that of the paraboloid

Archimedes, an example from *The Method*.

Any segment of a right-angled conoid (a paraboloid of revolution) cut off by a plane at right angles to the axis is $1\frac{1}{2}$ times the cone which has the same base and the same axis as the segment.

This can be investigated by our method, as follows.

Let a paraboloid of revolution be cut by a plane through the axis in the parabola BAC; and let it also be cut by another plane at right angles to the axis and intersecting the former plane in BC. Produce DA, the axis of the segment, to H, making HA equal to AD [see Figure 5.9].

Imagine that HD is the bar of a balance, A being its middle point.

[32] *The Works of Archimedes*, 24 (second numbering, after p. 326), in F&G 4.A9(c).

[33] The diagrams in this section are taken from Heath.

The base of the segment being the circle on BC as diameter and in a place perpendicular to AD, imagine (1) a cone drawn with the latter circle as base and A as vertex, and (2) a cylinder with the same circle as base and AD as axis.

In the parallelogram EC let any straight line MN be drawn parallel to BC, and through MN let a plane be drawn at right angles to AD; this plane will cut the cylinder in a circle with diameter MN and the paraboloid in a circle with diameter OP.

Now, BAC being a parabola and BD, OS ordinates,[34]

$$DA : AS = BD^2 : OS^2, \text{ or } HA : AS = MS^2 : SO^2.$$

Therefore $HA : AS$ = (circle, radius MS) : (circle, radius OS) = (circle in cylinder): (circle in paraboloid).

Therefore the circle in the cylinder, in the place where it is, will be in equilibrium about A with the circle in the paraboloid, if the latter is placed with its centre of gravity at H.

Similarly for the two corresponding circular sections made by a plane perpendicular to AD and passing through any other straight line in the parallelogram which is parallel to BC.

Therefore, as usual, if we take all circles making up the whole cylinder and the whole segment and treat them in the same way, we find that the cylinder, in the place where it is, is in equilibrium about A with the segment placed with its centre of gravity at H.

If K is the middle point of AD, then K is the centre of gravity of the cylinder; therefore $HA : AK$ = (cylinder):(segment).

Therefore cylinder = $2\times$(segment).

And cylinder = $3\times$(cone ABC) [Euclid's *Elements*, Book XII Prop. 10];

therefore segment $= \frac{3}{2}$ (cone ABC).

We now consider just the paraboloid and the cylinder (Figure 5.10). The upshot of the method is that Archimedes took the paraboloid away, hung it from the bar at H, and observed that the paraboloid and the cylinder are then in balance. Since the weight of the cylinder acts through its centre, which is half-way along, the cylinder is twice the weight of the paraboloid. The result then follows, since the cylinder is three times the volume of the cone, by a result in Euclid.

Having seen where we are heading, let us go back and fill in the middle part of the argument. Archimedes took any slice through the objects, parallel to the base. The slice produces circular cross-sections of both cylinder and paraboloid.

Archimedes observed that the circular cross-section of the paraboloid, if moved to the far end H of the bar, would balance the circular cross-section of the cylinder (when left where it is). The reason for this is an argument in which the fact that we are dealing with a paraboloid (a rotated *parabola*) is critical. The property of the parabola needed is that mentioned earlier, that distances along the axis are as squares on the ordinates; so

$$DA : SA = BD^2 : OS^2.$$

[34]The ordinate of a point with respect to two coordinate axes at right angles is the perpendicular distance of the point from the horizontal axis; in modern terms, it is the y-coordinate. Similarly, the abscissa of a point with respect to two coordinate axes at right angles is the perpendicular distance of the point from the vertical axis; in modern terms, it is the x-coordinate.

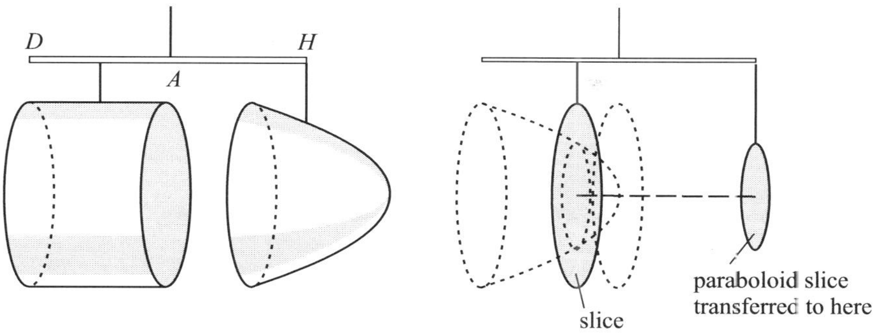

Figure 5.10. Weighing or balancing the slices

Since $HA = DA$ and $BD = MS$, we have

$$HA : SA = MS^2 : OS^2.$$

Now, by *Elements*, Book XII, Prop. 2, circles are as the squares on their diameters[35] — or as the squares on the radii, which is what we have here, so

$$HA : SA = \text{cross-section of cylinder : that of the paraboloid.}$$

At this stage, the equation is precisely that of the law of the lever. Archimedes proved in *On the Equilibrium of Planes* that:[36]

> Two magnitudes, whether commensurable or incommensurable, balance at distances reciprocally proportional to the magnitudes.

This is just what we have here. So the circular cross-section of the paraboloid (taken at distance HA) balances the circular cross-section of the cylinder (taken where it is, at distance SA). Since this conclusion applies to any slice, the effect is that all slices of the paraboloid (that is, the whole paraboloid) can be piled up at H and balances the cylinder as it is.

This proposition is a good example of Archimedes at his most versatile and flamboyant. If *The Method* was not generally taken up, that could be because no-one else could follow it so well. Plutarch remarked on the difficulty of discovering Archimedes' synthetic proofs beforehand, and his remarks seem equally applicable to *The Method*, despite the hopes that Archimedes expressed in his letter to Eratosthenes that others could make discoveries through it.

Archimedes wrote to Eratosthenes[37] that the results found through this method

> had to be demonstrated by geometry afterwards because their investigation by the said method did not furnish an actual demonstration.

There were two difficulties, and historians have disagreed over which was uppermost in Archimedes' thoughts. The first difficulty was that, following both Plato and Aristotle, there could be unease about the intrusion of 'mechanical' concepts. The very notion of weights balancing is mundane and ungeometrical in the Platonic view, and Aristotle had emphasised that purely geometrical properties such as *area* could not be properly proved by reference to concepts such as *weight* or *centre of gravity* arising in a subsidiary science such as mechanics.

[35] In F&G 3.E3.

[36] *The Works of Archimedes*, 192, and F&G 4.A4.

[37] *The Works of Archimedes*, *The Method*, 13 (second numbering, after 326, in F&G 4.A9(a)

The second difficulty was that the technique of slicing and moving the slices around appears to reawaken the logical difficulties we discussed earlier. For an elementary paradox appears very quickly, in conjunction with what Archimedes wanted the slices to effect. In order for the law of the lever to apply, the slices must have positive weights — zero magnitudes cannot enter the process; but to consider the cylinder and paraboloid as made up of all their slices, each slice must be of zero thickness and thus have zero weight: although the cylinder can be considered as made up of thin cylindrical discs, the paraboloid certainly cannot be. Thus we are back to a solid analogue of the problem that may have earned Antiphon's scorn — a paraboloid made up of circular slices is no *logical* advance on a circle made up of an indefinite number of straight lines — or, to put it another way, much the same heuristic intuitions as may have guided Antiphon are evident, in a more sophisticated framework, in the work of Archimedes 150 or more years later — and they were to be rediscovered, without the benefit of *The Method*, by mathematicians of the early 17th century.

5.3 Conics

Although several different curves were known to the Greeks, only one family of curves was studied in depth by many: these were the *conic sections*. As we shall see, there are many definitions of these curves, and it can be tricky to pass convincingly from one definition to another and see that they are equivalent. We therefore break with our usual practice and introduce the curves before discussing their history, in the hope that by deliberately being anachronistic we are also being helpful.

A modern mathematician would say that there are three types of (non-degenerate) conic section — the *ellipse* (including the *circle*), the *parabola*, and the *hyperbola* — and that they all derive from sections of a cone. What does that mean? First, the cone is what is more properly called a *double cone*, and is obtained as follows. We take a circle C in a plane, and a point O not in that plane, and draw all the straight lines through O that also pass through a point on C: note that these are whole lines, not half-lines starting at O. The resulting shape is a double cone. This double cone can be slanted or a right double cone.

A *section* is defined as the those points of this double cone that also lie in a plane; planes that pass through the chosen point are excluded. As we have mentioned, three different kinds of curve can be obtained in this way (see Figure 5.11):

- If the plane meets only one half of the double cone and is not parallel to any of the lines making up the cone, then the resulting conic section is an *ellipse*; note that circles are special cases of ellipses.
- If the plane meets only one half of the double cone but is parallel to one of the lines making up the cone, then the resulting conic section is a *parabola*.
- If the plane meets both halves of the double cone, then the resulting conic section is a *hyperbola*.

The three types of curve were given their names by Apollonius, as we shall describe later.

Most modern mathematicians do not like to work in three dimensions, and prefer to deal with the conic sections as curves in the plane. Moreover, they prefer to work with the curves using algebra. So the curves are often presented via their standard (canonical) equations. In this presentation:

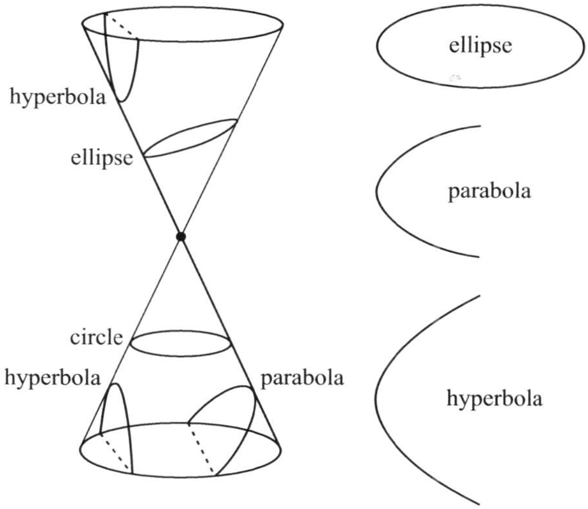

Figure 5.11. Three sections of a double cone

- an ellipse has equation $\frac{x^2}{a^2} + \frac{y^2}{b^2} = 1$
- a parabola has equation $y^2 = 4ax$
- the hyperbola has equation $\frac{x^2}{a^2} - \frac{y^2}{b^2} = 1$,

where a, b, and c are arbitrary non-zero real numbers.

Plainly it is quite a jump from the sections of a three-dimensional cone to these algebraic equations.[38] On the other hand, it is reassuring to know that these curves have fairly simple equations. It suggests (correctly) that they can be systematically studied.

Our questions, as historians of mathematics, are rather different from those of mathematicians. When were the conic sections first studied, and why? In what form were they introduced? How were they studied? How did they pass from their discoverers to later mathematicians?

The discoveries of Menaechmus. The conic sections may have arisen in connection with the work of Menaechmus on the problem of doubling the cube, studied in Chapter 4.

The historical record offers us some remarks about what Menaechmus did, and also about some sophisticated work by Archimedes and Apollonius more than a century later. We are also fortunate to have an account in some detail by Diocles, a contemporary of Apollonius, whose work *On Burning Mirrors* was written shortly before Apollonius wrote his account of conic sections. As the last-known discussion of conics written before the influence of Apollonius spread, this work is a guide to the developments reached earlier.

The name of Menaechmus, 'a student of Eudoxus who also was associated with Plato' as Proclus described him,[39] is traditionally given in connection with the discovery and early study of the conic sections, somewhere around 360–350 BC, but as you might expect, the early history of conic sections is somewhat obscure. Towards the end of the next century a

[38] We discuss the transition from geometry to algebra in Chapter 13.

[39] See (Morrow 1970, 55).

work appeared that became the definitive text, the *Conics* of Apollonius. Earlier treatises by Euclid and a certain Aristaeus are mentioned by the commentator Pappus, but are now lost. We shall try nevertheless to see what happened between the times of Menaechmus and Apollonius.

First, what was it that was discovered some time in the mid-4th century BC? The name 'conic section' refers to a curve and also reflects a particular way of defining that curve, as a section of a cone. Historically these two aspects may be distinct, however. It is generally accepted that it was in trying to double the cube that mathematicians of the Academy constructed curves later perceived as sections of cones. But it is one thing to tackle a problem, and quite another to study the way that the problem was solved. The perception of these curves as conic sections may indeed be from later in the century, and due perhaps to Aristaeus.

The historian Wilbur Knorr has argued that:[40]

> Menaechmus and the geometers in the decades immediately before and after him initiated not a theory of the conic sections, but a body of geometric problems solved according to a form of the method of analysis. When the resolving loci turned out to be curves which were later known as conic sections, these were at first constructed, if at all, via point-wise procedures.[41] Only late in the fourth century, near the time of Euclid, did one conceive of generating a class of curves via the sectioning of cones and begin their geometric investigation.

It was thus by the end of the 4th century BC that the conic sections were defined as sections of cones, the sections being made perpendicular to a generator of the cone;[42] Euclid's definition of a cone is given in the *Elements*, Book XI.[43] Knorr observed that Archimedes, when referring to 'what has been demonstrated in the *Conic Elements*' must have been referring to Euclid's or Aristaeus's treatments, but both are lost. He surmised that they were superseded by Apollonius's account, noting that Apollonius said that most of the first two Books of his own work were already familiar. Knorr concluded[44]

> The theory at Euclid's time may be viewed as consisting essentially of the first two books and some portions of the third.

Depending on the shape of the cone, the three different kinds of curve can be produced by a plane section perpendicular to a generator. If the angle at the top is less than a right angle, the section gives rise to an ellipse; if a right angle, to a parabola; if greater than a right angle, to a hyperbola (see Figure 5.12).

The definitions of conics are of little use in themselves unless further mathematical properties can be deduced. In fact, from these definitions it is not a difficult application of elementary plane geometry to find what the Greeks called their *symptom*: the condition that the points on the curve satisfy.

The symptom of the parabola is the simplest to describe. It is

$$(\text{ordinate})^2 = (\text{abscissa}) \times (\text{constant parameter})$$

— or, in the form the symptom arises in the work of Archimedes,

for any two points on the parabola, the abscissae are as the square on the ordinates.

---

[40] See (Knorr 1982, 7).

[41] A procedure constructs a curve point-wise if it supplies points that lie on the curve one by one.

[42] A generator of a cone is a line drawn through the vertex of the cone and lying entirely on its surface.

[43] F&G 3.E2.

[44] See (Knorr 1986, 111).

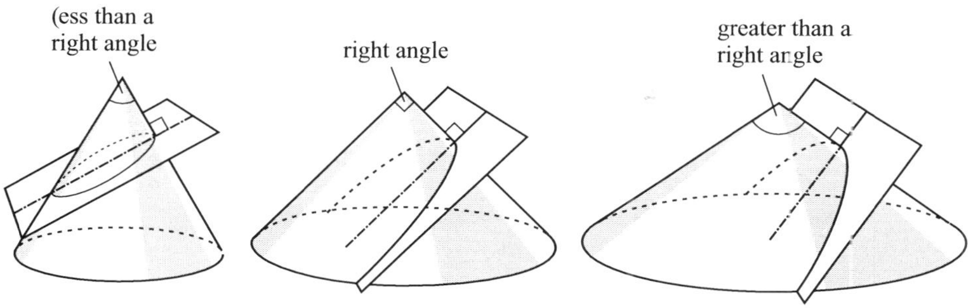

Figure 5.12. Conic sections from three different cones

By the time we meet conic sections in the work of Archimedes, they generally make their appearance directly through the symptom – that is, Archimedes worked from this condition, taking it to be something evidently known and not needing to be derived afresh, in such a way that the definition involving cones had in effect become a 'definition by property'. It is the symptom of the curve that now serves as its defining property.

The symptom of the ellipse is of the form

$$(\text{ordinate})^2 < (\text{abscissa}) \times (\text{constant})$$

by an amount that depends on the abscissa, and the symptom of the hyperbola is of the form

$$(\text{ordinate})^2 > (\text{abscissa}) \times (\text{constant})$$

by an amount that depends on the abscissa.

Diocles, cones, and burning mirrors. Before looking at Apollonius's work on conics, we consider Diocles' *On burning mirrors.* The introduction describes the problem, which is associated with the familiar names of Archimedes' correspondents Conon and Dositheus, and concerns reflections from mirror surfaces (see Figure 5.13). According to Diocles, the problem of finding a mirror that reflects the Sun's rays to a single point was solved practically by Dositheus. Diocles proved this result by considering properties of the parabola.

The property derived theoretically is that every ray coming in parallel to the axis of the parabola is reflected to a point, which we now call the *focus* of the parabola; thus, parallel sunlight is focused to a point by a parabolic mirror. The focus is easily related to the symptom of the parabola; its distance from the vertex of the parabola is one quarter of the 'constant parameter', in the above formulation of the symptom.[45]

Diocles's introduction to *On Burning Mirrors*.

He said: Pythion the Thasian geometer wrote a letter to Conon in which he asked him how to find a mirror surface such that when it is placed facing the Sun the rays reflected from it meet the circumference of a circle. And when Zenodorus the astronomer came down to Arcadia and was introduced to us, he asked us how to find a mirror surface such that when it is placed facing the Sun the rays reflected from it meet a point and thus cause burning. So we want to explain the answer to the problem posed by Pythion and to that posed by Zenodorus; in the course of this we shall make use of the premises established by our predecessors. One of those

[45]Diocles, in (Toomer 1976) and F&G 4.C1.

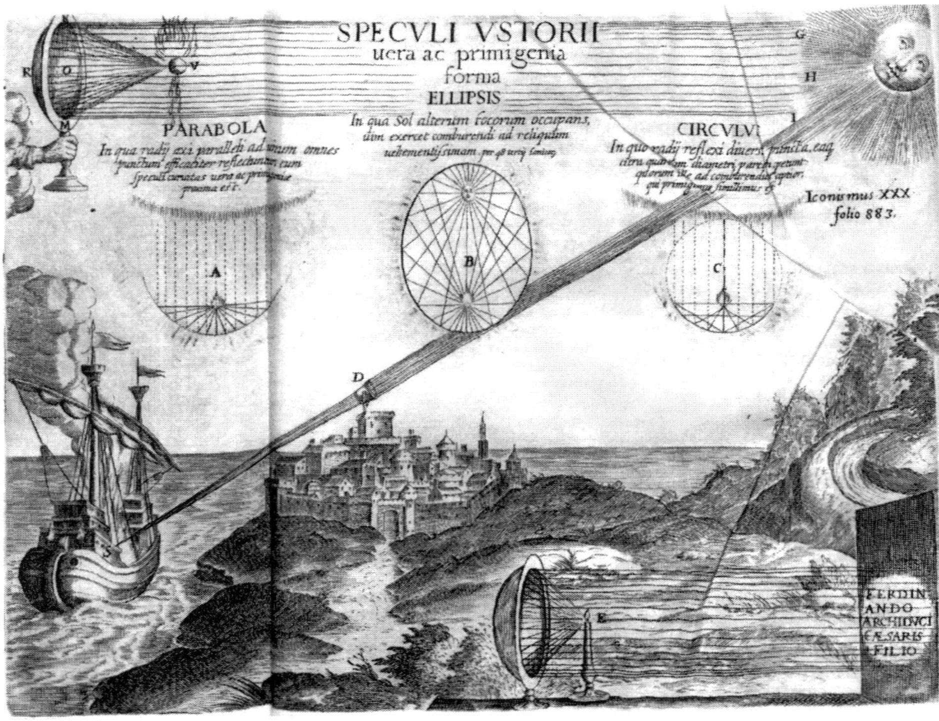

Figure 5.13. How Archimedes might have set fire to the Roman fleet using parabolic mirrors, from Athanasius Kircher's *Ars Magna Lucis* (1646)

two problems, namely the one requiring the construction of a mirror which makes all the rays meet in one point, is the one which was solved practically by Dositheus. The other problem, since it was only theoretical, and there was no argument worthy to serve as proof in its case, was not solved practically. We have set out a compilation of the proofs of both these problems and elucidated them.

The burning-mirror surface submitted to you is the surface bounding the figure produced by a section of a right-angled cone [parabola] being revolved about the line bisecting it [its axis]. It is a property of that surface that all the rays are reflected to a single point, namely the point [on the axis] whose distance from the surface is equal to a quarter of the line which is the parameter of the squares on the perpendiculars drawn to the axis [the ordinates]. Whenever one increases that surface by a given amount, there will be a [corresponding] increase in the above-mentioned conic section. So the rays reflected from that additional [surface] will also be reflected to exactly the same point, and thus they will increase the intensity of the heat around that point. The intensity of the burning in this case is greater than that generated from a spherical surface, for from a spherical surface the rays are reflected to a straight line, not to a point, although people used to guess that they are reflected to the centre; the rays which meet at one place in that [a spherical] surface are reflected from the surface [consisting] of a spherical segment less than half the sphere, and [even] if the mirror consists of half the sphere or more than half, only those rays reflected from less than half the sphere are reflected to that place.

We see that there were two problems, of which the practical one asks for the surface that reflects the Sun's rays to a point. The answer to the second problem is that the surface is a parabolic mirror, obtained by rotating a parabola about its axis. Not only does Diocles solve the problem, he also discovers where the point is at which the rays are focused: it is one quarter of the way along the axis, in terms of the parameters used to define the original parabola. Diocles contrasts this with the effect of a spherical mirror, which, he says incorrectly, reflects all the incoming rays to a straight line. Diocles is plainly pleased to have solved this problem theoretically, guided as he may have been by Dositheus's actual construction, and with justice, because properties of the conic sections were little known in his day.

We next take a look at how he proceeded. His strategy was first to work with a parabola in a plane, and only at the end to appeal to the symmetry of the rotated figure. Now he proceeded as follows (see Figure 5.14).

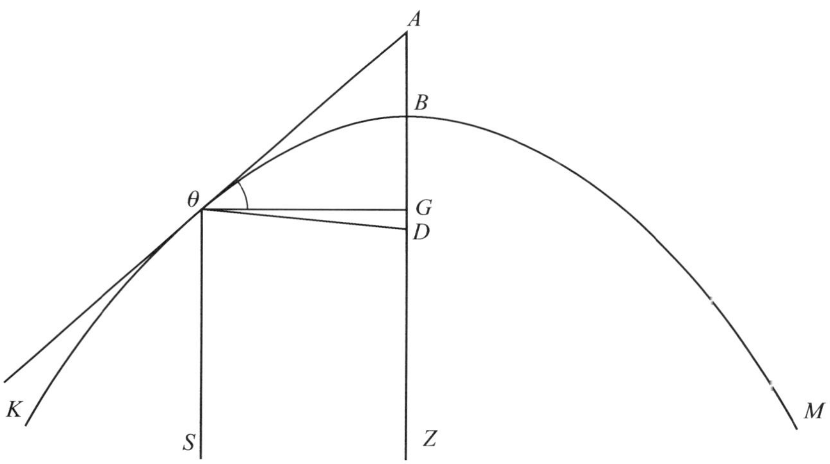

Figure 5.14. The focal property of the parabola (simplified)

Diocles

- described a parabola KBM and its axis ABZ
- located the point D where the rays will be focused
- chose an arbitrary point θ on the parabola
- drew the tangent θA to the parabola at the point θ
- made the plausible assumption that a ray of light $S\theta$, parallel to the axis and hitting the parabola at the point θ, is reflected as if by the tangent line.

Then he appealed to the idea that when light is reflected from a straight line the incident and reflected rays make equal angles with the line, and now it is a matter of calculation. Diocles' argument is an ingenious manipulation of angles, which rests on a fact about tangents to parabolas that he merely states: the distances AB and BG are equal.

Even if you do not work in detail through Diocles' proof, which you will find below, you can observe his geometrical style. Indeed, it is remarkable how much we can infer from the state of mathematics by simply observing what he took for granted. As we will see from his proof,[46] the parabola comes equipped with a line BH called 'half the parameter of the squares on the ordinates': this is half the constant parameter of the symptom, which is the distance from the original cutting plane to the point of the cone (see Figure 5.15).

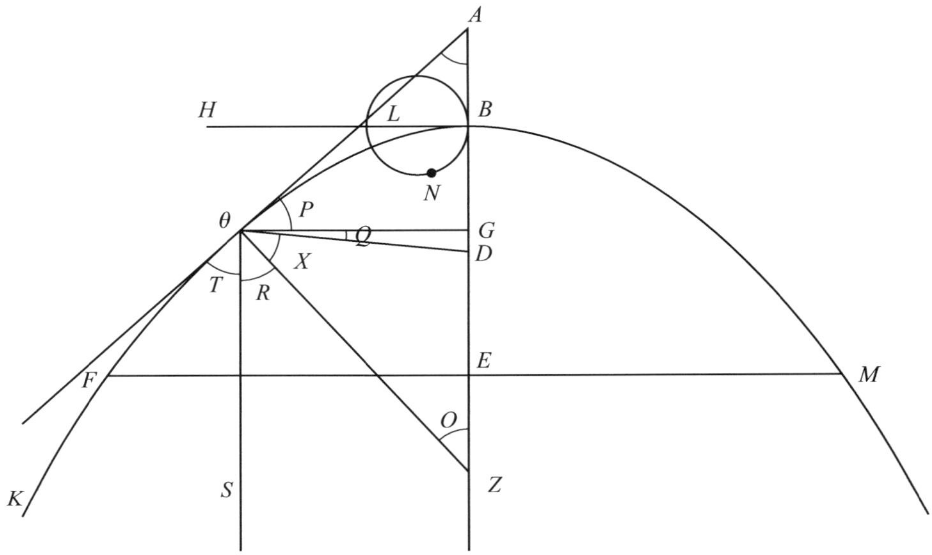

Figure 5.15. The focal property of the parabola

However, Diocles did not start from the symptom of the curve. Instead, he produced a series of properties of parabolas as though they were familiar truths generally known to experts in geometry at the time: these properties are that $AB = BC$, that the line drawn from θ perpendicular to θA meets AX beyond E, and that $GZ = BH$. Given his effortless and confident handling of the tangent θA and of the line θZ perpendicular to it, we can judge that this is a sophisticated treatment by an accomplished research mathematician.

The main body of this proof reads as an efficient contribution to the theory of conic sections.[47] But its last paragraph and the introduction place this theoretical knowledge in the practical context of mirror construction, so the overall enterprise has slipped beyond the grasp of the strict Platonic canons. There are several profound physical assumptions embedded in his procedure: that light rays can be treated like geometrical lines, that rays from the Sun are parallel, and so on. Note that the surface he described at the end, obtained by revolving the parabola around its axis, is just the paraboloid of revolution whose investigation by Archimedes we studied above.[48]

Diocles proves the focal property of the parabola.

Let there be a parabola KBM, with axis AZ, and let half the parameter of the squares on the ordinates be line BH. Let BE on the axis be equal

[46] F&G 4.C2.

[47] Diocles, in (Toomer 1976) and F&G 4.C1 and F&G 4.C2.

[48] Diocles, in (Toomer 1976) and F&G 4.C2.

to BH, and let BE be bisected at point D. Let us draw a line tangent to the section at an arbitrary point, namely the line θA, and draw the line θG as ordinate to AZ. Then we know that $AB = BG$ and that the line drawn from θ perpendicular to θA meets AZ beyond E. So let us draw $Z\theta$ perpendicular to θA, and join θD.

Then $GZ = BH$ and $HB = BE$, so $GZ = BE$.

We subtract GE, common (to GZ and BE), then the remainder $GB = EZ$.

But $GB = BA$, so $AB = EZ$.

And $BD = DE$, because BE is bisected at D, so the sum $AD = DZ$.

And because triangle $A\theta Z$ is right-angled and its base AZ is bisected at D, $AD = D\theta = DZ$.

So $\angle O = \angle X$ and $\angle A = \angle PQ$.

So let a line parallel to AZ pass through θ, namely line θ.

Then $\angle O = \angle R$, which is alternate to it, and $\angle O = \angle X$, so $\angle X = \angle R$ also.

And $\angle PQX = \angle RT$, right angles, so $\angle T = \angle PQ$, remainders.

So when line $S\theta$ meets line $A\theta$ it is reflected to point D, forming equal angles, PQ and T, between itself and the tangent $A\theta$. Hence it has been shown that if one draws from any point on KBM a line tangent to the section, and draws the line connecting the point of tangency with point D, e.g. line θD, and draws line $S\theta$ parallel to AZ, then in that case line $S\theta$ is reflected to point D, i.e. the line passing through point θ is reflected at equal angles from the tangent to the section. And all parallel lines from all points on KBM have the same property, so, since they make equal angles with the tangents, they go to point D.

Hence, if AZ is kept stationary, and KBM revolved [about it] until it returns to its original position, and a concave surface of brass is constructed on the surface described by KBM, and placed facing the Sun, so that the Sun's rays meet the concave surface, they will be reflected to point D, since they are parallel to each other. And the more the [reflecting] surface is increased, the greater will be the number of rays reflected to point D.

Apollonius. When we turn to Diocles' contemporary Apollonius, we are unequivocally back in the high geometrical research tradition. To attempt to work through Apollonius's *Conics* is an unnerving experience, and we shall have to content ourselves with enough of a glimpse of this extraordinary work to appreciate its considerable influence in later centuries.

Little is known of Apollonius's life beyond what can be gleaned from the prefatory letters to the various Books that comprise the *Conics* (see Figure 5.16).[49] The impression given is like that of other members of the mathematical research community over the previous two centuries, of someone who travelled about and spent some time in various places around the Mediterranean, meeting people of like interests with whom he kept in touch. Born in Perga on the south coast of Asia Minor, north-west of Cyprus, Apollonius mentioned Alexandria, Ephesus, and Pergamum as places he visited or stayed at.

[49] In F&G 4.D1, 4.D2.

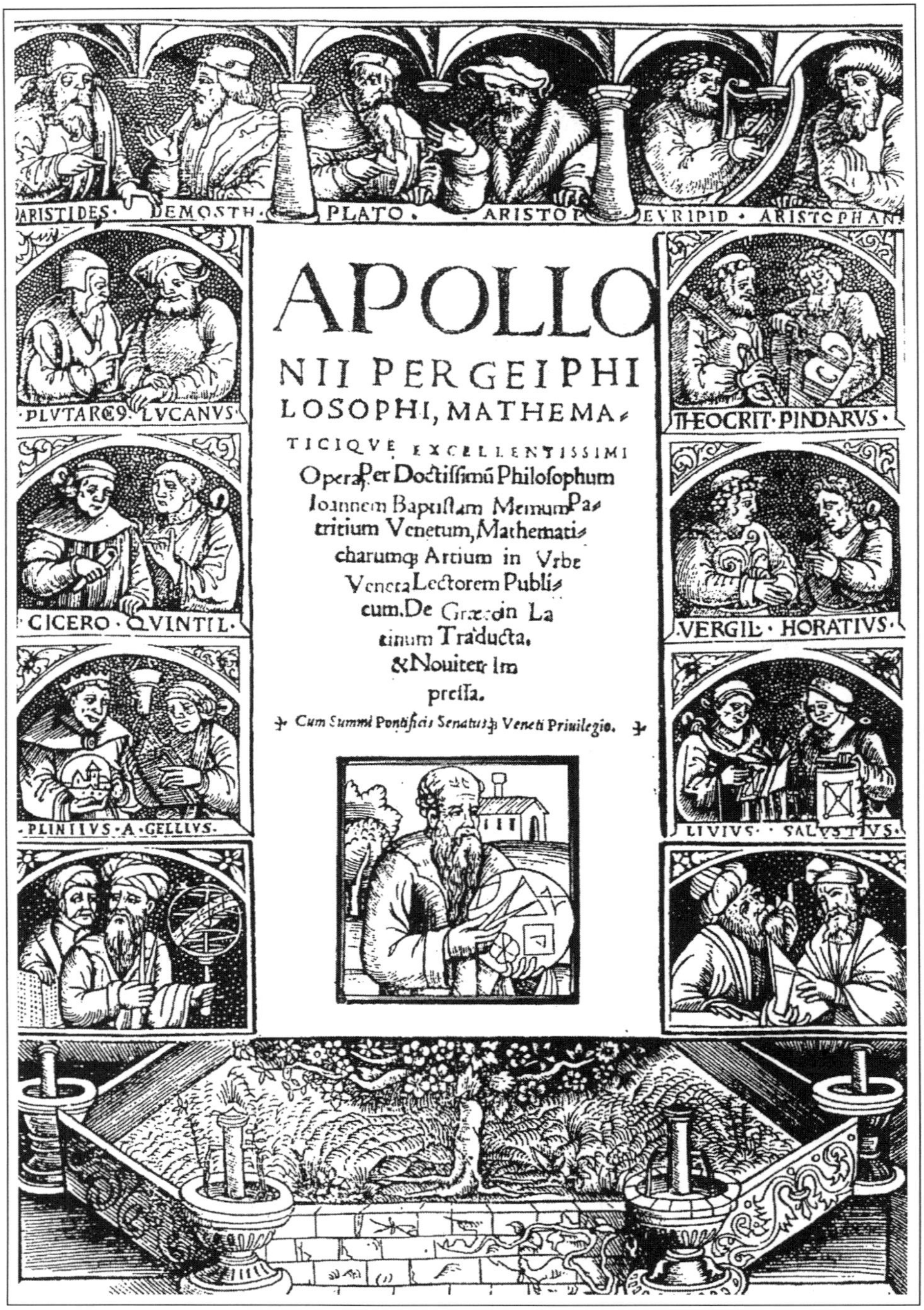

Figure 5.16. Frontispiece of Giovanni Baptista Memmo's edition of Apollonius's *Conics* (Venice, 1537), the first Latin translation

Apollonius's *Conics* presents an inexorable piling-up of 387 propositions in the seven surviving Books (the eighth being lost), icily austere in forbiddingly rigorous geometrical language, the rationale for the pattern of the whole being dimly perceptible. Many subsequent writers have commented on the difficulty of grasping his principles of selecting and ordering his material. Descartes seemed to suggest that Apollonius wrote down whatever came into his head, while B. L. van der Waerden (a major geometer of the 20th century who later turned to the history of mathematics) observed:[50]

> Apollonius is a virtuoso in dealing with geometric algebra, and also a virtuoso in hiding his original line of thought. This is what makes his work hard to understand; his reasoning is elegant and crystal clear, but one has to guess at what led him to reason in this way, rather than in some other way.

Book I of Apollonius's *Conics*. The Preface to Book I is of particular interest, as Apollonius there described the plan of the whole work.[51] It may be from here that Descartes got the idea that Apollonius put down whatever came into his mind; even Apollonius seems aware that some confusion could arise from the circulation of different versions of his work.

Apollonius to Eudemus, from the General preface to *Conics*.

> If you are restored in body, and other things go with you to your mind, well and good; and we too fare pretty well. At the time I was with you in Pergamum, I observed you were quite eager to be kept informed of the work I was doing in conics. And so I have sent you this first book revised, and we shall dispatch the others when we are satisfied with them. For I don't believe you have forgotten hearing from me how I worked out the plan for these conics at the request of Naucrates, the geometer, at the time he was with us in Alexandria lecturing, and how on arranging them in eight books we immediately communicated them in great haste because of his near departure, not revising them but putting down whatever came to us with the intention of a final going over. And so finding now the occasion of correcting them, one book after another, we publish them. And since it happened that some others among those frequenting us got acquainted with the first and second books before the revision, don't be surprised if you come upon them in a different form.
>
> Of the eight books the first four belong to a course in the elements. The first book contains the generation of the three sections and of the opposite branches, and the principal properties in them worked out more fully and universally than in the writings of others. The second book contains the properties having to do with the diameters and axes and also the asymptotes, and other things of a general and necessary use for limits of possibility. And what I call diameters and what I call axes you will know from this book. The third book contains many incredible theorems of use for the construction of solid loci and for limits of possibility of which the greatest part and the most beautiful are new. And when we had grasped these, we knew that the three-line and four-line locus had not been constructed by Euclid, but only a chance part of it and that not very happily.

[50]See (Van der Waerden 1961, 248).

[51]Apollonius, *Conics* in (Taliaferro 1952, 603) and F&G 4.D1.

> For it was not possible for this construction to be completed without the additional things found by us. The fourth book shows in how many ways the sections of a cone intersect with each other and with the circumference of a circle, and contains other things in addition none of which has been written up by our predecessors, that is in how many points the section of a cone or the circumference of a circle and the opposite branches meet the opposite branches. The rest of the books are fuller in treatment. For there is one dealing more fully with maxima and minima, and one with equal and similar sections of a cone, and one with limiting theorems, and one with determinate conic problems. And so indeed, with all of them published, those happening upon them can judge them as they see fit.

We concentrate here on topics from the first four books which he described in the Preface as 'an elementary introduction' — recall our earlier discussion in Chapter 3 of the meaning of the word *Elements*, for Apollonius was not claiming that the books were simple or easy.

Book I, he said, contains the fundamental properties of the conic sections 'worked out more fully and generally than in the writings of others': this is a crucial clue to his achievement. It seems that Apollonius achieved successfully for conic sections what Euclid did not quite manage (nor perhaps intend) for elementary geometry — to rework the elements from scratch as a unified theory. It is in large measure the generality of Apollonius's treatment that makes it difficult to comprehend, and yet it is very rewarding once one has done so. His generality was firmly announced right from the start. The first Definitions immediately confront us with a more general framework than we have met before.[52]

Apollonius, *Book I*: first definitions.

(1) If from a point a straight line is joined to the circumference of a circle which is not in the same plane with the point, and the line is produced in both directions, and if, with the point remaining fixed, the straight line being rotated about the circumference of the circle returns to the same place from which it began, then the generated surface composed of the two surfaces lying vertically opposite one another, each of which increases indefinitely as the generating straight line is produced indefinitely, I call a conic surface, and I call the fixed point the vertex, and the straight line drawn from the vertex to the centre of the circle I call the axis.

(2) And the figure contained by the circle and by the conic surface between the vertex and the circumference of the circle I call a cone, and the point which is also the vertex of the surface I call the vertex of the cone, and the straight line drawn from the vertex to the centre of the circle I call the axis, and the circle I call the base of the cone.

(3) I call right cones those having axes perpendicular to their bases, and I call oblique those not having axes perpendicular to their bases.

(4) Of any curved line which is in one plane, I call that straight line the diameter which, drawn from the curved line, bisects all straight lines drawn to this curved line parallel to some straight line; and I call the end of the diameter situated on the curved line the vertex of the

[52] Apollonius, *Conics* in (Taliaferro 2000, 3–4) and F&G 4.D3.

curved line, and I say that each of these parallels is drawn ordinatewise to the diameter.

(5) Likewise, of any two curved lines lying in one plane, I call that straight line the transverse diameter which cuts the two curved lines and bisects all the straight lines drawn to either of the curved lines parallel to some straight line; and I call the ends of the [transverse] diameter situated on the curved lines the vertices of the curved lines; and I call that straight line the upright diameter which, lying between the two curved lines, bisects all the straight lines intercepted between the curved lines and drawn parallel to some straight line; and I say that each of the parallels is drawn ordinatewise to the [transverse or upright] diameter.

(6) The two straight lines, each of which, being a diameter, bisects the straight lines parallel to the other, I call the conjugate diameters of a curved line and of two curved lines.

(7) And I call that straight line the axis of a curved line and of two curved lines which being a diameter of the curved line or lines cuts the parallel straight lines at right angles.

(8) And I call those straight lines the conjugate axes of a curved line and of two curved lines which, being conjugate diameters, cut the straight lines parallel to each other at right angles.

Apollonius's definition of a cone differs from that given by Euclid in two major ways and one minor one. The minor one is that Euclid's cone is a figure of revolution of a right-angled triangle, whereas Apollonius has a straight line fixed at a point and tracing round a circle (as in Figure 5.11). But they could still amount to the same thing, were it not for the two striking features of Definition 1: the line is of indefinite length in both directions, and the point need not be vertically above the centre of the circle (assuming the latter to be horizontal), but can be anywhere except in the plane of the circle. So Apollonius's general starting object, the conic surface, is a doubly extended cone that can be quite skewed. His *cone* is half that, from the vertex down as far as the circle, and Euclid's cone is a special case of that — in Definition 3, Apollonius called this a *right cone*.

After a few ground-clearing propositions, Apollonius unleashed three propositions (Propositions 11–13)[53] that show how successfully he could work at this very general level. These show that for any cone and a plane section through it, the resultant curve has the same symptom (or defining attribute) as one of the previously known conic sections. So what previously were sections perpendicular to a side of three different cones can now be arrived at by three different sections of just one double cone.

Apollonius called the three curves the *parabola*, *hyperbola*, and *ellipse*. His reason for this is interesting. Whereas the previous names for the curves referred to their *construction* ('section of an acute-angled cone', etc.), Apollonius's names for them referred to their *symptoms*, the geometrically expressed property satisfied by points on the curves. To see what this property is, recall that Apollonius was working in as general a way as possible, so any point on the curve can be considered in relation to the segments that it cuts off along some diameter (his Definition 4) and are parallel to the tangent at the end of that diameter: these are the distances x and y in Figure 5.17. Now, conic sections have constant parameters associated with them, related in the old definition to the distance of the cutting plane from the vertex of the cone.

[53] F&G 4.D4.

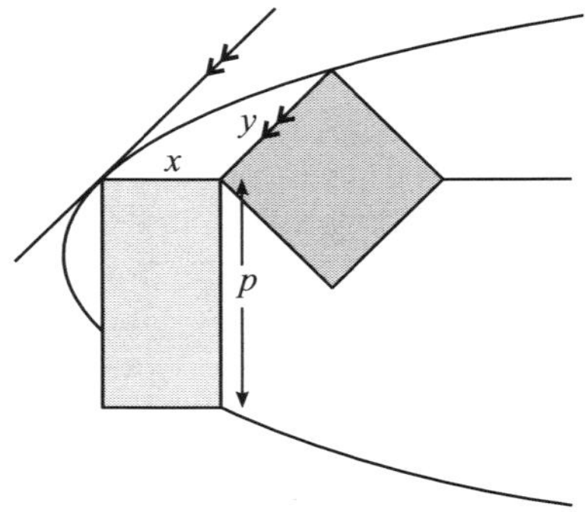

Figure 5.17. The parameter p for the parabola

Apollonius showed that for each curve the two shaded areas on each diagram are equal, one being the square on the length y, and the other being a rectangle one of whose sides is the length x. It is how the other side of the rectangle relates to the parameter p that determines the shape and name of the curve.

If the other side is p itself, then the relation is what we would express as $y^2 = px$, and the curve is a parabola. When the other side is less than p (by an amount depending on x) then the curve is an ellipse, and when it is greater than p, the curve is a hyperbola.[54] We omit these cases.

We shall not go through the proofs of these results. But to gain a further insight into Apollonius's style, consider the statement of the proposition for the parabola (Prop. 11). Here it is, in Taliaferro's translation:[55]

Apollonius introduces the parabola.

> If a cone is cut by a plane through its axis, and also cut by another plane cutting the base of the cone in a straight line perpendicular to the base of the axial triangle, and if, further, the diameter of the section is parallel to one side of the axial triangle, and if any straight line is drawn from the section of the cone to its diameter such that this straight line is parallel to the common section of the cutting plane and of the cone's base, then this straight line to the diameter will equal in square the rectangle contained by (a) the straight line from the section's vertex to where the straight line to the diameter cuts it off and (b) another straight line which has the same ratio to the straight line between the angle of the cone and the vertex of the section as the square on the base of the axial triangle has to the rectangle contained by the remaining two sides of the triangle. And let such a section be called a parabola.

It is an interesting exercise to try to match up what is said there with the description of the parabola given above. The ordinate y is

> any straight line ... drawn from the section of the cone to its diameter ... parallel to the common section of the cutting plane and of the cone's base.

[54] We can remember which is which because of the cognate words in English: a *parable* is a story in which a comparison of similar things is implied; an *elliptic* remark falls short of what is meant; and *hyperbole* is rhetorical exaggeration, exceeding what is meant.

[55] See (Taliaferro 2000, 19) and F&G 4.D4.

The abscissa x is

> the straight line from the section's vertex to where the straight line to the diameter cuts it off.

The parameter p is

> another straight line which has the same ratio to the straight line between the angle of the cone and the vertex of the section as the square on the base of the axial triangle has to the rectangle contained by the remaining two sides of the triangle.

So if we incorporate the shorter symbolism in place of the Apollonian terms, his proposition reads:

> If ... (the sectioning is done in such a way),
> then y will equal in square the rectangle contained by x and p.
> And let such a section be called a *parabola*.

The middle line of this is the relation $y^2 = px$ above.

There are two morals to this. One is that Apollonius makes good sense if we can find an appropriate language to understand him. The other is that Apollonius's own understanding and geometrical intuition must have been extraordinarily powerful and deep, even at a time when verbal expression of mathematics was the norm. That said, there were both gains and losses in Apollonius's new and highly general approach. Notice, for instance, how complicated the parameter p is in Apollonius's formulation, compared with the simple distance along the cone previously. This price must be set against the benefits of the new unified theory.

Apollonius's interests. What properties of curves was Apollonius interested in? He did not engage with the things investigated by Archimedes and the Eudoxan tradition (areas, volumes, filling things up with polygons, and so forth), but rather with other aspects of geometry: What lines go where? How do they cut? How do the lengths relate? What are the different properties and relations of tangents, diameters, and normals? A host of auxiliary lines and definitions came into play, leaving readers in no doubt about the richness and diversity of properties evinced by the conic sections.

Let us look briefly at one example to see the kind of result he obtained. In Book IV, Prop. 9 he explained how to draw tangents to a conic section from a point outside it (see Figure 5.18).[56] We draw through the point D any two lines that cut the conic section, and divide that part of each line within the curve in a particular ratio (the 'harmonic ratio' $EL : LH = ED : HD$), and similarly for the other line. The tangents from D meet the conic where the line through the two new points cuts the conic, but his *reductio* proof fails to illuminate what this curious but significant ratio has to do with tangents. Note, though, that Apollonius proved the result for any conic section, and that he did not need to consider each type separately. This is one of the benefits of his generality of treatment, that he was able to construct a unified body of techniques and concepts in place of a previously more diverse subject.

Finally, what was the point of all this? Had the massive edifice of Apollonius's *Conics* any particular use? This is a question that could be asked about other things too, but we raise it here because Apollonius himself did so. The Prefaces to *Conics*, Books IV and

[56] F&G 4.D6(d).

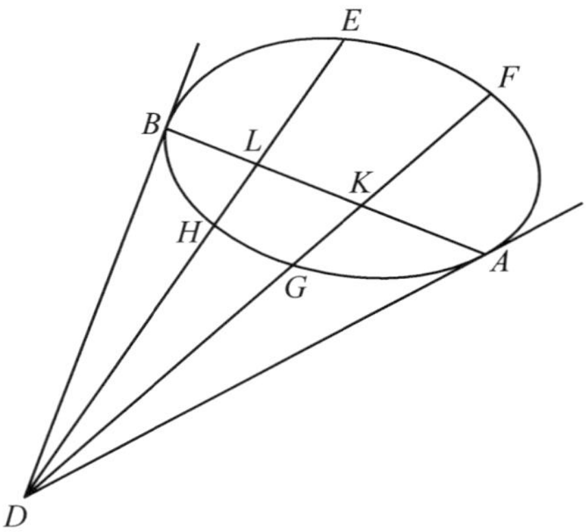

Figure 5.18. How to draw tangents to a conic section from a point outside it

V display the answers that Apollonius gave to this question.[57] In fact, towards the end of each preface he gives two kinds of answer: the use that other mathematicians have, in order to further their own studies, and an interesting 'art for art's sake' argument that his theorems are intrinsically worthwhile. This aesthetic response has been echoed by many mathematicians down the ages, and is entirely convincing to those who see it thus, but incomprehensible to those who do not.

Apollonius's high opinion of the value of mathematics was evidently shared by the English mathematician and astronomer Edmond Halley (see Figure 5.19) — the quotation from Vitruvius's *De Architectura*, Book VI reads: 'Aristippus the Socratic philosopher was shipwrecked on the coast at Rhodes and, observing geometrical diagrams drawn on the sand, is said to have cried out to his companions: 'Be of good cheer! For I see the traces of man.'

5.4 Further reading

We give here some suggestions for further reading covering Chapters 3–5.

Artmann, B. 1999. *Euclid: the Creation of Mathematics*, Springer. Useful and readable account of the thirteen Books in Euclid's *Elements*.

Cuomo, S. 2001. *Ancient Mathematics*, Science of Antiquity, Routledge. This book is remarkable for its breadth, and covers topics not included here such as mathematics among the Romans.

Fowler, D.H. 1999. *The Mathematics of Plato's Academy*, 2nd edition, Oxford University Press. An illuminating attempt to reconstruct the development of 4th-century BC ideas of ratio and proportion, attentive to sources and to papyral evidence.

Heath, Sir T.L. 1981. *A History of Greek Mathematics*, Dover (1st edition, Clarendon Press, Oxford, 1921). The standard two-volume general history, which is thorough and informative although some judgements need revision in the light of subsequent scholarship.

Knorr, W.R. 1975. *The Evolution of the Euclidean* Elements, Reidel. A substantive study of the development and significance of incommensurable magnitudes, which we only touch on in this book.

[57] F&G 4.D2(b),(c).

Figure 5.19. Frontispiece of Halley's 1710 edition of Apollonius's *Conics*

———. 1986. *The Ancient Tradition of Geometric Problems*, Dover. This book develops the discussion of the problem-solving aspect of the geometrical research tradition (notably, the three classical problems) further and more fully than we do here.

Lloyd, G.E.R. 1979. *Magic, Reason and Experience*, Cambridge University Press. This book discusses the development of mathematics and other Greek sciences in the context of social and intellectual history.

Tuplin, C.J. and Rihll, T.E. (eds.) 2002. *Science and Mathematics in Ancient Greek Culture*, Oxford University Press. A very readable survey by several authors of the many aspects of Greek mathematics and science.

Van Brummelen, G. 2009. *The Mathematics of the Heavens and the Earth: The Early History of Trigonometry*, Princeton University Press. An account of Greek and Indian trigonometry that reaches to the European renaissance.

———. 2013. *Heavenly Mathematics: The Forgotten Art of Spherical Trigonometry*, Princeton University Press. A companion volume to the previous one, it not only teaches but motivates and excites.

Van der Waerden, B.L. 1961. *Science Awakening*, Oxford University Press. Livelier than Heath's book, and only slightly less thorough.

6

Greek Mathematics: Later Years

Introduction

This chapter looks at some topics in Greek mathematics that took place several centuries after the work of Euclid and Archimedes. After a general overview, we investigate three topics: the work of Ptolemy the astronomer, number problems discussed by the mathematician Diophantus (mid-3rd century AD), and the commentating tradition.

Ptolemy's astronomy was to prove definitive for the next 1400 years, but Diophantus's problems have a much more complicated history. Some have seen these as closer to the Mesopotamian tradition of problem solving, while others see them as among the earliest works we have in algebra. However these issues may be resolved, Diophantus's work was to prove of major importance to the European mathematicians of the 16th and 17th centuries, and we pick up that theme in Chapter 12. Finally, the commentating tradition is important for the way it shaped the subsequent reception of Greek mathematics in Europe over the centuries.

6.1 The Hellenistic world

By the beginning of the 3rd century BC, the intellectual focus of the Greek-speaking world was no longer Athens, but Alexandria (in present-day Egypt). This involved more than a mere shift of locale, in that it coincided with a change of intellectual style so marked that historians refer to it by a fresh name, the Hellenistic period. It ended with the Roman conquest of 146 BC, although many traditions persisted well beyond that date, and some scholars speak of Hellenism to cover any predominantly Greek-influenced aspect of life.

Typical of the new Hellenistic style were two institutions established by the Emperor Ptolemy, the Museum and the Library. The Museum (its name means the House of the Muses, a sisterhood of nine goddesses who supported various forms of creativity) was a scientific research institute (in our terms), founded in the spirit of Aristotle's Lyceum; it proved extraordinarily productive in the sciences of astronomy, mathematical geography, anatomy, and physiology, as well as in mathematics. The Library at Alexandria developed rapidly to become the greatest library of antiquity, containing at its peak perhaps 10,000–15,000 or more rolls of papyrus; by about 250 BC, the library catalogue alone occupied 120 rolls.

During the 3rd century BC the three most famous Greek mathematicians — Euclid, Archimedes, and Apollonius — had associations with Alexandria. In their different ways, these three mathematicians form the culmination of the Greek geometrical research tradition. Their works were preserved to a greater extent than those of their predecessors, but were not significantly advanced, developed, or built upon for nearly two thousand years.

But the shift in intellectual style, mentioned above as characteristic of the new Hellenistic age, is revealed in their other interests. Euclid wrote a book called *Optics*, Archimedes had a high reputation in practical and theoretical aspects of mechanics, and Apollonius was noted for his astronomical studies. Although the later Hellenistic age was generally somewhat uneventful from the viewpoint of 'pure' mathematics, the mathematical sciences in a broader sense were pursued vigorously and to good effect.

In particular, the high Platonic distinction between mathematics and practical computation became rather blurred as time went on, and we find works in which great geometrical sophistication is combined with high computational skill. It is convenient to trace this through two routes. The first route stays close to the geometrical tradition and is associated here with the names of Heron, Ptolemy, and Pappus. The second route will take us to people who were intellectually more diverse and set a higher store by the study of numbers — neo-Pythagoreans such as Nicomachus, and the entirely unexpected figure of Diophantus.

Heron. To exemplify the practical mathematics tradition in Alexandria, we first consider briefly the work of Heron, who lived in the 1st century AD. Heron wrote many works on mechanics, as outlined by Pappus in Book VIII of his *Mathematical Collection*.[1]

Pappus on mechanics.

> The science of mechanics, my dear Hermodorus, has many important uses in practical life, and is held by philosophers to be worthy of the highest esteem, and is zealously studied by mathematicians, because it takes almost first place in dealing with the nature of the material elements of the universe. For it deals generally with the stability and movement of bodies [about their centres of gravity], and their motions in space, inquiring not only into the causes of those in virtue of their nature, but forcibly transferring [others] from their own places in a motion contrary to their nature; and it contrives to do this by using theorems appropriate to the subject matter. The mechanicians of Heron's school say that mechanics can be divided into a theoretical and a manual part; the theoretical part is composed of geometry, arithmetic, astronomy and physics, the manual of work in metals, architecture, carpentering and painting and anything involving skill with the hands. The man who had been trained from his youth in the aforesaid sciences as well as practised in the aforesaid arts, and in addition has a versatile mind, would be, they say, the best architect and inventor of mechanical devices. But as it is impossible for the same person to familiarize himself with such mathematical studies and at the same time to learn the above-mentioned arts, they instruct a person wishing to undertake practical tasks in mechanics to use the resources given to him by actual experience in his special art.

[1]See (Thomas 1980) Vol. 2, 615–621, and F&G 5.A2.

> Of all the [mechanical] arts the most necessary for the purposes of practical life are:
>
> (1) that of the *makers of mechanical powers*, they themselves being called mechanicians by the ancients — for they lift great weights by mechanical means to a height contrary to nature, moving them by a lesser force;
>
> (2) that of the *makers of engines of war*, they also being called mechanicians — for they hurl to a great distance weapons made of stone and iron and such-like objects, by means of the instruments, known as catapults, constructed by them;
>
> (3) in addition, that of the men who are properly called *makers of engines* — for by means of instruments for drawing water which they construct, water is more easily raised from a great depth;
>
> (4) the ancients also describe as mechanicians the *wonder-workers*, of whom some work by means of pneumatics, as Heron in his *Pneumatica*, some by using strings and ropes, thinking to imitate the movements of living things, as Heron in his *Automata* and *Balancings*, some by means of floating bodies, as Archimedes in his book *On Floating Bodies*, or by using water to tell the time, as Heron in his *Hydria*, which appears to have affinities with the science of sun-dials;
>
> (5) they also describe as mechanicians the *makers of spheres*, who know how to make models of the heavens, using the uniform circular motion of water.

Pappus then expanded his comments on Archimedes. His discussion is interesting, both for what it tells us about the breadth of Archimedes' interests and reputation (five centuries after his death), and also for his slightly nervous justification for applying geometry to everyday life. His final sentence makes no sense unless there were people around who protested, perhaps on Platonic grounds, against the sullying of pure geometry by its applications to everyday arts and sciences.

> Archimedes of Syracuse is acknowledged by some to have understood the cause and reason of all these arts; for he alone applied his versatile mind and inventive genius to all the purposes of ordinary life, as Geminus the mathematician says in his book *On the Classification of Mathematics*. Carpus of Antioch says somewhere that Archimedes of Syracuse wrote only one book on mechanics, that on the construction of spheres, not regarding any other matters of this sort as worth describing. Yet that remarkable man is universally honoured and held in esteem, so that his praises are still loudly sung by all men, but he himself on purpose took care to write as briefly as seemed possible on the most advanced parts of geometry and subjects connected with arithmetic; and he obviously had so much affection for these sciences that he allowed nothing extraneous to mingle with them. Carpus himself and certain others also applied geometry to some arts, and with reason; for geometry is in no way injured, but is capable of giving content to many arts by being associated with them, and, so far from being injured, it is obviously, while itself advancing those arts, appropriately honoured and adorned by them.

Of particular interest to us is Heron's *Metrica*, which was rediscovered only in 1896 in Constantinople (now Istanbul), in a manuscript from the 11th or 12th century. It is an intriguing blend of practical mensuration and methodological treatment more akin to the pure geometrical research tradition.

We illustrate Heron's work with his famous demonstration of how the area of a triangle can be found from the lengths of its three sides. That there must be such a relationship is obvious from the fact that the sides determine the triangle completely (up to congruence), but Euclid had not given such a result; in fact, as you will see, it is not easy to prove and cannot have been easy to discover. In modern terms, if the sides of the triangle have lengths a, b, and c, and we set the *semi-perimeter* $s = \frac{1}{2}(a + b + c)$, then Heron's formula for the area A of the triangle is

$$A = \sqrt{s(s-a)(s-b)(s-c)}.$$

The passage is in two parts. The first paragraph gives a sequence of instructions for determining the area of a triangle from the lengths of the sides, while the rest is a geometrical proof of the formula that is essentially being used. The first part is indicative of a problem-solving milieu: Heron presented the successive steps for a particular example, while stating that there is a general method; the justification that follows is then in the pure geometrical tradition. Note that his method hinges on constructing a segment equal in length to s.

There is an unusual feature in the first paragraph that is not paralleled in the later geometrical justification. The method consists of two parts: reaching some number whose square root is the answer, and applying an independent set of recipe-like instructions for working out an approximation to the square root if you cannot immediately find an exact value. Heron asserted that if this gives as accurate an answer as you want, that is fine; if not, then you do it again. Thus we see that the practical uses of this method were at the forefront of his attention.[2]

Heron on geometrical mensuration.

> There is a general method for finding, without drawing a perpendicular, the area of any triangle whose three sides are given. For example, let the sides of the triangle be 7, 8 and 9. Add together 7, 8 and 9; the result is 24. Take half of this, which gives 12. Take away 7; the remainder is 5. Again, from 12 take away 8; the remainder is 4. And again 9; the remainder is 3. Multiply 12 by 5; the result is 60. Multiply this by 4; the result is 240. Multiply this by 3; the result is 720. Take the square root of this and it will be the area of the triangle.
>
> Since 720 has not a rational square root, we shall make a close approximation to the root in this manner. Since the square nearest to 720 is 729, having a root 27, divide 27 into 720; the result is $26\frac{2}{3}$; add 27; the result is $53\frac{2}{3}$. Take half of this; the result is $26\frac{1}{2} + \frac{1}{3} (= 26\frac{5}{6})$. Therefore the square root of 720 will be very nearly $26\frac{5}{6}$. For $26\frac{5}{6}$ multiplied by itself gives $720\frac{1}{36}$; so that the difference is $\frac{1}{36}$. If we wish to make the difference less than $\frac{1}{36}$, instead of 729 we shall take the number now found, $720\frac{1}{36}$, and by the same method we shall find an approximation differing by much less than $\frac{1}{36}$.

[2]This and the next extract are from Heron, *Metrica* I, 8, in (Thomas 1939) Vol. 2, 471–475, in F&G 5.A5.

Heron then gave a proof of his claim that these computations work in every case. He drew heavily on Euclid's *Elements* — there are no fewer than seven references to it in this short passage, which is evidence of the growing importance of that work. He began by breaking the triangle into several right-angled triangles, and drawing a segment CH, equal to the semi-perimeter of the triangle whose area he wishes to find (see Figure 6.1).

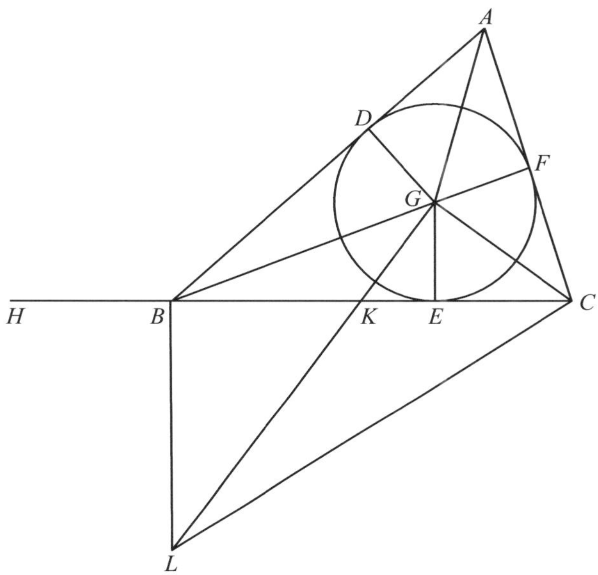

Figure 6.1. Heron on the area of a triangle

The geometrical proof of this is as follows: *In a triangle whose sides are given to find the area.* Now it is possible to find the area of the triangle by drawing one perpendicular and calculating its magnitude, but let it be required to calculate the area without the perpendicular.

Let ABC be the given triangle, and let each of AB, BC, CA be given; to find the area. Let the circle DEF be inscribed in the triangle with centre G [Euclid, Book IV, Prop. 9], and let AG, BG, CG, DG, EG, FG be joined. Then by [Euclid, Book I, Prop. 41]

$$BC.EG = 2.\text{triangle } BGC,$$
$$CA.FG = 2.\text{triangle } AGC,$$
$$AB.DG = 2.\text{triangle } ABG.$$

Therefore the rectangle contained by the perimeter of the triangle ABC and EG, that is the radius of the circle DEF, is double of the triangle ABC. Let CB be produced and let BH be placed equal to AD; then CBH is half the perimeter of the triangle ABC because $AD = AF$, $DB = BE$, $FC = CE$ [Euclid, Book III, Prop. 17]. Therefore $CH.EG =$ triangle ABC. But $CH.EG = \sqrt{CH^2.EG^2}$; therefore (triangle $ABC)^2 = HC^2.EG^2$.

Heron next constructed a right-angled triangle with base CH and area equal to that of the given triangle ABC, by specifying the height BL of this new triangle. He then established the existence of several similar triangles in the accompanying figure.

> Let GL be drawn perpendicular to CG and BL perpendicular to CB, and let CL be joined. Then since each of the angles CGL, CBL is right, a circle can be described about the quadrilateral $CGBL$ [Euclid, Book III, Prop. 31]; therefore the angles CGB, CLB are together equal to two right angles [Euclid, Book III, Prop. 22]. But the angles CGB, AGD are together equal to two right angles because the angles at G are bisected by AG, BG, CG and the angles CGB, AGD together with AGC, DGB are equal to four right angles; therefore, the angle AGD is equal to the angle CLB. But the right angle ADG is equal to the right angle CBL; therefore the triangle AGD is similar to the triangle CBL.

He next used these similar triangles to exhibit segments equal in length to the other ingredients in his computation: the difference between the semi-perimeter and each of the three sides of the original triangle.

> Therefore $BC : BL = AD : DG = BH : EG$, and so [Euclid V. 16] $BC : BH = BL : EG = BK : KE$, because BL is parallel to GE; hence [Euclid Book V, Prop. 18] $CH : BH = BE : EK$. Therefore $CH^2 : CH.HB = BE.EC : CE.EK = BE.EC : EG^2$, for in a right-angled triangle EG has been drawn from the right angle perpendicular to the base; therefore $CH^2.EG^2$, whose square root is the area of the triangle ABC, is equal to $(CH.HB)(CE.EB)$. And each of CH, HB, BE, CE is given; for CH is half of the perimeter of the triangle ABC while BH is the excess of half the perimeter over CB, BE is the excess of half the perimeter over AC, and CE is the excess of half the perimeter over AB, inasmuch as $EC = CF$, $BH = AD = AF$. Therefore the area of the triangle ABC is given.

Heron's work tempts us to speculate on what the history of mathematics would look like if more of his work had survived and less of Euclid's. It suggests that there might have been more of an interest in the applications of mathematics in the Hellenistic period than our sources allow us to describe, not only because this result of his is useful but also because of his evident interest in finding actual numbers in given cases — something of no concern in Euclid's *Elements*.

Claudius Ptolemy. Nowhere was this blurring between mathematics and computation more true than in astronomy. Disregarding Plato's injunction to 'ignore the visible heavens', Alexandrian astronomers developed the mathematical means to explain and predict celestial movements and phenomena to a high degree of precision. This culminated in a work of Claudius Ptolemy, in about AD 150, called the *Mathematical Syntaxis*; it is usually known by its Arabic name of *Almagest*, a translation of its popular name 'The greatest compilation'. This work was as effective as Euclid's *Elements* in summarising and making redundant the works of his predecessors, and was the definitive astronomical text for the next 1500 years.[3]

There was certainly considerable Mesopotamian influence on the work of Ptolemy and his predecessors, not least in their adoption of the sexagesimal system for astronomical computation. Mesopotamian astronomers had developed a systematic mathematical astronomy from about 500 BC onwards, which was fully worked out by the time of Alexander the Great's conquests. Although there had been many earlier contacts between Greek

[3] We study Ptolemy's work in greater detail in the next section.

and Mesopotamian astronomers, which made Mesopotamian observations distantly accessible within the Greek cultural orbit, Greek mathematical astronomy, as developed up to Ptolemy four centuries later, is different in principle: in the Greek systems calculations were based on a geometrical model of the planetary movements, rather than on the more strictly arithmetical procedures of the Mesopotamians.

One aspect of Ptolemy's work that provides evidence for the sophisticated computational skills in Alexandria is his trigonometry. To determine numerical relationships between sides and angles of triangles — which are fundamental to astronomy, mathematical geography, navigation, etc. — he worked with chords in circles, rather than our sines and cosines (a later Indian invention), although they amount to the same thing. Besides the *Almagest*, Ptolemy's other works testify to the range of mathematical sciences of his period: he wrote works entitled *Optics* (on vision, light, and the mathematics of reflection and refraction), *Harmonics* (on music theory), and *Geography*. This last work was influential: it was the first to employ latitude and longitude in a systematic way, and included two important map projections.

To the dismay of some historians of science, Ptolemy also wrote a work on astrology, his *Tetrabiblos*. The distinguished historian George Sarton wrote 'It is a great pity that Ptolemy wrote it', while Toomer described it as 'a specious "scientific" justification for crude superstition'.[4] However this may be, the *Tetrabiblos* was an influential astrological work for many years, and acts as good evidence of the breadth of Hellenistic mathematical activity.

Ptolemy also wrote a work on mechanics, which has not survived. Perhaps this loss is an indication that his work in this area was less highly regarded.

Pappus. Pappus lived in Alexandria between about AD 290 and 350 and wrote a number of works that, taken together, are called the *Mathematical Collection*. They are on a variety of subjects, and seem to have been written as a guide to other works, but with the passing of the centuries the *Mathematical Collection* has sometimes become our only source of information (see Figure 6.2). Happily for us, Pappus also supplied historical commentaries that bring together other accounts of particular problems and add further solutions of his own.

The first book of the *Mathematical Collection*, and part of the second book, are lost, but highlights from the rest give some idea of its range. Book IV has accounts of squaring the circle and trisecting an angle. Book V deals with isoperimetric problems: 'find the largest area enclosed by a curve of given length'; here, Pappus famously praised bees for their sagacity in being so economical in the way they construct their hexagonal honeycombs.[5] Book VII discusses several works on geometry by Apollonius that are now lost, and here Pappus wrote an account of the locus 'to three or four lines' that was to stimulate Descartes, as we shall see in Chapter 13.

Pappus's account of mechanics, which we quoted from above, occurs in Book VIII of the *Mathematical Collection*. He defended it as being a truly mathematical subject and not just one of utilitarian value, and it is tempting to interpret his Heronian division of mechanics into theoretical (sciences) and manual (arts) as a distinction analogous to Plato's for mathematics, but they are not quite the same thing. Plato's distinction hinged on greater or lesser approaches to truth, by which criterion the 'lower' level was effectively unnecessary, whereas the Pappus/Heron distinction is more of two complementary and

[4]See (Sarton 1959, 60) and (Toomer 1975, 198).

[5]See F&G 5.B5.

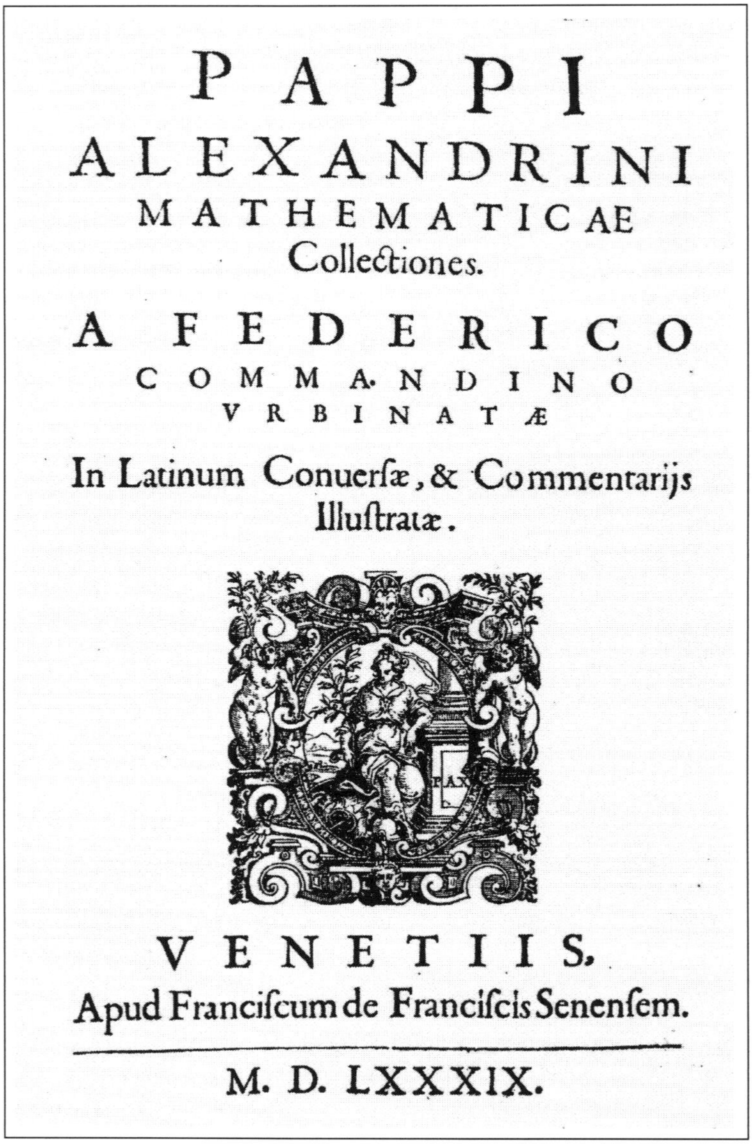
PAPPI
ALEXANDRINI
MATHEMATICAE
Collectiones.
A FEDERICO
COMMANDINO
VRBINATÆ
In Latinum Conuerſæ, & Commentarijs
Illuſtratæ.
VENETIIS.
Apud Franciſcum de Franciſcis Senenſem.
M. D. LXXXIX.

Figure 6.2. Pappus's *Mathematical Collection* (Venice, 1589)

equally necessary aspects of mechanics, split between different people for purely pragmatic reasons.

The Neo-Pythagoreans and Nicomachus. We saw earlier that the Greeks of the 4th century BC believed, rightly or wrongly, that their mathematics was derived from the Egyptians. In a similar way, Hellenistic and later Roman thinkers attributed the things they valued to some distant golden age.

Throughout the Graeco–Roman world, the early centuries AD saw a flourishing of many religious cults and philosophical beliefs. One such was what we call neo-Pythagoreanism, a revival (as they saw it) of the knowledge and beliefs of Pythagoras. Especially influential was the work of Nicomachus of Gerasa, who probably lived in Judaea around AD 100, and whose *Introduction to Arithmetic* was a basis for arithmetic teaching in the West for the next 1500 years. The opening of this work has a tribute to Pythagoras, leading to an explanation of the quadrivium classification of mathematical studies.[6]

We get a good idea of how much the neo-Pythagorean approach differed from Euclid's style of mathematics by comparing Nicomachus's treatment of perfect numbers with a glimpse of Euclid's, four centuries earlier.[7]

A *perfect number* is an integer that is the sum of all its proper divisors — that is, its divisors other than the number itself. For example, the proper divisors of 6 are $1, 2$, and 3, and $1+2+3=6$, so 6 is perfect; similarly, $28 = 1+2+4+7+14$ is perfect. The next two perfect numbers are 496 and 8128, and then there are no more until 33,550,336.

In Book IX of the *Elements*, Euclid proved that every number of the form $2^{k-1} \times (2^k - 1)$ is perfect if $2^k - 1$ is prime. For example,

$$6 = 2 \times 3 = 2^1 \times (2^2 - 1), \quad 28 = 4 \times 7 = 2^2 \times (2^3 - 1),$$

$$496 = 16 \times 31 = 2^4 \times (2^5 - 1), \quad \text{and} \quad 8128 = 64 \times 127 = 2^6 \times (2^7 - 1).$$

Now we look more closely at Euclid's method for producing some perfect numbers in *Elements*, Book IX, Prop. 36. Euclid began by letting A be the double of the unit length (he called it a *dyad* in the course of the proof), $B = 2A$, $C = 2B$, $D = 2C$, and so on until the sum $1 + A + B + C + D + \cdots = E$ is prime; then he claimed that the product DE is perfect.

If we set $A = 2$, then $B = 4$, $C = 8$, and so on up to $D = 2^{k-1}$, then

$$E = 1 + 2 + \cdots + 2^{k-1} = 2^k - 1,$$

and the claim is that when $2^k - 1$ is prime, $2^{k-1} \times (2^k - 1)$ is perfect. In Euclid's presentation, the argument begins:

Perfect numbers.

> *If as many numbers as we please beginning from a unit be set out continuously in double proportion, until the sum of all becomes prime, and if the sum multiplied into the last make some number, the product will be perfect.*
>
> For let as many numbers as we please, A, B, C, D, beginning from an unit be set out in double proportion, until the sum of all becomes prime, let E be equal to the sum, and let E by multiplying D make FG; I say that FG is perfect.
>
> For, however many A, B, C, D are in multitude, let so many E, HK, L, M be taken in double proportion beginning from E; therefore, *ex aequali*, as A is to D, so is E to M [VII. 14].
>
> Therefore the product of E, D is equal to the product of A, M [VII. 19].
>
> And the product of E, D is FG; therefore the product of A, M is also FG.
>
> Therefore A by multiplying M has made FG; therefore M measures FG according to the units in A.
>
> And A is a dyad; therefore FG is double of M.

[6]See (Nicomachus 1926) in F&G 2.D4.

[7]See (Nicomachus 1926) in F&G 3.D3(a) and 3.D3(b).

Euclid then continued in this heavily — one might say obscurely — geometrical fashion, showing that the number that he wanted to show is perfect is equal to the sum of the divisors that enter its construction. He then showed that there are no other divisors — more precisely, he showed that the divisors A, B, C, D, E, HK, L, and M of FG are the only divisors of FG. It follows that FG is the sum of its proper divisors and is accordingly perfect.

Now we look at Nicomachus's account.[8] You may find it helpful to take his example of 8128, and record its successive factorisations as 1, 2 and 4064, 4 and 2032, ..., to 64 and 127.

Nicomachus on perfect numbers.

> Now when a number, comparing with itself the sum and combination of all the factors whose presence it will admit, neither exceeds them in multitude nor is exceeded by them, then such a number is properly said to be perfect, as one which is equal to its own parts. Such numbers are 6 and 28; for 6 has the factors half, third, and sixth, 3, 2 and 1, respectively, and these added together make 6 and are equal to the original number, and neither more nor less. Twenty-eight has the factors half, fourth, seventh, fourteenth, and twenty-eighth, which are 14, 7, 4, 2, and 1; these added together make 28, and so neither are the parts greater than the whole nor the whole greater than the parts, but their comparison is in equality, which is the peculiar quality of the perfect number.
>
> It comes about that even as fair and excellent things are few and easily enumerated, while ugly and evil ones are widespread, so also the superabundant and deficient numbers[9] are found in great multitude and irregularly placed — for the method of their discovery is irregular — but the perfect numbers are easily enumerated and arranged with suitable order; for only one is found among the units, 6, only one other among the tens, 28, and a third in the rank of the hundreds, 496 alone, and a fourth within the limits of the thousands, that is, below ten thousand, 8128. And it is their accompanying characteristic to end alternately in 6 or 8, and always to be even.
>
> There is a method of producing them, neat and unfailing, which neither passes by any of the perfect numbers nor fails to differentiate any of those that are not such, which is carried out in the following way.
>
> You must set forth the even-times-even numbers from unity, advancing in order in one line, as far as you please:
>
> $$1, 2, 4, 8, 16, 32, 64, 128, 256, 512, 1024, 2048, 4096, \ldots .$$
>
> Then you must add them together, one at a time, and each time you make a summation observe the result to see what it is. If you find that it is a prime, incomposite number, multiply it by the quantity of the last number added, and the result will always be a perfect number. If, however, the result is secondary and composite, do not multiply, but add the next and

[8] See (Nichomachus 1926) in F&G 3.D3(b).

[9] A number was called *superabundant* if it is greater than the sum of its divisors, and *deficient* if it is less. The modern definition of these terms is different.

observe again what the resulting number is; if it is secondary and composite, again pass it by and do not multiply; but add the next; but if it is prime and incomposite, multiply it by the last term added, and the result will be a perfect number; and so on to infinity. In similar fashion you will produce all the perfect numbers in succession, overlooking none.

Nicomachus then explained that 6 and 28 are perfect numbers, before continuing:

When these have been discovered, 6 among the units and 28 in the tens, you must do the same to fashion the next. Again add the next number, 8, and the sum is 15. Observing this, I find that we no longer have a prime and incomposite number, but in addition to the factor with denominator like the number itself, it has also a fifth and a third, with unlike denominators. Hence I do not multiply it by 8, but add the next number, 16, and 31 results. As this is a prime, incomposite number, of necessity it will be multiplied, in accordance with the general rule of the process, by the last number added, 16, and the result is 496, in the hundreds; and then comes 8128 in the thousands, and so on, as far as it is convenient for one to follow.

It may seem to you that Nicomachus is much easier to follow than Euclid. Both tell you how to produce an example of a perfect number: just follow the instructions, as in a recipe book. Where their enterprises differ is that Euclid proved his result, whereas Nicomachus did not. Euclid's proof, moreover, is couched in geometrical terms and is difficult to follow.

Next, and no doubt connected with this lack of proof, Nicomachus made some audacious claims about perfect numbers that range from the true (every even perfect number ends in 6 or 8) to the false (the endings 6 and 8 alternate; there is one perfect number in the units, one in the 10s, one in the 100s, and so on), and the still unproved (all perfect numbers are even, and his (Nicomachus's) method produces every perfect number). Euclid made a more restricted claim, but one capable of proof. Finally, unlike Euclid's, Nicomachus's discussion invokes ethical analogies ('fair and excellent', 'ugly and evil').

So their treatments are very different, and it is tempting to see in Nicomachus a debased popularisation of Euclidean number theory, an assimilation into 'low' mathematics of the high classical past. But wherever we place him, the topic of perfect numbers has little to do with everyday computational mathematics, and in that respect Euclid and Nicomachus agreed.

Nicomachus was particularly emphatic about his debt to Pythagoras, and certainly his somewhat theological conception of number epitomises what tradition associates with the Pythagoreans, as in this passage:[10]

All that has by nature with systematic method been arranged in the universe seems both in part and as a whole to have been determined and ordered in accordance with number, by the forethought and the mind of him that created all things: for the pattern was fixed, like a preliminary sketch, by the domination of number pre-existent in the mind of the world-creating God.

But many other classical Greek concepts enter Nicomachus's exposition, especially ideas from Plato. For our purposes in gaining an impression of Roman culture, it does not matter precisely what sources Nicomachus drew upon. The things to notice are that there were several of them, they came from the fairly distant past, and Nicomachus may have

[10]See (Nicomachus 1926, Chap. 6).

been little better able than we are to discriminate between them. One final remark is in order: the attention paid to Nicomachus's work, despite his flimsy grasp of mathematics, testifies eloquently to the power of transmission. Whatever its merits, his work survived and gathered the lustre of antiquity — we shall never know how much was lost.

6.2 Ptolemy and astronomy

In this section we present some of the intricacies that astronomers discovered, and sought to describe, of the motion of the Sun, the Moon, and the planets, concentrating mainly on the work of the Greek astronomer Claudius Ptolemy. It is not intended as a history of astronomy covering a period of several centuries, but as a way of suggesting how issues in astronomy were formulated. It is divided into three parts: an introduction to naked-eye astronomy; an account of Ptolemy's work and some of its relations to its predecessors; and a brief account of some of Ptolemy's mathematics.

The observable phenomena. In all latitudes near the equator the Sun rises and sets every day, the Moon goes through a cycle of phases roughly once a month, and the seasons repeat more or less once a year. But this rough information has to be refined before it is of use, because farmers need to know when to plant their crops. To this end a remarkable amount of detailed information can be collected by naked-eye observation of the heavens.

In the course of a year the amount of daylight in a day varies with the seasons, from less than half the day to more than half and back. There are two days when the daylight fills exactly half the day: these are the equinoxes (*equi* for equal, *nox* for night) with equal days and nights. After the spring equinox the daylight hours last more than half the day and the weather becomes noticeably and consistently warmer, while after the autumn equinox the weather turns colder, so the spring equinox is a good time for planting. If you need to predict its arrival, you need to know the number of days in a year and how to count them. This is where the trouble starts and astronomy begins to get interesting.

There are about 365 days in a year, but this number is not accurate. A closer approximation is $365\frac{1}{4}$, which is why once every four years we add an extra day to the year, making it a leap year: if you work on the basis that there are only 365 days in a year, then in as few as 100 ($= 25 \times 4$) years your predictions will be 25 days too early, with possibly catastrophic results.

Can the phases of the Moon provide a better guide? The Mesopotamian and Greek civilisations knew that the Moon goes through its phases every 29 or 30 days, making a lunar calendar of 354 days. This gives a slippage of 11 days a year, which seems terrible — after only 17 years the spring equinox would occur where the autumn one does — so a new month needs to be inserted into the year every three years or so. But when, exactly? The Mesopotamians and the Greeks also knew that 235 lunar months make almost exactly 19 years (the error is about two hours). So, for example, the full Moon on 9 February 2020 will be followed by a full Moon on 9 February 2039 in almost the same place in the sky.

This cycle is nowadays known as the *Metonic cycle*, after the Greek astronomer Meton, and is accurate to one day in 219 years. Meton used it in 432 BC to argue for a way of calculating the timing of the extra month, and his rule was inscribed on public buildings in Athens. It seems that this cycle was known earlier to the Mesopotamians, who also knew of yet more cycles that predict very accurately when an eclipse will take place, whether it will be a lunar or a solar eclipse, full or partial, and so on. These cycles can accommodate the fact that the Moon does not move with uniform apparent motion across the sky. Eclipses tend to be of greater importance in cultures where the Sun and the Moon play a major role.[11]

Days and years, and the precession of the equinoxes. What exactly do we mean by a 'day' and a 'year'? There are many definitions, corresponding to different ways that these terms can be interpreted; our account is a deliberate over-simplification.

A *day* can be defined as the time it takes for the Earth to rotate once upon its axis.[12] A *year* is the time for the Earth to return to its original position in its orbit around the Sun. Those cultures that took the 'common-sense' view that it is the Earth that is fixed and the heavens that move could, and did, make comparable definitions — but how can they be measured? These facts are easier to measure at night: a day is the time it takes for the night-time sky to return exactly to the position that it had the night before.

A year is measured in much the same way: we use the fact that in the course of the night many stars either rise or set, and over the course of the year many stars have times when they are invisible because they never rise. What is called the *heliacal rising* of a star is the first time that it is visible above the Eastern horizon at dawn: before the rising the star is below the horizon; thereafter it rises four minutes earlier each morning, rising higher until it is lost in the brightness of the morning sunlight. The Egyptians started their calendar year with the heliacal rising of Sirius, which occurs reliably just before the annual flooding of the Nile.

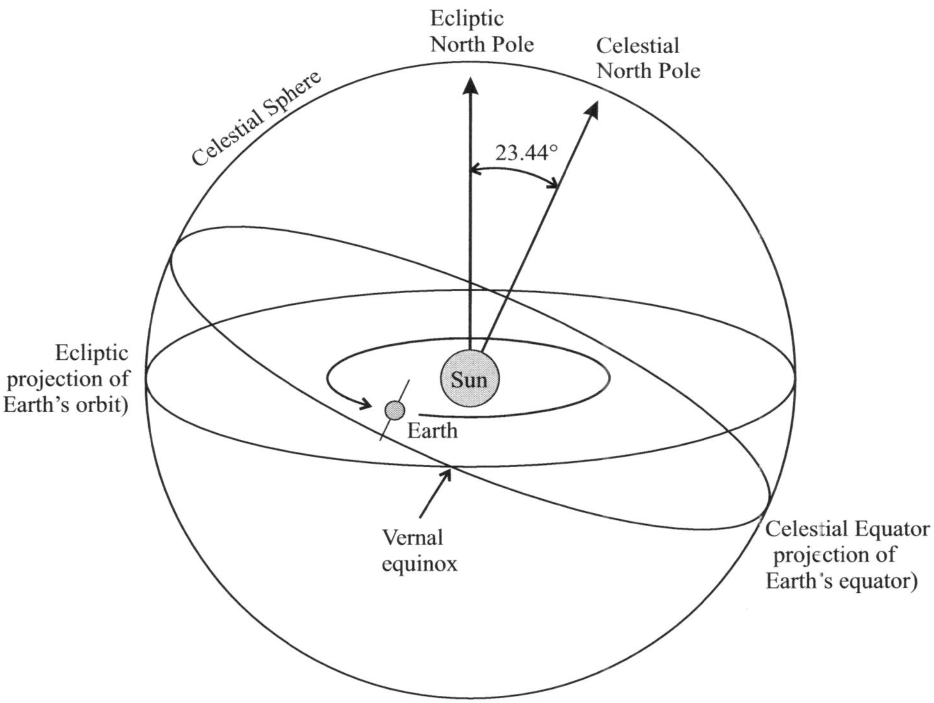

Figure 6.3. Plane of the ecliptic

[11]The Metonic cycle is still used by the Christian churches to keep track of Easter, and for many centuries it essentially solved the problem of the calendar, by enabling people to keep track of the seasons.

[12]Technically, this is the *sidereal day*, and is about four minutes shorter than the *solar day* which measures the average time from noon to noon.

The night-time sky varies from night to night through the year, because the Earth orbits the Sun in a plane, called the *plane of the ecliptic* (see Figure 6.3).[13] The Earth's axis is inclined to this plane, so it is possible to identify the stars that seem to sit on the ecliptic. This belt of stars, known as the *zodiac*, was conventionally divided up into twelve blocks and given the names that still form the so-called 'houses' of astrology, also known as the signs of the zodiac (Taurus, Leo, etc.), as the astrologers called the constellations.

Now imagine that it is sunset. As the stars come out, we can place the Sun in its position on the zodiac. This is another pattern that you expect to repeat every year, and so it does — or very nearly so. However, if you track the heliacal rising of a star accurately, you find that this position shifts slowly, by about 1 degree every 71.6 years. This phenomenon is the *precession of the equinoxes*: it means that if the Spring equinox first occurs when the Sun is 'in' Taurus (say), then it will do so for a further 71.6×30 $(= 2148)$ years, after which the Spring equinox occurs when the Sun 'enters' another 'house'.

The honour of discovering the precession of the equinoxes is much debated. The Greek astronomer Ptolemy credited his predecessor Hipparchus with it, while some historians credit the Mesopotamian astronomer Kidinnu of the 4th century BC; Kidinnu's name is also attached to the discovery of the Metonic cycle.

While we are still looking at the night-time sky and pondering issues astrological, we should notice the *planets*. The word is Greek and means 'wandering star': it refers to the fact that these objects move relative to the stars in the sky, which to naked-eye astronomers seem to move as if rigidly attached to the heavens.

Two of the planets, Mercury and Venus, are visible only at dawn and dusk, and disappear altogether at various times of the year. The remaining three that are visible to the naked eye, and were therefore known to ancient astronomers, are Mars, Jupiter, and Saturn, collectively called the *outer planets*. They move more or less in the plane of the ecliptic, and their brightness waxes and wanes as the years go by. The planets, along with the Sun and the Moon, are the only astronomical bodies that vary much from day to day, so they attracted the attention of astrologers and to this day have the names of ancient gods. Astrologers gave them considerable significance, and ascribed properties to them that influenced people when they were born in ways that reflected where the planets were at the time of their birth, so understanding their movements was particularly important.

But there were also difficulties. Although the three outer planets generally go round in the same direction from one night to the next, there are times when they appear (as seen from the Earth) to go into reverse. This is the phenomenon of *retrograde motion* (see Figure 6.4), and it is hard to describe accurately. This was a problem for astrologers, who needed to work out not only where the planets will be at some future time (auspicious for a birth or a battle), but also where they had been at some past time.

Ptolemy's account. We turn now to Ptolemy and his achievements. We know disappointingly little about Ptolemy himself, except that he worked in Alexandria: even his dates are conjectural, although we can deduce from some of his observations that he was alive between AD 125 and 150.

As we noted in the introduction to this chapter, Ptolemy was the author of the only full-length Greek astronomical work to have survived, later called the *Almagest*. His work is therefore our best source of information for what had been done by earlier Greek mathematicians and astronomers, although it should not be used uncritically. It can also be mined for information about what the Greeks learned from astronomers in the Mesopotamian tradition, and had become available to them ever since the Alexandrian conquest.

[13]The word *eclipse* is derived from the same root.

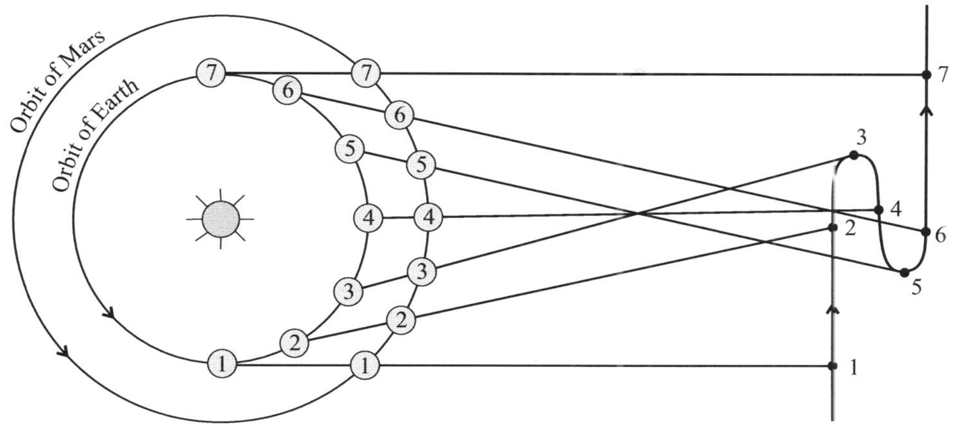

Figure 6.4. The retrograde motion of Mars

Among the first of the Greek astronomers were Eudoxus and Apollonius. They were followed by Hipparchus (c. 190–120 BC), and then over two centuries later by Ptolemy who seems to have taken over some of Hipparchus's ideas without acknowledging them. Eudoxus and Apollonius had used the Mesopotamians' ideas of the ecliptic and the division of the zodiac into twelve parts, but their theories were aimed at giving only qualitative descriptions of the motion of the heavens. Hipparchus, however, had access to Mesopotamian data: this shows up in his theories of the motion of the Moon and the planets. How he acquired this data is uncertain: Toomer[14] has suggested that Hipparchus may have obtained some of it himself from Mesopotamian sources, and the sheer amount of information in those sources makes it unlikely that they were ever translated systematically into Greek texts. In particular, Mesopotamian astronomers had centuries of data on eclipses, from which they had constructed very accurate predictive theories, essential for the purposes of astrology.

So the Greeks inherited from ancient Mesopotamia much more than the division of the unit into 60 parts, from which we still have 60 minutes in an hour, 60 seconds in a minute, and the division of the circle into 360 degrees. They could build on what Mesopotamian astronomers had described, and they sought to explain it through the construction of various elaborate geometrical accounts, and perhaps even with physical models. Here we have space to describe only a little of what Ptolemy did, but we should remember that the complexities in the motion of the Sun, Moon, Mars, Jupiter, and Saturn were well known to every competent astronomer and astrologer, and also to many a ruler and priest.

Ptolemy did not attempt a physical explanation by investigating the cause of the motion. He aimed instead at an accurate geometrical description of the motion that would enable good long-term predictions and fitted with what was, by then, quite a substantial amount of data. Where he differed most markedly from Hipparchus, upon whose work he often naturally relied, was in his rejection of the use of Mesopotamian predictive schemes based solely on extrapolation from prior observations (see Figure 6.5).[15]

[14] See (Toomer 1988).

[15] Probably no Greek astronomer knew how the Mesopotamian predictive models had been arrived at.

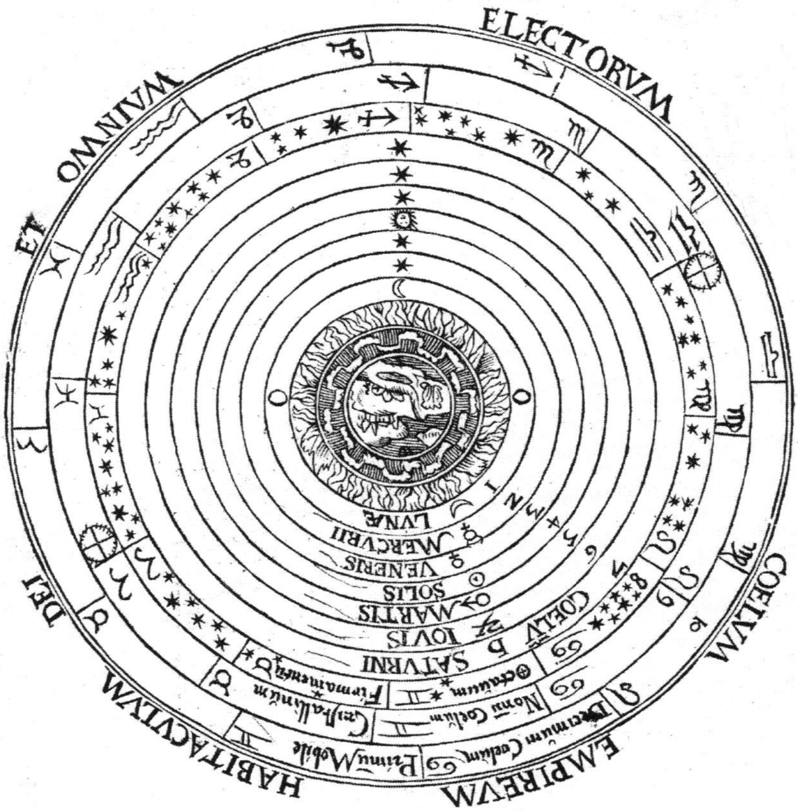

Figure 6.5. The Ptolemaic system of planetary astronomy

Ptolemy wanted to fit his data to geometrical models, and based his work on five preliminary statements:

- the heavens are spherical and rotate as a sphere
- the Earth is a sphere
- the Earth is at the centre of the cosmos
- the distance from the Earth to the heavens is so large that the Earth can be taken as a mathematical point
- the Earth does not move.

He then set out a well-organised account, starting with two books of general preliminaries (some mathematical, some astronomical). In Book III Ptolemy considered the motion of the Sun and the length of the year, in Book IV he considered the motion of the Moon, and in Books V and VI he made comparisons of the motion of the Sun and Moon, including the determination of eclipses. Then came accounts of the motion of the fixed stars, heliacal risings and the like, and then the motion of the planets.

The organisation is cumulative: the solar theory rests on observations of solstices and equinoxes; the lunar theory uses the solar theory; the account of the precession of the equinoxes uses the solar and lunar theories; and finally the account of planetary motion uses all of the previous work.

Ptolemy's first approximation was that all seven bodies (Sun, Mercury, Venus, Moon, Mars, Jupiter, and Saturn) orbit the Earth in circles at uniform speed. But he did not place the centres of these circles exactly at the centre of the Earth. To take care of the fact that no account of that kind can be accurate enough if the circles are supposed to rotate at uniform speed, he modified the motion in ways specific to each body. He imagined that each body (let us take Mars) is carried round on a small circle anchored to the original circle (called the deferent): he said that Mars rotates on a small circle whose centre is on the large circle that itself revolves with uniform speed around the earth. Remarkably complicated curves can be constructed in this way, and these can (for example) display retrograde motion. Because the smaller circles are mounted on the original circle, they are called *epicycles* (from *epi*, upon). To obtain a more accurate description, Ptolemy made Mars move on its epicycle with uniform speed as seen, not from the centre of the epicycle, but from another point called the *equant*, situated on the other side of the centre of the deferent from the Earth (see Figure 6.6).

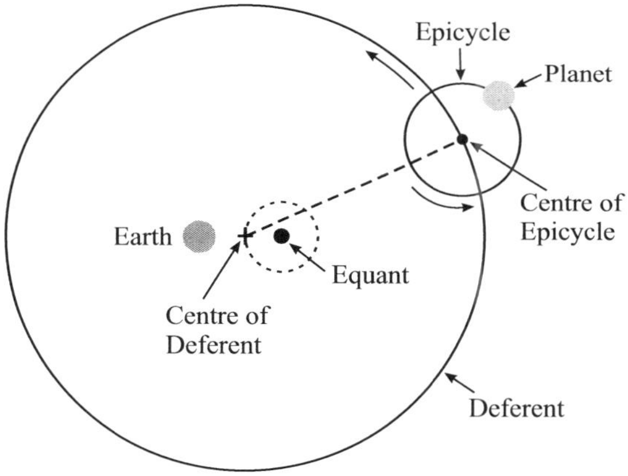

Figure 6.6. An epicycle

Arguments rage to this day about the accuracy of Ptolemy's work and its unexpected errors, and about his debt to his predecessors, but there is no question that it became the definitive account. It is a daunting work, one that no-one would idly contemplate setting aside, and it became the basis for all future work in astronomy in every culture that knew of it: the Islamic world, and later the reviving West. Four Arabic translations of his *Almagest* were made, of which one has survived, and from one of them Gerhard of Cremona made a Latin translation in 1175 during his stay in Toledo (it was not published until 1515). A 9th-century Greek manuscript of the work survived, and can be found in the Vatican Library, but it was never published; a Latin translation of it was published in 1451 and re-published in 1528. The first scholarly edition of the work was brought out by Grynaeus in Basel in 1538, and is based on the Greek text.

Trigonometry. Astronomical measurements are full of angles. Stars and planets are located with reference to their latitudes and longitudes on the celestial sphere. From these observations, and the times at which they were made, a mass of other quantities must be calculated (mostly other angles) which relate the observations to their theoretical explanations. In this way the observed departures from simple circular motion, the so-called *anomalies*, are explained. These calculations were done by a sophistication of the methods of Euclidean geometry that we call *trigonometry*.

Today, trigonometry relates the three sides and the three angles of a right-angled triangle by three fundamental formulas involving three functions of an angle: *sine*, *cosine*, and *tangent*. For the triangle in Figure 6.7, these are

$$\sin A = \frac{a}{c}, \quad \cos A = \frac{b}{c}, \quad \tan A = \frac{a}{b}.$$

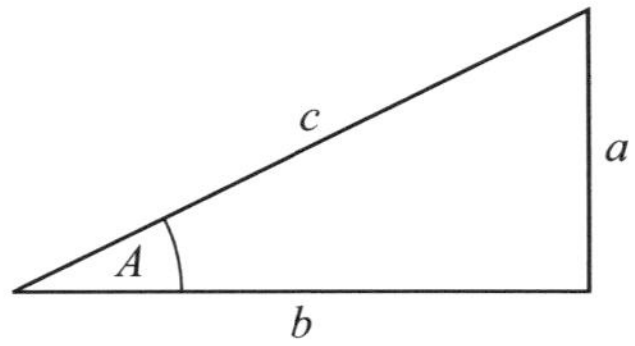

Figure 6.7. The trigonometry of a right-angled triangle

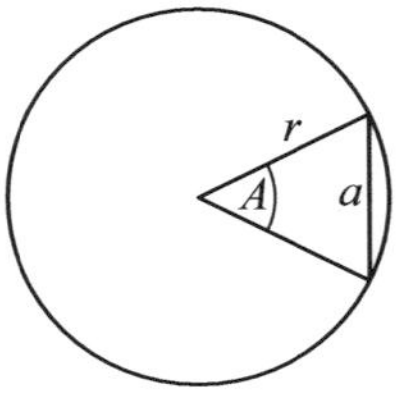

Figure 6.8. The trigonometry of chords in a circle

Ptolemy used trigonometry to find the sizes of chords in a circle, given the corresponding angles: this corresponds to calculating values for sine and cosine, in modern terminology. In this case, the relevant formula is

$$a = 2r \sin\left(\frac{A}{2}\right),$$

where r is the radius of the circle and the chord of length a subtends an angle of A at the centre (see Figure 6.8).

Ptolemy took a circle of radius 60 and divided its circumference into 360 parts (which correspond to *degrees*), and set himself the task of finding the lengths of chords as the corresponding angle increases in steps of $\frac{1}{2}$ degree from $0°$ to $180°$. 'In general', he said, 'we shall use the sexagesimal system because of the difficulty of fractions' (Book I, Section 10). As he observed, a few of these quantities are easy to find: an angle of 60 degrees can be found at the centre of a regular hexagon, so it subtends a chord of 60 degrees, and the corresponding chord also has length 60 (see Figure 6.9).

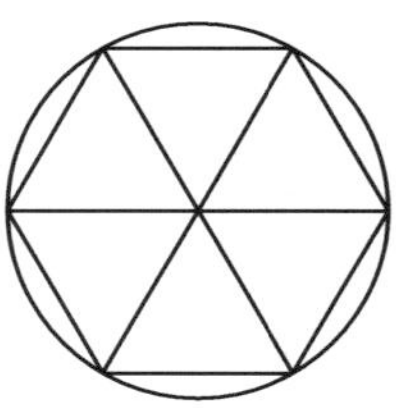

Figure 6.9. Ptolemy on the length of the side of a regular hexagon

What about an angle of 90 degrees? Ptolemy remarked (Book I, Prop. 10) that 'the side of the inscribed square ..., when squared, [is] double the square on the radius' and since 'the square of the radius is 3600, the square on the side of the square will add up to 7200, ... and so in length, chord of arc $90°$ approximately = $84°51'10''$ (see Figure 6.10). This is not too clear, although we can see that we are back briefly with the *Meno* example. Ptolemy was calculating the length of the side of a square inscribed in a circle of radius

60, so he was calculating 60 times the square root of 2, which, in sexagesimal fractional notation, is nearly $84 + \frac{51}{60} + \frac{10}{3600}^\circ$.

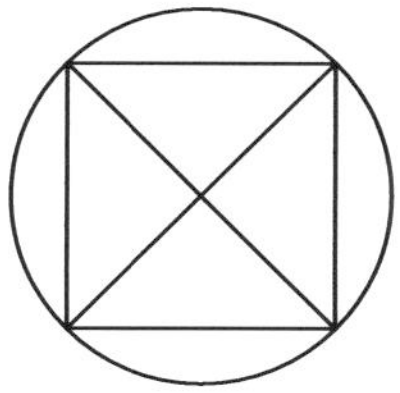

Figure 6.10. Ptolemy on the length of the side of a square

It is clear that only a few angles can be handled directly. What Ptolemy needed, and produced, was a systematic way of proceeding by seeing what happens when angles are successively halved: as he put it,

> from the chord of an arc of 12° there can be found the chord subtending an arc of 6°, and those subtending arcs of $3^\circ, 1\frac{1}{2}^\circ$, and $\frac{3}{4}^\circ$, respectively.

However, all of the arcs whose chords can be found easily are multiples of $\frac{3}{4}^\circ$, and repeated halving can never produce an angle of 1° from such angles. Ptolemy did not consider even hypothetically trisecting an angle since, as he put it, 'the chord of a third of the arc is in no way geometrically given'. So he resorted to an approximate procedure, trapping the chord corresponding to an angle of 1° between those for an angle of $\frac{3}{4}^\circ$ and $1\frac{1}{2}^\circ$. A single application of his halving procedure then gave him the chord corresponding to an angle of $\frac{1}{2}^\circ$.

Many calculations, all of them suppressed, gave Ptolemy his table of chords.[16] Answers were expressed in sexagesimal fractions, in the form of 'an integer + so many 60ths + so many 3600ths'. His table occupies almost four pages in a modern edition.

Conclusion. We cannot do better than sum up Ptolemy's contribution with two quotations. The first comes from a scholar of Hellenistic mathematics, Alexander Jones, who wrote in 1991 after a detailed analysis of Ptolemy's writings and his sources:[17]

> Ptolemy seems to have been the first to understand the importance of working out the entire system of the heavenly bodies according to a logical deductive plan and a consistent methodology. The *Almagest* was so successful in this respect that it is only with much difficulty that we are coming to appreciate its originality.

The second comes from Ptolemy himself, in the *Greek Anthology*, and belongs to a tradition of writings by astronomers on the romance of astronomy:

> I know that I am mortal and the creature of a day; but when I search out the massed wheeling circles of the stars, my feet no longer touch the Earth, but, side by side with Zeus himself, I take my fill of ambrosia, the food of the gods.

[16] Notice that although an angle of 5° is 10 times an angle of $\frac{1}{2}^\circ$, it does not follow that a chord of 5° is 10 times a chord of $\frac{1}{2}^\circ$.

[17] See (Jones 1991, 441–453).

6.3 Diophantus

The roving searchlight we have been playing across the Graeco–Roman scene has exhibited mathematicians of great versatility and application, but from the 'pure' mathematical perspective none was of the calibre of Archimedes and Apollonius, or of their 5th- and 4th-century BC predecessors. The fame and works of these giants lived on, as in Pappus's tribute to Archimedes, but the high mathematical research tradition seems by this time to have petered out.

There is, however, one Alexandrian mathematician of great subsequent influence, whose work is a puzzle for historians who like neat categorisations. He is Diophantus, who probably lived in the middle of the 3rd century AD, perhaps around AD 250. The puzzle has been neatly put this way:[18]

> if Diophantus had not existed, no historian of ancient Greek mathematics would have invented him

— in other words, but for the evidence of Diophantus's book, the *Arithmetica*, there would be no reason to suppose that anyone would have created a body of mathematics like it, at any time throughout the Greek period.

Diophantus's *Arithmetica* is not part of the geometrical research tradition. It is not about geometry, and there is none of the formal proposition/proof form we associate with that, although traces of the geometrical language of his predecessors can be found in his work. Nor does it look like a development of Euclidean number theory, even though the name *Arithmetica* might lead us to expect this. But it is also difficult to place the work firmly within the practical/computational tradition. Compared with earlier problem traditions (such as in the Rhind Papyrus), Diophantus is concerned with numbers simply in themselves, not as numbers of something-or-other, and his problems can be complicated and subtle. It seems to be 'straight' mathematics, carefully thought through and with no apparent neo-Pythagorean overtones: there is little sign of the influence of Nicomachus here. In short, Diophantus's work does not seem to fit in with anything we have seen before – but this might be a reflection of how much we have lost. It might not seem so out of place if we were less ignorant of some aspects of early mathematics.

So the work of Diophantus is strikingly different from that of Euclid's *Elements*, Archimedes' *The Sand-reckoner*, or Apollonius's *Conics*. Indeed, as we shall see when we look at some of his problems, it is much easier and more tempting for us to transcribe the problems of the *Arithmetica* into our present-day algebraic notation, without feeling that we are misrepresenting them. For this reason, some historians have referred to Diophantus as 'the father of algebra', inspired perhaps by Sir Thomas Heath who gave his book *Diophantos of Alexandria* the subtitle *A Study in the History of Greek Algebra*. In this section, we explore to what extent such views of Diophantus are justified.

Unfortunately, the *Arithmetica* seems to stand alone, without precedent or successor in Greek mathematics. It is unlike surviving contemporary works, but it seems implausible that it was unique in this way. It was written at the time of the neo-Platonic and neo-Pythagorean revival, and the study of whole-number or fractional solutions of problems that it contained was in keeping with such ideas, so it is likely that surviving sources represent only a fraction of the works originally written. Indeed, fewer than half of the original thirteen books of the *Arithmetica* have come down to us in the original Greek, although ten of them are extant if we include Arabic versions. Two other known works by Diophantus are completely lost and only a small fragment of a third work remains. So any

[18] See (Knorr 1985, 150).

judgement on the 'uniqueness' of the *Arithmetica* is based only on surviving sources from a period from which few works have survived.

Some of Diophantus's problems. Let us now look at some of the 189 problems to be found in the *Arithmetica*. We first look at just the statements of the problems and rather sketchily at his solutions, to give ourselves a feeling for the types of questions he discussed. We mainly use Heath's translation,[19] but the two versions of Book I, Prop. 7 illustrate the extent to which his versions of particular problems and solutions may be condensations of a more extensive original text.[20]

***Arithmetica*, Book I, Prop. 7.**

(a) A literal translation by I.E. Drabkin

From the same number to subtract two given numbers so that the remainders will have a given ratio to one another.

Let the numbers to be subtracted from the same number be 100 and 20, and let the larger remainder be three times that of the smaller.

Let the required number be $1x$. If I subtract from it 100, the remainder is $x - 100$ units; if I subtract from it 20, the remainder is $x - 20$ units. Now the larger remainder will have to be three times the smaller. Therefore three times the smaller will be equal to the larger. Now three times the smaller is $3x - 300$ units; and this is equal to $x - 20$ units.

Let the deficiency be added in both cases. $3x$ equal $x + 280$ units. If we subtract equals from equals, $2x$ equals 280 units, and x is 140 units.

Now as to our problem. I have set the required number as x; it will therefore be 140 units. If I subtract from it 100, the remainder is 40; and if I subtract from it 20, the remainder is 120. And the larger remainder is three times the smaller.

(b) A summary translation by Sir Thomas Heath

From the same [required] number to subtract two given numbers, so as to make the remainders have to one another a given ratio.

Given numbers 100, 20, given ratio $3 : 1$.

Required number x. Therefore $x - 20 = 3(x - 100)$, and $x = 140$.

It is evident that Heath supplied 19th-century school algebra in place of what Diophantus actually wrote. There can be good reasons for doing this, but it is obvious that it introduces its share of distortions. Nonetheless, we shall often use Heath's translation because it takes us directly to some of the difficulties we have with Diophantus's mathematics however it is translated.

Now look at the problem in Book I, Prop. 27 (in Heath's translation).[21]

To find two numbers such that their sum and product are given numbers.

Necessary condition. The square of half the sum must exceed the product by a square number.

---

[19](Heath 1885).

[20]For (a) see (Cohen and Drabkin 1945, 29) and for (b) see (Heath 1885, 133), both in F&G 5.D1.

[21]See F&G 5.D2.

Given sum 20, given product 96.
$2x$ the difference of the required numbers.
Therefore the numbers are $10 + x, 10 - x$.
Hence $100 - x^2 = 96$.
Therefore $x = 2$, and the required numbers are $12, 8$.

This invites comparison not with Euclid but with the Mesopotamian problems we studied in Chapter 2. As we saw there, solving quadratic problems in which the sum (or difference) and product of two unknown numbers are given was standard in Mesopotamian mathematics, if only in the special case of igûm–igibûm problems. Diophantus's text, however, first states the problem generally and then gives a necessary condition for the existence of a solution. It is only after doing this that he turns to a particular case, such as might be found on a Mesopotamian tablet — here, to find two numbers whose sum is 20 and whose product is 96.

How should we read this problem? In modern algebraic notation, it is equivalent to solving the simultaneous equations

$$x + y = 20 \quad \text{and} \quad xy = 96.$$

A Mesopotamian formulation of this problem might be:

I have added length and width: 20. I have multiplied length and width: 96.

Diophantus cleverly avoids having to denote two unknowns explicitly — there are no equivalents of the Mesopotamian terms 'length' and 'width'. Instead, as a look at the surviving text makes clear, Diophantus used just one word or symbol for 'the unknown' and rephrased the problem accordingly. The actual calculation leading to the solution is, however, similar to Mesopotamian methods. The novel twist, observing explicitly that such problems need have a solution under suitable conditions,[22] derives from a more overtly Greek formulation of such questions.

A similar scheme is found in Book II, Prop. 8: 'To divide a given square number into two squares'.[23] Here we notice a characteristic feature of the problems Diophantus posed and the solutions he offered. He was not aiming at maximum generality, such as a formula that would express all possible solutions to the problem. He did not answer this question by saying something — or indeed anything — like: let n^2 be the given square, then the required numbers are of the form

$$n\,\frac{p^2 - q^2}{p^2 + q^2} \quad \text{and} \quad n\,\frac{2pq}{p^2 + q^2}.$$

Nor does he seem to have been teaching a general method that the reader should see beneath the specific numbers, although we shall see that historians do try to find some system in his work. But, as did everyone at the time, he required of his solutions that they be what numbers were taken to be: positive whole numbers or fractions. This feature of his work can sometimes mean that his problems have definite answers that might otherwise have had an empty generality, much as the problem

$$\text{Find } x \text{ and } y \text{ such that } 3x + y = 10$$

has infinitely many solutions — there is always a value of y for each value of x — but there are only three pairs of solutions in positive integers:

$$x = 1 \text{ and } y = 7;\ x = 2 \text{ and } y = 4;\ \text{and } x = 3 \text{ and } y = 1.$$

[22]For example, we cannot find two numbers whose sum is 20 and whose product is 104.
[23]F&G 5.D3.

Arithmetica, Book II, Prop. 8.

> *To divide a given square number into two squares.*
>
> Let it be required to divide 16 into two squares.
>
> And let the first square $= x^2$; then the other will be $16 - x^2$; it shall be required therefore to make $16 - x^2 =$ a square.
>
> I take a square of the form $(mx - 4)^2$, m being any integer and 4 the root of 16; for example, let the side be $2x - 4$, and the square itself $4x^2 + 16 - 16x$. Then $4x^2 + 16 - 16x = 16 - x^2$. Add to both sides the negative terms and take like from like. Then $5x^2 = 16x$, and $x = \frac{16}{5}$.
>
> One number will therefore be $\frac{256}{25}$, the other $\frac{144}{25}$, and their sum is $\frac{400}{25}$ or 16, and each is a square.

Once again, we first find a general statement of the problem, then a particular square number (in this case 16) is chosen for the worked example and the problem is re-expressed so as to involve one unknown only; in this problem, however, a 'necessary condition' is not required. Diophantus then proceeded, much as the Mesopotamians did, by working with a specific number (here, 16) while illustrating a general method. He arranged for the unknown to appear in an equation in a simple form, manipulated the equation according to some simple rules, and gave his answer.

This particular problem has a remarkable later history, as Figure 6.11 indicates.[24] Famously, in the 17th century Pierre de Fermat wrote the following against Book II, Prop. 8, in his copy of a recent translation of Diophantus's *Arithmetica*:

> On the other hand it is impossible to separate a cube into two cubes, or a biquadrate into two biquadrates, or generally *any power except a square into two powers with the same exponent*. I have discovered a truly marvellous proof of this, which however the margin is not large enough to contain.

The figure shows the Greek text, its Latin translation by Claude-Gaspar Bachet, a French mathematician, and the note that Fermat wrote in the margin of his copy of an earlier edition, indicating that he had proved that the equation $x^n + y^n = z^n$ has no positive whole-number solutions when $n > 2$.

In Book III, Prop. 10 Diophantus asked for three numbers such that the product of any two, when added to a given number, yields a square.[25] As usual, a special case was chosen, taking 12 as the given number, and in the course of the calculation a particular square was chosen, namely 25. We are invited to understand that these numbers are typical, and that the same method would also work for other choices, although perhaps not so smoothly. After some further work, which we omit, Diophantus arrived at an expression of a quadratic relationship ($52x^2 + 12 =$ a square), at which point he interestingly did not mention the particular 'obvious' solution $x = 1$. Instead, his eventual solution of the problem is reached after a gap in the working covered by the remark 'This is easy'.[26]

In Book V, Prop. 9 presents a problem of a different kind, being an example of what Heath called Diophantus's 'method of approximation to limits'.[27] As before, we shall not follow it in detail, but extract two interesting features of his solution. The problem asks for two numbers whose sum is unity and such that if the same number is added to each of

[24] We take this story up again in Chapter 12.

[25] F&G 5.D4.

[26] At least, the remark is in the text as it has come down to us.

[27] F&G 5.D8.

Arithmeticorum Liber II. 61

interuallum numerorum 2. minor autem 1 N. atque ideo maior 1 N. + 2. Oportet itaque 4 N. + 4. triplos esse ad 2. & adhuc superaddere 10. Ter igitur 2. adscitis vnitatibus 10. æquatur 4 N. + 4. & fit 1 N. 3. Erit ergo minor 3. maior 5. & satisfaciunt quæstioni.

ς^ο ἑνός. ὁ ἄρα μείζων ἔσται ς^ο ἑνὸς μ^ο β̄. δεήσει ἄρα ἀριθμοὺς δ̄ μονάδας δ̄ τριπλασίονας εἶναι μ^ο β̄. καὶ ἔτι ὑπερέχειν μ^ο ῑ. τρὶς ἄρα μονάδες β̄ μετὰ μ^ο ῑ. ἴσαι εἰσὶν ςς^οις δ̄ μονάσι δ̄. καὶ γίνεται ὁ ἀριθμὸς μ^ο γ̄. ἔσται ὁ μὲν ἐλάσσων μ^ο γ̄. ὁ δὲ μείζων μ^ο ε̄. καὶ ποιοῦσι τὸ πρόβλημα.

IN QVAESTIONEM VII.

CONDITIONIS appositæ eadem rattio est quæ & appositæ præcedenti quæstioni, nil enim aliud requirit quàm vt quadratus interualli numerorum sit minor interuallo quadratorum, & Canones iidem hic etiam locum habebunt, vt manifestum est.

QVÆSTIO VIII.

PROPOSITVM quadratum diuidere in duos quadratos. Imperatum sit vt 16. diuidatur in duos quadratos. Ponatur primus 1 Q. Oportet igitur 16 − 1 Q. æquales esse quadrato. Fingo quadratum à numeris quotquot libuerit, cum defectu tot vnitatum quod continet latus ipsius 16. esto à 2 N. − 4. ipse igitur quadratus erit 4 Q. + 16. − 16 N. hæc æquabuntur vnitatibus 16 − 1 Q. Communis adiiciatur vtrimque defectus, & à similibus auferantur similia, fient 5 Q. æquales 16 N. & fit 1 N. 16/5. Erit igitur alter quadratorum 256/25. alter verò 144/25 & vtriusque summa est 400/25 seu 16. & vterque quadratus est.

ΤΟΝ ἐπιταχθέντα τετράγωνον διελεῖν εἰς δύο τετραγώνους. ἐπιτετάχθω δὴ τὸν ῑς̄ διελεῖν εἰς δύο τετραγώνους. καὶ τετάχθω ὁ πρῶτος δυνάμεως μιᾶς. δεήσει ἄρα μονάδας ῑς̄ λείψει δυνάμεως μιᾶς ἴσας εἶναι τετραγώνῳ. πλάσσω τὸν τετράγωνον ἀπὸ ςς^ων. ὅσων δήποτε λείψει τοσούτων μ^ο ὅσων ἐστὶν ἡ τῶν ῑς̄ μ^ο πλευρά. ἔστω ςς^ο β̄ λείψει μ^ο δ̄. αὐτὸς ἄρα ὁ τετράγωνος ἔσται δυνάμεων δ̄ μ^ο ῑς̄ λείψει ςς^ων ῑς̄. ταῦτα ἴσα μονάσι ῑς̄ λείψει δυνάμεως μιᾶς. κοινὴ προσκείσθω ἡ λεῖψις, καὶ ἀπὸ ὁμοίων ὅμοια. δυνάμεις ἄρα ε̄ ἴσαι ἀριθμοῖς ῑς̄. καὶ γίνεται ὁ ἀριθμὸς ῑς̄ πέμπτων. ἔσται ὁ μὲν σνς̄ εἰκοστοπέμπτων. ὁ δὲ ρμδ̄ εἰκοστοπέμπτων, καὶ οἱ δύο συντεθέντες ποιοῦσι ῡ εἰκοστοπέμπτα, ἤτοι μονάδας ῑς̄. καί ἐστιν ἑκάτερος τετράγωνος.

OBSERVATIO DOMINI PETRI DE FERMAT.

CVbum autem in duos cubos, aut quadratoquadratum in duos quadratoquadratos & generaliter nullam in infinitum vltra quadratum potestatem in duos eiusdem nominis fas est diuidere cuius rei demonstrationem mirabilem sane detexi. Hanc marginis exiguitas non caperet.

QVÆSTIO IX.

RVRSVS oporteat quadratum 16 diuidere in duos quadratos. Ponatur rursus primi latus 1 N. alterius verò quotcunque numerorum cum defectu tot vnitatum, quot constat latus diuidendi. Esto itaque 2 N. − 4. erunt quadrati, hic quidem 1 Q. ille verò 4 Q. + 16. − 16 N. Cæterum volo vtrumque simul æquari vnitatibus 16. Igitur 5 Q. + 16. − 16 N. æquatur vnitatibus 16. & fit 1 N. 16/5 erit

ΕΣΤΩ δὴ πάλιν τὸν ῑς̄ τετράγωνον διελεῖν εἰς δύο τετραγώνους. τετάχθω πάλιν ἡ τοῦ πρώτου πλευρὰ ς^ο ἑνός, ἡ δὲ τοῦ ἑτέρου ςς^ων ὅσων δήποτε λείψει μ^ο ὅσων ἐστὶν ἡ τοῦ διαιρουμένου πλευρά. ἔστω δὴ ςς^ο β̄ λείψει μ^ο δ̄. ἔσονται οἱ τετράγωνοι ὃς μὲν δυνάμεως μιᾶς, ὃς δὲ δυνάμεων δ̄ μ^ο ῑς̄ λείψει ςς^ο ῑς̄. βούλομαι τοὺς δύο λοιπὸν συντεθέντας ἴσους εἶναι μ^ο ῑς̄. δυνάμεις ἄρα ε̄ μ^ο ῑς̄ λείψει ςς^ο ῑς̄ ἴσαι μ^ο ῑς̄. καὶ γίνεται ὁ ἀριθμὸς ῑς̄ πέμπτων.

H iij

Figure 6.11. Fermat's last theorem

them the result is a square. If the two numbers are x and $1-x$ and the given number is A then we have to arrange for $A+x$ and $A+1-x$ to be squares. These conditions imply that $2A+1$ must be a sum of two squares. Diophantus began by observing a necessary condition: the given number A cannot be odd and $2A+1$ cannot be divisible by a prime number of the form $4n-1$. These remarks show a deep, if mysterious, insight into the fact that primes of the form $4n-1$ cannot be a sum of two squares. Diophantus next took the given number A to be 6, which meets his conditions, and wrote

> Therefore 13 must be divided into two squares each of which > 6. If we then divide 13 into two squares the difference of which < 1, we solve the problem.

In the course of the subsequent working we find the request to 'divide 13 into two squares such that their sides may be as nearly as possible equal to 51/20'. This arises because Diophantus has noticed that $6\frac{1}{2}+\frac{1}{400}$ is a square — in fact, $\left(\frac{51}{20}\right)^2$. A solution here is possible only because of his careful selection of a specific numerical example. His eventual solution was that the numbers x and $1-x$ must be $\frac{4843}{10201}$ and $\frac{5328}{10201}$.

Some of the problems are even more challenging. Book VI of the *Arithmetica* is concerned with right-angled triangles, where expressions involving their sides have to be squares or cubes. For example, Problem 19 is typically ingenious: 'To find a right-angled triangle such that its area added to one of its perpendiculars gives a square, while the perimeter is a cube'.[28]

Was there a method in Diophantus's work? How should we view what Diophantus did? The above examples give a fair impression of the books as a whole, which consist largely of problems. The *Arithmetica* is not a work of geometry, unlike the other Greek works we have looked at, and is not even a work of theory. Instead of theorems and constructions, there are problems and some unsystematic methods.

Historians have often looked for a system unifying these books and have come away empty handed: as the German mathematician and historian Hermann Hankel wrote in 1874:[29]

> It is difficult for a modern mathematician, even after studying 100 Diophantine solutions, to solve the 101st problem.

This seems to resonate with other historians and is often quoted. There is general agreement that Diophantus indicated his line of reasoning for individual problems, but there is no logical development within the work as a whole — he did not develop a subject systematically from first principles.

More recently, however, a Greek historian, Yannis Thomaidis, has sought to contest this claim.[30] Before we examine his argument, let us note that there is no reason for Diophantus to have been systematic: that may be an entirely modern requirement of an author. In fact, as noted, his problems are often reminiscent of the Mesopotamian arithmetical tradition. Diophantus may have become familiar with this tradition, since there is evidence of trade and cultural contacts over this period between Alexandria and Mesopotamia. We shall see in Chapter 7 that later Arabic writers sometimes set and solved problems with methods reminiscent of Diophantus, and this has led some historians to speculate that there may have been a tradition of such problems, of which we have only a few isolated fragments.

---

[28]F&G 5.D9.

[29]Quoted in (Heath 1885, 55).

[30]See (Thomaidis 2005, 591–640).

To understand Thomaidis's argument, let us follow his analysis of Book II, Prop. 8, which sticks much more closely to the Greek than did Heath.[31] As noted above, Diophantus chose 16 as the given number that is to be divided into two squares. Then if one of the squares is x^2, the other must be $16-x^2$. So $16-x^2$ must itself be a square, and Diophantus argued that one can arrange this by taking an arbitrary multiple of x minus the side (4) of the given square (16). Without any explanation, he took the arbitrary multiple to be 2, so he looked at $2x - 4$, whose square is $4x^2 - 16x + 16$. This gave him the equation $16 - x^2 = 4x^2 - 16x + 16$. At this point you should suspect that this equation has been contrived, and we shall shortly see that this is one of the kernels of Thomaidis's insight. For, as Diophantus immediately argued, we can always add or subtract the same kinds of numbers from both sides of an equation. In this case we can add x^2 and $16x$ to each side to get $16 + 16x = 5x^2 + 16$. Then we can subtract 16 from each side to get $16x = 5x^2$. If we ignore the solution $x = 0$ (which Diophantus would not have considered a solution), we get $x = \frac{16}{5}$. A simple calculation now shows that the sought-for squares are $\frac{256}{25}$ and $\frac{144}{25}$.

To summarise Thomaidis's argument briefly, he claims two things. One is that there is method in Diophantus's choice of what look like arbitrary numbers. In this case, the 2 is arbitrary, but the choice of constant (the 4 in the $2x - 4$) is not: it is deliberately chosen to make the resulting equation as simple as possible. It would take us too long to examine this, but we can show that it has some claim to being a general feature of Diophantus's solutions.

Consider again, for example, the solution of Book I, Prop. 27, which we looked at earlier. The problem asks for two numbers such that their sum and product are two given numbers, and Diophantus found that the numbers are 8 and 12. In line with what we have seen, Diophantus took the particular case where the sum is 20 and the product 96, and let the sought-for numbers be $10 - x$ and $10 + x$. This gave him the equation $100 - x^2 = 96$ to solve, from which he deduced that $x = 2$. Again, it is the choice of expression for the sought-for numbers that is a clue. Had Diophantus let the numbers be x and $20 - x$, he would have come up with the equation $20x - x^2 = 96$, which is not so easy to solve. Thomaidis's first claim is that Diophantus repeatedly chose these auxiliary numbers (the 2 in Book II, Prop. 8 and the 10 in Book I, Prop. 27) to make the problem as easy as possible, and that this deliberate policy amounts to a general approach, if not a systematic method. Note also that the 20 and the 96 are chosen to meet the 'necessary condition', for $10^2 - 96$ is a square (in this case, 4).

Thomaidis's other claim is more straightforward: it is that Diophantus took as a fundamental principle for solving equations the idea that we can always add or subtract the same kinds of number from both sides of an equation. Since this idea was presented in the introductory part of the *Arithmetica* we can readily agree with Thomaidis here. Thomaidis's position, based on these two claims, is that Diophantus had a systematic idea of how to deal with equations, one that worked well each time that the auxiliary numbers could be chosen so as to make the final equation suitably simple.

Was Diophantus an algebraist? The historian of mathematics Carl Boyer argued that it is on account of his notation that Diophantus 'has a good claim to be known as the father of algebra'.[32] Since working with symbols is a major feature of elementary algebra, this leads us to ask how Diophantus wrote his mathematics. He denoted the unknown by a special symbol (in some manuscripts looking like ς, possibly from the last letter of

[31] Heath's version is in F&G 5.D3; Thomaidis's is (Thomaidis 2005, 595–608).

[32] See (Boyer 1989, 204).

arithmos, number), which is omitted when its powers are given. Of these, the square (*dunamis*) and cube (*kubos*) of the unknown are written (in extant manuscripts) as Δ^γ and K^γ, so they are simply word abbreviations; there is no symbolic connection between the unknown ς and its square (Δ^γ) or cube (K^γ), any more than there is a symbolic connection between α (= 1) and ι (= 10) in Greek numerals.[33]

Addition is indicated solely by juxtaposition of terms, but there is a symbol for subtraction like an inverted ψ plausibly derived from the word *leipsos*, meaning 'deficient'. No symbol for multiplication or division appears, nor is there one for equality, which was asserted only verbally. When Diophantus wrote down his solutions to problems he incorporated this rudimentary symbolic notation into otherwise written prose, but he did not set out his calculations line by line as we do.

How substantial is this use of symbols? Certainly, we should not say that Diophantus's work should *not* be classified as algebra, just because it does not contain our notation for the unknown and its powers. It would be unreasonable to demand that the development of algebraic notation follows a neat succession of stages in the way that such distinctions might suggest. Likewise, the presence of particular numbers rather than letters for coefficients (the use of 20 and 96, where we might write a and b) was to remain a feature of mathematical work of this kind until at least the early 17th century, and cannot be held against Diophantus.

On the positive side, Diophantus clearly indicated manipulative methods for solving different types of problems, all of which we would now express as algebraic equations. His emphasis on numerical problems with some element of generality in the methods, gives his work an algebraic feeling. He occasionally resorted to geometrical language, but even in Book VI (where the problems involve right-angled triangles) such language seems to have been used simply as a vehicle for presenting numerical problems, much as it was with the Mesopotamian ones. His apparent lack of interest in the solution of a quadratic equation where the unknown can equal 1 reminds us of the Platonic or Pythagorean view that 1 is not a number but 'the source of numbers'. For all these reasons, most historians of mathematics agree with Boyer and consider Diophantus to be an early 'algebraist'.

More precisely, the problems might be said to belong to arithmetic, while Diophantus's methods are algebraic. More precisely still, it is the rules for manipulating equations that strike us as algebraic, but not Diophantus's abbreviated expressions, since they are not well adapted to algebraic manipulation.

However, there is also some validity in a contrary point of view, that there is no real justification to acclaim him as 'the father of algebra'. This claim is not on the strength of his notation, and it is not even justified by the mathematical content of his work. Rather, we could say, on the basis of the nature of the problems, that Diophantus was primarily a highly competent arithmetician working within the field that we today call number theory.

If we go down that route, there is another peril of anachronism that we must avoid. The most striking departure from earlier works is Diophantus's emphasis on what are called *indeterminate problems*: problems whose solutions are not uniquely determined by the information given, but for which there may be only finitely many solutions in integers. For this reason, the modern branch of number theory concerned with whole-number solutions to indeterminate equations is sometimes called *Diophantine analysis*. However, general solutions to indeterminate problems, indicating all possible answers, are not found in the work of Diophantus, so the modern use of the expression 'Diophantine analysis' to describe the study of such equations (often called 'Diophantine equations') within number

[33] Δ is the first letter of *dunamis* and K is the first letter of *kubos*.

theory can be misleading. This is not only because the term 'analysis' suggests a structured or systematic study, something rather lacking in Diophantus's work, but also because Diophantus contented himself with finding a single numerical solution in each case. This is entirely consistent with the Greek view of number, whereas in modern times the restriction to whole-number or fractional-number solutions must be deliberately made.

We should also remember that Diophantus is an isolated case. Earlier, we quoted Knorr's comment that if Diophantus had not existed it would have been impossible to invent him. Even from this preliminary summary we must ask ourselves what we are trying to do when we introduce terms like 'algebra' and 'algebraist'. There is no reason to avoid them when, as here, they have some plausibility and when we attempt to clarify how we intend to use them. When we can make clear what historical purpose is served in this way — by pointing to a shift in mathematical style and purpose, and by finding instructive similarities and differences — terms like 'algebra', however anachronistic, can play a useful role in the historian's toolkit.

6.4 The commentating tradition

From the 3rd century AD onwards we see the growth of another tradition with a different emphasis. It was pre-eminently the age of the commentator. The way to disseminate, teach, and actually do mathematics became that of writing commentaries, explanations, and expansions upon the work of great mathematicians of the past. This was not something completely new, but was more a shift of emphasis and balance. For even by the time of Aristotle, mathematics had a past: his pupil Eudemus wrote histories of geometry and arithmetic, and the great landmarks of Euclid's *Elements* and Ptolemy's *Almagest* incorporated and consolidated earlier work. The commentating tradition, in which the prime focus was upon preserving, clarifying, and transmitting the mathematics of the past, nevertheless represented a different way of conceiving mathematical and teaching activity, which continued through various cultures right up to the 17th century.

Why did this development take place? To observe merely that no longer were there mathematicians like Hippocrates or Apollonius is a statistical comment rather than an explanation. The answer, if there is one, may lie in the interplay of the mathematical research tradition with the political and social circumstances of the Hellenistic world.

One aspect of Greek mathematics to which we have so far given little thought is how it was taught and transmitted at the time. The fact that all we have are texts and other written sources about Greek mathematics should not lead us to overlook that there was a strong oral and face-to-face component to Greek mathematical communication and development. As we shall see, this was also connected with the place of women in education.

You have seen, in the work of Apollonius and even in parts of Euclid's *Elements*, that products of the high research tradition can seem almost meaningless without an accompanying explanation. The development of strict logical formulation was to ensure the truth-status of results, and not their motivation or understandability.

This is all very well so long as a teaching and research community remains in being, to transmit its values, practices, and understandings to each fresh generation, but conditions in the Roman world stopped this happening. Even the neo-Platonic Academy, of which Proclus became head in the 5th century AD, shared at best a name and distant ideals with what Plato had founded. There is no evidence that the mathematical research tradition of 4th-century BC Athens was sustained and maintained through the intervening centuries. To judge by what has come down to us, it had probably lasted but a century or two at most.

If that is so, then scholars such as Pappus and Proclus would have been faced, just as we are, with documents from the distant past but little by way of accompanying oral explanation. An important activity thus became the writing of commentaries that explain and expand obscure portions, add any new result that had since been found, cite any other document that seemed to cast light on the matter in hand, and so on. However, there is much evidence that the oral tradition continued in other fields, such as philosophy and Christian exegesis, so a decline in mathematical research is difficult to explain.

As we noted earlier, without Pappus's study of what remained of the Library of Alexandria in his day, our knowledge and understanding of Greek mathematics would be greatly impoverished. The later history of the Library is obscure: it seems to have fallen into gradual decay, perhaps from as early as the 1st century BC. It is a frustrating reflection for the historian that all Greek mathematical learning was once collected together in a single place, but Romans, Christians and the ravages of time ensured that by the time of the Moslem sack of Alexandria in AD 646 there was little of the Library left to destroy.

From about the 3rd century AD onwards, the set of beliefs known as *neo-Platonism* became prominent in intellectual circles of Alexandria and elsewhere. As its name suggests, it was thought of as drawing from the truths of Plato's philosophy, but, as with neo-Pythagoreanism, many other things were included. The end result was an idealistic religion of high ethical purity, which inspired many of the best scholars of the age.

One aspect of neo-Platonic beliefs relevant to our concerns is that it was a philosophy of revelation. Truth was revealed by the divine, not reachable by human thought alone, and such revelations had been granted to the wise men of old: Pythagoras, Plato, Nicomachus, etc. So we have another slant on the commentating tradition. Wanting to establish and understand what had been written in the past was not simply curiosity about past mathematics, but could also be important for philosophical or religious belief.

We should mention two other Alexandrian commentators: Proclus, whom we re-visit briefly below, and Theon of Alexandria, who lived in the 4th century (he mentions seeing two eclipses in 364 AD) and on whose edition of Euclid's *Elements* our present knowledge is essentially based. All but one of the extant Greek manuscripts of the *Elements* are copies of the version that Theon prepared for the use of his students at the Museum, of which he is the last attested member. Such a slender thread reminds us of how much must have been lost completely, and how partial is any picture of Greek mathematics that we construct. That Theon is our source is a mixed blessing, as he does not always seem to have understood the *Elements* very well.

Hypatia and the situation of women. If we set aside the tradition that Pythagoras's wife was a mathematician as being too slender to trust, then one of the first women mathematicians of antiquity was Theon's daughter, Hypatia of Alexandria, who was head of the neo-Platonic school in Alexandria.[34] Although none of Hypatia's work survives, she is reported to have written commentaries on Apollonius's *Conics*, Ptolemy's *Almagest*, and Diophantus's *Arithmetic*. Knorr has argued that her editorial hand may also be visible in the extant text of Archimedes' *Measurement of a Circle*. For such a prominent identification with learning and a Greek tradition of scepticism, she was brutally hacked to death by a Christian mob in AD 415.

Hypatia's story prompts us to investigate the general situation of women in classical times. Plainly her chances of learning mathematics were greatly increased by her father being a mathematician. Women intellectuals are seldom encountered in antiquity, and men's opinions differed about women's suitability for such tasks. Plato in his *Republic* argued

[34]Pappus mentions one earlier woman mathematician, Pandrosion, in the *Mathematical Collection*, vol. 3.1.

that women should be admitted to the elite of rulers; although they would remain subordinate to elite men, they could still rule over lesser men. In order to eliminate knowledge of paternity, as part of his programme for the eradication of private property, Plato proposed that women and children be shared. He even went so far as to deny that, childbearing and physical strength aside, there was any significant difference between men and women. By the time he wrote his *Timaeus*, Plato was surely implying a hierarchy when he described the Gods' creation of people in these terms:[35]

> Once more into the cup in which he had previously mingled the soul of the universe he poured the remains of the elements, and mingled them in much the same manner; they were not, however, pure as before, but diluted to the second and third degree. And having made it he divided the whole mixture into souls equal in number to the stars, and assigned each soul to a star; and having there placed them as in a chariot, he showed them the nature of the universe, and declared to them the laws of destiny, according to which their first birth would be one and the same for all ... and to come forth the most religious of animals; and as human nature was of two kinds, the superior race, was of such-and-such a character, and would hereafter be called man. ... He who lived well during his appointed time was to return and dwell in his native star, and there he would have a blessed and congenial existence. But if he failed in attaining this, at the second birth he would pass into a woman, and if, when in that state of being, he did not desist from evil, he would continually be changed into some brute who resembled him in the evil nature which he had acquired.

At the end of his *Timaeus* he spelled out his point even more bluntly:[36]

> Of the men who came into the world, those who were cowards or led unrighteous lives may with reason be supposed to have changed into the nature of women in the second generation.

By the same token, light-minded men are apparently born again as birds.[37] Finally, by the time Plato wrote the *Laws*, his views had shifted to one that was more accepting of Athenian reality.

Similarly, Aristotle was of the opinion that:[38]

> The slave is entirely without the faculty of deliberation; the female indeed possesses it but in a form which lacks authority, and children also possess it, but in an immature form.

Neither of these writers was a policy maker, and it is hard to determine to what extent they were forming opinions or passing on the conventional wisdom of their day. Aristotle can be seen as putting the wisdom of his day into a supposedly cosmologically ordained order. The historian Sarah Pomeroy has written that:[39]

> There is general agreement that politically and legally the condition of women in Classical Athens was one of inferiority,

adding that 'the question of her social status has generated a major controversy'.

---

[35] Plato, *Timaeus*, 41d–e, in (Plato 1970, 218).

[36] Plato, *Timaeus*, 90e, in (Plato 1970, 296).

[37] Plato, *Timaeus*, 91d, in (Plato 1970, 297).

[38] Aristotle, *Politics* Book I, Part XIII, 1260a.

[39] See (Pomeroy 1975, 58).

Opinions range from the view that women were despised and kept in seclusion, via the view that although secluded they were valued and ruled the home, to the view that they were respected and as free as any women before the 20th century. Such a range of attitudes doubtless reflects both the situation of the authors and the women of their day, and the charged role of Athens as the 'birthplace of democracy'. Such rhetoric finds it hard to accommodate the role of slavery and the status of women.

One important function for women was to provide heirs for the principal Athenian families. Early marriage was usual, but so important was procreation that divorce was available for either partner and re-marriage was socially acceptable. Women's dowries were protected by law, but women themselves had the legal status of children and never came of legal age. Education typically excluded women, being aimed chiefly at making citizens (male, by definition) and soldiers. So it seems that Aristotle was offering his explanation for what Athenian law and education made into a social fact: that relationships between men and women were necessarily unequal. The only public realm in which women could exert any influence was that of religion, through the practices of the many overlapping cults.

The situation in Hellenistic times was different. The first Egyptian queen reigned in 270 BC, and the famous Cleopatra, who came to the throne in 51 BC, was the seventh with that name. Women could hold other public offices — there were women magistrates, for example — and women from wealthy families entered public life in various ways. More modest gains were made in legal terms.

The opinions of philosophers shifted too. Epicurus, who lived at the time of Alexander and founded a school of philosophy known as Epicureanism (a form of materialism that emphasised the importance of leading a peaceful life and asserted that pain and fear were clear examples of what is wrong), admitted women to his school, and the first woman philosopher that we know by name (Hipparchia) dates from this time. There were also women poets, of whom Sappho has become the most famous.

Women in the Roman Empire also had these limited opportunities. The daughters of poorer families could attend school, and the wealthy might have private tutors at home. Women could inherit. If they had more than three children they could become legally autonomous, and many upper-class women were educated and became patrons of the arts, while some were doctors. There were women teachers and some women slaves were literate.

Intellectual accomplishment was not taken to be a disadvantage, rather the reverse. In a work now lost, Plutarch recorded that Cornelia, the last wife of Pompey, was charming because she knew geometry and philosophy. And in a reversal of fortunes, another writer said that Roman women had taken to carrying round copies of Plato's *Republic*, because it advocated that wives be held in common; this was taken to be an argument, he said, for their licentiousness. Juvenal, the last of the Roman satirists, who lived at the end of the 1st and the start of the 2nd centuries AD, even provided us with one of the first complaints about women who were found to be too well read and to discourse endlessly at dinner parties about grammar and obscure poets, thereby ensuring that 'one woman will be able to bring succour to the labouring moon!'.[40]

Hypatia therefore appears as one of a number of women intellectuals tolerated in the Roman Empire. The situation in early Christian times was much bleaker: the patriarchal

---

[40]See Juvenal, *Satire* 6, transl. G.G. Ramsay, available at `http://www.tertullian.org/fathers/juvenal-satires-06`. Eclipses were said to be caused by witches, whose unwelcome incantations could be drowned out by very loud noises.

hand of the Church and its hostility to learning joined to make women almost invisible until the time of Hildegard of Bingen in the 12th century. It is therefore no surprise that named women mathematicians about whom something is known reappear only in 17th-century Europe, although historians are now patiently discovering a steady number of women known in the intervening period for their intellectual abilities.

The end of the neo-Platonic movement. The grisly murder of Hypatia effectively brought the neo-Platonic movement in Alexandria to an end. The last great neo-Platonic philosopher in Athens was Proclus in the 5th century AD. Although our interest in him has been primarily for what we can learn from him about the history of Greek mathematics, his own influence on subsequent thought was considerable. In the words of two historians at the beginning of the 20th century:[41]

> The works of Proclus, as the last testament of Hellenism to the church and the middle ages, exerted an incalculable influence on the next thousand years. They not only formed one of the bridges by which the mediaeval thinkers got back to Plato and Aristotle; they determined the scientific method of thirty generations, and they partly created and partly nourished the Christian mysticism of the middle ages.

It is worth noting that references to Renaissance neo-Platonism later in the course mean a set of beliefs deriving ultimately from Plato but as seen through the eyes of Proclus who lived nearly half-way in time between Plato and the Renaissance.

Boethius. We mention one more writer of the period, whose philosophy was neo-Platonic and whose work was of great historical influence: the Roman aristocrat Boethius (see Figure 6.12).

We have said little about the Romans, as their interest in mathematics seems to have been largely confined to elementary practical mathematics. Boethius, however, appears to have set out to translate into Latin, and comment upon, as much Greek learning as he could gain access to. He was imprisoned for political reasons, and then killed in AD 525 before he could put all of his plans into effect, but he wrote an *Arithmetic* based on that of Nicomachus, and a *Music* derived from Nicomachus, Ptolemy, and Euclid.

It was Boethius who, as far as we know, first used the word *quadrivium* for the classification of the mathematical sciences into arithmetic, geometry, astronomy, and music.[42] It is not certain whether he also compiled works on geometry and astronomy, although these were attributed to him in the Middle Ages. That these and other works were given Boethius's name testifies to the fact that his role in transmitting knowledge of the quadrivium to the West caused him to become revered, for a time, as one of the great mathematical giants of the distant past.

By the time of Boethius, the Western Roman Empire had collapsed. Some four years after Boethius's death in AD 529, the Eastern Emperor Justinian ordered the closure of the Academy, as the neo-Platonism it taught was seen as antagonistic to Christianity. The teachers moved east to the court of the Persian King Chosroes. Although they came back to Athens after a few years, the living succession of neo-Platonic learning had been impaired. One of these teachers was the commentator Simplicius, to whom we owe the preservation of the fragment of Eudemus about Hippocrates' work on lunes which you met in Section 3.2.

---

[41] See (Harnack and Mitchell 1929), Vol. 16, p. 219.

[42] See (Boethius 1983, 71, 73) in F&G 2.D5.

Figure 6.12. Boethius, Pythagoras, Plato, and Nicomachus, from a medieval manuscript.

Journey to Byzantium. So we can see the 6th century as the one in which the Greek tradition of learning and commentating on mathematics ceased. As a coda, we sketch briefly how the texts and knowledge of this tradition survived, in part, until the revival of classical learning in Renaissance Europe nearly a millennium later.

In simple terms we can think of there being three streams. The first was the continuing tradition within western Europe, initially in monasteries and later in universities. While the European 'Dark Ages' were by no means as dark as they are sometimes represented, it cannot be claimed that a tradition headed by the works of Boethius transmitted much of mathematical consequence beyond a simplified version of the *Arithmetic* and the music theory of Nicomachus. Second, we have the much richer and more complicated creative transmission via the world of Islam. This spread with astonishing speed and energy after the death of Mohammed in AD 632, so that within thirty years all of the Greek-speaking communities along the southern and eastern shores of the Mediterranean had been incorporated, as well as Mesopotamia and Persia. As we shall see in Chapter 8, the subsequent

Islamic assimilation of peoples and cultures included the translation into Arabic of vast numbers of Greek mathematical, scientific, and philosophical texts.

The third stream of transmission or preservation was in that part of the old Eastern Roman Empire which held off the expansion of Islam for 800 years: the Byzantine civilisation, centred on Constantinople (now Istanbul). Only here, by and large, were original texts preserved through copying in the original language, but what survived in this way was only what was thought worth copying, for pedagogical or other reasons. Many ancient texts, and perhaps whole research traditions or mathematical topics, did not survive this winnowing process generation by generation.

Compared with Islamic scholars, and also with the earlier Hellenistic commentator tradition, Byzantine scholars seem on the whole to have copied faithfully, rather than developing the mathematics or creatively rewriting the texts. This is an advantage from our point of view, while giving the impression of a curiously inanimate style of scholarship. One historian has written, half-sympathetically:[43]

> The dead hand of academicism was laid upon them from the start by their determination to write — and, one suspects, to think — in a dead language, the Attic dialect of the fifth century BC; the weight of their classical inheritance.

However, we should not overlook some Byzantine developments beneficial to later scholarship. One such was the introduction, around AD 800, of spaces between words, and the introduction of upper-case and lower-case letters. Previously, Greek texts had no such differentiation or spacing. Because almost all the surviving Greek texts from Byzantium are written in the new style, we know that they date from the 9th century or later.

The Byzantine transmission of texts was described by Sir Thomas Heath with reference to the works of Archimedes.[44] It emerges that only in 9th-century Constantinople were 'the works' of Archimedes collected together, and from that activity stem the text and knowledge of all subsequent generations. That manuscript found its way to the West (to Sicily) in the 12th century, and today all but one of the existing Archimedean manuscripts are copies or translations of two original manuscripts, both of which are lost. The first of these was used by William of Moerbeke in 1269 when he made his Latin translation of a number of the Archimedean treatises; the original vanished some time after 1311. The second of Moerbeke's sources was copied several times during the Italian Renaissance, but it disappeared during the 16th century, although some of the copies survive.

The exception, as we saw in Chapter 5, was *The Method*, the very important work by Archimedes that was discovered only in 1906. It is possible that other lost texts are still to be found, either in Istanbul or elsewhere. From the various dates given by Heath, we can see that the 15th and 16th centuries were particularly active in the translating, editing, and printing of the works of Archimedes. This process of the gradual spreading of knowledge of Archimedes' works is fairly typical in general outline: a manuscript or two filters through to western Europe in the later Middle Ages, and then the trickle turning into a flood in the half-century leading up to (and after) the eventual fall of Constantinople to Islam in 1453. The excitement among scholars, and the additional knowledge stimulated by the wealth of texts in Byzantine custodianship, were critical factors in what we call the Renaissance.

Conclusion. Here our survey of Greek mathematics, its history, and its transmission, ends. You have seen that the Byzantine civilisation preserved that culture (or those aspects

[43] See (Jenkyns 1983); the Attic dialect was that of Plato and other Athenians of his time.

[44] In (Heath 1921, Vol. 2, 25–27) and F&G 4.B4.

of it that we are interested in) until the 15th century. Later we pick up the story of what happened in the Renaissance and onwards, when we shall see that it was the mathematics from the time of Euclid and Archimedes that the later Renaissance mathematicians devoted their energies to learning, digesting, and generally coming to grips with, and which provided the springboard for the great flourishing of European mathematics in the 17th century.

So we have at least a two-fold interest in classical Greek mathematics: it is of great importance in its own right, as a different style of mathematics from anything that we know of previously, and it was the study of, reaction to, and development of these works that helped to catalyse and stimulate the development of 'modern' mathematics during the Renaissance. But as scholarship advances, we have a third interest: in the mathematicians of the Islamic world, their own mathematics, and their responses to the Greek legacy that they inherited. It is to that story that we turn in Chapter 8, after first looking at two other mathematical cultures that flourished while intellectual life in Europe declined: those of India and China.

6.5 Further reading

Cuomo, S. 2007. *Pappus of Alexandria and the Mathematics of Late Antiquity*, Cambridge University Press.. A detailed analysis of this important mathematical text and an examination of the work's wider cultural setting.

Heath, Sir T.L. 1885. *Diophantus of Alexandria: A Study in the History of Greek Algebra*, Cambridge University Press. Still the only version in English, but more of a reworking than a literal translation.

Pomeroy, S.B. 1975. *Goddesses, Whores, Wives, and Slaves: Women in Classical Antiquity*, Schocken Books. The title doesn't quite say it all, but this is a scholarly, wide-ranging, and thorough examination of its theme.

7

Mathematics in India and China

Introduction

In this chapter and the next we consider three important mathematical cultures, those of India, China, and the Islamic world. We try to find out how and why mathematics was studied at various times in those different places, and to analyse some aspects of the mathematics that was done.

The difficulties facing anyone who tries to investigate these cultures are enormous, and we are fortunate to be able to follow a number of distinguished historians who have worked, or are working today, in these fields. First of all, there are the difficulties raised by geography and chronology. The territorial expanses of India, China, and the Islamic world are vast, and what unity there was varied with time and the rise and fall of empires. Second, there was never a single common language spoken in any of these regions, even though many learned Indian documents were written in Sanskrit, the Arabic language dominated the Islamic world, and Chinese in one form or another was the principal written language in China itself. Worse, the very terms can be misleading. What is meant by 'India', for example, is usually not the country marked out as India on modern maps. It might be larger — and, more importantly, it might be less coherent and composed of several independent, even warring, principalities.

We should rather think of the terms 'India', the 'Islamic world', and 'China' as catch-all terms like 'Europe'. Europe is not the right geographical unit for thinking about mathematics at the time of ancient Greece: the Alexandrian Empire extended well into the east but did not reach into the north of modern Europe. 'Europe' is a better term, and its inhabitants came to think of it as more of a single place, when it became a largely Christian entity that was bordered to the south and east by the territories of Islam. But we must also note how much boundaries shifted, most notably in Spain in the 15th century. That said, European countries fought each other, spoke different languages, came to divide bitterly over the interpretation of the Christian religion, and gave up using their one common tongue, Latin.

Similar complexities affect the three cultures we shall be looking at. In the case of India, there are a rough geographical location, a common learned language (Sanskrit), and some sense of cultural continuity, which we can trace more in the Hindu community than in the Buddhist or Jainist ones. For the Islamic world, which came to include large parts of India in the period from the 7th to 13th centuries AD, the unifying features were Islam and the Arabic language, but different traditions existed, of which the division into Sunni and Shiite (centred on Persia, where the non-Arabic language Persian was spoken and written) is the best known. Islamic culture is not only the religion, it is also a legal system, and this too unified a region that extended from Morocco in the West across North Africa, through Turkey and Afghanistan into India and beyond. At various times different parts of the Islamic world seemed to be dominant — Baghdad, Egypt, and Turkey — but in the era of horse and sail the extent to which a ruler in Baghdad could influence events in Morocco or Spain was slight. China presents the greatest degree of unity, for it became recognisably one country with a ruling court, in what is called the Western Han period (206 BC–AD 8). Thereafter its boundaries did not shift very much, but it was periodically convulsed by dynastic wars and invasions, and these are marked by the succession of names that distinguish the phases of Chinese art and culture, among them the Tang (618–907), Sung (960–1269), and Ming (1368–1644) dynasties.

This complicated and extensive history raises problems for historians, and for us as students. Our sources are few, and their transmission was varied (copies of copies, etc.). Even when studied by those comfortable with the language they can be difficult to read, and they may be entirely silent on why the work was done at all, or why in this particular way. We may at best glimpse matters that are otherwise lost. We may see examples of profound discoveries but be unable to see how they were made, or how (and whether) they were proved to be correct. The temptation to sweep the few things we have into tidy generalisations about Indian, Chinese, or Islamic mathematics can be hard to resist, and we must remember that even when the surviving documents are richly informative much has been lost. This problem is particularly acute when the earliest periods of Indian and Chinese cultures are concerned, when sources are few and especially when great claims are made for the antiquity and profundity of what some writers would have us believe was there.

A further difficulty attends the study of these cultures for those coming from the West. It is hard for us not to see them as alien or exotic, even if we have tried our best not to see them as inferior or hostile. We must remember that our view (as students who cannot read the original sources) is brought to us by historians and writers with a variety of agendas. They include explorers and colonial administrators as well as scholars from several centuries. It is possible to detect several distinct ways in which these cultures have been analysed, and the evidence selected accordingly. Some writers have looked at them to see what they had that 'we' (Europeans and, later, Americans) have, and to see who had it first. This is a very Western-centric approach, in which mathematics is defined, often tacitly, as the mathematics 'we' do, and the other culture scrutinised for what is has in common with ours. Evidence of difference is suppressed.

Other writers emphasise the exotic and the different, whether or not they present it as better or as weird. This can be a genuine effort to see the other culture more on its own terms, but it can pander to stereotypes of that culture that are often pernicious, and it can be blind to similarities in our 'own', Western, culture, past or present. Finally, because much of the writing about these cultures was done originally by Western scholars with their own agendas, Indian, Chinese, and Islamic writers have sometimes fought back too strongly, over-stating the originality of their own cultures. Maoist China was not the only place to make exaggerated claims for its ancient ways.

The greatest problem facing us as students, however, is the neglect of the history of mathematics in these cultures. To be intelligible, sources have to be accessible, collected, catalogued, edited, and perhaps transcribed. Only then can they speak to us, and the sad truth is that for much of the time, and in most of the territories we are studying, this has simply not been the case. Many losses are due to the ravages of time: documents are fragile and easily destroyed. The Indian climate is particularly unsuited to the survival of paper. Other losses are the consequences of wars and fires, some (as in China) the result of deliberate destruction. But not until the 1950s were there any attempts to collect, preserve, catalogue, and make available the documents that do survive, and the results are still patchy. There are undoubtedly major discoveries still to be made in libraries across Asia and elsewhere.

Here we present some arguments and evidence that shed light on the history of mathematics in these cultures. It is entirely refracted through the work of expert historians in those fields, and it draws on only a part of their work. In each case it is possible to make some remarks about what we have looked for and tried to present.

First, we have tried to give some impression of the chronological range of what was done, to give some indications of the earliest surviving evidence and some sense of what it can tell us. We have also tried to reach to the 16th or 17th centuries AD, but we have been less interested in contacts between these cultures and the West, fascinating though those stories are.

Second, we have looked at why mathematics was done in each culture. There are broadly three reasons. Astronomy is one, including computations of the calendar, astrology, and cosmology. Navigation does not seem to have been an issue, but for the Islamic world calculating the direction of Mecca and the time (for the purposes of prayer, as well as agriculture) were vital and difficult tasks. Other reasons were literally more down to earth: land surveying, estimating areas and volumes, taxation and division of estates, and the teaching of numeracy to an elite. Finally, in each culture the mathematicians themselves (whatever the word 'mathematician' might mean) often studied mathematics for its own sake, pursuing questions and methods beyond any practical need.

Then we look at how this mathematics was done, and this will be our main interest. We have selected themes that are rich and interesting in each culture. One is the writing of numbers and number systems. We look at how numbers were written, with special emphasis on zero, fractions, negative numbers, and very large numbers. We consider how numbers were calculated, for example by such means as counting rods (in China), and how rules and methods for working directly with numbers developed (long multiplication and division, root extraction, etc.). Then we look at the study of geometry in each culture, the rise of trigonometry, and at evidence for the use of algebra (another term that we try to define carefully).

Our study of how this mathematics was done leads us to a consideration of whether and why it was accepted. A rough and ready distinction can be made in the history of 'European' mathematics between Euclidean geometry and algebra, according to which geometry was concerned with proof and algebra with calculation, perhaps according to a set of rules that formed an algorithm. Indeed, mathematicians did sometimes turn to Euclidean geometry to give their methods the security of a decent proof, and the shift at the start of the 17th century to regard algebra as itself being capable of proof was a momentous one in the history of 'European' mathematics. We shall see that in India, China, and the Islamic world there was always a greater emphasis on number, calculation, and algorithms than was the case in the West, and this has given rise to a controversy as to the extent to which those cultures possessed a concept of proof.

It is evident that we are plunged into issues of historians' presumptions. Those who assert that mathematics began in Greece and have proof at its core are likely to think that other cultures must have lacked a proper concept of proof, and to that extent be deficient. Those looking for the exotic may agree that proof was lacking, but elevate some other virtue instead. Yet other historians may believe that some concepts, such as science or mathematics, are not at all culturally dependent, while allowing for any amount of regional variation: they may expect to find proofs in the mathematics of India, the Islamic world, and China, but not Euclidean-style geometrical proofs. So we have to ask what we take proof to be, and this leads to an interesting debate among historians as to whether, and to what extent, an understanding of algorithms implies the existence of a proof.

Finally, we offer some examples of the mathematics of each of these cultures at its best, different but remarkable achievements that we hope will give an impression of what they could, and did, do. At every stage we have tried to present good original sources in translation.

7.1 Indian mathematics

Much of what we know of early mathematics in the Indian subcontinent survives in the form of the Vedas (the word *veda* means ‘knowledge’), written in Sanskrit at various stages in the first millennium BC. The Vedas gave instructions for religious purposes.

Among the Vedas were the Śulba-sūtras, or cord-rules, that explained how to make shapes of various kinds (altars, fireplaces, etc.): stakes and marked cords were used to make right-angled triangles, squares, and so forth. It is evident that some form of the Pythagorean theorem was known, although not under that name, and there is no reason to suppose a Greek influence, although a Mesopotamian influence can be neither affirmed or denied. The most important branch of the Vedas for the history of mathematics was the part concerned with calendars and astronomy, and we shall see that a number of important discoveries made by Indian mathematicians derive from a flourishing astronomical tradition.

However, our most detailed written sources for mathematics in India are much later, around the time of Āryabhaṭa, who was born in AD 476 and whose *siddhānta* (or treatise), the *Āryabhaṭīya*, survives. He seems to have taken the occasion to write not just on astronomy but on many computational issues, and it is with calculation that our brief survey here is chiefly concerned. His work is also one of the principal sources for the notation for numbers that is in international use today, the decimal place-value system with its essential use for a zero, and we shall discuss that too. The *Āryabhaṭīya* survives partly because it was the subject of numerous commentaries – indeed, these old Sanskrit works are remarkably cryptic and difficult to understand without considerable help, so one supposes that there was an oral tradition explaining them, although little is known about this. We shall refer to one commentary, written by Bhāskara I in AD 629.

We conclude our account with a few words about another famous work, the *Brāhma-sphuṭa-siddhānta* (Corrected Treatise of Brahman), by Brahmagupta, who was born in AD 598. His treatise became one of the Indian astronomical works that taught the Muslim world astronomy. We shall see that it also digresses into some remarkable and entirely mathematical problems that are well worth our attention.

The Indian number system. There were many ways of naming numbers in Sanskrit. One system recorded in the early Vedas is visibly Indo-European and decimal: the numbers begin

eka, dva, tri, catur, panca, sad, sapta, asta, nava, dasa.

Another system records numbers as the result of a calculation: for example, *trisapta* represents 21 (three times *seven*), as in the French *quatre-vingt* for 80.

The system of writing numbers that Āryabhaṭa used, explained, and valued for its simplicity, was the system then coming into use in India: a decimal place-value system with distinct symbols for the numbers from 1 to 9 and a special symbol for an empty place (see Figures 7.1 and 7.2). This Hindu system for writing whole numbers involved symbols that gradually changed as they migrated, one variant ending up as our Western symbols 0, 1, 2, ..., 9. Later Islamic writers extended them to include decimal fractions, thus creating what is now justly known as the *Hindu–Arabic system* of numeration.[1]

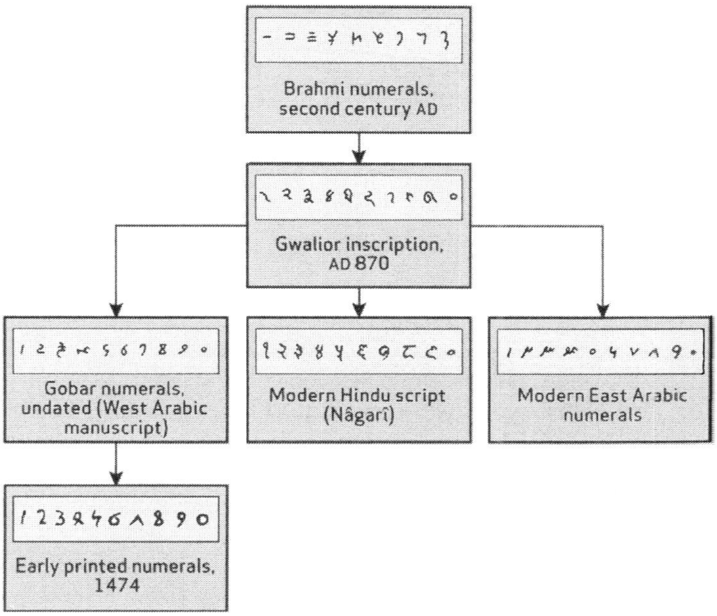

Figure 7.1. The evolution of decimal number systems

The clever idea underlying the decimal, or any, place-value system is that the value of a symbol depends on its place in the numeral: so when we write 99 the two 9s do not stand for 'two of some thing', but for *ninety-nine*. Āryabhaṭa explicitly stated that 'a place should be ten times the previous place', which tells us that his numerals were to be read from right to left, and not from left to right as we do.

This raises the question of what to make of a numeral such as 205 (two hundred and five). It has no tens, but cannot be written as 25, because that would be twenty-five. So a symbol has to go in to indicate that there are no tens. This seems to be the origin of the zero symbol, which acquired currency with the spread of the decimal place-value system and eventually acquired equal rights with the numerals 1, 2, ..., 9 and could stand on its own. However, it should be noted that Ptolemy's table of chords[2] gives the chords and their sixtieths for all the arcs from $\frac{1}{2}^{\circ}$ to 180° and ends with this entry: 120 00 00 000 — Ptolemy surely knew very well that the chord of this arc is simply zero.

[1] Some authors use the phrase 'Indo–Arabic'.

[2] Toomer edition (1984, 57–60).

Figure 7.2. The Bakhshali manuscript in which a symbol for zero (in the form of a dot) occurs numerous times. The manuscript was redated in 2017 to the 3rd or 4th century AD, making it currently the world's oldest recorded origin of the zero symbol that we use today.

Could Indian mathematicians have been influenced by contemporary astronomers? We have already seen that in Mesopotamia it was customary to use a modified decimal system in ordinary life, and the full-blooded sexagesimal system for mathematical and astronomical work. The sexagesimal system for recording astronomical data already allowed for expressions like 205, which could mean (in this notation) 2 + (0 60ths) + (5 3600ths); there are many such entries in Ptolemy's trigonometrical tables, for example. But the sexagesimal system quickly becomes clumsy in everyday life, and one can imagine that a decimal system with the same flexibility would have been adopted, leaving sexagesimals for the experts — where they remain to this day in latitudes, longitudes, minutes, and seconds.

Āryabhaṭa. Āryabhaṭa was one of the earliest Hindu astronomers and mathematicians that we know about. He was born in AD 476, in what is nowadays called Patna in India, and at the age of 23 he wrote his Sanskrit text the *Āryabhaṭīya*. It was much studied by scholars, especially in southern India, and eighteen commentaries on it are known. This helped to ensure the survival of the text, and several modern editions exist, as well as translations into English, French, and Hindi.

The *Āryabhaṭīya* was an astronomical work, and necessarily involved him in a considerable amount of mathematics. As he put it:[3]

> By the grace of Brahma, the sunken jewel of the best of true knowledge has been brought up by me out of the ocean of true and false knowledge by means of the boat of my intellect.

Like many later mathematical works written in Sanskrit, it was written in verses (in this case 121 of them, called *slokas*), which does not make it easy to read. After a lengthy introduction the *Āryabhaṭīya* opens with a section of thirty-three verses, called the *Ganita* (or 'calculation'), that is purely mathematical. Āryabhaṭa showed how to find square roots

[3]See (Āryabhaṭa 1930, 81).

Box 7.1. Fractions. As we have seen, Egyptians used unit fractions, as did Greek mathematicians, except in astronomy, where they used sexagesimal fractions. Sexagesimal fractions were the common property of the astronomers from Mesopotamian times, and are found, for example, in Ptolemy's *Almagest*.

Common fractions were used by, and may be due to, Āryabhaṭa. He wrote

$$\begin{array}{c} a \\ b \\ c \end{array}$$

to denote $a + \frac{b}{c}$. He could use $\frac{b}{c}$ with $b > c$ in the course of a calculation, but would write the final answer as an integer plus a fractional part that was less than 1.

Āryabhaṭa also wrote about the numerator and denominator of a fraction and gave the standard rules for combining two fractions and reducing them to a common denominator. In particular, if he had to find x such that $a : b = x : d$ he would answer that $x = ad/b$.

Decimal fractions are an invention of the Islamic world, specifically of Abu'l-Hasan al-Uqlīdīsi in AD 952, who used a vertical line for what we would call the decimal point. He used them unsystematically, and the first systematic account of them is by al-Samaw'al in 1172. Al-Uqlīdīsi also wrote fractions in the manner of Āryabhaṭa, and in a widespread acknowledgement of the source of decimal arithmetic, including the use of fractions, Arabic and Islamic mathematicians called it Hindu Reckoning.

The refinement that separates the numerator and denominator of a common fraction by a horizontal bar was introduced by Abū Bakr Muhammad ibn 'Abdallāh ibn Ayyash al-Hassar in the 12th century, and in this form they spread to the West — Fibonacci wrote fractions with the numerator and denominator separated by a horizontal bar in his *Liber Abaci* (1202).

and cube roots of numbers, solve some geometrical problems involving areas of plane rectilinear figures and circles, use the Pythagorean theorem, and sum arithmetic progressions of numbers. He also gave general rules for solving quadratic equations, similar to the Mesopotamian process, and introduced a notation for fractions (See Box 7.1).

Āryabhaṭa is also well remembered for his method for solving problems that we solve today with what is called the *Chinese remainder theorem*. This is because the earliest solution to problems of this type is credited to the Chinese mathematician Sunzi (Sun Tzu) of the 3rd century AD, as we shall see.[4]

> Find a number that leaves a remainder of 3 on division by 5, and a remainder of 4 on division by 7.

The smallest such number is 18, and all solutions are of the form '18 plus a multiple of 35 ($= 7 \times 5$)'. He was led to this type of problem by a question in astronomy:

> Given the times that two planets take to complete their orbits, when will they first return to the positions that they presently occupy?

[4]The mathematician of this name is not to be confused with the much earlier Sunzi (or Sun Tzu) who lived some nine centuries earlier and is the author of a famous manual *The Art of War*. No connection between Sunzi and Āryabhaṭa is known or thought likely.

Such problems are not trivial, and here a little modern algebra may help. The general question of this type is:

> Find a number that leaves a remainder of a on division by m, and a remainder of b on division by n.

If the required number is N, then the first statement gives the equation $N = mx + a$, while the second one is $N = ny + b$, for some whole numbers x and y. Subtracting the second equation from the first gives $mx - ny = b - a$, and this is the problem, expressed verbally, that Āryabhaṭa solved.

Note that we have one equation but two unknowns, which is why it is called an *indeterminate equation*. Since the unknowns occur only to the first power, it is an *equation of the first degree* or a *linear equation*. When the unknowns have to be whole numbers then it is the smallest positive solution that is sought.[5]

Āryabhaṭa's method, called the *kuttaka* or *pulveriser*, involves successive division and finding remainders (see Box 7.2); we include it with Bhaskara I's commentary as an example of just how much assistance his work needs!

Āryabhaṭa extended his method to cope with such problems as finding the smallest number that leaves a remainder of 1 on division by 2, 3, 4, 5 and 6, and is exactly divisible by 7. As he put it: 'Say quickly what it should be, mathematician!' We shall not describe how he came to his solution, but leave his ability to find it as an illustration of his insight into such problems.[6]

Geometry. We next look at two other topics that Āryabhaṭa considered. The first is the Pythagorean theorem, which we quote from Bhaskara I's commentary.[7]

> That which is the square of the base and the square of the upright side is the square on the hypotenuse.

This is a perfectly general statement — we have left behind the world of specific values for the sides (such as 3, 4, and 5). Unfortunately for historians, it was not accompanied by a proof. Earlier in his commentary, Bhaskara I had divided a square of side 7 into eight triangles forming four rectangles with sides of 3 and 4 with a 1 by 1 square in the middle, but he did not suggest that the figure stood for a proof of the general theorem; he was merely interested in explaining about squares. So it does not seem possible to decide whether Āryabhaṭa possessed a proof of the general Pythagorean theorem, still less what proof he had, even though he clearly knew that the theorem was correct.

Another geometrical problem that Āryabhaṭa considered concerned the measurement of circles. He gave the area of a circle as half the circumference multiplied by half the diameter. For actual numerical problems, such as the compilation of a sine table preparatory to its use in astronomy, the ratio of the circumference to the diameter is needed, and Āryabhaṭa gave this value for what we call π:[8]

> A hundred increased by four, multiplied by eight, [added to] sixty-two thousands [is] the approximate circumference of a circle with diameter two *ayuta* [that is, 20,000].

---

[5] A similar view of mathematical problems was taken earlier by Diophantus (see Chapter 6), but there is no evidence of any connection, nor any need to suppose one.

[6] The answer is 301.

[7] See (Keller 2006, 83).

[8] Quoted in (Plofker 2009, 128), from the *Āryabhaṭīya*, verses 9–12.

Āryabhaṭa's prolix expression, written in verse, is equivalent to the value of 3.1416 for π. It is remarkably accurate, being very close to the value 3.14159 given by the Chinese geometer Liu Hui, who obtained it by considering the area of a polygon with 3072 sides that he produced from a hexagon by repeatedly bisecting the sides.[9]

Box 7.2. The pulveriser.
Āryabhaṭa's pulveriser is devoted to solving problems such as the following, which he stated explicitly (in verse form):[a]

> [A quantity when divided] by twelve has remainder which is five, and furthermore it is seen by me [having] a remainder which is seven, when divided by thirty-one. What should one such quantity be?

Bhaskara I's explanation of the solution was as follows:

> One should divide the divisor of the greater remainder by the divisor of the smaller remainder. The mutual division [of the previous divisor] by the remainder [is made continuously. The last remainder] having a clever [thought] for multiplier is added to the difference of the [initial] remainders [and divided by the last divisor].
>
> The one above is multiplied by the one below, and increased by the last.
>
> When [the result of this procedure] is divided by the divisor of the smaller remainder, the remainder, having the divisor of the greater remainder for multiplier, and increased by the greater remainder is the [quantity that has such] remainders for two divisors.

Bhaskara's commentary is some help, but it still leaves us well aware of how different their way of expressing their ideas was from ours.

A modern explanation, that captures the sense of dividing and taking remainders until a good guess (the 'clever thought for multiplier') can be made, is as follows. We are asked to find a number N such that

$$N = 12x + 5 = 31y + 7,$$

where x and y are whole numbers. We deduce that:

$$x = (31y + 2)/12 = 2y + w \text{ (say)},$$
$$\text{so} \quad 31y + 2 = 24y + 12w,$$
$$\text{giving} \quad y = (12w - 2)/7 = 1.w + v \text{ (say)}.$$

It follows that

$$12w - 2 = 7w + 7v,$$
$$\text{and so} \quad w = (7v + 2)/5 = 1.v + u \text{ (say)}, \text{ giving } v = (5u - 2)/2.$$

At this point we make a clever guess: $u = 2, v = 4$ is a solution of the last equation, and on working backwards we eventually find that $N = 317$. This is the smallest possible solution; all other solutions differ from 317 by multiples of 372, the least common multiple of 31 and 12.

[a] See (Keller 2006, 128).

[9] We look at Liu Hui's method in Section 7.2.

Because Liu Hui wrote his commentary on the *Nine Chapters on the Mathematical Art* in the 3rd century AD, some earlier historians have suggested that Āryabhaṭa learned his estimate for π from this Chinese source. Modern scholars find no evidence that Āryabhaṭa had direct access to Chinese sources, although they concede that this value for π could originally have come into India from China via some unknown earlier transmission. So even this apparently straightforward topic hints at the possibility of connections between Indian mathematicians and those in China, while, as with the Chinese remainder theorem, there is no compelling reason to believe that there was any contact at all.

Brahmagupta. We do not know when Brahmagupta, who was born in AD 598 in Rajasthan in north-west India, first came into contact with the tradition inspired by the *Āryabhaṭīya*, but he seems to have liked the mathematics and disliked the way that it was applied in astronomy.

His *Brāhma-sphuṭa-siddhānta*, written when he was 30, sought to correct the *Āryabhaṭīya* and other older astronomical works, and it too dealt on occasion with purely mathematical topics. It was one of the Indian astronomical works that taught astronomy to the Islamic world; others may have reached Islamic scholars even earlier, but less is known about them. Both the *Brāhma-sphuṭa-siddhsiddānta* and Ptolemy's *Mathematical Syntaxis* were later translated into Arabic, and this is when Ptolemy's great work acquired its modern name, the *Almagest*.

The Caliph al-Mansūr, who founded Baghdad in AD 762 and reigned there until his death in 775 at the age of 63, saw the need for astronomy in determining the direction of Mecca and managing the lunar calendar in use in the Islamic world. He ordered a number of Indian works to be translated into Arabic, and in this way the Hindu numeral system entered the Islamic world, along with the rules for calculating with them. These rules included the tricky ones involving zero that Brahmagupta had laid down, and which show that Indian mathematicians knew how to calculate with zero — it was no longer only a place holder. Equally important, the rules for calculating with *negative* numbers were also given. They are among the more mathematical parts of Brahmagupta's work, from which we now quote:[10]

Brahmagupta's rules for arithmetic.

> 18.30. [The sum] of two positives is positive, of two negatives negative; of a positive and a negative [the sum] is their difference; if they are equal it is zero. The sum of a negative and zero is negative, [that] of a positive and zero positive, [and that] of two zeros zero.
>
> 18.31. [If] a smaller [positive] is to be subtracted from a larger positive, [the result] is positive; [if] a smaller negative from a larger negative, [the result] is negative; [if] a larger [negative or positive is to be subtracted] from a smaller [positive or negative, the algebraic sign of] their difference is reversed — negative [becomes] positive and positive negative.
>
> 18.32. A negative minus zero is negative, a positive [minus zero] positive; zero [minus zero] is zero. When a positive is to be subtracted from a negative or a negative from a positive, then it is to be added.
>
> 18.33. The product of a negative and a positive is negative, of two negatives positive, and of positives positive; the product of zero and a negative, of zero and a positive, or of two zeros is zero.

[10]Quoted in Plofker, in (Katz 2007, 429), and in (Plofker 2009, 151).

> 18.34. A positive divided by a positive or a negative divided by a negative is positive; a zero divided by a zero is zero; a positive divided by a negative is negative; a negative divided by a positive is [also] negative.
>
> 18.35. A negative or a positive divided by zero has that [zero] as its divisor, or zero divided by a negative or a positive [has that negative or positive as its divisor]. The square of a negative or of a positive is positive; [the square] of zero is zero. That of which [the square] is the square is [its] square-root.

We should not let his misguided remarks about dividing by zero obscure the main point: here for the first time we have a clear record of how zero can be treated as a number, and how to calculate with it.

Elsewhere in his book Brahmagupta wrote about arithmetic and on geometry, but we stay with his Chapter 18, which was much more algebraic, and look at his discussion of how to solve some equations of the form

$$Nx^2 + k = y^2,$$

(in modern notation) for positive integers N and small values of k (such as $\pm 1, \pm 2$, or 4); for example, when $N = 5$ and $k = 1$, the equation is $5x^2 + 1 = y^2$ and the smallest solution in positive integers is $x = 4$ and $y = 9$, since

$$(5 \times 4^2) + 1 = 81 = 9^2.$$

It is easy to exhibit values of N that make this problem impossible to solve by trial and error, so Brahmagupta's success is a formidable achievement. Particularly striking was his success with $N = 61$ and $k = 1$: here the smallest positive solution, which he succeeded in finding, is $x = 226{,}153{,}980$ and $y = 1{,}866{,}319{,}049$. It is clear that he must have had a general method, but all that have come down to us are hints that allow one to take solutions to the problem and produce others.[11]

A great advance by Bhāskara II (not the Bhāskara we met earlier) in his *Bīja-gaṇita* (Seed Computation), written in 1150, was to discover a general method for solving this problem. It is the *chakravala* or cyclic method, and starts with a guess at a solution to a 'nearby' equation. In Box 7.3 we give his general method (which you may find hard to follow), followed by a specific example. In this connection, Plofker has written:[12]

> It is not known how, or whether, Indian mathematicians showed that this method would always provide a solution after a finite number of cycles.

Trigonometry. Much more could be said about Indian mathematics. The historian David Pingree wrote:[13]

> Indian astronomers in the third or fourth century, using a pre-Ptolemaic Greek table of chords, produced tables of sines and versines, from which it was trivial to derive cosines. This new system of trigonometry, produced in India, was transmitted to the Arabs in the late eighth century and by them, in an expanded form, to the Latin West and the Byzantine East in the twelfth century.

Here, the *versine* of an angle α is defined to be $1 - \cos\alpha$.

---

[11] See the discussion in (Plofker 2009, 154–156).

[12] See Plofker in (Katz 2007, 474) and, for more detail on the mathematics, (Plofker 2009, 194–195).

[13] See (Pingree 2003, 45–54).

Box 7.3. Bhāskara II's cyclic method.

Suppose that $x = a, y = b$ is a solution to the problem $Nx^2 + k = y^2$. Certainly $x = 1, y = m$ is a solution of $Nx^2 + (m^2 - N) = y^2$. Bhāskara then used a method derived by Brahmagupta's of combining solutions to deduce that, for any number b,

$$N(am + b)^2 + (m^2 - N)k = (bm + Na)^2,$$

so that

$$N(am + b)^2/k^2 + (m^2 - N)/k = (bm + Na)^2/k^2.$$

This implies that $x = (am + b)/k, y = (bm + Na)/k$, is a solution of

$$Nx^2 + (m^2 - N)/k = y^2.$$

Bhāskara now chose a number m such that $am + b$ is divisible by k. This implies that $m^2 - N$ and $bm + Na$ are also divisible by k, and so

$$x = (am + b)/k, \ y = (bm + Na)/k$$

are integer solutions of the equation $Nx^2 + (m^2 - N)/k = y^2$, where $(m^2 - N)/k$ is also an integer. The insight is to try to make $(m^2 - N)/k$ smaller than k and then hope to repeat this argument until a solution of the equation $Nx^2 + 1 = y^2$ is obtained.

We now give one of Bhāskara's examples. To solve the equation

$$61x^2 + 1 = y^2,$$

he took $a = 1$, $b = 8$, and $k = 3$, and sought a number m for which $(m + 8)/3$ is an integer and $m^2 - 61$ is as small as possible (ignoring sign). So he took $m = 7$ and obtained $x = 5$, $y = 39$, as a solution of the equation $61x^2 - 4 = y^2$. Using Brahmagupta's method he then found $x = 226{,}153{,}980, y = 1{,}866{,}319{,}049$ as the smallest solution of $61x^2 + 1 = y^2$.

In fact, Bhāskara always stopped when $(m^2 - N)/k$ is one of the numbers ± 1, ± 2, or ± 4, for then he could apply Brahmagupta's method to find a solution to the equation $Nx^2 + 1 = y^2$. It seems clear that he knew that his method always worked, although he gave no proof.

In particular, the fundamental trigonometrical function, the sine function, is due to Indian astronomers of the 4th or 5th century AD. The *Surya Siddhānta*, an Indian astronomical handbook of this time, gives in verse form the values of the sine function for every multiple of $3\frac{3}{4}^\circ$ from 0° to 90°: this value $3\frac{3}{4}^\circ$ is obtained from 60° by repeated bisection.

One way that this might have been done — and the text is not clear — is by exploiting the connection between $\sin A$ and versine $2A$, which is

$$\sin^2 A = \tfrac{1}{2}(1 - \cos 2A) = \tfrac{1}{2}(\text{versine } 2A).$$

On this interpretation the sines are found by taking a succession of square roots and versines, starting from the known value of $\sin 30^\circ = \frac{1}{2}$. Rules for finding square roots had been given earlier by Āryabhaṭa, who also computed a sine table for these values, apparently by starting with the smallest sine and working upwards. Intermediate values of the sine function were usually found by linear interpolation, but more accurate methods were also used in later tables.

The sine of an angle is closely related to the corresponding chord, as Figure 7.3 makes clear. The angle $2A$ at O, the centre of a circle of radius r, subtends the chord PQ of length $2a$. The sine of the angle A is a/r, and what Ptolemy computed for the angle $2A$ is the ratio $2a/r$, which is therefore $2 \sin A$ — note that this not what we would call $\sin 2A$.

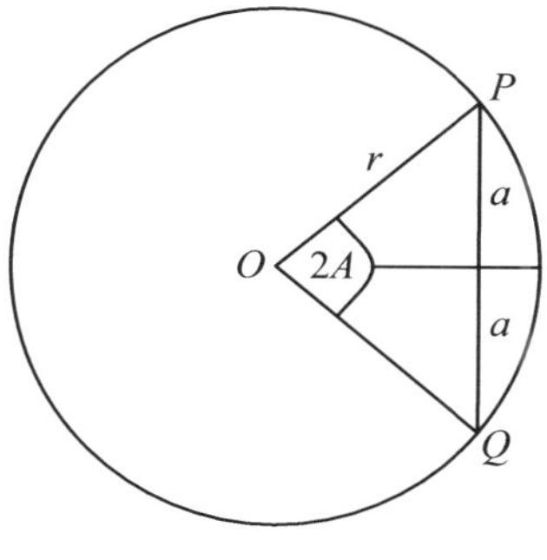

Figure 7.3. Sines and chords

So it is crucial to Pingree's claim that the Indian work, although later than Ptolemy's, was nonetheless independent of it. It is also important to note the sine, cosine, and tangent functions were liberated from geometry and allowed to form equations with each other to each other, as in the formula $\sin^2 A + \cos^2 A = 1$, where A is an angle in a right-angled triangle. This opened the way to a study of them as *functions* — which could in turn assist in the computation of their values. Among the Indian mathematicians who promoted this work in the century after the *Surya Siddhānta* was Āryabhaṭa.

Inevitably a claim of this magnitude has been contested. Those who support it point to the differences between the underlying geometrical models of Indian astronomy and those of Ptolemy's *Almagest*; those who oppose it point to the similarities. Both sides are hindered by the remarkable lack of collections of Indian astronomical data, which would help us to understand how theory and observation had been made to interact. But we also lack knowledge of Hellenistic Greek astronomical methods other than Ptolemy's, which might have been transmitted to the Indians, as well as Mesopotamian ones. There is also the fact that no surviving Indian astronomical text specifically refers to Ptolemy and the *Almagest*, unlike the later practice of Islamic astronomers. Plofker's tentative assessment is that:[14]

> medieval Indian mathematical astronomy was originally largely inspired by the models and parameters of its Hellenistic counterpart, but developed many original methods as well.

Concluding remarks. We conclude this section with a lengthy passage in which Plofker summarises her investigations into the contributions of Indian mathematicians, their roots, and their reception in the Islamic world:[15]

> Indian mathematics, of course, did not develop in entire isolation, but we know little about its early interactions with other mathematical traditions. We have seen that some of its topics were apparently inspired by contact with Babylonian and Greek astronomy and astrology (such as the trigonometry of chords which Indian astronomers recast as sines), and foreign scholars in their turn doubtless encountered and absorbed some Indian discoveries ...

[14] See (Plofker 2009, 120).
[15] See (Plofker 2007, 434–435).

More direct transmissions from India to West Asia followed the rapid expansion of Islam. Parts of northwest India were under Muslim political control by the early eighth century, when Muslim merchants were already present on the Malabar coast in the southwest and in other areas. In the second half of the eighth century, a delegation from India arrived at the new caliphate in Baghdad with a *siddhānta* belonging to the school of Brahmagupta, which scholars of the caliph's court translated or adapted into Arabic. The early translations from Sanskrit inspired several other astronomical/astrological works in Arabic ... They reveal that the emergent Arabic astronomy adopted many Indian parameters, cosmological models, and computational techniques, including the use of sines.

These Indian influences were soon overwhelmed — although it is not completely clear why — by those of the Greek mathematical and astronomical texts that were translated into Arabic under the Abbasid caliphs. Perhaps the greater availability of Greek works in the region, and of practitioners who understood them, favored the adoption of the Greek tradition. Perhaps its prosaic and deductive expositions seemed easier for foreign readers to grasp than elliptic Sanskrit verse. Whatever the reasons, Sanskrit-inspired astronomy was soon mostly eclipsed by or merged with the "Graeco–Islamic" science founded on Hellenistic treatises.

Decimal place-value arithmetic, however, retained its status as a crucial mathematical tool, as well as its name 'Indian computation'. Arabic algebra too may have been influenced by Indian techniques for calculation with unknowns, although the links are far from clear, for example, we would expect to see in an Indian-inspired algebraic tradition an explicit discussion (such as Brahmagupta provided) of the relations between positive and negative quantities, but Islamic algebra generally does not admit negative numbers.

In any case, Muslim mathematicians soon came to identify their subject strongly with the methodology and philosophy of Greek mathematics, and presumably grew less receptive to different approaches.

7.2 Chinese mathematics

It would be an interesting exercise to write a history of mathematics that started with China. If we knew only what Chinese mathematicians of two millennia and more had done, and followed only later Chinese developments up to some date like AD 500, and if we did not know what Western mathematics was like or modern mathematics has become, we would form a certain impression of how advanced mathematics got started: what were its main concerns, its methods, its interest in computation, in proof, in utility, and in mathematics 'for its own sake'. We might then become informed about Greek mathematics of the period, and find it exotic, and the surviving sources only too full of holes. Above all, we would not see the same modern concerns in embryo in the early material — although we might see others — and we could not claim a cultural continuity from the earliest times to today.

This hypothetical exercise invites us to approach Chinese mathematics without expecting it to be 'Greek', and it might suggest that we should be less confident in some of our assertions about the history of Greek mathematics. Each culture tells us that the other one did not need to evolve as it did. Let us see what the sources permit us to say about the origins and development of mathematics in a culture that was not influenced by the Greek example.[16]

The earliest examples of written numerals in China are known to us from inscriptions on oracle bones (happily, bones survive being buried quite well) dating from the late Shang period, around the 12th and 11th centuries BC. The numbers range from 1 to 30,000 with special signs for 10, 100, 1000 and 10,000 (Figure 7.4). By the Han period (206 BC– AD 220) a decimal system for numbers had been established that survives almost unchanged today.

1	2	3	4	5	6	7	8	9	10
一	二	三	四	五	六	七	八	九	十

10^2	10^3	10^4
百	千	萬

Figure 7.4. Chinese numerals 1 to 10, 100, 1000, 10,000

The Chinese at this time wrote the *rank* of a number after the decimal digit, so 5263 was written as 5 thousands 2 hundreds 6 tens 3. In this system a number like 503 is written as 5 hundreds 3, which is different from 53 (5 tens 3), so there was at this stage no special symbol for zero.

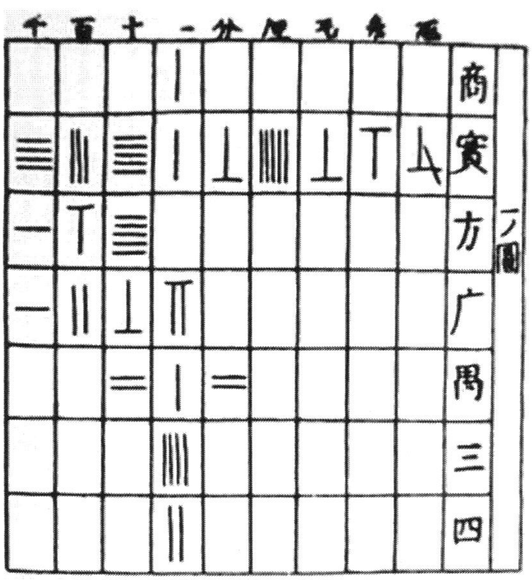

Figure 7.5. A Chinese counting board

The way that computations were done is explained in some surviving texts from a later period, but there is no reason to doubt that they describe an older practice still current

[16]We must be careful: the Chinese state observatory was sometimes staffed by Indian astronomers, who brought with them the Greek division of the circle into 360°, 'although Chinese astronomers continued to favor a circle divided into $365\frac{1}{4}$ degrees', to quote Dauben, in (Katz 2007, 376).

when the texts were written. Computations were done on a counting board (see Figure 7.5), which can be thought of as columns arranged in powers of 10 (ascending from right to left, as we do today). This made it easy to do elementary arithmetic, including the extraction of roots, and ultimately to solve systems of linear equations. Numbers on the board were represented by a simple system of bamboo rods, which could be moved around easily as the computation required. A zero in a column stands for none of the corresponding power of ten, and multiplication by ten was very easy: just move all the rods one column to the left. Each of these numerals appears in two different forms, vertical and horizontal (see Figure 7.6). This is to make the numerals in consecutive columns of the counting board easier to distinguish.

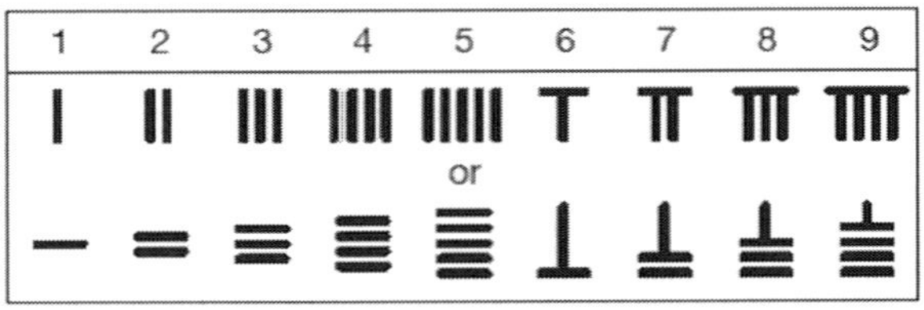

Figure 7.6. How numerals appear on a Chinese counting board.

For a long time the oldest Chinese text known was the *Jiu zhang suan shu* (Nine Chapters on Mathematical Procedures), which was produced around 200 BC; we shall look at it in detail below. But in 1983–1984 Chinese archaeologists discovered an older work, comparable in its importance for the history of mathematics to the much better known tomb of the warriors in Xi'an that dates from 235 BC (and since then a number of other similar sources have been found in other tombs). They excavated a tomb and found a number of books written on bamboo strips (see Figure 7.7), including one on mathematics called the *Suan shu shu* (Book of Numbers and Computations).[17] The tomb can be dated to around 186 BC, but what makes the find so significant is that it is an original source and not a later copy; the *Nine Chapters*, for example, survives only in a copy made in AD 1213.

The *Book of Numbers and Computations* begins by explaining multiplication and how to write and manipulate fractions. The problems in the book are always presented with particular values for the entities involved. The book makes it clear that mathematics is useful for tax collectors, for finding how long it will take groups of people to do a job of work, for computing ratios of different types of grain for the purposes of bartering, and generally for dealing with mixtures. Then it turns from arithmetic to geometry and tackles surveying problems to do with areas and volumes. Unfortunately for historians, but like many a manual to this day, it is short on names and does not tell us who discovered these ideas and when.

A typical arithmetical problem is Problem 12:[18]

> *Fox goes through customs*
> A fox, raccoon, and hound go through customs, and (together) pay tax of 111 *qian*. The hound says to the raccoon, and the raccoon says to the fox: since your fur is worth twice as much as mine, then the tax you pay should be twice as much! How much should each one pay? The result says: the hound pays $15\frac{6}{7}$ *qian*, the raccoon pays $31\frac{5}{7}$ *qian*, and the fox pays 63 and $\frac{3}{7}$ *qian*. The method says: let each

[17]This work has already been given three different titles in Western studies; we follow (Dauben 2008, Appendix I). Figure 7.7 is taken from Dauben, in (Katz 2007, 202).

[18]See (Dauben 2008, 116).

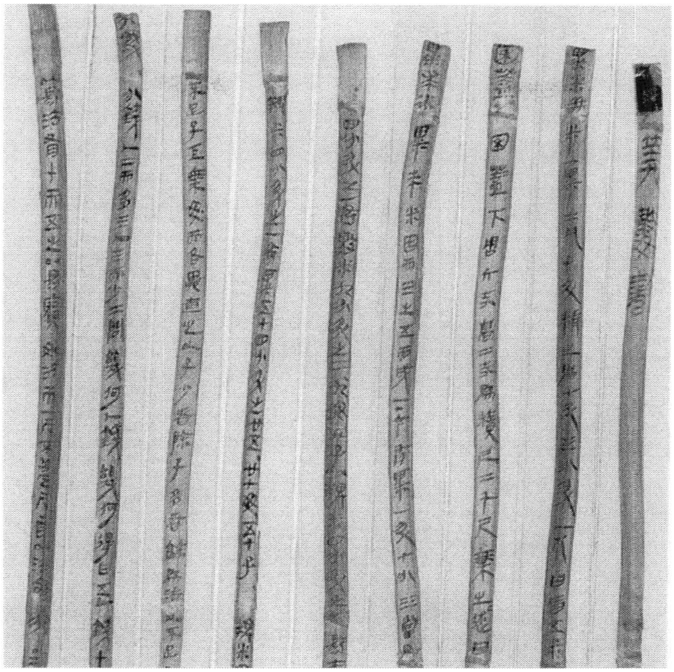

Figure 7.7. Bamboo strips of the *Suan shu shu*

> one double the other; adding them together (1 + 2 + 4), 7 is the divisor; taking the tax, multiplying by each (share) is the dividend; dividing the dividend by the divisor gives each one's (share).

A somewhat more complicated example is Problem 15 about taxes, where the use of a counting board is indicated:[19]

> *Combined taxes*
> (The tax on) 3 (square) *bu* of millet is 1 *dou*; (on) 4 (square) *bu* of wheat is 1 *dou*; (and on) 5 (square) *bu* of small beans is 1 *dou*. If the combined tax (on all of them together) is 1 *shi* (capacity), then how much is the tax (on each one)? The result says: the tax on millet is $4\frac{12}{47}$ *dou*; the tax on wheat is $3\frac{9}{[47]}$ *dou*; (and) the tax on beans is $2\frac{26}{[47]}$ *dou*. The method says: put down (on the counting board the amount of) millet 3 *bu*, wheat 4 *bu*, and beans 5 *bu*; let the product of the millet and wheat be the dividend for the beans; the product of the beans and the millet be the dividend for the wheat; (and the product of the wheat and the beans be the dividend for the millet); for each of the different (amounts) put down (on the counting board) one *shi* multiplied by each (of the amounts for beans, wheat, and millet) as the dividends; (taking) 47 as the divisor gives the result in *dou*.

These problems are typical of mercantile mathematics: one can imagine transactions like this happening every day in the China of the time. Problem 51 (below) is about another distinctively Chinese concern, one step removed from the hurly-burly of the market place

[19] See (Dauben 2008, 119).

and the tax office – simultaneous equations (systems of linear equations with more than one unknown).[20]

> *Dividing coins*
> [When] dividing coins, if each person gets 2 coins, then there are three coins too many; if each person gets 3 coins, then there are 2 coins too few. How many people and how many coins are there? The answer says: 5 people and 13 coins. Cross-multiply the excess and deficiency by the denominators (and combine the products) as the dividend, add the numerators together as the divisor; if there is always an excess or likewise a deficiency, cross-multiply the numerators and denominators and put each down (on the counting board) separately; subtract the smaller numerator from the larger numerator; the remainder is the divisor; use the deficiency as the dividend.

The geometrical problems are much more varied: some are genuinely difficult, in that they would have stretched any mathematician from any culture at the time, and it is not at all clear in every case that the scribe has expressed himself correctly. There are problems that give the correct method for finding the volume of a cone (on the tacit assumption that the circumference is 3 times the diameter) and from this a method for finding the volume of a frustum of a cone is given (Problems 57–59).

Nine Chapters on Mathematical Procedures. We turn now to the most famous book on mathematics from ancient China, the *Nine Chapters on Mathematical Procedures*, commonly known as the *Nine Chapters*. We do not know who wrote it, nor even precisely when it was written; the date of 200 BC is only an estimate. Most likely it survived when other texts did not because it quickly became a 'Canon' (*jing*). This is how Liu Hui, the first commentator whose comments on it have come down to us, referred to it in AD 263, and its canonical status meant that it was to be preserved and explained to successive generations because of its perceived merits. As a result, several commentaries were written about it. Liu Hui's preface mentions two other scholars who at various dates had edited the text: Zhang Cang (ca 250–152 BC) and Geng Shouchang (fl.50 BC).

Tradition selected two of these to be handed down with the Canon: Liu Hui's completed commentary, and the commentary composed by a group under the supervision of Li Chungfeng, and presented to the throne in AD 656. Every surviving edition of the *Nine Chapters* contains both of these commentaries, and historians agree that these commentaries are generally much more interesting than the original text if we want to understand what the texts are saying and begin to understand ancient Chinese mathematics.

The *Nine Chapters* is mainly composed of 246 problems, together with their answers, and algorithms for solving them. As commentators have pointed out, several of the topics discussed are taken from the *Book of Numbers and Computations*.[21] These include the computation of areas, problems about grains, and proportional division. Both books share the algorithmic approach that became fundamental in much of later mathematics in China and was emphasised by many mathematicians. The way in which these algorithms were performed, according to the descriptions of the *Nine Chapters* and to the testimony of later writings, can be seen to have shaped the approach to mathematics, and hence to algorithms themselves, in ancient China.

The contents of each of the nine chapters can be briefly indicated.[22]

---

[20] See (Dauben 2008, 150).
[21] See (Cullen 2004) and (Dauben 2008).
[22] We follow (Chemla 2001).

Chapter 1: Rectangular field. Algorithms are given for the basic computations on fractions, and to compute the areas of the fundamental figures, or 'fields': the rectangle, the triangle, the circle, the annulus, the spherical sector, and so on.
Chapter 2: Millet and husked grains. The rule of three[23] is stated and applied. It is mainly used for computing equivalences between grains, according to official rates enacted by the National Treasury and in problems relating to the payment of taxes in kind.
Chapter 3: Parts weighted according to the degree. An algorithm is given for dividing a given quantity into unequal parts, and applied to the problem of officials with different statuses.
Chapter 4: Small lengths. Algorithms are given for various sorts of division, including division by fractions and the extractions of square roots and cube roots, which are conceived of as divisions, as well as other methods of root extraction.
Chapter 5: Discussing works. Algorithms are given for computing the volumes of basic solids: parallelepipeds and cylinders, trapezoidal prisms, pyramids and truncated pyramids with square bases, tetrahedra, cones, and truncated cones, with a view to the carrying out of civil works.
Chapter 6: Paying taxes in a fair way according to transportation. The title of the chapter comes from its opening problems, which concern a fair distribution of taxes among various administrative units. Problems in this chapter often require combining rules of three.
Chapter 7: Excess and deficit. The rules of double false position are given.
Chapter 8: Measures in square. An algorithm for solving systems of n linear equations with n unknowns is described and progressively extended throughout the chapter.
Chapter 9: Base and height. 'Base' and 'height' are the technical terms for the sides of a right-angled triangle, the subject of this section. An algorithm that corresponds to the Pythagorean theorem opens the chapter. One problem is solved by a quadratic equation.

The *Nine Chapters*, like the *Book of Numbers and Computations*, carries out calculations with specific numbers even when the explanation is entirely general. For example, after a general algorithm for multiplying fractions has been given, a problem in the *Nine Chapters* reads:

> Suppose one has a field which is $\frac{4}{7}$ *bu* wide and $\frac{3}{5}$ *bu* long. One asks how much the field makes.

Specific numbers are used, even when the setting is abstract, as in this problem involving the addition of fractions:

> Suppose that one has $\frac{1}{3}$, $\frac{2}{5}$. One asks how much one gets if one gathers them.

The given algorithm is perfectly general and abstract:

> The denominators multiply the numerators that do not correspond to them; add up; take this as the dividend (*shi*). The denominators being multiplied by one another make the divisor (*fa*) ... If the denominators are equal, one adds them (the numerators) directly with each other.

In our notation, this is

$$\frac{a}{b} + \frac{a'}{b'} = \frac{ab' + a'b}{bb'},$$

or a direct addition when $b = b'$. Note, moreover, that the wording of the algorithm allows it to be applied to any number of fractions added together, not just two.

---

[23]In mathematics, the rule of three allows one to deduce the value of the remaining quantity in a proportion $a : b = c : d$ when three of the quantities are known.

Some of the problems in the *Nine Chapters* are geometrical, such as this procedure for computing the area of a circle:

> Half of the circumference and half of the diameter being multiplied one by the other, one obtains the *bu* of the product.

The *Nine Chapters* is often sophisticated and difficult in its treatment of algorithms. As an example, consider one of the novelties in the book, its treatment of simultaneous equations in Chapter 8. We quote from Liu Hui's commentary on the rule for solving such equations, showing how it is applied to a specific problem. The rule he is explaining is given in italics.[24] It is clear that the method is entirely general. It is remarkable for the use of negative numbers, as you will see.

Liu Hui explained step by step how to set out the coefficients on a counting board and to eliminate them systematically until the entries above the diagonal are zero and the result is obtained.[25] His explanation, as he admitted, is not easy to follow, so we have interpolated a modern commentary based on one provided by the historian Karine Chemla and her co-author.[26]

The first problem of Chapter 8 is about the price of grain of three different qualities, to which Liu Hui attached a general rule:[27]

> Suppose that 3 *bing* of top quality millet, 2 *bing* of medium quality millet, [and] 1 *bing* of low quality millet produce 39 *dou* of grain; 2 *bing* of top quality millet, 3 *bing* of medium quality millet, [and] 1 *bing* of low quality millet, yield 34 *dou.* 1 *bing* of top quality millet, 2 *bing* of medium quality millet, [and] 3 *bing* of low quality millet, yield 26 *dou*. One asks how much is produced respectively by one of high-, medium-, and low quality millet. Answer: Top quality millet yields $9\frac{1}{4}$ *dou* [per bundle]; medium quality millet $4\frac{1}{4}$ *dou*; [and] low quality millet $2\frac{3}{4}$ *dou.*
>
> The Measures in Square Procedure: ...Lay down in the right column three bundles of top quality millet, 2 bundles of medium quality millet, [and] 1 bundle of low quality millet. Yield: 39 *dou* of grain. Similarly for the middle and left columns.

Commentary: Writing x, y, and z for top, medium, and low quality millet, we obtain these equations:

$$3x + 2y + 1z = 39$$
$$2x + 3y + 1z = 34$$
$$1x + 2y + 3z = 26$$

> *Liu Hui*: Things from different groups are put together in wholes that are mixed up. For each type, in a row, there are quantities, and one expresses their products/dividends [that is, their constant term] globally. One does it in such a way that each column is made of *lü*. If there are two things, one measures [each] twice, if there are three things; one measures [each] thrice; all types of things] are measured [as many times as] the number of things. One brings rows together to make columns, this is why one calls this [procedure] 'measures in square'. To the right and to the left of a column, no [other column] exists that is the same. Moreover, one ensures that they are formulated only if there is something on which to ground

[24] In the 19th century the same problem became known in the West as the method of Gaussian elimination.

[25] This problem is also discussed by Dauben in (Katz 2007, 277–279).

[26] See Karine Chemla and Guo Shuchun, *Les Neuf Chapitres*, Dunod, 2004, 616–662.

[27] These translations have been provided by Chemla.

> them. This is a universal procedure. However, with abstract formulations it would be difficult to understand. This is why the [authors] linked it on purpose to millet to eliminate the obstacle. Further, one lays down the middle and the left columns like the right one.

Commentary: The algorithm 'measures in square' (*fangcheng*) is introduced within the framework of the first problem, and in this case amounts to the above system of equations:

$$\begin{aligned} &\text{(I)} \quad 3x + 2y + z = 39 \\ &\text{(II)} \quad 2x + 3y + z = 34 \\ &\text{(III)} \quad x + 2y + 3z = 26 \end{aligned}$$

The first step of the algorithm is to describe an arrangement of all the coefficients in an array on the counting surface as follows:

1	2	3
2	3	2
3	1	1
26	34	39

(Note that the equations are read vertically, and from right to left.)

> *Use [the number of bundles of] top quality millet in the right hand column to multiply the middle column then with [the right column] eliminate vertically the top number of the middle column.*
>
> *Liu Hui*: The intention (*yi*) that [guided) the making of the procedure is that if one subtracts the inferior column from the greater column and if they are subtracted one from the other repeatedly, then the head position must be exhausted first. When, above, a position does not exist [that is, is zero], then in this column, there is also a kind of thing missing. If this is so, since one displayed sets of *lü* to subtract the ones from the others, this does not affect the confrontation [that is, the relationship] between the remaining quantities [*shu*]. If thus one makes the head position disappear, then, below, one has eliminated the production [*shi*] related to one thing. In this way, when one subtracts adjacent columns from one another, if one examines in depth what relates to positive and negative, then one can know the results. The fact that one first makes the multiplication of the middle column by the top quality millet of the right column has the intention of carrying out a 'homogenization-equalization'. To carry out a 'homogenization-equalisation' is to subtract the right column from the middle column, between quantities that are in front of one another. Even though, in conformity with the [norms] of simplicity and ease, one does not express the procedure in terms of homogenization-equalization if one does consider the procedure from the view point of the intention (*yi*) of homogenizing and equalizing, its meaning (*yi*) is actually so.

Commentary: Each equation corresponds to a column, and all the coefficients associated with the same unknown are placed in the same line. The algorithm is described — albeit rather obscurely by Liu Hui — by reference to this table. It first prescribes multiplying the central column (equation II) by the highest coefficient (associated with x) of the right column:

(I)	$3x + 2y + z = 39$	1	6	3
(II)	$6x + 9y + 3z = 102$	2	9	2
(III)	$x + 2y + 3z = 26$	3	3	1
		26	102	39

"Moreover, [with [the amount of] top quality millet in the right column], one also multiplies [the whole of] the column following that one [that is, the left column(s) which follow(s) the middle one, since one goes from right to left] and, with it [the right column], one likewise eliminates".

Liu Hui: Again one eliminates the head of the left column.

Then, with the [amount of] medium-quality millet of the middle column, if it has not vanished, one multiplies the whole of the left column, and with [the middle column], one eliminates vertically.

Commentary: Then, one subtracts the right column from the central one as many times as needed to eliminate the highest position (i.e. the term in x) in the middle column:

(I)	$3x + 2y + z = 39$	1		3
(II)	$5y + z = 24$	2	5	2
(III)	$x + 2y + 3z = 26$	3	1	1
		26	24	39

Liu Hui: One likewise makes the two columns eliminate, with one another, the medium quality millet of one column.

*If the [amount of] low-quality millet on the left hand side has not vanished, the upper [position] is taken as divisor and the lower as dividend. The dividend is thus the dividend [*shi*] of the low-quality millet.*

Commentary: The last two operations are repeated, first between the right and left columns to eliminate the highest position [the term in x] of the left column, then between the central and the left columns to eliminate the highest left position [the term in y] of the left column. At the end of this process, the array of numbers on the table is as follows:

(I)	$3x + 2y + z = 39$			3
(II)	$5y + z = 24$		5	2
(III)	$36z = 99$	36	1	1
		99	24	39

The completion of the algorithm is now easy: the left column (equation III) yields z, then the central column (equation II) yields y, and the right column (equation I) yields x, as Liu Hui went on to explain. In fact,

$$36z = 99 \text{ implies that } z = \tfrac{99}{36} = 2\tfrac{3}{4},$$

so the equation from the middle row becomes

$$5y + 2\tfrac{3}{4} = 24, \text{ which implies that } y = 4\tfrac{1}{4},$$

and so the final equation becomes

$$3x + 9\tfrac{1}{2} + 2\tfrac{3}{4} = 39,$$

giving $x = 9\frac{1}{4}$, as Liu Hui had claimed.

He noted in particular that working with fractions on a counting board is difficult and remarked that

However, if one counts the computations, the calculations with rods are too complex and not economical. The reason for making other methods is to simplify. However, even if it is not worth still using the old method, this develops different methods.

What would happen if the process asked for a number to be subtracted from a smaller one? In such cases, Liu Hui advised that red counting rods be used for gains and black counting rods be used for losses, and called positive and negative respectively, so the rule for measuring in square involves operations on red and black entries. He also carefully spelled out how they are to be combined, as follows. [28]

Like signs subtract

Liu Hui:

This is red subtract red, black subtract black. Subtract one column from another to eliminate the first entry. So if the first entries are of like sign, this rule should be applied. If the first entries are of opposite sign the rule below should be applied.

Opposite signs add

Liu Hui:

Subtracting one column from another depends on appropriate entries with the same signs. Opposite signs [entries] are from different classes. If from different classes, they cannot be merged but are subtracted. So merging red by black is to subtract black, merging black by red is to subtract red. Red and black merge to the original [color], this is "adding"; they are mutually eliminating. This is by eliminating [the top entries] using addition and subtraction to achieve the bottom constant. The prime purpose of the rule is to eliminate the first entry; the magnitudes of entries in other positions are of no concern, either subtract or merge them. The reasoning is the same, not different.

Positive without extra, make negative; negative without extra, make positive

Liu Hui:

"Without extra" is "without merging." When nothing can be subtracted, put the subtrahend in its place [with color changed], subtracting the resulting constant from the bottom constant. This rule is also applicable to columns whose entries are of mixed signs. In the rule, entries with the same signs subtract their constant terms, entries with opposite signs add their constant terms. This is positive without extra, make it negative; negative without extra, make it positive.

Opposite signs subtract; same signs add; positive without extra, make positive; negative without extra, make negative

Liu Hui:

This rule uses "opposite signs subtract" as an example to illustrate how, with the above rule, they complement each other. Positive and negative are used to express that they are opposites to permit operating on these two classes. To say negative does not necessarily mean less, to say positive does not necessarily mean more. Thus interchanging the red and black rods in any column is immaterial. So one can make the first entries to be of opposite sign. These Rules are in fact but one. So applying these two rules, interchange every operation and sign in carrying out the calculation processes from the top entries to the bottom entries. This is the

[28] The next extract comes from Dauben's essay in (Katz 2007, 280–281).

> method. Also subtract to eliminate [the leading entry] of each column. There is no limit on how many [columns] until [there remains] one entry and one constant.

These red and black marks correspond to positive and negative numbers, but they appear only in the working: problems were never set that would produce negative numbers as answers.

The Pythagorean theorem (gou-gu). The ninth chapter of the *Nine Chapters* is devoted to what we would call right-angled triangles (although there was no specific term for 'triangle' in Chinese at that time). Here Liu Hui explained and used the Pythagorean theorem, which was certainly known to Chinese mathematicians independently of the Greeks. As usual, the explanation of the result — its proof — was couched in terms of explicit numbers, but the argument is entirely general. The square on the hypotenuse is placed inside a larger square and 'emptied out' so as to fill the region between the two squares. His explanation is not at all clear, and several later Chinese mathematicians sought to amplify it. It is preceded by a rule that explains how the theorem was to be used.[29]

We label the sides of a right-angled triangle a, b, and c with $a \leq b < c$, so $a^2+b^2 = c^2$. The rule explains that $c^2 - a^2 = b^2$, and this is the form in which it is proved. The initial figure shows a square with side $a + b$ and a square of side c inside it whose four corners lie on the sides of the first square. The figure is dissected and the pieces moved around until they form a square of side $2c$ with a square inside it of side $2a$ bordered all round by a strip of width $c - a$. The proof that this can be done is equivalent to a proof that $c^2 - a^2 = b^2$.

> *Gou-gu rule*: Add the squares of the *gou* [b] and the *gu* [a], take the square root [of the sum] giving the hypotenuse. Further, the square of the *gu* is subtracted from the square on the hypotenuse. The square root of the remainder is the *gou*. Further, the square of the *gou* is subtracted from the square on the hypotenuse. The square root of the remainder is the *gu*.
>
> *Liu*: The shorter side [of the perpendicular sides] is called the *gou*, and the longer side the *gu*. The side opposite to the right angle is called the hypotenuse [*xian*]. The *gou* is shorter than the *gu*. The *gou* is shorter than the hypotenuse. They apply in various problems in terms of rates of proportion. Hence [I] mention them here so as to show the reader their origin. Let the square on the *gou* be red in color, the square on the *gu* be blue. Let the deficit and excess parts be mutually substituted into corresponding positions, the other parts remain unchanged. They are combined to form the square on the hypotenuse. Extract the square root to obtain the hypotenuse.

The diagram that has become famous in discussions about Chinese knowledge of the Pythagorean theorem is called the hypotenuse diagram. The oldest extant copy of it is much later than Liu Hui's account, and is due to the 6th-century commentator Zhen Luan (see Figure 7.8).[30]

Unfortunately, the colours in the second diagram do not match what Liu Hui wrote, but a link between the two figures is supplied by Zhao Shuang's commentary, which was written in the 3rd century AD.[31] First, he stated the Pythagorean theorem.

> The base and the altitude are multiplied by themselves. Add to make the hypotenuse area. Divide this to open the square (i.e. take the square root), and this is the hypotenuse.

---

[29] Dauben, in (Katz 2007, 286). There is another helpful account in (Straffin 1998).

[30] Dauben, in (Katz 2007, 222).

[31] Dauben, in (Katz 2007, 221–222).

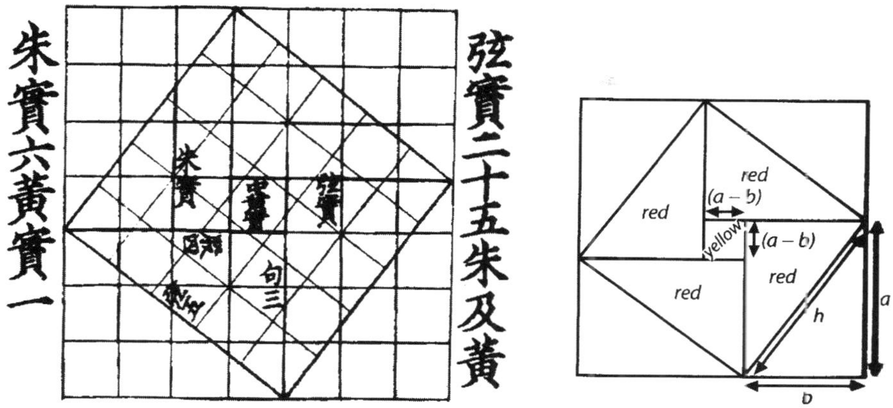

Figure 7.8. A Chinese diagram of the Pythagorean theorem

Then he discussed it, as follows:

> In accordance with the hypotenuse diagram, you may further multiply the base and altitude together to make two of the red areas. Double this to make four of the red areas. Multiply the difference of the base and the altitude by itself to make the central yellow area. If one [such] difference is added [to the four red areas], the hypotenuse area is completed.

In the figure on the right, we have labelled the sides of one of the (red) corner triangles a, b and h. As the lower right corner of the figure shows, two of these triangles fit together to form a rectangle of area ab, so four of these triangles have area $2ab$. The side of the small square in the middle is of length $a - b$. The area of the inner square (the square on the hypotenuse) is made up of this inner square and four red triangles, so the inner square has area $2ab + (a - b)^2 = h^2$.

The area of the big square is $(a + b)^2$, and this is made up of 8 red triangles and the yellow square. So $(a + b)^2 = 4ab + (a - b)^2$. But the large square is made up of a square of side a, a square of side b (in opposite corners), and four red triangles, so $(a + b)^2 = a^2 + b^2 + 2ab$. Therefore

$$4ab + (a - b)^2 = a^2 + b^2 + 2ab$$

and so

$$a^2 + b^2 = 2ab + (a - b)^2 = h^2$$

and the Pythagorean theorem is proved.

Once in possession of the result, Liu Hui and others after him applied it to solve problems that earlier Mesopotamian mathematicians had tackled, such as ladders that have slid down walls. But he also applied it to a problem about the height of a mountain, and to find the radius of a circle inscribed in a right-angled triangle and touching all of its sides. Curiously, given that this account is the last of the nine chapters, he had already used it in the first chapter to find the ratio of the circumference of a circle to its diameter, as we shall shortly see.

The *Nine Chapters* is interesting in a number of ways. It shows how intimately Chinese mathematical culture was sustained by the bureaucracy. It demonstrates a great interest in methods for solving problems involving numbers, methods that we are tempted to call 'algebraic', and it displays these with such confidence that it is hard to deny that

Chinese mathematicians understood them, knew that they worked, and regarded them as proved. This has generated a controversy among present-day historians of mathematics who are accustomed to the proof styles of Euclidean geometry — but why should proofs in 'algebra' resemble proofs in geometry? In Dauben's view:[32]

> Chinese mathematicians were clearly concerned about justifying their methods and establishing the validity of their results. Their proofs were not axiomatic proofs, but they were proofs nevertheless, and they were clearly able to establish the truth of correctness of the solutions they proffered.

The comparison with the Greek situation is striking. There are fewer names in the Chinese tradition, and they do not seem to have been what we take most, if not all, Greek mathematicians to have been: men with a private income and enough leisure to pursue purely intellectual interests at some stage in their lives. The absence in the Mediterranean of an overarching society with its court and bureaucracy makes us wonder whether that is the explanation for the disparity of interest in abstract proof in the two cultures. The attention here to weights, measures, prices, and taxes points us to another interesting difference that might reflect the nature of our sources. We have nothing like this amount of evidence that the Greeks bought and sold on this scale — but surely they did, so why has almost nothing come down to us? Was it because it was never written down, or because it has been lost in transmission?

Approximations to π. In the *Nine Chapters* Liu Hui rejected as too naive and obviously incorrect the idea that the ratio of the circumference of a circle to its diameter is exactly 3. He then set out a method for obtaining successively better estimates, starting with a regular inscribed hexagon as an approximation to the circle and replacing it successively with regular figures having 12, 24, 48, 96, and 192 sides. At each stage he computed the area of the polygon that he obtained, and regarded it as a better approximation to the area of a circle of radius 1 (see Figure 7.9, which comes from the *The Sea Island Mathematical Manual*, Liu Hui's extension to Chapter 9 of the *Nine Chapters*).

Liu Hui worked with the areas of the inscribed polygons. At the first stage the polygon is made up of 6 triangles, at the second stage $6 \times 2 = 12$ triangles, ..., and at the nth stage $6 \times 2^{n-1}$ triangles. So, let S_n be the area of the nth polygon, s_n the length of its side, and a_n the length of the so-called *apothem* of the nth triangle (see Figure 7.10).[33] Then, by the formula 'area $= \frac{1}{2}$ (base $\times$ height)', we have

$$S_n = 6 \times 2^{n-1} \times \tfrac{1}{2} \times s_n \times a_n.$$

The relationship between s_n and a_n is given by the Pythagorean theorem as $(\frac{1}{2}s_n)^2 + a_n^2 = 1$.

At each stage, we obtain the new polygon by bisecting the triangles that make up the previous polygon. Liu Hui related the $(n+1)$th figure to the nth figure by applying the Pythagorean theorem to the small right-angled triangles with hypotenuse s_{n+1} and sides $\frac{1}{2}s_n$ and $1 - a_n$. The procedure is the same at each stage, but numerical values are used, facilitated by the simple Chinese decimal system.[34]

---

[32] Dauben, in (Katz 2007, 377).

[33] The *apothem* is the line segment drawn from the centre to the midpoint of one of the sides of the regular polygon.

[34] The word *gu*, used here to mean a polygon, is the word for an angular sacrificial vase, see Martzloff (1997, 278, n.24). The translation of this extract from the *Nine Chapters* is taken from Dauben's essay in (Katz 2007, 235–236).

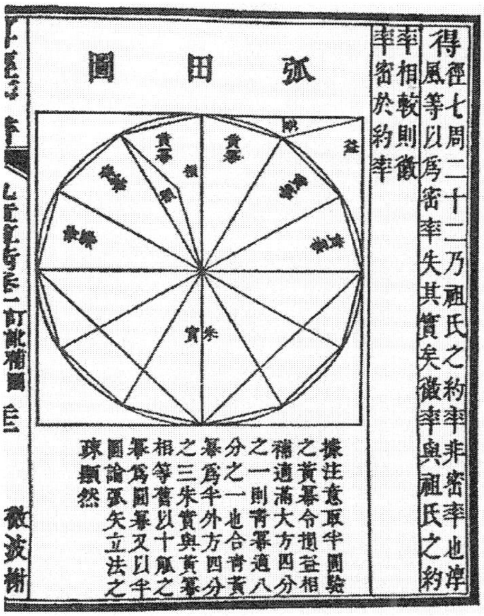

Figure 7.9. Liu Hui's calculation of π

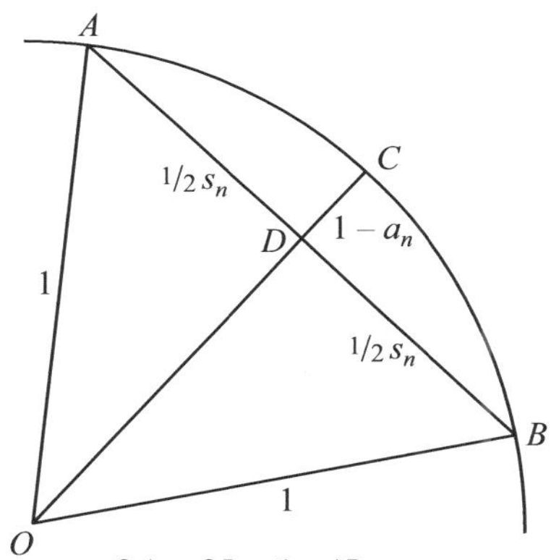

$OA = OB = 1,\ AB = s_n$

apothem $OD = a_n$

co-apothem $DC = 1 - a_n$

Figure 7.10. The division at the nth stage

Liu Hui explained how to find the area of a regular figure with 6, 12, 24, . . . , 192 sides by an iterative process that amounts to bisecting the sides at each stage and calculating the areas of the new triangles in terms of the old ones. His explanation is written out in specific numbers at each stage, but it is the same argument every time. He then asserted that the process can be carried out indefinitely, yielding steadily better approximations to the area of the circle.

> *Rule [for circular fields]*: Multiplying half the circumference by the radius yields the area of the circle in [square] *bu*.

> *Liu*: Taking half the circumference as length and the radius as width; hence multiplying the two gives the area in [square] *bu*. Given a circle of diameter 2 *chi*, consider its inscribed regular hexagon. The rate of the diameter of the circle to its perimeter is 1 to 3.
>
> As in the figure, 3 times the product of the side of the hexagon by the radius yields the area of the inscribed regular dodecagon. Dividing again; then 6 times the product of the side of the dodecagon by the radius yields the area of a 24-gon. The larger the number of sides, the smaller the difference between the area of the circle [and that of its inscribed polygons]. Dividing again and again until it cannot be divided further yields a regular polygon coinciding with the circle, with no portion whatever left out.

Liu Hui did not say whether the process would end after finitely many stages, or whether it could be carried out indefinitely, producing ever-increasing accuracy in its result, but the latter alternative is surely the more likely. He next turned to the details of the construction.

> Outside a regular polygon there are co-apothems. Multiply the sides by the co-apothem giving the area [of the polygon plus these co-apothem rectangles] is larger than that of the circle. As the number of sides increases [without limit] the polygon coincides with the circle and no co-apothem exists, so that [the polygon plus those co-apothem rectangles] is no longer larger than the circle. Take the perimeter [of the inscribed n-gon] to multiply the radius; compare that with the area of the inscribed [$2n$-gon]; it is twice as large. Hence the Rule "Multiplying half the circumference by the radius yields the area of the circle in [square] *bu*." According to this method, the more precise rate is different from a rate of 3 to 1, which is the perimeter of the [inscribed] hexagon [to the radius].

Next Liu Hui criticised earlier writers on the topic, in a way that further illuminates his opinion about how, if ever, the process concludes.

> The difference between a polygon and a circle is just like that between the bow (arc) and its chord, which can never coincide. Yet such a tradition has been passed down from generation to generation and no-one cares to check it. So many scholars followed the tradition that their error has persisted. It is hard to accept without a convincing derivation. Planar forms are either curvilinear or rectilinear. A study of the rates for a square and its inscribed circle may be trivial, but it will lead to far-reaching consequences. By elaborating on this point, one gets wide applications. Let us verify the facts by diagrams for a more precise rate. Assertions without facts are not sound, so notes and comments in detail are given as follows.

Next he described how the calculation of the area of the dodecagon is carried out, and then the 24-gon.[35]

> The rule for calculating the dodecagon from the hexagon: Let the diameter of a circle be 2 *chi*; and 1 *chi* the radius [one side of the hexagon] be the hypotenuse, and half the side, 5 *cun* be the *gou* (base) of a right-angled triangle, now find the *gu* (altitude). From the square on the hypotenuse subtract the square on the *gou*, 25 [square] *cun*, and the remainder 75 [square] *cun*. Extract the square root up to *miao*, or *hu* and one digit down once more to the tiny decimal, which is taken as

[35] The words *cun, fen, li, miao,* and *hu*, stand for successive decimal parts of the appropriate unit: a 'tenth', a 'hundredth', and so on.

numerator with 10 as its denominator, i.e. $\frac{2}{5}$ *hu*. Now the root is the *gu*, 8 *cun* 6 *fen* 6 *li* 2 *miao* and $5\frac{2}{5}$ *hu*, which is 1 *cun* 3 *fen* 3 *li* 9 *hao* 7 *miao* $4\frac{3}{5}$ *hu* less than the radius, the latter number is called the small base; half a side of a hexagon is called a small altitude. From these one gets the [small] hypotenuse. The square on it would be 267, 949, 193, 445 [square] *hu*, where the residue has been neglected. Extracting the square root, get the side of the inscribed dodecagon.

The rule for calculating the 24-gon from the dodecagon: as above, let the radius be the hypotenuse, half a side is the *gou*, now find the *gu*. Quarter of the square on the small hypotenuse is 66,987,298,361 [square]*hu*, where the residue has been neglected. Take it as the square on the *gou*, and subtract it from the square on the hypotenuse. Extract the square root of the remainder, giving a *gu* of 9 *cun* 6 *fen* 5 *li* 9 *hao* 2 *miao* $5\frac{4}{5}$ *hu*. Subtracting it from the radius, the remainder is 3 *fen* 4 *li* 7 *miao* $4\frac{4}{5}$ hu, i.e. the small *gou*. Take half a side [of the dodecagon] as the small *gu*, and one gets the small hypotenuse. The square on the small hypotenuse would be 68,148,349,466 [square] *hu* where the residue has been neglected. Extracting the root, get the side of the [inscribed] 24-gon.

Liu Hui continued through three more iterations of the method, but we have seen enough to understand what he did. His method bears some similarities to the method used earlier by Archimedes, but there are important differences. Liu Hui worked with the area of a circle, Archimedes with its perimeter. Liu Hui took the successive approximations as obviously better and gave no argument to show that they converged; nor did he consider circumscribed polygons but rather the little rectangles on each side that stick out beyond the circle, so there is no hint of the method of exhaustion. On the other hand, Liu Hui was greatly helped by the Chinese method of writing numbers and computing square roots.

He then discussed the accuracy of his result:

The difference of areas $\frac{105}{625}$ [between the 96-gon and the 192-gon] is a small amount. ... take $\frac{36}{625}$ from the difference of areas and add it to the area of a 192-gon. Take the sum as the area of the circle, i.e. $314\frac{4}{25}$ [square] *cun*. The diameter squared yields the area of the [circumscribed] square as 400 [square] *cun*. Reducing it by the area of its inscribed circle, get rates of 5000 to 3927, which means that if the area of a square is 5000, that of its inscribed circle will be 3927; and if the area of a circle is 3927, that of its inscribed square will be 2500. Dividing the area of the circle $314\frac{4}{25}$ [square] *cun* by its radius 1 *chi* and doubling it, we obtain 6 *chi* 2 *cun* $8\frac{8}{25}$ *fen* as the circumference. Reducing the diameter 2 *chi* with the circumference we obtain the sought rate of the circumference to its diameter: 3927 to 1250. Such a rate should be precise enough. However, such a method is still too rough to use. So calculate the side of a 1536-gon for the area of a 3072-gon. With the residue omitted, obtain the same result as estimated, which is proved to be true.

Liu Hui's discussion drew the attention of later commentators. We quote briefly from the commentary composed by Li Chungfeng, where it is hard not to detect the cautious written style of the civil servant.

Li Chungfeng.: In antiquity, the rate was always taken as 3 to 1; with this, however, one would have the circumference too short but the diameter too long, though it is correct for a hexagonal field. Why is this so? Suppose the side of the hexagon is 1 *chi*, its diameter [diagonal] would obviously be 2 *chi*, i.e., with the perimeter rate [being] 6, [or] 1 and 3.

> We would like to give another example to elucidate this point. Make six equal equilateral triangles, with 1-*chi* sides and arrange them round a common point forming a regular hexagon with diagonal 2 *chi*. With the common point as center, and a side as radius, we draw a circle. The circumference passes through all the six corners of the hexagon, and encompasses the six sides, for the distance between two opposite sides is shorter than the diagonal. Here, the perimeter is 6 *chi*, the diameter 2 *chi*, and each side 1 *chi*, so that the distance between two opposite sides is less than 2 *chi*, therefore rates of 3 and 1 would mean a shorter circumference, and a longer diameter. This indicates that rates of 3 and 1 are not precise, but a simple shortcut. Liu Hui considered this as too crude, so sought new rates. But it is hard to determine a number exactly coincident with the product of the circumference and diameter, and neither of the rates Liu gave us seems to be exact. Zu Chongzhi considered them as rough enough to be studied further. In compiling and commenting on the book we have collected all the relevant views, examined which is right and which is wrong, and come to the conclusion that Zu's is the most precise. So we add our humble opinion to Liu's comments for the study of future scholars.

The search for accurate approximations to the value of π may owe more to mathematical curiosity than to any application, but they are nonetheless likely to be found when the mathematically adept have some reason to look for them, and one such reason was calendar reform. The astronomer Zu Chongzhi lived between AD 429 and 500, when there were intense debates about the need for a new calendar. Influential court officials sought to oppose the move arguing that tradition was secure and novelty amounted to blasphemy. Zu Chongzhi supported the reformists and publicly proclaimed that he was, and one should be, 'Willing to hear and to look at proofs in order to examine truth and facts', and that 'Floating worlds and unsubstantiated abuse cannot scare me'. In the end, calendar reform was postponed until after his death, when the calendar he had advocated was accepted, in no small part thanks to the advocacy of his son Zu Geng.[36] Zu Chongzhi's value for the ratio of the circumference of a circle was between 3.1415926 and 3.1415927 (in our decimal notation), which is accurate to six decimal places. Unfortunately, we do not know the method that he used, because his *Method of Interpolation* was reported lost as early as 1127.

The readership of the *Nine Chapters* can be guessed at. Many problems in the *Book of Numbers and Computations* and the *Nine Chapters* read like the sorts of problems that faced the bureaucracy of the Han dynasty generally: paying civil servants, managing granaries and civil works, and carrying out standard grain measurements. Geng Shouchang, mentioned earlier as one of the presumed authors of the *Nine Chapters*, worked for the National Treasury at some stage in his life, and Zhang Cang, another possible author mentioned above, also dealt with accounting and finance at high levels of the bureaucracy. Indeed, several other scholars known in Han times for their mathematical ability worked in similar jobs. It seems clear, therefore, that mathematics developed historically in Han-dynasty China to assist the administration that was in charge of economic matters.[37]

But economics was not the only master: astronomy was another major influence. The 7th-century mathematician Wang Xiaotong, a commentator on the *Nine Chapters*, is typical in interpreting many problems as being about astronomical questions, and sources record that both Zhang Cang and Geng Shouchang worked in astronomy. Moreover, a

[36]See (Li and Du 1987, 80–82).

[37]See (Chemla 1997).

comparison of the *Nine Chapters* with a Han Canon on astronomy and mathematics shows that parts of the *Nine Chapters* that are not to be found in the *Suan shu shu* grew out of procedures used in astronomical activities — for example, the problems dealing with root extraction and the study of right-angled triangles. But astronomy, astrology, and calendrical matters were the other main concern of the bureaucracy.

The Chinese remainder theorem. The problems that go by the name of the Chinese remainder theorem today are typically of the form:

> A number when divided by 3 gives a remainder of 2, when divided by 5 gives a remainder of 1, and when divided by 7 gives a remainder of 6. What is the number?

The smallest positive number satisfying these requirements is the solution: in this case it is 41.

These problems are first discussed in the surviving Chinese literature by Sunzi, who lived in the late 4th and early 5th centuries, and at greater length by Qin Jiushao in the 13th century, who also gave a valid general method for tackling such problems. By then, such problems had cropped up in the Islamic literature — ibn al-Haytham tackled a special case and showed how to solve it completely. They also turn up in the work of Leonardo of Pisa (Fibonacci). It is clear he was stating Sunzi's problem and tackling it in the same way, which suggests that it had travelled a long way without ever being so well understood that authors felt willing to depart from the original script.

Here is Sunzi's problem, from the *Sunzi suan jing* (The Mathematical Classic of Master Sun):[38]

> *Sunzi's rule*
>
> Suppose we have an unknown number of objects. If they are counted in threes, 2 are left, if they are counted in fives, 3 are left and if they are counted in sevens, 2 are left. How many objects are there?
>
> Answer: 23.
>
> Rule: If they are counted in threes, 2 are left: set 140. If they are counted in fives, 3 are left: set 63. If they are counted in sevens, 2 are left: set 30. Take the sum of these [three numbers] to obtain 233. Subtract 210 from this total; this gives the answer.
>
> In general: For each remaining unit when counting in threes, set 70. For each remaining unit when counting in fives, set 21. For each remaining unit when counting in sevens, set 15. If [the sum obtained in this way] is 106 or more, subtract 105 to obtain the answer.

Sunzi's rule would also solve the problem:

> If they are counted in as, r are left, if they are counted in bs, s are left and if they are counted in cs, t are left. How many objects are there?

by the following method. When counting in as, set bcR, where R will be explained shortly. When counting in bs, set caS, and when counting in cs, set abT. Consider the number $bcRr + caSs + abTt$, and subtract the largest multiple of abc that is less than this number, say $nabc$. The solution is $x = bcRr + caSs + abTt - nabc$, and indeed this is correct. Unfortunately, Sunzi gave no explanation for why this works. Let us see why it does, and explain the mysterious numbers R, S, and T.

[38] See (Martzloff 1997, 310).

The number $bcRr + caSs + abTt$ leaves the same remainder as $bcRr$ on dividing by a, because the numbers $caSs$ and $abTt$ are divisible by a. So the number $bcRr$ and the number we want give the same remainder on dividing by a, and for this we find a small number R such that bcR is 1+ (a multiple of a). In Sunzi's problem, with $a = 3$, $b = 5$, and $c = 7$ we find that $bc = 35$, and $35 \times 2 = 70 = 69 + 1$. So Sunzi set $R = 70$ (not 35, as you might have expected). But in his problem, it transpires that $S = 1 = T$. We remark that just because Sunzi did not explain his method does not mean that he did not know that it always works — it might simply be a reflection on what he thought the mathematical culture of his day required from a book.

Eight centuries later, Qin Jiushao did offer a general method, and his algorithm bears striking similarities to ones later found and used in the West. It is a complicated object at this level of generality, and we have chosen not to describe it,[39] but at its core it amounts to an explanation of how the numbers corresponding to our R, S, and T are found in general. There is also an explanation of how to proceed if a division goes exactly, and a reminder that it is necessary to work in units such that all the numbers that appear are integers (because in specific problems a man might sell the Chinese equivalent of twelve pounds four ounces of rice, and so forth).[40]

Later Chinese mathematics. The surviving Chinese mathematical books have come down to us because in 656 Li Chungfeng assembled, the *Suanjing shishu* (The Ten Canons of Mathematics) into a collection and presented it the Tang Emperor Gaozong. Li Chungfeng supervised the writing of commentaries on each canon, and the collected work became the basis for mathematical education in China, providing the topics for examinations. Under the Sung dynasty the Department of the Imperial Library went further and printed the collection in 1084, which made it the first ever printed book on mathematics. By then one canon was already missing (Zu Chongzhi's 5th-century *Method of Interpolation*) and the only canon from the Tang period is there because it replaced another lost canon. In 1213 Bao Huanzhi reprinted the collection, and added to it the *Shushu jiyi* (Memoir on the Methods of Numbering) by Xu Yue, which he considered to be one of the canons used in the Tang mathematics school.

Dauben has surveyed the transition from Han to Tang times in these words:[41]

> When the "Ten Classics" were finally canonized as a standard set of official works in the Tang dynasty, the fate of mathematics was mixed. Sometimes it was encouraged, sometimes not. At certain times it was taught at the Imperial Academy, the Institute of Astronomy, or the Institute of Records. In the third year of the Xian Qing reign (658 CE), having been reestablished only a few years earlier, it was again criticized and abandoned by official decree for reasons that we have already encountered:
>
>> Since mathematics ... leads only to trivial matters and everyone specializes in their own way, it distorts facts and it is therefore decreed that it shall be abolished ... At the beginning of the Tang a lot of the arguments in the *Mathematical Manual of the Five Government Departments* and others in the *Ten Mathematical Classics* were found to be contradictory. Li Chungfeng and others in the Imperial Academy were ordered to write commentaries, after which the Emperor decreed

[39] See (Martzloff 1997, 313–323) and Dauben in (Katz 2007, 310–319).

[40] For the details, see Dauben in (Katz 2007, 314–315).

[41] Dauben in (Katz 2007, 375–376).

> the new edition of the Ten Classics be used throughout the State Institutes (*Records of Official Books* (656 CE), quoted from [Li and Du 1987, 105]).

> Apparently, the collations of Li Chungfeng and his colleagues were not so successful that mathematics was valued with respect to the certainty to which the Greeks ascribed it. But later, in the Tang dynasty, mathematics was again rehabilitated ... Due to concerns about inconsistencies between the various ancient mathematical texts (and doubtless a conviction that there should be no disagreement about mathematical matters), when the Emperor called for an official compendium of classical works to be approved as the official texts for use at the Imperial Academy (the Guo Zi Jian) and as the basis for civil service examinations in mathematics, it was Li Chungfeng (with a retinue of other experts) who collated and annotated the "Ten Books of Mathematical Classics." These that have been surveyed here serve as a grand summary of the achievements of Chinese mathematics over the 1000 years or so spanning the period from the Han to Tang dynasties.

The chief novelty revealed by the collection is the emergence of books devoted to particular domains of mathematics, such as the representation of numbers and solving problems with algebraic equations. Perhaps for this reason, authors' names become more prominent. The use of the counting board was described in greater detail. The *Mathematical Canon by Master Sun*, the *Mathematical Canon by Zhang Qiujian* and others abound in details on how to organise data on the counting surface to make use of the algorithms given. These books are our best source of evidence for how the counting board was used in ancient China. Various computational practices, such as square and cube root extraction, were systematised, but in other areas something seems to have been lost: in the *Mathematical Canon by Master Sun* and the *Mathematical Canon by Zhang Qiujian*, the algorithms for solving systems of simultaneous equations are no longer as subtle and complete as they were in the Han canon.

As we have seen, the first occurrence of the 'Chinese remainder theorem' in a mathematical book appears in the *Mathematical Canon of Master Sun*, as one of a series of problems to do with calendar making. The problem circulated widely — it can be found in several Arabic writings of the period — and it is natural to ask to what extent the mathematics of China found its way to the West, into India, and to the Islamic world.[42] As we have seen, Indian mathematicians used the place-value decimal system, the concepts of fractional arithmetic, and quadratic irrationals. Although it seems impossible to say where this knowledge originated, so much is common that mathematical contacts between China and India must have been important to both civilisations.

There seems also to have been direct contact between China and the Islamic world, by-passing India. The rules of double false position occur in Arabic sources from the 9th century onwards, as they were to do later on in medieval Europe, but they have not been found in Indian sources. In the 12th century the Chinese theory of algebraic equations appears in an Arabic algebraic tradition, but again it has not been found in Indian sources. This suggests that there might have been a tradition of direct exchange of mathematical knowledge throughout Central Asia, even before the founding of the Mongol Empire in 1279.

The later history of mathematics in China is hard to follow. There seems to have been a revival of interest in the subject in the later part of the Sung dynasty in the 13th century,

[42]This is the subject of (Libbrecht 1973).

followed by a decline starting in the 14th century, during which much was forgotten. The *Nine Chapters* was much appreciated during this revival, and commentaries on it suggest that it was better understood than it had been for some time.

The best mathematician of the 13th century was Li Ye, also known as Li Zhi.[43] He gave up his job as a civil servant in northern China when Mongol armies started to make life difficult there, and supported by local officials and friends he devoted himself to the study of various topics, including history, poetry, and mathematics. His book *Measuring the Circle* (1248) was printed around 1282, which helped it to survive.

Measuring the Circle entirely revolves around a new method — the study of polynomials in one indeterminate, represented readily enough with a place-value notation. Polynomials are represented by a column of numbers that list the coefficients of the successive powers (positive as well as negative) of the indeterminate. These coefficients are marked, either positively or negatively, or else set equal to zero. Li Ye developed an algebra for them, showing how they could be multiplied and divided and brought to bear on problems about right-angled triangles. In a sense, his book was a continuation of the *Nine Chapters* and the later work of Wang Xiaotong (see below).

Binomial coefficients were also known to Chinese mathematicians, notably Yang Hui in the 13th century, who reminded his readers that the 11th-century mathematician Jia Xian had already discussed them. Their work was in turn extended by Zhu Shijie, another 13th-century mathematician. We have time for only a small selection of these.

The *binomial coefficients* are the numbers that appear when a two-term expression (a 'binomial') is expanded to a power. For example,

$$(a+b)^4 = a^4 + 4a^3b + 6a^2b^2 + 4ab^3 + b^4,$$

so the numbers $1, 4, 6, 4, 1$ are the binomial coefficients associated with the fourth power. They arise naturally in many problems of an algebraic or a combinatorial kind, and they satisfy a variety of rules, so it is not surprising that many different mathematical cultures have encountered them.

Chinese interest in the binomial coefficients arose in the context of extracting roots. For example, any rule for extracting the fourth root of a number will be based on the expansion of a binomial of the form $(a+b)^4$, where the number a is one approximation to the root and $a+b$ is intended to be a better approximation. In 1261 Yang Hui took up this problem and reminded his readers that the 11th-century mathematician Jia Xian had given a display of binomial coefficients in his *Shi suo suan shu* (The Key to Mathematics). This table was extended to the coefficients of the 8th power in Zhu Shijie's *Precious Mirror of the Four Elements* (see Figure 7.11).[44]

This tabular display does more than list the coefficients: it strongly suggests that the author knew that each number in the table is the sum of the two numbers immediately above it, and this is indeed what Yang Hui says in the next extract, which comes from his book *Xiangjie jiuzhang suanfa* (A Detailed Analysis of the Mathematical Methods in the *Nine Chapters*).[45]

Yang Hui on the binomial expansion.

The original method for the [binomial] expansion. This is found in the arithmetic book *Shi suo* (Unlocking coefficients) and this method has been

[43] He changed 'Zhi' to 'Ye' to avoid confusion with the name of the third Tang emperor: the names are only one stroke apart in Chinese.

[44] Notice that the fourth entry in the penultimate row is incorrect: it says 34 and should agree with the fifth entry, which correctly says 35.

[45] See Dauben, in (Katz 2007, 331–332).

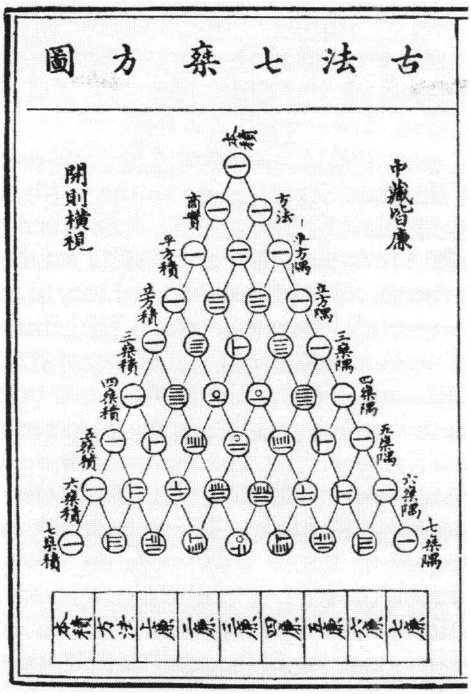

Figure 7.11. Zhu Shijie's table of binomial coefficients

used by Jia Xian. The unit coefficients of the absolute terms are on the left. The unit coefficients of the highest powers are on the right.

$$
\begin{array}{ccccccccccccc}
 & & & & & & 1 & & & & & & \\
 & & & & & 1 & & 1 & & & & & \\
 & & & & 1 & & 2 & & 1 & & & & \\
 & & & 1 & & 3 & & 3 & & 1 & & & \\
 & & 1 & & 4 & & 6 & & 4 & & 1 & & \\
 & 1 & & 5 & & 10 & & 10 & & 5 & & 1 & \\
1 & & 6 & & 15 & & 20 & & 15 & & 6 & & 1
\end{array}
$$

The unit coefficients of the absolute terms lie along the left.

The unit coefficients of the highest powers lie along the right.

The centre contains all the other coefficients.

When the coefficients are multiplied by their respective terms, [the sum] forms the expansion which may then be removed.

The working for finding the coefficients of the binomial expansion: Origin of the method of finding the coefficients by 'unlocking'. Arrange the positions for the powers of the expansion. For instance, in the binomial to the sixth power arrange five positions as shown above and place the unit coefficient of the highest power of the expansion on the outside. Let the coefficients of the highest powers be 1, and add the numbers in the [two] places above in order to find the number in the place below. [Continue in this manner] till the first position is reached. The number obtained for the first position is 6, with the number for the second place above it as 5, the number for the third place above as 4, the number for the fourth place above as 3, and the number for the next place above as 2. Repeat as before with the unit coefficients of the highest powers, adding the [two] numbers above in order to find the number in the position below.

To find the number in the second position:

6 the previous number 5 add 10 and stop 4 add 6 to give 10
3 add 3 to give 6 2 add 1 to give 3.

To find the number in the third position:

6 15 the previous number 10 add 10 and stop 6 add 4 to give 10 3 add 1 to give 4.

To find the number in the fourth position:

6 15 20 the previous number 10 add 5 and stop 4 add 1 to give 5.

To find the number in the fifth position:

6	15	20	15 the previous number	5 add 1 to give 6
first coefficient	*second coefficient*	*third coefficient*	*fourth coefficient*	*fifth coefficient*

Notice that the method is an iterative one: to find the entries in a particular row you have first to find the entries in the previous row. It does not provide a rule or a formula for computing an arbitrary entry in an arbitrary row. This differs from the way that the binomial coefficients are commonly presented today.

Concluding remarks on Chinese mathematics. Our emphasis has been on the utilitarian character of Chinese mathematics, although we have also described some of its major discoveries, especially in the fields of algebra and algorithmic mathematics. So it seems appropriate to end with Dauben's concluding words from his account of Chinese mathematics:[46]

> What is perhaps most compelling about the examples of Chinese mathematics presented here is how similar in tone is the wonder mathematics inspires, not only in the myriad varieties of applications it brings to all aspects of life, but how it also enables the human mind to transcend physical limitations and even measure the immeasurable. As Wang Xiaotong, who criticized Zu Geng's method of interpolation as wrong or incomplete, bragged to the Emperor about his own accomplishments in the treatment of the volumes of irregular solids:
>
>> Your subject [Wang Xiaotong] has now produced new methods that go further ... He progressed from level surfaces and extended that to narrow and sloping objects ... Your subject thought day and night, studying all the Classics, afraid that any day his eyes would close forever and the future not be seen; he progressed from level surfaces and extended that to narrow and sloping objects [yielding] twenty methods in all in the *Classic Continuation of Ancient Mathematics*. Please request

[46]Dauben in (Katz 2007, 378).

> capable mathematicians to examine the worth of the reasoning. Your subject will give a thousand gold coins for each word rejected.[47]

In this sense, the best Chinese mathematicians sought to find ever better applications of what past masters had accomplished. And it was work that could be done with a sense of certainty in its perfection, for Wang Xiaotong must have been confident that he was not about to lose thousands of gold coins for any errors that might be discovered in his results. But the spirit of mathematics was perhaps best summarized by Xu Guangqi in his reflections on the significance of the subject both East and West:

> In truth mathematics can be called the pleasure-garden of the myriad forms, the Erudite Ocean of the Hundred Schools of philosophy.

7.3 Further reading

Chemla, K. 2012. *The History of Mathematical Proof in Ancient Traditions*, Cambridge University Press. A profound analysis of proofs in many different cultures, showing how much was done outside the Greek tradition.

Cullen, C. 2007. *Astronomy and Mathematics in Ancient China: The 'Zhou Bi Suan Jing'*, Cambridge University Press. Provides a lucid account, accessible to non-specialists, of how Chinese astronomers did their work in ancient times, together with a translations of a first century AD Chinese book on astronomy.

Dauben, J.W. 2007. Chinese mathematics, in (Katz 2007, 187–384) is surely the first thing to read for anyone taking up the study of mathematics in China: it is lucid, wide-ranging, and up-to-date.

Li Yan and Dù Shírán. 1987. *Chinese Mathematics: A Concise History*, Clarendon Press, Oxford, is a valuable and thorough survey.

Plofker, K. 2007. Mathematics in India, in (Katz 2007, 385–514) performs the same service as Dauben's essay in Katz's book, but for India. It also reaches much further into the modern period than we have tried to do.

———. 2009. *Mathematics in India*, Princeton University Press. A wide-ranging survey carefully grounded in the sources.

[47]See (Li and Du 1987, 101–102).

8

Mathematics in the Islamic World

Introduction

Intellectual enquiry in many fields flourished for several centuries in the Islamic world. Here we open a window on the important and original contributions of Islamic mathematicians, recognising that we cannot possibly do justice to them here. They did much more than acquire and transmit Greek mathematics: they were original mathematicians with critical and inventive minds, and they had important links to mathematicians and astronomers in India. Islamic mathematical activity is important, both for the transmission of earlier mathematics to the West and because of significant Islamic contributions to both algebra and geometry.

We shall explore two topics: the Arabic and Persian study of quadratic and cubic equations, and the foundations of geometry. The former topic shows how far Arabic and Islamic mathematicians could differ from the Greek approach to mathematics, and the latter illustrates how thoroughly they had mastered it.

8.1 The Islamic intellectual world

The period of Islamic mathematics lasted for some eight hundred years, from about 730. There is much still to be learned about it, because there are obvious difficulties with the sources, which are not always accessible and are not written in languages that most historians of mathematics know, and because of the usual differences in working across cultures. The last twenty to thirty years have seen a revival of interest in the subject, and it is possible here to report on only a fraction of what is known.[1]

We have chosen to refer to this topic as *Islamic mathematics* because it was the rise of Islam (the civilisation that sprang from the religious teachings of Mohammed) that created an empire stretching from the Iberian peninsular, via North Africa, to the Indian sub-continent. The term *Arabic mathematics* is also used by some historians, loosely if

[1]This is most accessibly presented in (Berggren 1986) and in his essay in (Katz 2007).

interchangeably, although strictly the phrase means only 'mathematical works written in Arabic'.[2] In the Islamic world, works were written in Arabic by mathematicians of different countries, not all of them Islamic, and not all of them Arabs, who were able to communicate with each other because of their widely distributed common language. Other works were written by Islamic mathematicians who were Persian and wrote in that language (also sometimes called Farsi).

The subject of Arabic names is a complicated one, and we can do no better than quote an expert, Lennart Berggren.[3]

> Arabic names will be unfamiliar to many readers, so we give a short explanation here of how to read (and remember!) an Arabic name. Thus, a child of a Muslim family will receive a name (called in Arabic *'ism*) like Muḥammad, Ḥusain, Thābit, etc. After this comes the phrase 'son of so-and-so,' and the child will be known as Thābit ibn Qurra (son of Qurra) or Muḥammad ibn Ḥusain (son of Husain). The genealogy can be compounded. For example, Ibrāhīm ibn Sinān ibn Thābit ibn Qurra carries it back to the great-grandfather. Later in life one might have a child and then gain a paternal name (*kunya* in Arabic) such as Abū 'Abdullāh (the father of 'Abdullāh). Next comes a name indicating the tribe or place of origin (in Arabic *nisba*), such as al-Ḥarrānī, 'the man from Ḥarrān.' At the end of the name might come a tag (*laqab* in Arabic), it being a nickname such as 'the goggle-eyed' (al-Jāḥiẓ) or 'the tent-maker' (al-Khayāmī) or a title such as 'the orthodox' (al-Rashīd) or 'the blood-shedder' (al-Saffāḥ). Putting all this together, we find that one of the most famous Muslim writers on mechanical devices ... had the full name Badī' al-Zamān Abū al-'Izz Isma'īl ibn al-Razzāz al-Jazarī. Here the *laqab* Badī' al-Zamān means 'prodigy of the Age,' certainly a title a scientist might wish to earn, and the *nisba* al-Jazarī signifies a person coming from al-Jazarī, the country between the upper reaches of the Tigris and Euphrates Rivers.

The early Islamic Empire or Caliphate. In the early period of the creation of the Islamic empire, after around AD 650, centres of learning were established with libraries for the collection and translation of Greek and Indian works. The most important of these was Baghdad, which was established as the capital by Caliph al-Mansur;[4] Cairo and Damascus were also important. In Baghdad the Abbasid Caliphs encouraged the collection and translation of Greek works.

Initially, the works to be translated were those dealing with local economic and legal systems. It was important for the Islamic conquerors to keep existing trading and governmental institutions functioning effectively, and this required a knowledge of the underlying laws and procedures. This, in turn, led to a desire to comprehend their philosophical basis, and hence to the translation of philosophical manuscripts. Attention turned finally to the scientific and mathematical literature. In addition to this knowledge of the Greek heritage, from about the mid-8th century Islamic mathematicians became aware of earlier Indian astronomical and mathematical works. In the early 8th century Caliph al-Mansur established a royal library at Baghdad, later known as the Bayt al-Hikma or House of Wisdom. It grew to acquire its own observatory, and it was here that some of the most eminent Islamic

[2]This exactly parallels the situation with the phrase 'Greek mathematics', which we discussed in Chapter 3.

[3]Berggren, in (Katz 2007, 520). We have adopted his conventions as far as possible.

[4]*Caliph* is the Arabic term for the leader of the community of Islam; the territory ruled by the Caliph is the *Caliphate*. The Abbasids were an Arab family descended from an uncle of Muhammad; they ruled from 749 to 1258, although their rule was not recognised west of Egypt after 787.

scholars were based and given the opportunity to seek out all that survived of the Greek scholarship.

The change from Greek to Islamic culture (with an incorporation of Indian knowledge) is marked by a change of geographical focus from Alexandria to Baghdad, as well as by the linguistic change from Greek to Arabic. Arabic translations of scientific works were undertaken by scholars of high quality, and were frequently revised. However, as with almost all translations, the change of language resulted in distortions of meaning, particularly since Arabic did not have a comprehensive scientific vocabulary available and one had to be invented. Greek mathematical terms had to be expressed in the nearest appropriate Arabic terms of the day, and this often led to interpretation rather than strict translation.

Thus, despite the intrinsic competence and zeal of many Islamic scholars, the earliest translations were of variable quality, and it was not until the 9th century that Islamic translators really began to penetrate the ideas of Greek mathematicians: they were especially attentive to Greek conceptions of proof. The most important Greek works translated into Arabic were Euclid's *Elements* and *Data*, Archimedes' *On the Sphere and Cylinder* and *On the Measurement of the Circle*, Apollonius's *Conics*, writings of Diophantus and Heron, and those of the neo-Pythagorean Nicomachus.

This Greek heritage inspired the writing of many original Islamic works. As the Islamic empire expanded and developed, there was an increasing need for explicitly Islamic learning. Wealthy patrons encouraged mathematicians to interpret the Greek writings, and also to engage in research that would lead to treatises of their own. We can gather something of this general milieu from the writings of the *Banu Musa*, three wealthy brothers who lived in 9th-century Baghdad and whose name translates as 'the sons of Moses'.[5]

The *Banu Musa*'s book.

Because we have seen that there is a fitting need for the knowledge of the measure of surface figures and of the volume of bodies, and we have seen that there are some things, a knowledge of which is necessary for this field of learning but which — as it appears to us — no one up to our time understands, and [that] there are some things we have pursued because certain of the ancients who lived in the past had sought understanding of them and yet knowledge has not come down to us, nor does any one of those we have examined understand, and [that] there are some things which some of the early savants understood and wrote about in their books but knowledge of which, although coming down to us, is not common in our time — for all these reasons it has seemed to us that we ought to compose a book in which we demonstrate the necessary part of this knowledge that has become evident to us.

And if we consider some of those things which the ancients posed and the knowledge of which has become public among men of our time but which we need for the proof of something we pose in our book, we shall merely call it to mind and it will not be necessary for us in our book to describe it [in detail], since knowledge of it is common; for this reason we seek only a brief statement. On the other hand, if we consider something which the ancients posed and which is not well remembered nor excellently known but the explanation of which we need in our book, then we

[5]This survives in a Latin translation by Gerard of Cremona. This English translation is from (Clagett 1964, 239–241), in F&G 6.A1.

shall put it in our book, relating it to its author. It will be evident from what we shall recount concerning the composition of our book that one who wishes to read and understand it must be well instructed in the books of geometry in common usage among men of our time.

And everything which we have put in our book is our own teaching except the knowledge of finding the measure of the circumference of a circle from its diameter, for that is the work of Archimedes, and except the knowledge of placing two quantities between two quantities so that they are [all] in continued proportion. For although we have posed in our book in regard to the matter [of the two mean proportionals] the method that Menelaus fashioned, we put forth in addition our own method concerning it. And further we posed how to trisect an angle.

And indeed the understanding of all these things we have recounted in our book is of great moment for all those who seek a knowledge of geometry and computation, and the use [of these things] is vital and they are necessary for those who seek this knowledge. For the knowledge of the surface and volume of a sphere which is one of the things [presented here] is properly a part of those things which no one of our time, as far as I have seen, knows how to compute by a method according with the truth [which is] in one who claims to know the demonstration of his method. This book has been completed with the help of God.

It would seem that at this early stage in the Islamic world people were aware that some potentially useful mathematics had been lost and that there were results that no-one understood — presumably no-one knew how to prove them and therefore did not know why they were true. The Banu Musa set out to pull all this knowledge together, concentrating on those parts that were not well known and assuming a familiarity with elementary geometry. They claimed credit for the difficult material, except for a few topics. One was the relationship between the circumference of a circle and its diameter, which they ascribed to Archimedes — evidently they valued the polygon method that Archimedes had provided for evaluating this ratio to any level of accuracy. Another was the method of inserting two mean proportionals between two given magnitudes, which they credited to Menelaus. They also regarded the theory of the area and volume of a sphere as something that no-one in their own mathematical culture knew how to do. The attitudes of these scholars toward ancient Greek mathematics were in part practical — knowledge of areas and volumes would be useful — and in part theoretical — there were things that the 'ancients' (the Greeks) had understood and which needed to be understood again.

We turn now to look at some of the achievements of Islamic mathematicians in more detail.

8.2 Islamic algebra

First we look at what has come to be known as *algebra*. The word 'algebra' is used today in many different ways. As you will see, such use is linguistically anachronistic, inasmuch as it derives from an Arabic word introduced around the 8th century AD. Its core use today may well refer to the school subject, in which equations (such as $ax^2 + bx + c = 0$) are solved by the manipulation of symbols. There is also an extended use of the term, which refers to the manipulation of abstract quantities according to rules. In this spirit, some historians have also spoken of 'Greek geometrical algebra' to refer to a hybrid of algebra and geometry, while others even speak of 'Babylonian algebra' referring to problems we think of as solving linear/quadratic equations.

There is considerable disagreement about the validity of these terms, and we need to think carefully about what we mean when we use the word 'algebra' in an ancient context. A worthwhile concern is to ask what one gains by calling any piece of mathematics 'algebra' (as opposed, say, to 'arithmetic', or 'geometry')? Does the increase in our understanding of the original material outweigh the losses created as we 'translate' it from one kind of mathematical language to another? We shall see that, like Indian writers, Arabic writers went about their work in ways that were recognisably different from those that Greek mathematicians had employed, and the interesting question becomes how we can best capture and characterise this difference.

In fact, for cultural reasons to do with pride in their own achievements, or perhaps for personal reasons, there soon were scholars who claimed to be the equal of the Greeks, if not their outright superiors. Such claims did not please everyone with enough knowledge to have a reasoned opinion, as these critical comments by a 10th-century scholar Aḥmad ibn Muḥammad ibn 'Abd al-Jalīl al-Sijzī make clear:[6]

> I am astonished that anybody who pursues and occupies himself with the art of geometry, even though he acquires it from the excellent Ancients, thinks that there are weakness and shortcomings in them; and especially when he is a beginner and a student, with so little knowledge of it that he imagines that he can achieve with very little effort things, which he believes to be easy to handle and easily understood, although that was far beyond the understanding of those who are trained in this art and skilled in it . . .
>
> What makes it necessary to believe in the weakness of the excellent Archimedes, with his superiority in geometry over the rest of the geometers? He reached such a high level in geometry that the Greeks called him 'the geometer Archimedes'. None among the Ancients nor any of the later geometers were called by his name because of his excellence in geometry. He took great pains to find out useful things. By his power he completed the tools, the instruments and the mechanical procedures. He established the lemmata for the heptagon and followed a path leading to success. By his power we have understood the heptagon just as Heron understood the machines by his [Archimedes'] power and his hard work in mathematical matters.

Al-Sijzī then berated his contemporary Abū al-Jūd for making errors in his attempted construction of a regular heptagon and for then having the gall to claim that his method was simpler than Archimedes' 'ugly' method. By contrast, he continued:

> It is a wonderful achievement, the proof which Archimedes discovered for the lemmata of the heptagon. But he did not write it down in his book to prevent anyone, such as this outcast, who is not worthy of it, from making use of it. When I had acquired instruction from the knowledge of Archimedes, the lemmata of Apollonius, and in particular from my contemporaries such as al-Ala' ibn Sahl, I was also eager [to know] of this noble, abstruse proposition, and the division of the rectilineal angle into three equal parts, which I achieved with very little effort by means of the first treatise of the Book of Apollonius on Conics.
>
> Now I shall describe the affair. I shall quote the words of this person who misleads himself, so that it may serve as an education for beginners. I shall describe the wickedness in his own words and the mistake in what he constructed.

[6]Quoted in (Hogendijk 1984, 305–308), in F&G 6.A2.

> Then I shall follow it with the lemmata of the heptagon. I shall follow that with the construction of the heptagon. I shall finish the book with the division of the rectilineal angle into three equal parts. To God belongs success.

This argument, which led to mutual accusations of plagiarism, shows that, while the importance and excellence of Greek works were coming to be widely recognised, there were nevertheless Islamic writers who were critical of the ancients and less inclined to adopt the Greek heritage. It is interesting, too, that interest in the classical problems (not just the famous three, but also the construction of the heptagon, which yields a cubic equation in modern terms) was strong among the ambitious mathematicians of the Islamic world.

Muḥammad ibn Mūsā al-Khwārizmī. The most important 9th-century Islamic mathematicians was Muḥammad ibn Mūsā al-Khwārizmī, who flourished at the House of Wisdom around the year 825. He compiled astronomical tables and wrote a number of mathematical works, two of which, on arithmetic and algebra, have come down to us.

The *Arithmetic* is of particular significance, because it was through its Latin translations that the Hindu numerals and the arithmetical methods of calculation associated with them became known in the West (see Figure 8.1). Indeed, it has survived only in a Latin translation with the title *De Numero Indorum* (On the Hindu Reckoning). Al-Khwārizmī made no claim to originality in this work, and although it was clearly based on Indian writings, in Europe the techniques became particularly associated with his name in the form of the word *algorism* (from which our modern term 'algorithm' is derived).

The precise title of his work on the solution of equations, or *algebra* as it could be called, was *Al-kitab al-mukhtasar fi hisab al-jabr wa'l-muqabala*, which translates as *The Compendium on Calculation by Completion and Restoration* (al-jabr) *and balancing* (al-muqabala). *Al-jabr* (completion or restoration) refers to adding the same thing to both sides of an equation — that is, restoring all the terms to a standard form; *al-muqabala* (balancing or opposition) refers to subtracting the same term from both sides. The key word of the Arabic title, *al-jabr*, eventually came into common usage in the West as 'algebra', meaning something like 'calculating rules for solving equations'.

The book was dedicated to Caliph al-Ma'-mūn. It is doubtful whether al-Khwārizmī's *Algebra* (as we shall refer to it) was the first Islamic work bearing that name. The existence of several near-contemporary works containing similar material, at least one of which seems to have pre-dated his, suggests a tradition whose origins in the Islamic world remain to be traced.

Al-Khwārizmī's *Algebra*. This work is in three parts: rules for solving equations, practical measuring (areas and volumes), and problems arising mainly from the complicated Islamic laws of inheritance. It opens with a brief account of the decimal place-value principle, and then (in six chapters) deals with the following six 'standard' types of equation, where all the numbers are positive: no negative numbers were allowed:

al-Khwārizmī's description	*modern version*
squares equal to roots	$ax^2 = bx$
squares equal to numbers	$ax^2 = c$
roots equal to numbers	$ax = c$
squares and roots equal to numbers	$ax^2 + bx = c$
squares and numbers equal to roots	$ax^2 + c = bx$
roots and numbers equal to squares	$bx + c = ax^2$

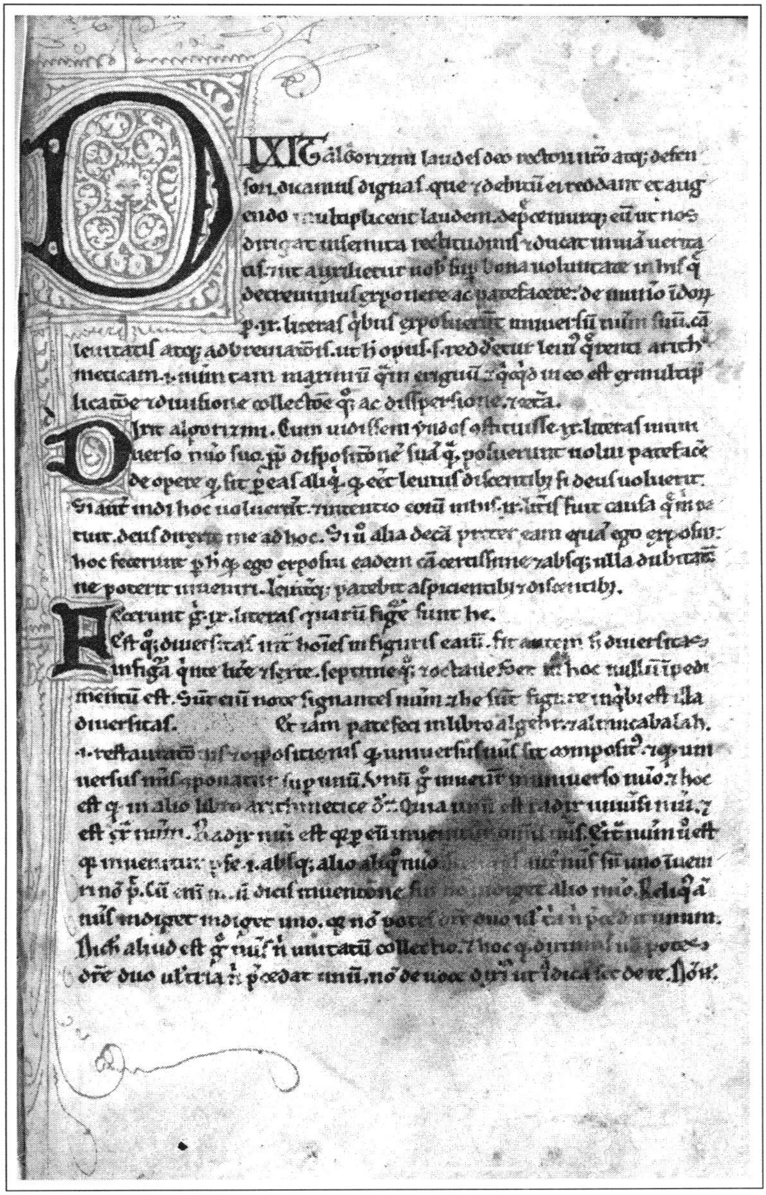

Figure 8.1. A 13th-century manuscript of al-Khwārizmī's *Arithmetic*.

In the first three of these chapters, al-Khwārizmī showed how to solve particular examples in the different cases where the coefficient a is unity, a fraction, and a whole number greater than 1, and he carefully indicated how quite complicated equations can be simplified into one of these standard forms.

The role of geometry in this work is interesting. Having defined the terms 'number' as the simple numbers that enter the problem (the 'coefficients' in a modern formulation), 'root' as the unknown in a problem, and 'square' as the square of the unknown – to which he gave the Arabic name *mal* meaning wealth — he went on to say:[7]

> A number belonging to one of these three classes may be equal to a number of another class. You may say, for instance, 'squares are equal to roots' or 'squares are equal to numbers' or 'roots are equal to numbers'.

This is quite unlike classical Greek procedures in which areas could only be compared with areas, lengths with lengths, and so on, and invites comparison with the earlier Mesopotamian formulation of similar problems, studied in Chapter 2.

We can learn more by following one of his examples in greater detail. He introduced his methods of solving equations produced by means of specific numerical examples. Our example is of 'roots and squares equal to numbers'.[8]

> for instance, 'one square, and ten roots of the same, amount to thirty-nine dirhems'; that is to say, what must be the square which, when increased by ten of its own roots, amounts to thirty-nine? The solution is this: you halve the number of the roots, which in the present instance yields five. This you multiply by itself; the product is twenty-five. Add this to thirty-nine; the sum is sixty-four. Now take the root of this, which is eight, and subtract from it half the number of the roots, which is five; the remainder is three. This is the root of the square which you sought for; the square itself is nine.

This example, which we would write as $x^2 + 10x = 39$, appears again and again in later Islamic works and in early Western treatises on algebra.[9] There are marked similarities with the Mesopotamian approach to such questions – indeed, the procedure itself seems as close as it can be, given that the Mesopotamian problem is not of al-Khwārizmī's form here (but is 'squares equal to roots and numbers'). They come out when we follow al-Khwārizmī's explanation of the steps to be followed in solving these equations. As you will see in the next extract, which expands on the one above, he described in terms of a particular numerical example with the clear implication that any other example of this form is to be treated analogously (see Figure 8.2).[10]

Al-Khwārizmī's completion of a square.

> We have said enough so far as numbers are concerned, about the six types of equations. Now, however, it is necessary that we should demonstrate geometrically the truth of the same problems which we have explained in numbers. Therefore our first proposition is this, that a square and 10 roots equal 39 units.
>
> The proof is that we construct a square of unknown sides, and let this square figure represent the square (second power of the unknown) which together with its root you wish to find. Let the square, then, be *ab*, of which any side represents one root. When we multiply any side of this by a number (or numbers) it is evident that that which results from the multiplication will be a number of roots equal to the root of the same number (of the square). Since then ten roots were proposed with the

---

[7]See (Al-Khwārizmī, 1831, 5–6), in F&G 6.B1(a).

[8]See (Al-Khwārizmī 1831, 8–11), in F&G 6.B1(a).

[9]The historian of mathematics L. C. Karpinski (1915, 19) said that this particular equation 'runs like a thread of gold through the algebras for several centuries'.

[10]Al-Khwārizmī (1930, 77–81) in F&G 6.B1(b).

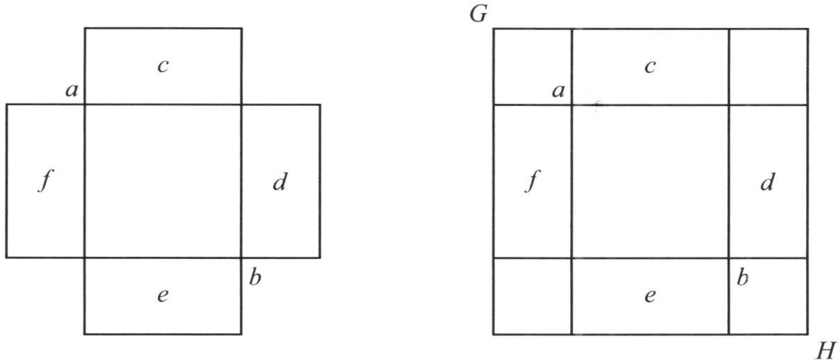

Figure 8.2. al-Khwārizmī's completion of a square

square, we take a fourth part of the number ten and apply to each side of the square an area of equidistant sides, of which the length should be the same as the length of the square first described and the breadth $2\frac{1}{2}$, which is a fourth part of 10. Therefore four areas of equidistant sides are applied to the first square, *ab*. Of each of these the length is the length of one root of the square *ab* and also the breadth of each is $2\frac{1}{2}$, as we have just said. These now are the areas *c, d, e, f.* Therefore it follows from what we have said that there will be four areas having sides of unequal length, which also are regarded as unknown. The size of the areas in each of the four corners, which is found by multiplying $2\frac{1}{2}$ by $2\frac{1}{2}$, completes that which is lacking in the larger or whole area. Whence it is [that] we complete the drawing of the larger area by the addition of the four products, each $2\frac{1}{2}$ by $2\frac{1}{2}$; the whole of this multiplication gives 25.

And now it is evident that the first square figure, which represents the square of the unknown $[x^2]$, and the four surrounding areas $[10x]$ make 39. When we add 25 to this, that is, the four smaller squares which indeed are placed at the four angles of the square *ab*, the drawing of the larger square, called *GH*, is completed. Whence also the sum total of this is 64, of which 8 is the root, and by this is designated one side of the completed figure. Therefore when we subtract from 8 twice the fourth part of 10, which is placed at the extremities of the larger square *GH*, there will remain but 3. Five being subtracted from 8, 3 necessarily remains, which is equal to one side of the first square *ab*.

This 3 then expresses one root of the square figure, that is, one root of the proposed square of the unknown, and 9 the square itself. Hence we take half of 10 and multiply this by itself. We then add the whole product of the multiplication to 39, that the drawing of the larger square *GH* may be completed; for the lack of the four corners rendered incomplete the drawing of the whole of this square. Now it is evident that the fourth part of any number multiplied by itself and then multiplied by four gives the same number as half of the number multiplied by itself. Therefore if half of the root is multiplied by itself, the sum total of this multiplication will wipe out, equal, or cancel the multiplication of the fourth part by itself and then by 4.

Let us follow what he did step by step but with modern notation. We are given a square of unknown side, and we call this side x. Al-Khwārizmī's method asks us to draw a square with side x and area, therefore, x^2. We then have to put four rectangles around the square, each of sides x and $\frac{10}{4} = 2\frac{1}{2}$ (the 10 is the given number of roots). The area of each rectangle is thus $\frac{10x}{4}$, so their combined area is $10x$, and we are told that the area of the square and these rectangles together is 39. This gives us the cross-shaped figure of width $x + \frac{10}{2} = x + 5$. Now we are asked to *complete the square*, by adding in the four corner squares whose sides are each $\frac{10}{4} = 2\frac{1}{2}$. The area of each of these squares is $(2\frac{1}{2})^2 = \frac{25}{4} = 6\frac{1}{4}$, so their combined area is 25. So the area of this completed square is $39 + 25 = 64$, and so the square must have side 8. This side is made up of the unknown x in the middle and two lengths of $2\frac{1}{2}$, making a total of $x + 5$, so $x = 3$.

We now repeat the calculation but with arbitrary coefficients: we solve the equation 'a square and b roots are equal to c numbers'. We construct a square with side x and area x^2. We add to it four rectangles with sides x and $\frac{b}{4}$. Each rectangle has area $\frac{bx}{4}$ and their combined area is bx, so the cross-shaped figure has area $x^2 + bx = c$ and width $x + \frac{b}{2}$. We add to this cross-shaped figure the four corner squares of side $\frac{b}{4}$ and area $\frac{b^2}{16}$, whose total area is $\frac{b^2}{4}$, and obtain a square of area $\frac{b^2}{4} + c$ and side $\sqrt{\frac{b^2}{4} + c}$. But the side of the square is also $x + \frac{b}{2}$, so we deduce that $x + \frac{b}{2} = \sqrt{\frac{b^2}{4} + c}$ and so $x = -\frac{b}{2} + \sqrt{\frac{b^2}{4} + c}$. This simplifies to

$$x = \frac{-b + \sqrt{b^2 + 4c}}{2}$$

which is equivalent to the familiar modern formula in the case when $a = 1$.

Al-Khwārizmī also considered equations of the form 'a square and numbers are equal to sides', such as the one that we would write as $x^2 + 21 = 10x$, that have two positive roots — this case $x = 3$ and $x = 7$. He observed that in this case one may, if one wishes, add or subtract the square root (because the result is positive in each case) and so such equations have two solutions. Abū Kāmil, who wrote a generation later than al-Khwārizmī, also noted the special case in which the two roots are the same — for example, $x^2 - 10x + 25 = (x - 5)^2$ — which happens when the polynomial is a perfect square, and the term under the square root is zero.[11]

In what ways does al-Khwārizmī's treatment mark an advance on what we read in Mesopotamian texts? The biggest difference is that al-Khwārizmī states explicitly that this equation is one of the general species of equation and that he is illustrating (via a specific example) the general solution method for this species. In contrast, the Mesopotamian tablet is specific; however, we do not know what the scribe said to his pupils while discussing a tablet's contents. Al-Khwārizmī was writing down a traditional method of solving such questions, but contributed an explicit awareness that they arise as examples of six different types. Because he was more explicit about what he was doing, one can say that his account amounts to a *proof* that the method works.

Al-Khwārizmī's approach is notable for its use of geometrical figures to justify equation-solving methods, and so it is not an algebraic but a geometrical one. That said, it is not a classical Greek proof either (and al-Khwārizmī never referred to Euclid): it is not presented in general terms, but as a geometrical demonstration of the validity of the method

[11] This happens, he remarked, when the product of half the roots equals the number, or, writing $x^2 + c = bx$ when $(b/2)^2 = c$. See (Abū Kāmil 1966, 38).

applied in a particular example. It is a geometrical validation of the solution $x = 3$, and is general in the sense that the same steps can be applied to other equations of the same type.

In this sense, therefore, it is a significant advance on 'proof by checking' — that is, by substituting numerical solutions into the original problem. But it finds only *positive* numbers as solutions, for only these can measure the side of something: al-Khwārizmī would not have considered negative numbers to exist, and so was blind to the 'other' solution of this equation, $x = -13$, which the modern quadratic equation formula produces. For how could he have begun by drawing a square on a side of *negative* length?

The mathematician who brought these two styles together was Thābit ibn Qurra, who lived in the second half of the 9th century. He was a pagan who belonged to a Sabian sect descended from Mesopotamian star worshippers with a long-standing interest in astronomy and astrology, and his native language was Syriac. This sect was tolerated by its Islamic rulers, and he is a good example of why it is so difficult to encompass the Islamic culture in a single word. According to tradition, he was discovered by the Banū Mūsā while working as a money changer in his native Ḥarrān, a town in northern Mesopotamia. They were impressed by his ability to speak several languages and brought him back to Baghdad to work with them at the House of Wisdom, where he soon distinguished himself as a mathematician of rare ability. In particular, he showed how the techniques of *Elements*, Book II, Props. 5 and 6 can be used to represent al-Khwārizmī's methods and prove that they are justified.

Thābit continued the work of the school of translators in the House of Wisdom. We owe to him the Arabic translations of many important Greek works by Apollonius, Archimedes, Euclid, Eutocius, and Ptolemy: without such Arabic translations, a number of Greek works (which have come down to us through Arabic translations only) would have been totally lost. In translating from Greek (and occasionally from Syriac translations of Greek), the Islamic mathematicians made contributions of their own, often extending and generalising the Greek work and providing alternative proofs of theorems.

Not all Islamic mathematicians were enthusiastic followers of the Greek tradition, and we must exercise special caution when considering al-Khwārizmī's sources. Any argument that he 'unified' Greek, Indian, and Mesopotamian methods presupposes some dependence on classical Greek mathematics. There is no doubt that he used Indian sources for his purely arithmetical and astronomical writings: we have already noted that he learned of the Hindu numerals from India, and that his astronomical tables were based on Indian and Persian texts — his value for π, for instance, corresponded precisely to that of the 5th-century Indian mathematician Āryabhaṭa. Some scholars have suggested that he drew on a native Syriac–Persian tradition. He was also aware of the Hebrew mathematical tradition, for he wrote a treatise on the Jewish calendar.

Be that as it may, in the preface to his *Algebra* al-Khwārizmī emphasised that his purpose was to write a popular and practical work, in contrast to Greek theoretical mathematics: this suggests that he did not wish to be counted among the 'Greek school' of contemporary Islamic mathematicians. Even though the geometrical demonstrations in the *Algebra* may superficially remind us of Greek geometrical procedures, they have considerable differences in intention from proofs in, say, the *Elements*. Indeed, the historian of mathematics Solomon Gandz has described al-Khwārizmī as 'the antagonist of Greek influence'.[12]

[12]See (Gandz 1936, 263–277).

The tradition of al-Khwārizmī was continued by various writers, such as Abū Kāmil ibn Aslam, who wrote an *Algebra* around AD 900 that was clearly based on that of al-Khwārizmī. In the introduction to his work, he referred explicitly to al-Khwārizmī's 'three species' from which quadratic equations are formed: roots, squares, and numbers.[13] The historian Jacques Sesiano has noted that Abū Kāmil's work, unlike al-Khwārizmī's, was written for mathematicians — 'that is to say, readers familiar with Euclid's *Elements*'.[14] He dealt with quadratic equations whose coefficients are themselves square roots of other numbers, and areas of some regular polygons.

Abū Kāmil also described a class of indeterminate problems that were to prove very popular down the ages, the so-called *bird problems*. In a typical one, you are asked to buy 100 birds for 100 *dirhams*, and whereas it costs 2 *dirhams* to buy a duck, one *dirham* buys two pigeons, three wood pigeons, four larks, or one chicken. If you wish to buy d ducks, p pigeons, w wood pigeons, l larks, and c chickens, you have to satisfy these equations:

$$d + p + w + l + c = 100,$$

and

$$2d + \frac{p}{2} + \frac{w}{3} + \frac{l}{4} + c = 100.$$

The implied condition that makes the problem a challenge is, of course, that the numbers d, p, w, l, and c must be natural numbers (and 0 is not admitted as a solution). Mathematician that he was, Abū Kāmil reasoned his way to a complete solution (missing only two cases): there are actually 1445 ways in which you can buy the birds under the stated conditions. Abū Kāmil wrote that, when he told people of his discoveries:[15]

> those who did not know me were arrogant, shocked, and suspicious of me. I thus decided to write a book on this kind of calculation, with the purpose of facilitating its treatment and making it more accessible.

It is easy to imagine a mathematician thinking of these problems while out shopping in the market, and indeed these problems did not originate with Abū Kāmil, but in China some centuries before. They were also current in India at the time that Abū Kāmil was writing, and similar problems are found in a book by Alcuin of York, the monk who was charged by Charlemagne around the year 900 with reviving education in his domains. Sesiano speculates that such problems originated in Greek Alexandria because some of them involve not birds but camels 'which are unlikely to have inspired a scholarly monk of Charlemagne's empire'.[16]

The early 11th century was another period of outstanding mathematical activity in the Islamic world. There was a new edition of Euclid's *Elements*, produced by ibn Sina, a Persian scholar. He is said to have written some two hundred books, including influential and original works on logic, philosophy and mathematics, and on medicine. These books helped to transmit Aristotle and Galen (the Greek writer on medicine) to the medieval West. Just as Euclid was often referred to as 'the Geometer' and Aristotle as 'the Philosopher' in later works, so ibn Sina (Latinised as Avicenna) was known as 'the Commentator'.[17]

In Egypt, ibn al-Haytham (better known in the West as al-Hazen or Alhazen) proposed a novel theory of vision, whereby light travels from objects to the eye. He wrote over fifty

[13]See (Abū Kāmil, 1966, 40, 52, 54) and F&G 6.B2 for Abū Kāmil's solution of the equation $x^2+21 = 10x$ and an indication of his treatment of two algebraic identities.

[14]See (Sesiano 2009, 64).

[15]Quoted in (Sesiano 2009, 76).

[16]See (Sesiano 2009, 78).

[17]See (Lindberg 1981, 60).

books, on subjects such as optics, perspective, astronomy, and astrology, and reconstructed Book VIII of Apollonius's *Conics*, which was already lost by that time. A celebrated problem known as *Alhazen's problem*, although it was originally asked by the astronomer Ptolemy, asks for the point on a spherical mirror where a ray of light is reflected from a given source to an observer. The solution that ibn al-Haytham gave is an ingenious exercise in the geometry of conic sections.

Omar Khayyām. As a result of political and ethnological changes within Islam, mathematical centres within the Islamic world flourished, died, and revived, and new centres arose. In the 11th century a new capital city arose in northern Persia around the irrigation centre of Merv. It was here that 'Umar ibn Ibrāhīm al-Khāyyamī (usually known in the West as Omar Khayyām) flourished for a while before moving to Isfahan, where a group of outstanding Islamic astronomers had gathered together to found an observatory.

As well as being an astronomer of note, Omar Khayyām was also a mathematician, philosopher, and poet. He encountered religious opposition, both to his reform of the Moslem solar calendar and to the atheistic tenor of his poem, the *Rubaiyat*, in which sceptical questions are asked about God, the Universe, and the passing of time.[18] It was reported that Omar Khayyām undertook a pilgrimage to Mecca, in order to clear himself of the charge of atheism. Although he contributed to mathematics, astronomy, law, history, and medicine, we concentrate here on his algebra and his work on the parallel postulate.

In his *Algebra* Omar Khayyām went well beyond al-Khwārizmī to treat certain cubic equations.[19] He noted that these ideas had not been discovered by the ancients (by which he meant the Greeks), but that he had several Arabic precursors in this area whom he respected. Specifically, he tells us that one Abū Ja'far al-Khāzin had solved a cubic equation derived from a problem of Archimedes by means of intersecting conic sections, but in the preface to his *Algebra* Omar Khayyām said that he liked to solve problems in complete generality and by making distinctions where necessary. As he put it:[20]

> I have always desired to investigate all types of theorem ... giving proofs for my distinctions.

Omar Khayyām then clarified cubic equations as al-Khwārizmī had done for quadratics: systematically, and in some sense geometrically. He observed al-Khwārizmī's rule that all numbers entering the equations must be positive. He then classified the equations accordingly into twenty-seven distinct types (see Box 8.1), of which two, for example, are 'cube and numbers equal to sides' and 'cubes and sides equal numbers': corresponding to $x^3 + c = bx$ and $x^3 + bx = c$ in modern notation. Of these twenty-seven types, one is not regarded as being an equation at all ($x^3 = 0$). Of the remaining twenty-six types, he showed that eight can be solved by Euclidean methods and a further four have no solutions at all (in positive numbers, a restriction tacitly supposed) while the remaining fourteen can be solved by the use of conic sections.[21] His arguments are an ingenious mixture of algebraic manipulations and classical constructions in three-dimensional Euclidean geometry, and he was careful to explain when a given type of equation has no solutions or more than one.

---

[18]It was recast as a splendid Victorian poem, bearing occasional resemblance to its original, by Edward Fitzgerald in 1859.

[19]See F&G 6.A3 and 6.B3.

[20]Quoted in (Berggren 1986, 118).

[21]An example of the latter type appeared in Chapter 4, in our discussion of the problem of doubling the cube.

Box 8.1. The different types of cubic equation.
There is a maximum of 27 types of cubic equation, corresponding to the ways that an expression of the form

$$x^3 + ax^2 + bx + c = 0$$

can be written so that all the coefficients are positive or zero – for example, as

$$x^3 + bx = ax^2 + c$$

(cubes and sides equal to squares and numbers). We can obtain this by allowing each of a, b, and c to be positive, zero, or negative. With three choices for each coefficient there are 27 possibilities.

Omar Khayyām enumerated these equations by considering how many non-zero terms occur. He excluded the case where no such coefficients occur (corresponding to the equation $x^3 = 0$).

There are six equations with only two non-zero terms; they are of the form $x^3 + ax^2 = 0$, $x^3 + bx = 0$ and $x^3 + c = 0$, where none of a, b, c can be zero. He regarded both cases of each of the first two types as being equivalent to a quadratic equation, and therefore solvable by Euclidean methods, but included two cases of the third type.

There are twelve equations with precisely three non-zero terms. He regarded the four cases where $c = 0$ as equivalent to quadratic equations. This left him with eight cases that are not equivalent to quadratic equations.

Finally, there are eight cases where none of a, b, or c are zero; none of these are equivalent to quadratic equations.

This gave him a total of 18 cubic equations that cannot be solved by Euclidean methods.

However, some of these equations cannot have any positive solutions and Omar Khayyām excluded them on those grounds. These are the four equations where all the terms are positive, and so he was left with 14 cubic equations that have to be solved.

The novelty of Omar Khayyām's complicated achievement is eloquent testimony to the vigour of the Islamic mathematical community at the time. Omar Khayyām himself commented on this striking mixture of geometry and algebra in one strange phrase that is worthy of particular note:[22]

> Algebras are geometric facts which are proved by Propositions 5 and 6 of Book II of the *Elements*.

By this he meant that when solving equations by algebra — that is, by carrying out the operations of *al-jabr* and *al-muqabala* — the results we obtain can be proved correct by geometry, because, appearances to the contrary notwithstanding, one is really applying these propositions. This argues for a deep appreciation of both the Greek style of mathematics and his own, that he can so casually translate the one into the other.

We can get an impression of the magnitude of Omar Khayyām's work if we first sketch an account in modern terms of how a cubic equation can be solved by intersecting a circle and a parabola (see Figure 8.3).

[22]See F&G 6.A3.

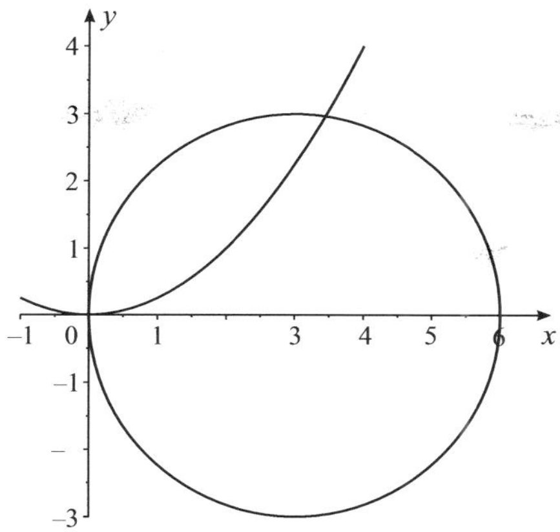

Figure 8.3. The circle $x^2 + y^2 - 6x = 0$ and the parabola $x^2 = 4y$

First we show how to obtain a cubic equation from a circle and a parabola.

We let the circle have equation $x^2 + y^2 - 2xa = 0$, so this is a circle with centre $(a, 0)$ and radius a. We let the parabola have equation $x^2 = by$, so it is symmetrical about the y-axis, and both curves pass through the origin. When we eliminate y from these equations we obtain

$$x^2 + \left(\frac{x^2}{b}\right)^2 - 2xa = 0,$$

which simplifies to

$$x^4 + b^2x^2 - 2b^2ax = 0.$$

This is an equation for the x-coordinates of the common points of the circle and the parabola. If we set aside the value $x = 0$ that corresponds to the origin, we are left with this cubic equation in x:

$$x^3 + b^2x - 2b^2a = 0.$$

We now show how to connect the above analysis with the solution of a given cubic equation

$$x^3 + \alpha x - \beta = 0,$$

for specific positive values of α and β. We have to determine the precise circle and parabola of the above types that are needed to solve the cubic equation, which requires that we express a and c in terms of α and β. But from $b^2 = \alpha$ we deduce that $b = \sqrt{\alpha}$, and from $2b^2a = \beta$ we deduce that $a = \beta/2\alpha$.

However, Omar Khayyām had to show how to find a geometrically, so he had to give constructions for a and b in terms of the given values (lengths) α and β. This gives his exposition a further layer that was more familiar to his audience than it is to us today. And he also had to explain in geometrical terms why his argument is valid, and this was probably difficult even for his most accomplished contemporaries.

If we also note that the cubic equations solved in this way are of a special kind (the number of squares is zero), and that to solve a more general cubic equation the parabola was replaced with the hyperbola, which makes the whole argument much more complicated, we can begin to see just what a tour de force Omar Khayyām's achievement was.

We now give an extract from Omar Khayyām's account of how to solve geometrically a cubic equation of the form *cube plus squares plus edges equal to numbers.*[23]

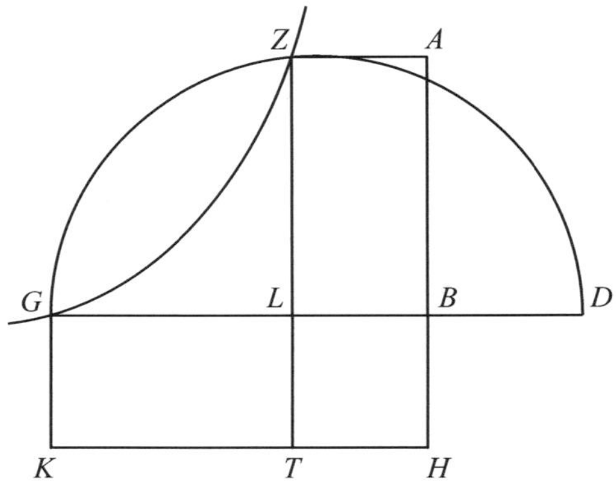

Figure 8.4. Omar Khayyām's geometrical solution of a cubic equation

Omar Khayyām's geometrical solution of a cubic equation.

We draw BH to represent the side of the square equal to the given sum of the edges, and construct a solid whose base is the square of BH, and which equals the given number. Let its height BG be perpendicular to BH. We draw BD equal to the given sum of the squares and along BG produced, and draw on DG as diameter a semicircle DZG, and complete the area BK, and draw through the point G a hyperbola with the lines BH and HK as asymptotes. It will intersect the circle at the point G because it intersects the line tangential to it [the circle], i.e., GK. It must therefore intersect it [the circle] at another point. Let it intersect it [the circle] at Z whose position would then be known, because the positions of the circle and the conic are known. From Z we draw perpendiculars ZT and ZA to HK and HA. Therefore the area ZH equals the area BK. Now make HL common. There remains [after subtraction of HL] the area ZB equal to the area LK. Thus the proportion of ZL to LG equals the proportion of HB to BL, because HB equals TL; and their squares are also proportional. But the proportion of the square of ZL to the square of LG is equal to the proportion of DL to LG, because of the circle. Therefore the proportion of the square of HB to the square of BL would be equal to the proportion of DL to LG. Therefore the solid whose base is the square of HB and whose height is LG would equal the solid whose base is the square of BL and whose height is DL. But this latter solid is equal to the cube of BL plus the solid whose base is the square of BL and whose height is BD, which is equal to the given sum of the squares. Now we make common [we add] the solid whose base is the square of HB and whose height is BL, which is equal to the sum of the roots. Therefore the solid whose base is the square of HB and whose height is BG, which we drew equal to the given number, is equal to the solid cube of BL plus [a sum] equal to the

[23]See (Khayyām 1950, 55–56), in F&G 6.B3.

given sum of its edges plus [a sum] equal to the given sum of its squares; and that is what we wished to demonstrate.

Thus this class has no variations, and none of its problems is impossible, and it has been solved by the properties of the hyperbola together with the properties of the circle.

In one respect, Omar Khayyām was dissatisfied with his achievements. The equations that required the use of conic sections for their solutions have answers that are given as line segments. Omar Khayyām looked for a solution as a formula involving the numbers that entered the problem; he was unsuccessful and concluded by expressing the hope that later mathematicians would solve the problem. It was to prove a long wait.[24]

The binomial coefficients. The binomial coefficients were studied by Persian mathematicians: al-Samaw'al ben Yahyā al-Mahribī, who died in 1174 reported on a rule known, he said, to Abū Bakr al-Karajī around 1007, and Jamshid al-Kāshī, who died in 1429, drew on work on Nasīr-al-Dīn al-Ṭusī from 1265 when he explained the rule for constructing the binomial coefficients in his *The Reckoners' Key*.

x	x^2	x^3	x^4	x^5	x^6	x^7	x^8	x^9	x^{10}	x^{11}	x^{12}
1	1	1	1	1	1	1	1	1	1	1	1
1	2	3	4	5	6	7	8	9	10	11	12
	1	3	6	10	15	21	28	36	45	55	66
		1	4	10	20	35	56	84	120	165	220
			1	5	15	35	70	126	210	330	495
				1	6	21	56	126	252	462	792
					1	7	28	84	210	462	924
						1	8	36	120	330	792
							1	9	45	165	495
								1	10	55	220
									1	11	66
										1	12
											1

Figure 8.5. Al-Samaw'al's binomial coefficients

Al-Samaw'al discussed the construction of the binomial coefficients in these terms (see Figure 8.5):[25]

Let us now recall a principle for knowing the necessary number of multiplications of these degrees by each other, for any number divided into two parts. al-Karajī said that in order to succeed, we must place "one" on a table and "one" below

[24]See Section 10.3.

[25]See Berggren, in (Katz 2007, 552–553).

> the first "one" , move the [first] "one" into another column, add the [first] "one" to the one below it. We thus obtain "two", which we put below the [transferred] "one" and we place the [second] "one" below two. We have therefore "one" , "two", and "one" . This shows that for every number composed of two numbers, if we multiply each of them by itself once — since the two extremes are "one" and "one" — and if we multiply each one by the other twice — since the intermediary term is "two" — we obtain the square of this number. If we then transfer the "one" in the second column into another column, then add "one" [from the second column] to "two" [below it], we obtain "three" to be written under the "one" [in the third column], if we then add "two" (from the second column) to "one" below it, we have "three" which is written under "three", then we write "one" under this "three"; we thus obtain a third column whose numbers are "one" , "three", "three", and "one". This teaches us that the cube of any number composed of two numbers is given by the sum of the cube of each them and three times the product of each of them by the square of the other.

Al-Samaw'al followed al-Karajī step by step in this fashion through fourth powers (called square-square), fifth powers (quadrato-cubes), and so on, until he had a table of binomial coefficients adequate for 12th powers.

Western interest in the binomial coefficients can be traced back to these Persian sources, and in the West they are often displayed in what is called *Pascal's triangle*, after their appearance in Blaise Pascal's *Traité du Triangle Arithmétique* (1665), but that was not their first appearance in Western literature — a table of them appears in Michael Stifel's *Deutsche Arithmetica* (1544), and in Niccolò Tartaglia's *General Trattato* (1556), although he seems to have known of it for some 20 years by then.

8.3 Islamic geometry

It is hard to select among a range of geometrical topics studied by Islamic mathematicians. We might choose, for example, to look at their work on mastering Greek works and reconstructing lost ones, on Ptolemaic astronomy, on patterns and designs, or on instruments such as the astrolabe. We have chosen a topic that well demonstrates their critical and investigative spirit: the parallel postulate.

The parallel postulate was mentioned briefly in Chapter 3 as playing a useful role in establishing many results in Euclid's *Elements*. It was used, as we saw, to prove that the angles of a triangle add up to two right angles and to prove the Pythagorean theorem.

As we saw in Chapters 3 and 4, the *Elements* is an impressively coherent and elaborately structured work. In concentrating on the parallel postulate, commentators in the Islamic world focused on what is perhaps its weakest point. They did not doubt the truth of the postulate, but they did not believe that it was necessary to make an assumption of that kind about parallel lines. Rather, they sought to derive the parallel postulate from Euclid's other initial assumptions. This sharp criticism led them into interesting discussions about the foundations of geometry.

There is evidence to suggest that the parallel postulate was contentious in Greek times. Indeed, it is possible that Aristotle knew of problems with it, even before Euclid decided to make it a postulate or assumption, and we saw in Chapter 4 that Proclus had declared that it ought to be struck from the postulates altogether, being in fact a theorem that followed from the other four postulates.[26]

[26] F&G 3.B1.

Be that as it may, neither Proclus nor anyone before him succeeded in deriving the parallel postulate from Euclid's other assumptions, and Islamic mathematicians had the field all to themselves. So it is interesting that from the start they seemed to have found the parallel postulate unsatisfactory and sought to give it the security of a proper proof. As we shall see, they engaged in a sustained and detailed critique of Euclid's *Elements* that surpassed anything they could have learned from the Greeks.

When studying these attempts at proving the parallel postulate you should ask yourself whether they have derived it validly, and if so, whether they have derived it from the other assumptions of Euclid's *Elements* alone, or whether some new assumption has crept in. It is best to read them pen in hand, drawing the necessary figures as you proceed. And since the parallel postulate makes claims about straight lines and parallel lines, you should first direct your critical gaze towards the use of those terms.

Al-Jawharī. One early commentator was al-ʿAbbās ibn Saʿīd al-Jawharī, who lived at the end of the 8th and the start of the 9th centuries and was an astronomer of the Caliph al-Ma'mūn. From an initial implicit assumption about angles he deduced that the lines l_1 and l_2 in Figure 8.6(a) are equidistant. He also deduced (see Figure 8.6(b)) that, given an angle and a point inside it, one can always draw a straight line cutting both arms of the angle.

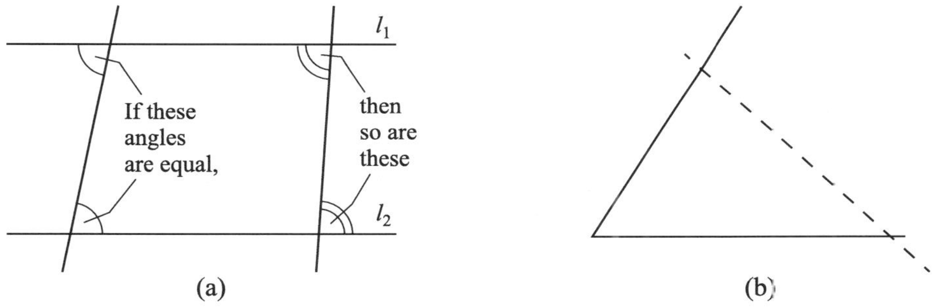

Figure 8.6. Al-Jawharī's figures

Is what al-Jawharī assumed true? Is what he deduced from it true? A natural response would be to say that these statements are obviously true — in which case, why is al-Jawharī deducing some of them from others? And indeed they are true — if you assume all of Euclid's postulates. But what al-Jawharī and his successors were investigating was whether the parallel postulate has to be assumed, as it had been by Euclid, or whether it can be proved from the other postulates. So they had to work without the parallel postulate, for obviously they could not assume what they wanted to prove.

But for many centuries this turned out to be easier said than done. You have just seen al-Jawharī make an assumption without realising it. He should have *proved* that you can draw the dashed line in al-Jawharī's second figure above — for assuming the existence of such a figure turns out to be equivalent to assuming the parallel postulate itself. It yields (by a valid argument that we omit) the result that every quadrilateral has an angle sum of four right angles, and from this the parallel postulate can be deduced. The word 'equidistant', which crept into al-Jawharī's argument, is indicative of his confusion, because it is by no means obvious that parallel lines must be equidistant.

Thābit ibn Qurra. Al-Jawharī was followed by Thābit ibn Qurra, who offered a 'proof of the celebrated postulate of Euclid' based on the theorem that:[27]

> If a straight line cuts two straight lines and if the alternating angles are equal, then the lines neither move apart nor draw closer on either side of the line.

This recalls Euclid's first use of the parallel postulate in the Euclid's *Elements*, Book I, Prop. 28, which says that, in Figure 8.7, *if the indicated angles are equal, then the straight lines are parallel.*

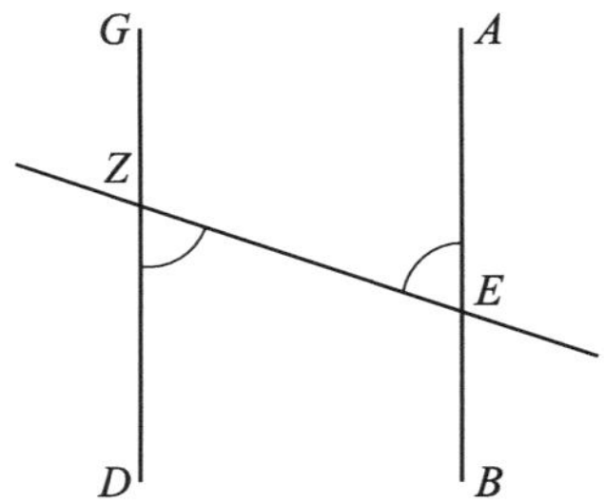

Figure 8.7. Thabit's symmetrical figure

In the course of his proof, Thābit asserted that if two straight lines that cross a third converge on one side of the line, then they diverge on the other, and in each case they do so without limit. He then argued by contradiction, because in the given case the figure is symmetrical about the midpoint of ZE. After his proof of this result, Thābit established the converse result, which says that if the lines neither move apart nor draw closer, then the stated angles are equal; this recalls Euclid's *Elements*, Book I, Prop. 29. So Thābit was building up a theory of straight lines that neither move apart nor draw closer, and hoped to show that it has all the features of the Euclidean theory of parallel lines. Accordingly, he next established the following result:[28]

> Given two equal straight lines which neither move apart nor draw closer to each other, if one joins the extremities of these straight lines by two straight lines, then these new straight lines neither move apart nor draw closer to one another.

This establishes the existence of parallelograms, from which the Euclidean parallel postulate was rapidly deduced.

Is this a proof of the parallel postulate? You may guess that the answer is 'no', and indeed it is not. In this case the mistake is to assume that there are such things as straight lines that neither move apart nor draw closer to one another. In fact, his assumption about lines neither converging and diverging is true only in Euclidean geometry.

For reasons like this, commentators have suggested that this attempt on the parallel postulate was an early work by Thābit, who normally achieved much higher standards. For exactly the opposite reasons, his second and much more sophisticated attempt is usually regarded as a mature work.

This time, Thābit began with a methodological reflection. Geometry, he said, is based on measurements of lengths, angles, and areas. Accordingly we imagine that an object can be made to coincide with a measuring instrument or another object; for example, we might lay down two lengths side by side to see which one is longer. Indeed, he went on:[29]

---

[27] See (Jaouiche 1986, 145).
[28] See (Jaouiche 1986, 146).
[29] See (Jaouiche 1986, 151–152).

> Many principles of proofs of propositions and theorems in geometry which have needed to be proved require the use of the operation we have mentioned: I would say the motion of one of two objects, of which one is the measure of the other ... and its movement, in our imagination, occurs without it being made to change its form until we put it, in that shape, on that by which it is measured. This is what Euclid had to do in the proof of proposition 4 of Book I of his book on the *Elements* and in the proof of proposition 8 of that Book, and these two propositions are among the most fundamental that one states and knowledge of them and their proofs prepares the way for the rest.[30]

In the proofs of these propositions, Euclid spoke of 'applying one triangle to another', which means placing one on top of the other and proving that the sides coincide exactly. So Thābit was not attributing a concept to Euclid that cannot be found in the *Elements*.

Thābit then observed that we are accustomed to imagining a straight-line segment of fixed length being rotated about one end so that the other end sweeps out a circle. All the radii of this circle are equal, and the moving line does not alter its length or cease to be straight. He then continued 'with something well known from the subject of solid bodies', as he put it:[31]

> Every point of a solid body which we imagine moved in its totality in a single direction of motion (one which is simple and rectilineal) moves along a straight line and traces in its motion the straight line along which it goes. And of the straight lines which the body contains, those which lie in the direction of movement pass likewise along a straight line; and as for those which do not lie along the direction of motion, they do not travel along themselves.

Not for the first time, or the last, a mathematician trying to explain a fundamental principle comes over as rather obscure. The basic idea, which he set out carefully, is that if one point of a solid body moves along a straight line, and the body does not rotate, then every point of the body moves along a straight line, and the lines in the body that are parallel to the direction of motion move along themselves. Put in that way, one begins to get nervous: Thābit has been at pains to keep the word 'parallel' out of his exposition, but has he avoided introducing the idea? We shall see, but first let us indicate his plan of attack.

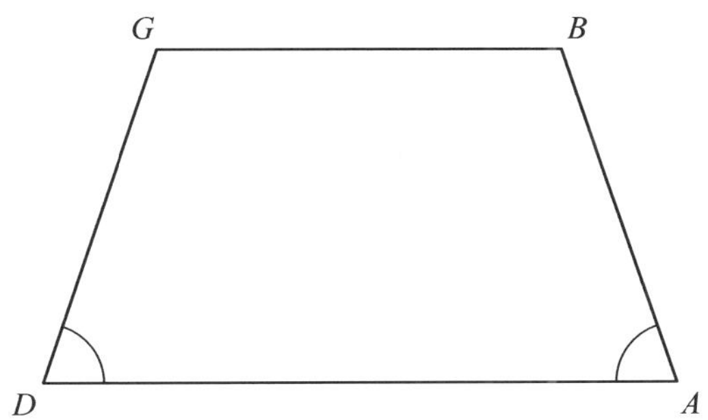

Figure 8.8. $GD = BA$ and $\angle GDA = \angle BAD$

On the basis of his assumption Thābit proceeded to show that, given a quadrilateral $GBAD$ with two opposite sides of equal length $(GD = BA)$ that meet the base at equal

[30] Euclid's *Elements*, Book I, Props. 4 and 8, present two of the congruence tests for triangles.

[31] See (Jaouiche 1986, 153).

angles ($\angle GDA = \angle BAD$), as in Figure 8.8, the other pair of angles are also equal ($\angle DGB = \angle ABG$). On the basis of a long and careful argument he then showed that if the base angles are less than a right angle, then the sides DG and AB must eventually meet as they are extended upwards.

The flaw in this argument is well hidden, and you might want to spend a few minutes trying to spot it. Thābit's hypothesis was that *if* a line segment moves along a straight line containing the segment, *then* a point not on the line also moves along a straight line. The moving point remains at the same distance from the points of the moving segment, and Thābit's hypothesis is that the moving point moves along a straight line. Using this hypothesis he deduced the truth of the parallel postulate. Euclid had already shown the converse result, that the parallel postulate allows one to deduce that parallel lines are equidistant. But what does that show? It certainly shows that, given all of Euclid's other assumptions, Thābit's hypothesis is equivalent to the parallel postulate, but does it show that the parallel postulate is true? The answer is *no*, unless you are prepared to grant the truth of Thābit's hypothesis. Now it is very plausible — perhaps more plausible than the parallel postulate. But it also lacks the self-evident character of the other Euclidean postulates.

After all Thābit's hard work, we are no further forward. We may base Euclidean geometry on a different hypothesis, but we have not derived the truth of the parallel postulate from the other postulates of the *Elements* alone.

Omar Khayyām and ibn al-Haytham. A century later, Omar Khayyām disagreed with his illustrious predecessor ibn al-Haytham over the use of motion in geometry.[32] Ibn al-Haytham believed that geometrical figures can legitimately be constructed by reasoning about what happens when something is moved, but Omar Khayyām disputed the validity of this.

This is worth looking at in more detail. Ibn al-Haytham's argument is straightforward and accessible if we take it gently and construct figures as he tells us to. The figure he described in the latter half of the first paragraph of the following extract is something like the right-hand half of Figure 8.9: a quadrilateral right-angled at A, B, and D. It is obtained by taking the line segment AB, drawing the lines AG and BD at right angles to it, and dropping a perpendicular from G to the line BD. Notice that ibn al-Haytham did not claim that the angle AGD in this quadrilateral is also a right angle.

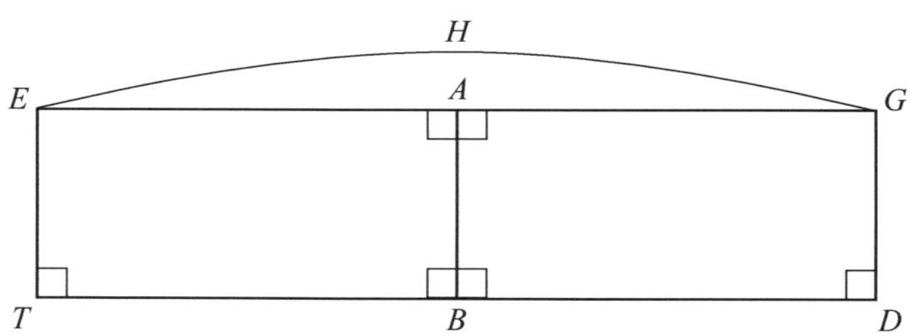

Figure 8.9. Ibn al-Haytham's construction

Ibn al-Haytham wanted to prove that GD is equal to AB, as that would indeed show that the angle AGD is a right angle, and so establish that AG and BD are parallel without invoking the parallel postulate. His strategy was the classical Greek one of *reductio ad*

[32]Ibn al-Haytham's 'proof' is in (Sude 1974, 94–96) and F&G 6.C1; Omar Khayyām's criticism of it appears in (Amir-Moez 1959, 276–277) and F&G 6.C2.

absurdum: to show that GD equals AB, he assumed that they were unequal and then tried to deduce a contradiction.

He assumed first that GD is *greater* than AB. Showing that this leads to a contradiction occupies the first half of his proof; the other half of the proof, showing that a contradiction arises when it assumed that GD is *less* than AB, proceeds similarly and will not be discussed here.

Ibn al-Haytham began by observing that, by the same construction as before, it is possible to construct a quadrilateral with three right angles on the other side of the line segment AB. This produces the whole of Figure 8.9, in which the left-hand side is the mirror image of the right-hand side because the line segments EA and AB are equal, but ibn al-Haytham, being a rigorous thinker, first proved that the two halves of the figure are congruent to each other.

He drew the diagonals GB and EB, and argued that the triangles GAB and EAB are congruent (that is, of equal shape and size) because they have two pairs of equal sides enclosing a right angle. From this it follows that the segments GB and EB are equal, and that the angles $\angle GBA$ and $\angle EBA$ are equal. Therefore the angles $\angle GBD$ and $\angle EBT$ are equal. It now follows that the triangles GDB and ETB are congruent, because they have two pairs of equal angles and a pair of equal sides (GB and BE). Therefore the segments GD and ET are equal. And since GD is greater than AB (by the *reductio* assumption), so too is ET greater than AB.

At last we come to his disputed motion argument. He imagined ET to move to the right — that is, T moves along the line TBD and ET stays perpendicular to that line and remains of the same length as it moves. He then said that EHG is straight because (as he claims to have proved earlier) when a point is moved in such a way it traces out a straight line. He also said that EAG is straight because he constructed it that way. He now deduced his sought-for contradiction — namely, that it is impossible for two straight lines, such as EHG and EAG, to enclose an area. He argued in detail as follows:

Ibn al-Haytham's motion argument.

> Let us imagine line ET moving along line TB, while, during its motion, it is perpendicular to it, so that angle $\angle ETB$, throughout the motion of ET, is always right. When point T, by the movement of line ET ends up at point B, line ET will coincide with line BA, since the two angles $\angle ETB$, $\angle ABD$ are equal (because each of them is a right angle). When line ET coincides with line BA, point E will be outside line AB and higher than point A, since line ET had been shown as being greater than line AB. Therefore, let line ET, while it coincides with line BA be line BH. Line BH can be imagined also after this position moving in the direction of GD, and it is in the equivalent of its first position. Then, when point B, by the motion of line BH ends up at point D, line BH coincides with line DG since the two angles $\angle HBT$, $\angle GDB$ are equal because they are right. When line BH coincides with line GD, point H coincides with point G, since line HB is line ET and line ET is equal to line GD. When line BH (which is line ET) arrives at line GD and coincides with it, line ET will have moved over line TD, and point T will have ended up at point D. Point E will have ended up at point G. It was shown above in defining parallel lines that the higher end of every line moving in this way traces a straight line; therefore point E traces a straight line during the movement of line ET over line TB. Let the line [that] point E traces be line EHG; thus, line

> EHG is a straight line, but line EAG is a straight line by assumption. Point H has been shown to be higher than point A, so line EHG is other than line EAG. But the two points E, G are common to these two lines and are straight, therefore two straight lines would contain a space; but this is impossible. The impossibility necessarily follows from our assumption that line GD is greater than line AB. Therefore, line GD cannot be greater than line AB.

Omar Khayyām's objection to this proof is that it appeals to the concept of motion. He did not deny that figures may be defined by motion, for that was done by Euclid, but he disputed that the curve traced out by the point E is necessarily a straight line, as ibn al-Haytham had claimed. Instead, he felt that such a claim required a proof, which he then went on to give. At all events, he went on to give his own proof of the same result. It was to turn out, in fact, that any proof of such a claim is equivalent to establishing the parallel postulate.

Omar Khayyām argued that in a quadrilateral with two equal sides AC and BD erected at right angles on a common base AB (see Figure 8.10), the remaining angles — those at C and D — are equal, and that to deny that they are both right angles leads to a contradiction. The conclusion is that this quadrilateral has an angle sum of four right angles, and hence that every quadrilateral does, and so the Euclidean parallel postulate must be true. These conclusions would indeed follow if Omar Khāyyam's arguments were not flawed.

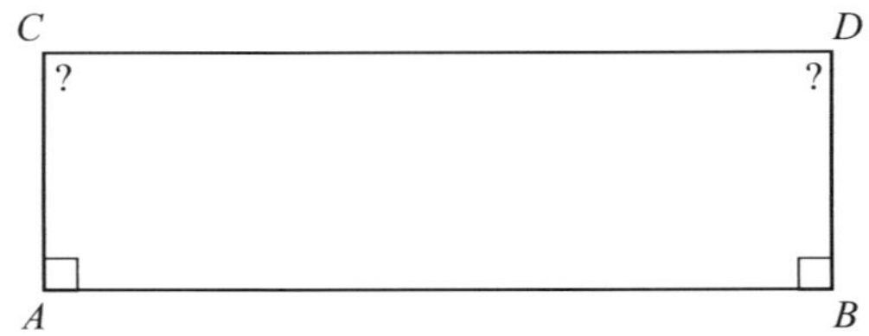

Figure 8.10. Omar Khayyām's quadrilateral

The impression gained from studying this material should not be of a number of mathematicians failing to prove the obvious, so much as of people recognising that an obvious and indeed attractive hypothesis might reasonably need to be proved. In point of fact they failed, and it was shown in the 19th century that they were bound to fail; but it is a sign of a lively and critical mathematical culture that they tried.[33]

Nasīr al-Dīn and Ṣadr al-Dīn ibn khwāja Naṣi al-Dīn. The last Islamic mathematicians whose work on the parallel postulate we mention are Nasīr al-Dīn, and his eldest son, Ṣadr al-Dīn ibn khwāja Naṣi al-Dīn. As well as studying the parallel postulate, Nasīr al-Ṭusī worked on plane and spherical trigonometry, treating this as a subject in its own right independent of astronomy, and making significant contributions to astronomy itself. He was appointed astronomer to Hūlāgū Khan (grandson of Genghis Khan and brother of Kublai Khan) and worked at a new centre of learning established at Maragha (in what is now northern Iran) by the Mongol rulers after they had sacked Baghdad in 1256, where in due course he was succeeded as director by his son.

[33] Among other discussions, see (Gray 1989).

The father's attempt at proving the parallel postulate, which we shall not discuss, was close to that of Omar Khāyyam. It was to become well known in the West when John Wallis discussed it in Oxford in 1663.

Fundamental to the son Ṣadr al-Dīn ibn khwāja Naṣi al-Dīn's much more forceful approach was the idea that lines cannot converge in one direction and then diverge. There could not, therefore, be a line joining them that marks their closest approach and from which they diverge in each direction. On the basis of this assumption, he too began by considering a quadrilateral $CDBA$ formed by two equal segments DC and BA perpendicular to a common base DB (see Figure 8.11). He showed, correctly, that the remaining pair of angles are equal: $\angle BAC = \angle DCA$. He then attempted to show that the angles $\angle BAC$ and $\angle DCA$ are right angles. It follows that the angle sum of the quadrilateral is two right angles, and from this Ṣadr al-Dīn speedily deduced the parallel postulate.

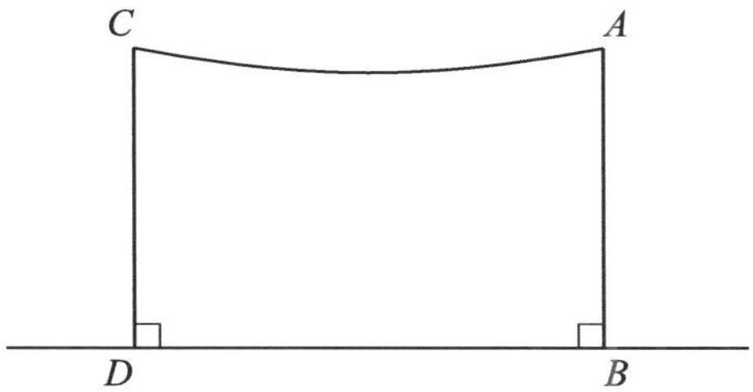

Figure 8.11. Ṣadr al-Dīn's quadrilateral

Ṣadr al-Dīn's defence of the parallel postulate is flawed (his fundamental assumption is equivalent to the parallel postulate), but to say so is not to dismiss its principal achievements, which are the clarity with which he thought through the fundamental, but elusive, concept of straightness and the rigour with which he marshalled his arguments.

Conclusion. We have been able to sketch only a small part of the mathematical activity in the Islamic world from the 7th to the 15th centuries. Nevertheless, it is clear that the tradition that Europe inherited through translation of the Arabic writings was extensive. In many respects, it was much more developed than any of the individual traditions that the Islamic conquerors had themselves assimilated over the preceding centuries. Islamic geometry, algebra, and astronomy did not end with al-Ṭusī, but by then something intangible and not always noticed at the time was under way: a simultaneous awakening of interest in mathematics and independent rational thought in Christian Europe and a hardening and loss of vigour in the Islamic world. No-one at the time would have correctly predicted how the balance of intellectual power would shift in the next few centuries, but the transformation was to be marked and it is to Europe that we now turn our attention.

8.4 Further reading

Berggren, L.J. 1986. *Episodes in the Mathematics of Medieval Islam*, Springer: an accessible and wide-ranging survey.

Gray, J.J. 1989. *Ideas of Space, Euclidean, non-Euclidean, and Relativistic*, 2nd ed. Oxford University Press. A history of the parallel postulate from its inception to its resolution, based on the work of Jaouiche (1986) and others for the Arabic material.

Katz, V. (ed.) 2007. *The Mathematics of Egypt, Mesopotamia, China, India, and Islam: a Sourcebook*, Princeton University Press.

Lyons, J. 2009. *The House of Wisdom*, Bloomsbury. A vigorous and readable defence of Islamic intellectual culture in the Middle Ages and its transmission to the West.

9

The Mathematical Awakening of Europe

Introduction

In this chapter we look at another part of the history of mathematics between the collapse of the Western Roman Empire and the early 16th century. This period is sometimes regarded as leading from Greek mathematics in the Graeco–Roman world to the rediscovery of Greek mathematics in the West, a rediscovery that forms part of a much larger rebirth of learning called the European *Renaissance*. As such, its history concerns the transmission of ideas between different cultures: the collecting of Greek mathematical works, their translation into other languages (specifically, Arabic and Latin), and their eventual impact on an increasingly autonomous and energetic European culture. Crucial to this story were the Arabic and Islamic mathematicians, translators, and commentators who worked over a period of some 700 years (from the mid-8th to the mid-15th centuries), from whom Western commentators and translators eventually acquired many of their texts.

The Western European response to Islamic scholarship and the Greek legacy came in two phases. In the earlier phase, centred on the 13th century, more was done with Arabic intermediaries (both texts and scholars), and in the second one, beginning in the 15th century, more translations were made directly from Greek originals. The reason for the separation was the catastrophe of the Black Death of around 1347–1351, in which one-third of the population of Europe died; figures for the Middle East and Asia are comparable.

When intellectual life resumed, a subtle shift had occurred. Europe was now more buoyant and self-confident intellectually and militarily, while the Islamic world was entering, imperceptibly at first, on a period of decline. So it was that in the later period Western scholars were better equipped to appreciate the significance of Greek thinking, to develop Greek mathematics in their own way, and so to forge their own mathematics. So the middle of the 14th century marks a natural separation point between this chapter and the next, which deals with mathematical developments in the Renaissance.

9.1 Mathematics in the medieval Christian West

In this section we move to the Christian West and trace the gradual awareness of Greek and Islamic learning that developed there between the 11th and the 14th centuries.

It is difficult to exaggerate the depths to which mathematics had sunk in Western Europe by the start of the 11th century. What remained of Platonic thinking on the subject had long since shrunk to the remnants recorded by Boethius (see Section 6.4), Church education followed Augustine in bracketing mathematics with astrology, which it disapproved of strongly, and confining its role to revealing God's design for the Universe. Euclid's *Elements* was unknown. Mathematical knowledge included little more than elementary arithmetic, some neo-Pythagorean number theory in the tradition of Nicomachus and Boethius, calculating rules for the Roman abacus, and the basic geometry needed in the construction of buildings. The contrast between the Christian West and Eastern Christianity in Byzantium on the one hand, and the Islamic world on the other, could not have been more stark.

It was unlikely that this state of affairs could have persisted for ever, but ironically the moves by some Western intellectually-minded Churchmen to find out what Islamic philosophers and natural scientists had to offer coincided with the violent incursions of Christian armies into Byzantium and the Holy Lands, known as the Crusades. The best-trained of these often ill-disciplined forces were the Normans, who had a strong base in Sicily, and some of their leaders took the opportunity to carve out small fiefdoms for themselves, which had the effect of forcing them to acquire the skills of accommodating a mixture of Muslims, Jews, and varieties of Christians. This in turn opened up opportunities for trade and the transmission of ideas. A similar story unfolded at the other end of the Mediterranean, as Christian kingdoms took to fighting the Islamic Caliphate in al-Andalus (Andalucia), the southern part of present-day Spain.

Trade played a crucial role in the awakening of Europe (see Figure 9.1). The two main routes by which Greek and Islamic learning penetrated the Christian West were via Sicily and Spain. These were important centres of trade between the Christian West and the Islamic countries, and by the latter part of the 11th century they had become places of direct contact between Western scholars (often monks) and Arab and Hebrew scholars settled in Europe. The first phase of translation was part of the European intellectual awakening of the 12th and 13th centuries. In Spain, Arabic texts were collected and translated, but in Sicily, many Greek works were also readily available.

Gerbert of Aurillac, who became Pope Sylvester II in the year 999, was an influential precursor in this respect. He was a Benedictine who travelled to Catalonia in 967 to study mathematics and music, before going to Rome in 970. In Catalonia he became interested in Arabic learning, and it is through him that Arabic science crossed the Pyrenees and reached as far as northern and eastern France — he taught in Rheims in 984. A number of texts are associated with his name or that of Boethius, although authorship is uncertain, and it seems that he promoted Pythagorean number theory in the spirit of Nicomachus, elementary geometry (without proofs), and the rudiments of astronomy. This shows a revival of interest in the West in the old quadrivium of mathematical subjects. He also taught the use of the abacus, and his symbols for the numerals may show an influence from Arabic, although there was no symbol for zero (see Figure 9.2).

Among the first great medieval translators and authors was an Englishman, Adelard of Bath, who was born around 1080. He studied at Tours, then taught at Laon (north-east of Paris), before travelling for seven years to places such as Sicily, Syria, and Palestine (the details of his journeying seem to be unknown). He had learned Arabic, and spent his time copying what he could. On his return he brought back two works that could only shock the

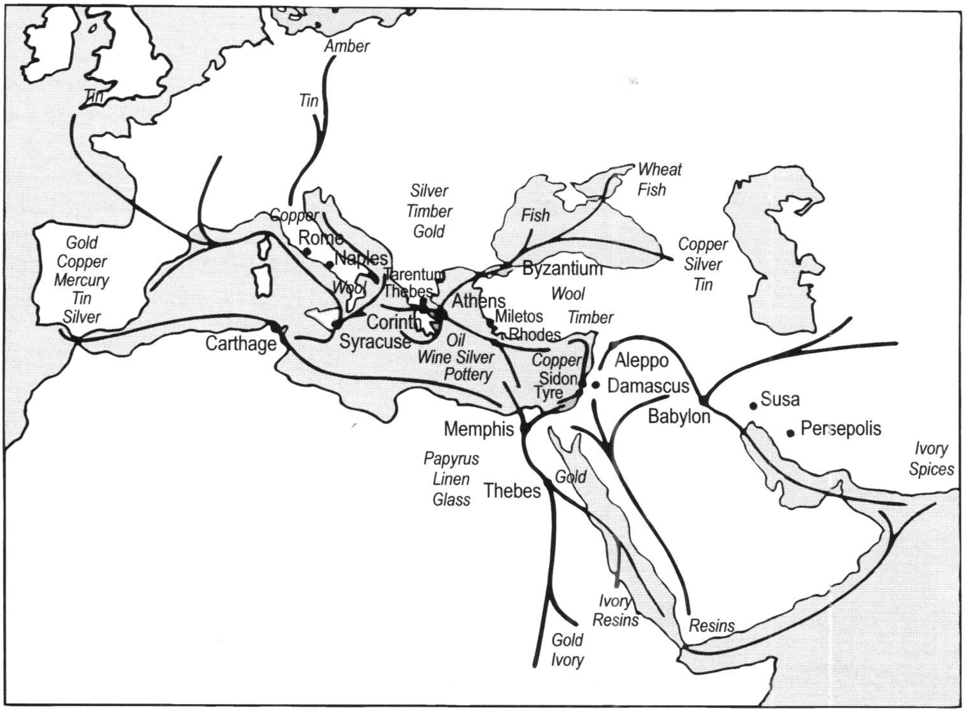

Figure 9.1. Mediterranean trade routes

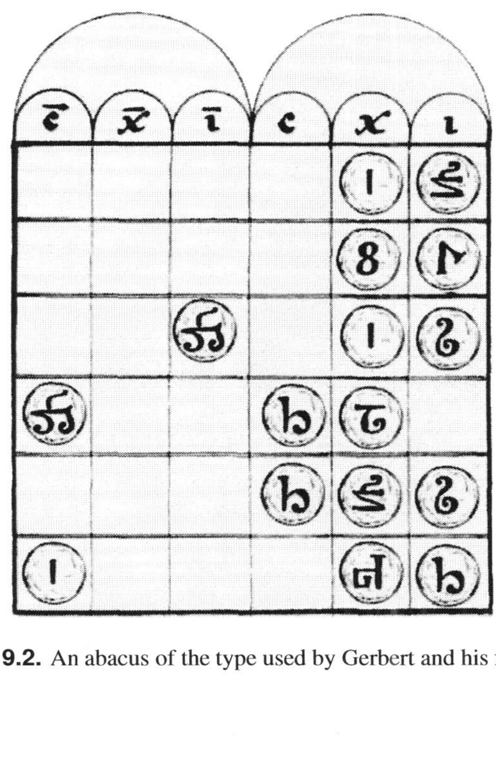

Figure 9.2. An abacus of the type used by Gerbert and his followers

Christian world out of its intellectual complacency: an edition of Euclid's *Elements* and a *zij* — a set of astronomical tables. This was a set of data giving the risings and settings of stars and the positions of the planets, and as such has to be adapted to the latitude where they are to be used; this *zij* had travelled and showed how the data had been recalculated for three different locations (we do not know where Adelard acquired it).

Such knowledge was completely beyond the ability of any European at the time. The impact of Euclid's *Elements* was comparable. For the first time the fundamentals of geometry were available, and it has been suggested that this is visible in the great change that swept over Christian architecture in the 12th century: the building of large cathedrals with their arched windows (themselves a Norman import from their contacts with the Arab world). Adelard's work has survived in three medieval editions of Euclid's *Elements*.

Adelard's example spurred others to follow him. Southern Spain was a largely tolerant and cultured Muslim caliphate embracing Toledo, Cordoba, Seville, and Granada. It had been created in the 720s when an Islamic army led by Tariq ibd Ziyad crossed from North Africa and defeated the Visigoths. (The name of the general survives in the name of his first fortress: Jabal Tariq, or Gibraltar.) When the tenth and last Umayyad Caliph was overthrown by the Abbasids, his grandson, Abd al-Rahman, continued Umayyad rule in Muslim Spain and extended it as far as Cordoba. By the late 10th century the Umayyads presented themselves as an alternative to Abbasid rule. Many Christians, known as Mozarebs, and Jews lived in their domains and may well have conducted much of their business in Arabic, making for a lively multilingual community. Contact with the Christian community to the north was sometimes fraught, but negotiations could take place.

Then at the end of the 11th century the Christian armies began to push steadily south. Toledo fell in 1085. Not surprisingly, one of the casualties of this shift in fortune was tolerance: al-Andalus came under the domination of the Almohads, conservative Islamic theologians with their own enthusiasm for strict adherence to a Holy Book. In 1170 they established their capital of al-Andalus in Seville. But Cordoba and Seville fell in the first half of the 13th century, thus reversing the flow of power that had seen a string of Arabic conquests in the 7th century. In 1492 the last Moorish outpost, Granada, fell to the armies of Ferdinand and Isabella, whose marriage had stood for a united Christian Spain (from which not only Muslims but also Jews were expelled).

Alfonso VI, the conqueror of Toledo, had recognised the considerable intellectual superiority of the Islamic world, and set about helping the Christian world to catch up. He copied his Islamic predecessors in recognising other faiths than his own: where they had granted limited rights to Jews and Christians as people of the Book, he did the same for Jews and Muslims. This atmosphere, which spread south for a time with the Christian 're-conquest', as it has controversially been called, created opportunities for Christian scholars to find and translate Arabic works, and several came to Toledo for that purpose.

Around 1150 Robert of Chester was one of the most prolific of these. He translated al-Khwārizmī's *Algebra* (and the *Koran* in 1143). At about this time Gerard of Cremona came looking for, and located, an Arabic edition of Ptolemy's *Almagest*, which he could not find in Greek. He translated it into Latin, and stayed in Toledo, producing translations of many other works. In a eulogy of him written shortly after his death his friends wrote of Gerard that[1]

> seeing the abundance of books in Arabic on every subject, and regretting the poverty of the Latins in these things, he learned the Arabic language, in order to be able to translate.

[1] Quoted in (Lindberg 1978, 65).

In the early 12th century, Sicily was under the rule of Frederick II, one of the great medieval emperors. He was a man of considerable learning, fluent in a number of languages including Arabic, and was described by his contemporaries as the 'Wonder of the World'. He was a great patron who encouraged the collection and translation of Greek and Arabic manuscripts. Eventually, Frederick II fell foul of the Vatican and was excommunicated by Pope Gregory II for delay in mounting a crusade, but the real problem lay in his plans to unify Italy. After his death in 1250 the struggle over Italy continued, culminating in a Papal triumph in 1266. As a result, Frederick's library was transferred to the Vatican, where the work of translation was resumed. We shall see that a number of important translations were based on sources in the Vatican, including some of the most useful editions of Euclid's *Elements*.

It is no easy task to catalogue all the translations made at this time; some of the most important of these are listed in Box 9.1.

The story of how the Greek inheritance was absorbed and developed shows that many of the same difficulties that Islamic culture had experienced were encountered again — problems of translation into a language ill-equipped to deal with it, for example. The whole intellectual climate in the Christian West was so different from that of the Greek world that many Greek ideas were inevitably misunderstood or not understood at all. The mathematical development in the West, stimulated by the Greek learning, nonetheless moved in directions that would have been foreign to Greek thought; there was virtually no existing contemporary knowledge into which the rediscovered Greek ideas could fit.

But if some translations were as literal as they could be made, many were far from faithful to their originals. There was a particular interest in the logical structure of past mathematics and the works were often adapted to meet the needs of the various schools of study — they were, in a word, *scholasticised.* One reason for this was the marked disparity between the Christian and Islamic cultures. Scholarship in the West, such as it was, had for some time been largely a matter of studying, copying and decorating ecclesiastical documents. The most intricate calculations were those needed for determining the date of Easter. The imaginative leap required to translate some of the most advanced Greek and Arabic works was to take several generations.

Two original mathematicians stand out in this period in the West: Leonardo of Pisa, known as Fibonacci (see Figure 9.3), and Jordanus de Nemore. Both of these were active in the early 1200s, and we look at them in turn.

Leonardo of Pisa (Fibonacci). Leonardo, whose father was an Italian working in North Africa, studied under a Moslem teacher and travelled widely in Egypt, Syria, and Greece. He became well acquainted with the Hindu–Arabic numerals and the methods of calculation associated with them, and in his *Liber Abbaci* of 1202 he strongly advocated their use.[2] The title of this work is misleading: it is not a book on the abacus, but one in which arithmetic and algebraic solutions of problems are investigated using the Hindu–Arabic numerals.

[2]The revised edition of the *Liber Abbaci* is dedicated to Robert of Chester and Michael Scot, who worked for Frederick II and translated texts by and commentaries on Aristotle from Arabic.

Box 9.1. Some 12th- and 13th-century translations of Arabic and Greek texts.

Spain, especially Toledo and Barcelona, but including Spanish scholars in Toulouse: translated from Arabic versions of Greek manuscripts and manuscripts except where indicated.
Adelard of Bath (there is uncertainty and disagreement about where Adelard worked): Euclid's *Elements*, Ptolemy's *Almagest* (from Greek), al-Khwārizmī's *Tables* and *Arithmetic*.
Gerard of Cremona: Archimedes' *On the Measurement of the Circle* and *On the Sphere and Cylinder* (fragments); Euclid's *Elements* and *Data*; Ptolemy's *Almagest*; works by Menelaus and others; al-Khwārizmī's *Algebra*, and Islamic works on mechanics, astronomy and optics.
Hermann of Carinthia: Euclid's *Elements* and Ptolemy's *Planisphere*.
Plato of Tivoli: Ptolemy's *Tetrabiblos*, and other works.
Robert of Chester: al-Khwārizmī's *Algebra*.

Sicily: translated from Greek manuscripts except where indicated.
Henricus Aristippus: Plato's *Phaedo*.
Eugenius of Palermo: Ptolemy's *Optics* (from Arabic).
Unknown translators: Euclid's *Elements*, *Data*, *Optics*, and *Catoptrics*.

The book is, perhaps, best remembered today for the 'rabbits problem' which gives rise to the Fibonacci sequence of numbers $(1, 1, 2, 3, 5, 8, 13, 21, \ldots)$, where each successive term is the sum of the previous two. Indian scholars had earlier found this sequence in their study of poetic metres, and it has since been discovered in remarkably diverse situations in nature.

The original version of the rabbits problem was as follows:[3]

> A certain man had one pair of rabbits together in a certain enclosed place, and one wishes to know how many are created from the pair in one year when it is the nature of them in a single month to bear another pair, and in the second month those born to bear also.

Here is another of his problems with solution, this time about a tree:[4]

> There is a tree $\frac{1}{4}$ and $\frac{1}{3}$ of which lie below ground, and are 21 *palmi*; it is sought what is the length of the tree; because the least common denominator of $\frac{1}{4}$ and $\frac{1}{3}$ is 12, you see that the tree is divisible into 12 equal parts; three plus four parts are 7 parts, and 21 *palmi* therefore as the 7 is to the 21, so proportionally the 12 is to the length of the tree. And because the four numbers are proportional, the product of the first times the fourth is equal to the second by the third: therefore if you multiply the second 21 times the third 12, and you divide by the first number, namely by the 7, then the quotient will be 36 for the fourth unknown number, namely for the length of the tree: or because the 21 is triple the 7, you take triple the 12, and you will have similarly 36.

[3]See (Fibonacci 2002, 404).

[4]See (Fibonacci 2002, 268) and, in a different translation, F&G 7.A1(b).

Figure 9.3. Leonardo of Pisa (c.1170–c.1250)

Leonardo's other works include the *Liber Quadratorum* (The Book of Squares), which includes indeterminate problems of the kind considered by Diophantus, and the *Practica Geometriae*, which uses algebraic methods to solve geometrical problems.

What could have been Leonardo's sources? Several Arabic authors are candidates, but their presence has to be inferred from the problems and methods that Leonardo used, for none is mentioned by name. Al-Khwārizmī is one, and Abū Kāmil is another, as has been shown by a careful comparison of problems in the *Liber Abbaci* with those in Arabic works. There are also sources deriving from Archimedes, Latin translations of the *Elements* from Arabic texts, and some Byzantine compilations of Diophantus.

It seems that the more advanced parts of Leonardo's work were little discussed, but many merchants and surveyors took up its practical aspects, and this led to a long tradition of practical arithmetic and algebra, clearly distinguishable from the scholastic tradition. Many of these problems had to do with changing money, and they can be quite complicated, whether because trade demanded it or because the author got carried away is not clear. Two features stand out. First, while many of these problems reduce to systems of linear equations, and as such belong to a much older tradition that started outside Europe, Leonardo's examples are among the most sophisticated and difficult. Second, he is occasionally almost willing to leave negative numbers as solutions — that is to say, while he rejected them as *solutions* he deliberately set problems in which they arise and then interpreted them as positive quantities to be subtracted. We can see this as a partial recognition that negative numbers — in this case, debts — can arise in real life. Leonardo's writings therefore provide an interesting insight into the kinds of problems and methods that were an intrinsic part of this practical tradition.

This practical mathematical tradition gradually also became a vernacular one — that is, using the language of everyday speech. Mathematics was increasingly useful in trade, not just between Italy and Egypt (say), but even between Italian towns that often had different currencies and different systems of weights and measures. A further significant impulse for this tradition was provided by the gradual spread of literacy in the West.

Jordanus de Nemore. So little is known about Jordanus de Nemore that the best anyone can say of his dates is that he flourished around 1230. We do not even know the country in which he was born, and the ascription 'de Nemore' is obscure: it means 'from Nemus' in Latin, but there is no identifiable place called Nemus from which he could have come; it has therefore been suggested that it is a corruption of 'de numeris', meaning 'of numbers', by association with his writings on arithmetic. He seems to have been a lay person who wrote on arithmetic, geometry and astronomy, and mechanics. He was best known for his work on the theory of weights, where he discussed how to resolve a static force into components in different directions, gave a proof of the law of the lever (explained earlier in Section 5.2), and discussed the concept of equilibrium.

His treatises on the subject (if indeed they are exclusively his — such was his influence that later writers may have attributed other people's work to him) are couched in a strikingly Euclidean framework, with postulates and theorems. The Euclidean example is generally maintained in his other writings. His *Algorismus*, which described the Arabic system of numbers and their uses, proceeds by definitions and theorems, so the basic operations up to the extraction of roots are presented formally and without examples, unlike in some other books of the period. His *De Numeris Datis* in four books is patterned on the style of formal statement of a proposition, its proof, and then a numerical example, so both Greek and Arabic influences are evident. The books are also notable for the system of single letters for quantities that Jordanus introduced (see Figure 9.4), although Fibonacci had done similar things a little earlier.[5]

Extensive use of symbols is evidence of the need to create a flexible mathematical language, and has led some authors to regard *De Numeris Datis* as a significant work in the history of algebra. Jordanus's *Arithmetica* is again modelled on the arithmetic books of Euclid's *Elements*, and differs markedly from the informal, popular, and sometimes inaccurate accounts of Nicomachus and Boethius. It became the standard work of reference on theoretical arithmetic in the Middle Ages, and was written in Euclidean style as a series of formal definitions and propositions, as though to remind his readers of the western classical tradition.

9.2 The rise of the universities

As we have seen, the 13th century was a time of renewed scholarship in Europe. Many Arabic and Greek texts were translated into Latin, and universities developed as associations of masters and scholars. The European world of learning was then more truly international than at any time since — indeed, 'the whole of educated Western Europe formed a single undifferentiated cultural unit'[6] and it benefitted from all those involved having learned Latin, the supra-national ecclesiastical and scholarly language, at the school associated with each cathedral and large church.

[5]See (Jordanus 1981, 127–132) and F&G 7.A2(a),(b).

[6]See (Knowles 1962, 80).

Jordanus De numeris datis.

fol. 188 v.

1) *Numerus datus est cuius quantitas nota est.* §. *Numerus ad alium datus est cum ipsius ad illum est proporcio data.* §. *Data est autem proporcio cum ipsius denominacio est cognita.* §. *Si numerus datus in duo diuidatur quorum differencia data est utrumque eorum datum.* §. Quia enim minor proporcio et differencia faciunt maiorem tunc minor porcio cum sibi equali et cum differencia facit totum sublata ergo differencia de toto remanebit Duplum minoris datum quo diuiso erit minor porcio data est et maior. §. Uerbi gracia . X . diuidatur . in duo quorum differencia duo que si auferatur de . X . relinquentur octo cuius medietas est quatuor et ipse est minor porcio altera sex.

2) *Si numerus datus diuidatur . per quodlibet quorum continue differencie date fuerint quodlibet eorum datum erit.* §. Datus numerus sit . a . qui diuidatur in . b . c . d . e . sit que e minimus et quisque eorum continue sit differencie date singulorum ad e . date erunt differencie . sit igitur . f . differencia . b . ad . e . et . g . h . differencie . c . ad . e . et. d . ad . e . et quia . e . cum singulis illorum facit singula istorum . manifestum est quod triplum ē . cum . f . g . h . facit illos tres . Quadruplum ergo . e . cum f . g . h . facit . a . singulis iis ergo demtis de . a . remanebit quadruplum . e . datum . quare . e . datum erit et per addicionem differenciarum erunt reliqua data. § hoc opus est . uerbi gracia . XL diuidantur per III quorum per . ordinem differencio sint IIII . III . duo. Differencia ergo primi ad ultimum . IX . et secundi ad illum . V . et tercij ad eum duo que simul faciunt . XVI . quibus demptis de . XL . remanebunt . XXIIII . quorum quarta pars . VI . et hoc erit minimus IIII . additis . a . IX . V . et duobus prouenient ceteri tres . VIII . XI . XV.

3) *Dato numero per duo diuiso si quod ex ductu unius in alterum producitur datum fuerit et utrumque eorum datum . esse necesse est.* § Sit numerus a b c . diuisus a b et . c . atque ex . a . b . in . c . fiat d . datus itemque ex a b c in se fiat e sumatur itaque quadruplum . d . qui sit . f . quo de e (sublato?) remaneat g . et ipse erit quadratum differencie a b ad . c . extrahatur ergo radix ergo et sit b . eritque b . differencia a b . ad c . tum quod sit b c datum erit et c et a b datum. § Huius opera facile . constabit huius modî uerbi gracia sit X . diuisus in numeros duos atque ex ductu unius eorum in alium fiat . XXI . cuius quadruplum et ipsum est . LXXXIIII . tollatur de

Figure 9.4. Jordanus's system of single letters for quantities, in *De Numeris Datis*, 1879 edition

This revival of learning invigorated the universities of Europe, but it had not, and was not to have, the same effect in the Islamic world. So before moving to look at the European situation we should look briefly at the Islamic case, bearing in mind that cross-cultural comparisons, although tempting, are always difficult and contentious. The Islamic system of *madrasas*, or colleges, was one of charitable endowments that specialised in teaching law. Unlike the later European universities, however, the transmission of knowledge and the granting of expertise was strictly personal. Reading and debate were part of the education, but each new scholar obtained his licence, so to speak, personally from his teacher. Not only did standards vary uncontrollably, but the madrasa itself had no collective authority. It has been argued that these factors weakened the Islamic universities.[7] Lacking a collective structure, and tied to a charitable status, madrasas were in a poor position to resist calls for orthodoxy and to provide autonomous centres of learning. One should not imagine a 13th-century European university as a seat of free speech and unfettered enquiry, but the Islamic system prescribed no book but the *Koran*, and recognised no higher authority than a well-articulated and defended tradition.

[7] See (Huff 1993).

The first two European universities, Bologna and Paris, grew up in the early 12th century as places where theology, law, and medicine were taught, and this syllabus, in whole or in part, came to characterise the early universities. The form that the teaching took was a mixture of lectures, reading, and public debate; in the Faculty of Law this was often of an adversarial kind. The university idea spread to other cities, for example to Oxford by the start of the 13th century, and the institutions acquired a degree of autonomy, both from the town where they were situated, and from the Church, through their expertise in law and also, increasingly, in theology. Crucially, only the university had the legal right to confer degrees, rising to the degree of doctorate, and as time went on it did so in a demanding and highly structured way in which, by the late 13th century, a close study of Aristotle was mandatory. One can only speculate, but it would seem that the widespread recognition in Europe that the Europeans had a great deal to learn was to foster universities, while a justifiable feeling in the Islamic world that they had understood many things led to a more restrictive attitude.

The two leading northern European universities of the 13th century were Paris and Oxford. At Paris the teaching was, on the whole, in the hands of Dominican friars, most notably Thomas Aquinas, who dominated intellectual life at the end of the 13th century. Aristotle's works, with their classical heritage, were emphasised and made compatible with Christian teachings; these featured a strong concern with logically clear rational argument, but were not especially mathematical.

Oxford, however, was more strongly a seat of Franciscan teaching; it was influenced by a tradition of Christian neo-Platonism stemming from Saint Augustine, who had died in the year 430. In this tradition mathematics played a more important role, and thus its study was the more encouraged. Robert Grosseteste — his name means large-head — who taught at Oxford during the first half of the 13th century and became its first Chancellor, wrote:[8]

> The usefulness of considering lines, angles and figures is the greatest, because it is impossible to understand natural philosophy without these ... For all causes of natural effects have to be expressed by means of lines, angles and figures, for otherwise it would be impossible to have knowledge of the reason concerning them ... Hence these rules and principles and fundamentals having been given by the power of geometry, the careful observer of natural things can give the causes of all natural effects by this method ... since every natural action is varied in strength and weakness through variation of lines, angles and figures.

Here we see in outline what might be thought of as a programme for investigating the world through mathematics, principles that scholars of succeeding generations would attempt to put into practice. Grosseteste's most famous follower was the Franciscan friar Roger Bacon who went even further (see Figure 9.5):[9]

> He who knows not mathematics cannot know the other sciences nor the things of this world ... And, what is worse, those who have no knowledge of mathematics do not perceive their own ignorance and so do not look for a cure. Conversely a knowledge of this science prepares the mind and raises it up to a well-authenticated knowledge of all things.

The enthusiasm of Grosseteste and Bacon for mathematics had its inspiration and clearest justification in the success of geometrical optics as a mathematical science — and

[8] In (Crombie 1953, 110).

[9] See (Molland 1968, 110).

Figure 9.5. Roger Bacon (c.1212–1294)

as the mathematisation of light, the most potent of neo-Platonic symbols. It was at Oxford in the next century that some of the most sophisticated mathematical discussions of the Middle Ages took place, in which mathematics was used and developed along the lines of the programme that Grosseteste and Bacon had advocated. Indeed, this went so far that one historian has spoken of 'the fourteenth-century habit of quantifying almost everything seen or unseen'.[10]

Most of the Oxford scholars concerned in this development were fellows of Merton College, and are often spoken of collectively as 'the Merton School', although their European contemporaries referred to them just as 'the English' (*Anglici*) or 'the British' (*Britannici*). Most prominent of this group was Thomas Bradwardine, who showed how the programmatic utterances of Grosseteste and Bacon could bear real fruit in the mathematical analysis of nature. He wrote:[11]

> It is mathematics which reveals every genuine truth, for it knows every hidden secret, and bears the key to every subtlety of letters; whoever then has the effrontery to study physics while neglecting mathematics, should know from the start that he will never make his entry through the portals of wisdom.

Bradwardine's treatise *On the Ratios of Velocities in Motions* (1328) was an especially influential application of ratio theory, along the lines of Euclid's *Elements*, Book V, to questions of how speeds of motion depend on the ratio of the force producing the motion to the resistance to movement. Bradwardine's work was very successful conceptually, as an abstract mathematical analysis: the empirical measurement of things like force and resistance was not the kind of enterprise he was engaged in. In the hands of Bradwardine and his fellow-scholars the mathematisation of magnitudes went far beyond anything

[10]See (Molland 1968, 123).

[11]See (Weisheipl 1959, 446).

thought of in Greek geometry. Other members of the Merton School attempted to quantify intensities of light, heat, colour, sound, hardness, and density, as well as magnitudes that mathematicians have still not managed to quantify effectively: knowledge, certitude, charity, and grace.

How far these developments would have gone we cannot know. Bradwardine, in company with many other scholars, died of bubonic plague (the 'Black Death') in 1349, shortly after he had been appointed Archbishop of Canterbury. Despite the ravages and devastation of the Black Death and other social and political miseries, teaching and learning of a sort did continue at Oxford, albeit in a rather depressed state.

Future ages would look back and see the early 14th century as a pinnacle of scholarly achievement, as what a recent historian has described as Oxford University's 'period of greatest glory', in which 'the schools of Oxford attained a reputation throughout the academic world that they never quite equalled before or after'.[12] When in the 16th and 17th centuries mathematicians and other scholars looked back towards a distant golden age of British learning, it was Oxford from Grosseteste to Bradwardine that came to mind, although specific knowledge of their achievements was tenuous by this time and verged on folklore; Roger Bacon, in particular, became in popular memory a magician and dabbler in the black arts.

Nicole Oresme. We mentioned earlier that many translations of Greek and Arabic work were done in the scholastic tradition. This too had its creative mathematicians, of whom the Norman scholar Nicole Oresme is one of the most illustrious. He was born near Bayeux in Normandy in 1320, and in 1348 he registered as a student of theology at the University of Paris, by then already established for nearly two centuries as the intellectual capital of Europe. This means that he would already have obtained a Master of Arts degree with the usual liberal arts training: the *trivium* (grammar, rhetoric, and dialectic) and the *quadrivium* (arithmetic, geometry, music, and astronomy).

In the course of developing his theological career (he eventually became Bishop of Lisieux in 1377, dying in office in 1382), Oresme became one of the first translators of Latin editions of Aristotle and of other scientific works into the French language, thus playing an important role in the vernacular educational tradition. It is a measure of the difficulty of this work, as well as his impact on the French language, that he had to coin some three hundred new French words by adapting Latin ones; medieval French did not have the technical vocabulary needed.

In his own work, Oresme blended powerful mathematics within an Aristotelian philosophical framework. An indication of his geometrical approach to the study of motion can be seen in Figure 9.6, taken from the 1505 printing of his *Treatise on the Latitude of Forms*, a work that dates from the 1350s. He extended the mathematical consideration of velocity and movement begun in earlier work at Merton College, Oxford. In particular, he gave an elegant graphical proof of what is called the mean speed theorem. This says that the distance covered by an object A moving with constant acceleration for a given period of time is the same as the distance covered by an object moving for the same period of time with the average (or mean) of the initial and final velocities of the object A. Oresme's work on ratio and proportion developed that of Bradwardine into a calculus of ratios (enabling fractional and irrational powers of quantities to be handled), which he then applied to the problem of whether heavenly motions are commensurable or incommensurable.

This is an important problem for anyone concerned to know whether certain planetary conjunctions will occur. Will two planets, assumed to be moving uniformly in circular

[12]See (Phillips 1985, 105).

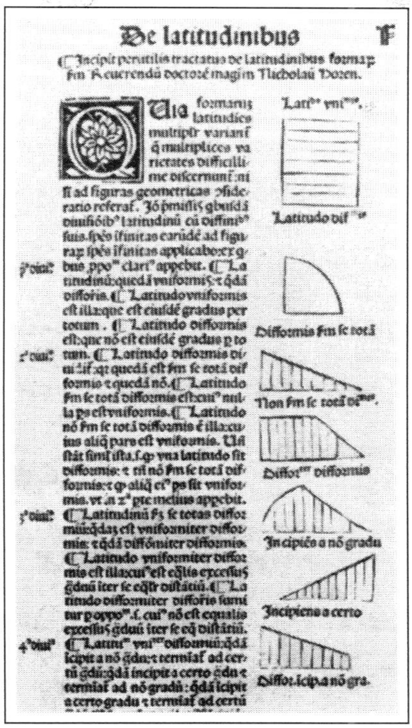

Figure 9.6. From Oresme's *Latitude of Forms*, 1505 edition

orbits, seem to meet? If so, will they do this once, or several times, or infinitely often? Oresme idealised the problem so that at issue are only exact coincidence of point-sized bodies along the line of sight from the centre of their orbits. He proved that if two bodies start from coincidence and move with commensurable velocities (where one velocity is a rational multiple of the other), then they will return to coincidence infinitely often, and he showed how to find the first time that they do so. Then he considered bodies starting from different positions, and then three bodies, showing that triple conjunctions (in which all three bodies lie in a line with the centre) may never occur.

Oresme then considered two bodies whose velocities are not commensurable, and showed that if they start from a given position then they will never conjunct at that position again, nor have they ever done so before. Moreover, if they ever conjunct again at some other point, then the angular separation between these two points of conjunction is incommensurable with the whole circle, and the times between any two successive conjunctions are also incommensurable. From this he deduced that if enough successive conjunctions are considered, then the distance between neighbouring points of conjunction can be made arbitrarily small. Oresme then went on to speak of the infinitely many points of conjunction that occurred in the past (all distinct and none to be repeated) and the infinitely many points of conjunction that will occur in the future (likewise all distinct and none to be repeated, and none of which have ever occurred before). This is a remarkable display of mathematics, directed to a novel purpose.

A dialogue between Arithmetic and Geometry. At the end of his *Treatise*, Oresme asked which of the conflicting hypotheses (commensurable or incommensurable motions) occurs in the heavens. He found himself unable to decide, and offered instead a report of a 'dream' in which Arithmetic and Geometry each sought to convince the judge, Apollo. Edward Grant, an editor of Oresme's work, has provided this helpful summary of the ensuing exchange of views, which has four aspects:[13]

Oresme on arithmetic and geometry.

1. ARITHMETIC: Rational ratios are better than irrational ratios because they produce pleasure, whereas irrational ratios cause offensive effects.

GEOMETRY: True, rational ratios do produce a certain beauty in the world; but if they were united harmoniously with irrational ratios this would be far better, since from such a union there would result a rich variety of wonderful effects.

2. ARITHMETIC: Arithmetic is the first-born of the mathematical sciences and her numbers are most worthy to represent the relationships between the celestial motions.

GEOMETRY: Geometry represents magnitude in general, which includes numbers as a special category. Thus, Geometry embraces not only numerical or rational ratios but much more, so that it not only furnishes as much beauty to the heavens as does Arithmetic, but adds much more splendour.

3. ARITHMETIC: Only if the celestial motions are commensurable and represented by rational ratios can the music produced by the moving spheres be harmonious concords; if these ratios were irrational, the music would be discordant.

GEOMETRY: If there really is celestial music, it would not vary as the velocities of the celestial motions but, rather, as volumes of the celestial spheres, or in some other way. But even if celestial music resulted from the celestial motions themselves, there is no evidence for assuming that the principal harmonic concordances would be produced. Furthermore, no one has yet determined whether celestial music is sensible or merely intelligible. But if it is sensible and created by fixed and rational ratios, it would be monotonous; only infinite variation is capable of producing interesting sounds.

4. ARITHMETIC: Unless the celestial motions are commensurable, astronomical prediction and tables, as well as knowledge of future events, would be impossible. Terrestrial effects would fail to repeat and a Great Year could not occur.

GEOMETRY: But if all celestial motions are commensurable, conjunctions and other astronomical aspects could occur only in a certain finite number of fixed and different places in the sky. Thus a peculiar consequence of commensurable celestial motions is that certain places in the sky would be preferred over other places. Would it not be better that such events be capable of occurring anywhere in the sky? Man cannot attain to exact knowledge of astronomical phenomena and must rest content with approximations. Indeed, acquisition of exact knowledge would serve to discourage

[13]See (Grant 1971a, 68–69).

> man from making continual observations; and if man had precise knowledge of future celestial positions he would become like the immortal gods themselves, a repugnant thought. Fortunately, this is not likely to pose problems since, as shown elsewhere by mathematical demonstration, it is probable that the celestial motions are incommensurable.

And what does Apollo decide, when he promises to 'announce the truth in the form of a judgement'? Alas, at that moment Oresme woke up, and the book ends with his confession that he does not know what Apollo actually said!

It is evident that ARITHMETIC and GEOMETRY are talking through each other, neither listening to what the other has to say. But it is also clear from his refusal to take sides that Oresme finds that each has a strong case. We can see that the musical analogy is vivid here, and that in some way Plato's organisation of the mathematical sciences has persisted. The inadequacy of the rational numbers for the representation of reality is strongly felt, but the price of admitting irrational numbers is held to be that knowledge becomes approximate where it had been exact — a high price to pay.

The second half of the 14th century saw a period of decline, partly due to the scourge of the Black Death. The subsequent period of recovery forms the topic of the next chapter.

9.3 Further reading

Crombie, A.C. 1969. *Augustine to Galileo*, Penguin, (1st edn. 1952): a full and helpful survey of Western science in the Middle Ages.

Fauvel, J., Flood, R., and Wilson R. (eds.) 2013. *Oxford Figures: Eight Centuries of The Mathematical Sciences*, second edition, Oxford University Press. A readable and stimulating account that will be worth coming back to in Chapter 11.

Goldstein, C., Gray, J.J., and Ritter, J. (eds.), 1996. *L'Europe Mathématique, Mathematical Europe*, Éditions de la Maison des Sciences de l'Homme, Paris. This book has a reasonably up-to-date account of many of the topics discussed here (some in English, some in French), with particular bearing on the construction of an idea of a European mathematical tradition.

Grant, E.N. 1971. *Science in the Middle Ages*, Wiley. A most readable short survey with a useful critical bibliography.

———. 1971a. *Oresme and the Kinematics of Circular Motion, edited with an introduction, English translation and commentary*, Wisconsin University Press. Scholarly and rich.

Hannam, J. 2009. *God's Philosophers: How the Medieval World Laid the Foundations of Modern Science*. Icon Books. This readable presentation does much to rescue medieval philosophers from the slurs of later generations. It gives an up-to-date account of the political and theological context in which they wrote and discusses Galileo's work in the same spirit.

Hay, C. (ed.), 1987. *Mathematics from Manuscript to Print: 1300–1600*, Oxford University Press. A collection of scholarly essays on some of the topics in this chapter.

Lindberg, D.C. (ed.), 1978. *Science in the Middle Ages*, University of Chicago Press. A collection of essays on different topics, including a lucid essay on medieval mathematics by Michael Mahoney.

Murdoch, J.E., 1984. *Album of Science: Antiquity and the Middle Ages*, Charles Scribner's Sons. There is much on medieval mathematics in this splendid volume, which comprises a lavish selection of illustrations with informative commentary by a leading scholar in the field.

10

The Renaissance: Recovery and Innovation

Introduction

Recovery from the Black Death was slow, and those who came after it found a world forever changed. The invention of printing in the mid-15th century in Europe, some eight centuries after wood-block printing had been invented in China, revolutionized culture and learning. The first book was printed in Germany by Johann Gutenberg in 1454: no longer had reliance to be placed on the copying of manuscripts, with its small number of copies, slowness, and inevitable errors. Standard printed texts began to circulate, and extensive personal libraries were built up in the humanistic studies.

Two traditions were enabled by the rise of the printed book. One is the renewed interest in the literature, values, and achievements of the ancient Roman and Greek world associated with what is called the Renaissance. In mathematics, new editions were made of those texts by ancient authors that had survived: works not only by Euclid and Archimedes, but by several others. The other was that of practical books, usually in the vernacular.

As with other turning points in cultural life, historians argue about the Renaissance at length. When was it? Where did it take place? Was it the same event in southern and northern Europe? Is its importance overstated? Did it begin, as many would say, with such men as Petrarch and other Italian humanists in the 14th century, or later with Leonardo da Vinci, the archetypal 'Renaissance man', at the end of the 15th century?

Such questions are never to be answered definitively, but the new resurgence of interest in ancient learning is conventionally dated from the fall of Constantinople to the Turks in 1453, when the Byzantine Empire came to an end and many Greek scholars moved west. They brought with them Greek texts that were very different from those studied earlier — texts that tended to be free of the Islamic additions and deletions, and of the commentaries that were often woven into Arabic translations. As a result, the Greek aspect of the texts stood out more clearly, and a changed attitude to translating could begin to spread, one more focussed on the rediscovery of the authentic ancient source and its study for its own sake.

This is not to say that translations were done always, or solely, with accuracy in mind, but rather that they could now be done in a more critical spirit, giving more weight to what were supposed to be the words of the original author, and less weight to accumulated tradition, however shrewd; we shall see examples of this in the work of Maurolico and Commandino. Furthermore, unlike in the 12th and 13th centuries, the intellectual level of European scholarship was now higher and the Islamic countries had embarked on a slow decline, invisible to people at the time but much debated by historians today. This helped to promote the idea of a Graeco–Roman tradition to which, quite unfairly, the Islamic contribution was considered marginal.

However, this high cultural Renaissance spread only slowly to mathematics and science, and it combined with a vigorous vernacular tradition — mathematics written in the native language of the region rather than in Greek or Latin — that was opened up by the new market in printed books.

The combined effect of printing and circulating new editions of mathematical texts for an elite and the production of original texts aimed at practical people (merchants and teachers) was a steady spread of mathematical learning throughout Europe. The university mathematical curricula developed from basic arithmetic and geometry to the study of Euclid and Archimedes, quite sophisticated astronomy, and a certain amount of civil, marine, and military engineering, but a further effect of printing was its uncontrolled nature, which put learning in the hands of many in a way that the powerful few found hard to control and curb.

The vernacular tradition of mathematics in 15th-century Europe was driven by trade. Trade promoted a need for new mathematics, such as double-entry bookkeeping, and with the spread of printing, energetic businessmen could begin to teach themselves the mathematics they needed. This was arithmetical and algebraic in flavour, and focussed more on problem solving than theory building. It mixed with the more theoretical classical tradition, and led to one of the high points in the history of algebra: the solution of cubic and quartic equations in 16th-century Italy. This story is notable not only for the mathematics involved, but also because of what it tells us about the life of mathematically minded scholars in those days. From some perspectives, the circumstances in which they worked may seem strange today, more reminiscent of modern industrial or military intrigue.

In the vernacular tradition, problems were initially expressed in words, and then in a gradually more symbolic language. We can regard this chapter as the 'birth process' of elementary algebra, in the sense that the methods were general, the solutions were properly derived, and algebraic rather than geometrical language was used.

This chapter starts by looking at the vernacular tradition and then at the recovery of ancient texts before turning to look in more detail at the new, Italian study of algebra. It ends with the merging of these traditions in the first influential rediscovery of Diophantus's *Arithmetica*.

10.1 Early European mathematics

Three men stand out as significant early figures in any analysis of the 'rebirth' of mathematics in the West, and they illustrate the complexity of the process. The first was Regiomontanus, one of a number of mathematicians in Europe in the latter part of the 15th century who were able to grasp the sophisticated technicalities of Euclid. But his contributions stalled, and after him came two figures, Nicolas Chuquet and Luca Pacioli, who between them capture the ambiguities involved in talking about the Renaissance.

Regiomontanus. The most influential mathematician of the 15th century was Johann Müller, known as Regiomontanus, the Latin name for his birthplace Königsberg (then the capital of East Prussia, now part of Russia) — an English term would be 'King's mountain'. He was a remarkable calculator, instrument maker, printer, translator, scientist, and mathematician. In a life lasting only forty years his achievements were quite exceptional. He was especially active in translating and publishing classical mathematical manuscripts, an activity in which he was supported by Cardinal Bessarion, the Papal Legate to the Holy Roman Empire. Bessarion's native language was Greek, and he was an ardent advocate of the translation enterprise.

Regiomontanus completed the translation of Ptolemy's *Almagest*, begun at Bessarion's instigation by his teacher in Vienna, Georg Peurbach. He then translated works of Apollonius, Heron, and Archimedes, and made original contributions to mathematics, of which the most important is his *De Triangulis Omnimodis* (On all Sorts of Triangles — see Figure 10.1), a systematic account of 'solving' triangles that, with his essay *On the*

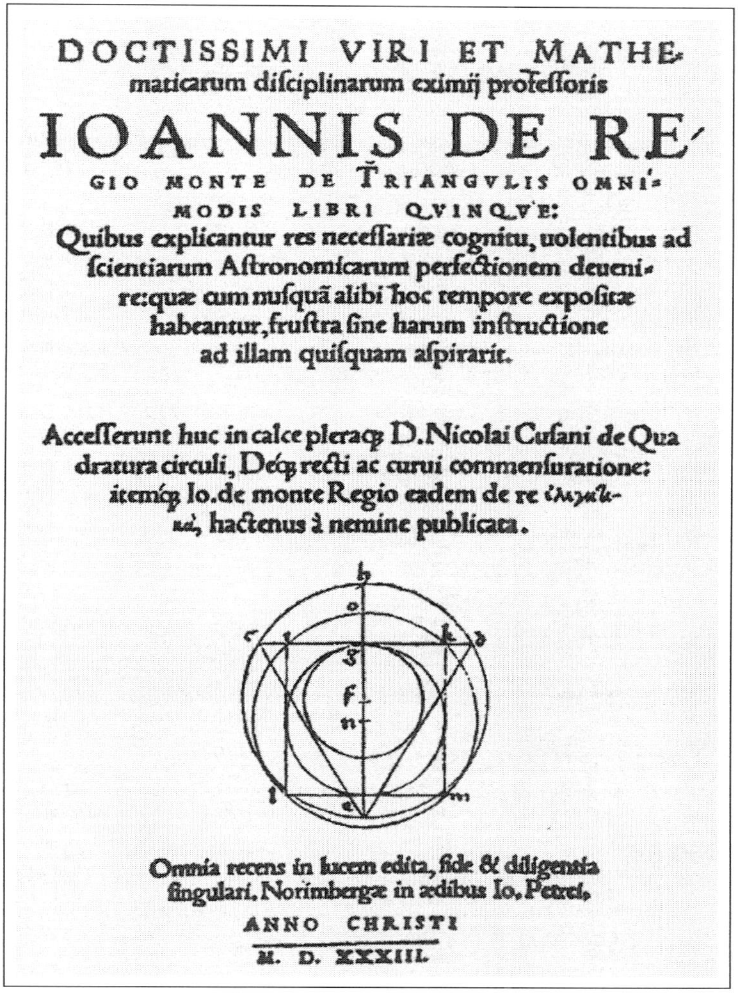

DOCTISSIMI VIRI ET MATHE-
maticarum diſciplinarum eximij profeſſoris
IOANNIS DE RE-
GIO MONTE DE TRIANGVLIS OMNI-
MODIS LIBRI QVINQVE:
Quibus explicantur res neceſſariæ cognitu, uolentibus ad ſcientiarum Aſtronomicarum perfectionem deueni-re: quæ cum nuſquã alibi hoc tempore expoſitæ habeantur, fruſtra ſine harum inſtructione ad illam quiſquam aſpirarit.

Acceſſerunt huc in calce pleraq; D. Nicolai Cuſani de Quadratura circuli, Deq; recti ac curui commenſuratione: itemq; Io. de monte Regio eadem de re ἐλεγκτικά, hactenus à nemine publicata.

Omnia recens in lucem edita, fide & diligentia ſingulari. Norimbergæ in ædibus Io. Petrei,
ANNO CHRISTI
M. D. XXXIII.

Figure 10.1. Regiomontanus's *De Triangulis Omnimodis*

Construction of Sine Tables (first published in 1541), formed in effect a complete introduction to the trigonometry learned from Islamic works, especially those of Nasīr al-Dīn (see Section 8.3). Written in manuscript around 1464, but printed only in 1533, *De Triangulis* formed the foundation of trigonometry in the Renaissance period. He also planned an edition of Jordanus de Nemore's *De Numeris Datis*, but died before this could be finished.

Regiomontanus also discovered a Greek text of the *Arithmetica* of Diophantus, but never found the time to translate it into Latin as he had wished. But when Bessarion moved to the Venetian Republic, Regiomontanus was able to give a series of lectures at the University of Padua, then under Venetian control. Among other sources, Regiomontanus drew on a 12th-century history of mathematics and astronomy by the Spanish translator Plato of Tivoli, stressing the continuity of mathematical tradition as it passed from culture to culture through the centuries and the crucial role played by translators. He argued that just as Diophantus was the source of Arabic algebra, and Islamic scholars had gained knowledge of algebra by translating Diophantus, so his own contemporaries should translate the same writer from Greek into Latin. This was not to happen for some time, however, as we shall see in Section 10.4.

Bessarion's death in 1472 robbed Regiomontanus of support for his work, and thus of further influence. Regiomontanus prepared a manuscript on algebra, but it remained unpublished for a century because, on his death, it came into the hands of a patron who did not make it available to the mathematicians of the day. Regiomontanus's untimely death may also have been significant for the limited development of algebra in the 15th century, but if no contemporary had quite his vision, there were a number of other mathematicians at work: in the 14th and 15th centuries there was an extensive Italian tradition with many surviving manuscripts on arithmetic and elementary algebra, and also an inventive German vernacular tradition.

Chuquet and Pacioli. The careers of two other mathematicians who flourished in the second half of the 15th century also illustrate the importance both of patrons and printing at the time: the Frenchman Nicolas Chuquet and the Italian Luca Pacioli.

Chuquet. Little is known about Chuquet, even his dates are uncertain. He seems to have been living in Lyons when he completed an extensive manuscript in 1484, and may have been working as a doctor or as a teacher of arithmetic. His *Triparty en la Science des Nombres*, as its name suggests, was a work in three parts: a section on arithmetic and algebra, a section on applications of these techniques to problems in geometry, and a commercial arithmetic. It was thus very much in the practical tradition, and Chuquet wrote in his vernacular (French), while his mathematics shows distinct traces of Italian influence. But it had little direct effect, mainly because it remained only in a single manuscript that lay hidden for centuries in various libraries and library vaults, and was not rediscovered until the 19th century. While much of it was incorporated into a printed work that was compiled by his pupil Estienne de la Roche and published in 1520, many of Chuquet's innovations were omitted or distorted by de la Roche, so Chuquet's influence on subsequent developments in algebra was at best slight and second-hand.

Among his innovations that could have had a valuable effect was his notation for exponents. The premier or unknown has exponent 1, written as a superscript to its coefficient: thus, our $3x$ would be written as 3^1. Similarly, squares have exponent 2, cubes have exponent 3, and so on. He also used the symbols $\overline{p}$ and $\overline{m}$ to stand for 'plus' and 'minus' respectively, so we would interpret the expression

$$3^3\overline{p}4^2\overline{m}1^1$$

as $3x^3 + 4x^2 - x$. A zero exponent was introduced for numbers (for example, 9^0 is 9), and negative exponents were also permitted (see Figure 10.2). The first example of an isolated negative number in an equation also appears in the *Triparty*.

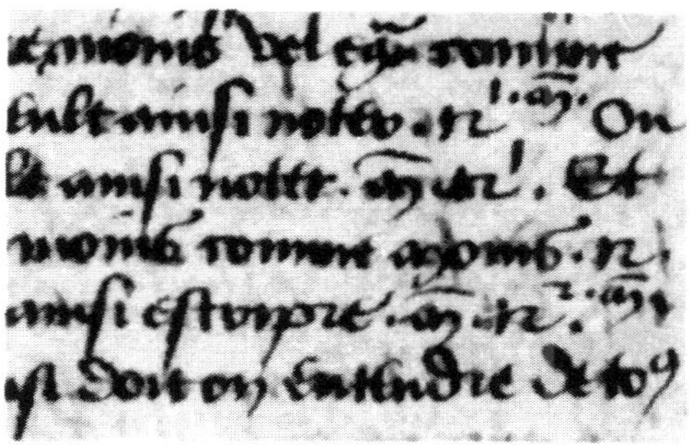

Figure 10.2. Chuquet's notation for negative exponents in his *Triparty*. The exponent for -1 appears in the second line as $1.\overline{m}$

It is possible that Chuquet's notation for exponents was inspired by the work of Oresme in the previous century. Chuquet's rules for the manipulation of exponents have in them the germ of the idea of logarithms, although they lack any concept of expressing all numbers to a common base.[1]

Chuquet then listed the first 21 powers of 2, from the one with 'denomination' 0 (that is, $2^0 = 1$) to the one with 'denomination' 20 ($2^{20} = 1{,}048{,}576$):[2]

Numbers	Denomination
1	0
2	1
4	2
8	3
16	4
32	5
...	...
262,144	18
524,288	19
1,048,576	20

and observed[3]

Chuquet on exponents.

> Now it is necessary to know that 1 represents and is in the place of numbers, whose denomination is 0. 2 represents ... the first terms, whose denomination is 1. 4 holds the place of the second terms, whose denomination is 2. And 8 is in the place of the third terms, 16 holds the place of the fourth terms, 32 represents the fifth terms, and so for the others. Now whoever multiplies 1 by 1, it comes to 1, and because 1 multiplied by 1

[1] The journey from Chuquet's relatively simple perception to the invention of logarithms is discussed in Chapter 11.

[2] Chuquet, *Triparty*, transl. in (Flegg, Hay, and Moss 1985, 151–153) and in F&G 7.B2.

[3] See (Chuquet 1985, 151–153) in F&G 7.B2.

> does not change at all, neither does any other number when it is multiplied by 1 increase or diminish, and for this consideration, whoever multiplies a number by a number, it comes to a number, whose denomination is 0. And whoever adds 0 to 0 makes 0. Afterwards, whoever multiplies 2, which is the first number, by 1, which is a number, the multiplication comes to 2; then afterwards, whoever adds their denominations, which are 0 and 1, it makes 1; thus the multiplication comes to 2^1. And from this it comes that when one multiplies numbers by first terms or vice versa, it comes to first terms. Also whoever multiplies 2^1 by 2^1, it comes to 4 which is a second number. Thus the multiplication amounts to 2^2. For 2 multiplied by 2 makes 4 and adding the denominations, that is, 1 with 1, makes 2. And from this it comes that whoever multiplies first terms by first terms, it comes to second terms. Likewise whoever multiplies 2^1 by 2^2, it comes to 8^3. For 2 multiplied by 4 and 1 added with 2 makes 8^3. And thus whoever multiplies first terms by second terms, it comes to third terms. Also, whoever multiplies 4^2 by 4^2, it comes to 16 which is a fourth number, and for this reason whoever multiplies second terms by second terms, it comes to fourth terms. Likewise whoever multiplies 4 which is a second number by 8 which is a third number makes 32 which is a fifth number. And thus whoever multiplies second terms by third terms or vice versa, it comes to fifth terms. And third terms by fourth terms comes to 7th terms, and fourth terms by fourth terms, it comes to 8th terms, and so for the others. In this discussion there is manifest a secret which is in the proportional numbers. It is that whoever multiplies a proportional number by itself, it comes to the number of the double of its denomination, as, whoever multiplies 8 which is a third number by itself, it comes to 64 which is a sixth. And 16 which is a fourth number multiplied by itself should come to 256, which is an eighth. And whoever multiplies 128 which is the 7th proportional by 512 which is the 9th, it should come to 65 536 which is the 16th.

At the end of the extract, Chuquet drew attention to the fact that if you want to multiply 128 by 512 you can proceed by observing that $128 = 2^7$ and $512 = 2^9$, so

$$128 \times 512 = 2^7 \times 2^9 = 2^{7+9} = 2^{16},$$

by the usual rule for exponents; since $2^{16} = 65{,}536$, you have: $128 \times 512 = 65{,}536$. In this way a multiplication has been reduced to an addition ($7 + 9 = 16$), via a knowledge of the powers of 2.

With the arrival of hand-held calculators, logarithms have almost disappeared from modern mathematical syllabuses and with them any experience of just how tedious and error-prone the operations of multiplication and division can be. For example, to compute $987{,}654 \times 23{,}456$ by hand we must perform thirty simple multiplications, starting with 4×6 and finishing with 9×2: this yields five rows of six- or seven-digit numbers that must then be added together to give the answer: $23{,}166{,}412{,}224$. Even a much smaller task, to divide $987{,}654$ by 97, is considered difficult by many people today (the answer is $10{,}182$). These are tasks that logarithmic tables made relatively easy, reducing the first to little more than an addition and the second to a subtraction. Chuquet's use of exponents was a modest step in that direction, although, as we shall see, the decisive step was taken much later.

Later writers made Chuquet's idea work in general. Suppose you know that $27 = 2^a$ and $79 = 2^b$, where a and b are some numbers (not necessarily integers). Then to multiply

27×79, you could argue as before:

$$27 \times 79 = 2^a \times 2^b = 2^{a+b}.$$

Having calculated $a + b$, if you also know that $2^{a+b} = 2133$, you would again have done multiplication by addition: $27 \times 79 = 2133$.

For this to work, somebody has to calculate a table expressing every number as a power of 2. The number a for which $2^a = m$ is called the *logarithm* of m to the base 2. So the idea is that to calculate $m \times n$ we proceed as follows. Suppose that $m = 2^a$ and $n = 2^b$: we say that $\log_2 m = a$ and $\log_2 n = b$, so $\log_2(m \times n) = a + b$. In the case above, working to four decimal places, we have $27 = 2^{4.7549}$ and $79 = 2^{6.3038}$, so we calculate

$$\log_2(27 \times 79) = 4.7549 + 6.3038 = 11.0587,$$

and look up in a table the number that has 11.0587 as its logarithm to the base 2. This, happily, turns out to be 2133.

Luca Pacioli. Much better known in its day than Chuquet's writings was the *Summa de Arithmetica, Geometrica, Proportioni et Proportionalità* by the Italian friar Luca Pacioli. This was published in Italy in 1494, although completed some years earlier, and was a compilation of arithmetic, algebra, elementary geometry, and the first account of double-entry bookkeeping.[4]

Although more comprehensive than Chuquet's work, the *Summa* contained less mathematical sophistication and innovation, and the notation and terminology were less generalised — for example, Pacioli's notation for roots was confined to square and cube roots, and his terminology for powers was limited. Like Chuquet, Pacioli included material directed towards the practical applications of arithmetic and algebra to problems of barter and currency exchange. The section on double-entry bookkeeping has become a classic, translated into many languages over the succeeding centuries.

Figure 10.3 shows Luca Pacioli as a geometry teacher with a pupil. Note the geometrical instruments (compasses and set-square), a Euclidean diagram, and a geometry text, as well as the Platonic solid (dodecahedron) on the table and a transparent solid suspended in the air. The original painting, traditionally attributed to Jacopo de Barbari, is in the Museo e Gallerie Nazionale di Capodimonte, Naples.

Pacioli is also notable for a work entitled *De Divina Proportione* (On Divine Proportion) on polygons and polyhedra. The outstanding illustrations in this book are due to his friend Leonardo da Vinci, with whom he lived for a time and to whom he taught some mathematics (see Figure 10.4).

But if Pacioli lacked mathematical sophistication by comparison with his French contemporary, he had a much more successful career, not least because he had a patron. Indeed, he had been able to dedicate his *Summa* to his pupil, the young Duke of Urbino. This association with the Court of Urbino at the height of its powers was only one of a number of useful contacts that Pacioli made throughout his life and that kept him and his work in the public eye.

Other writers on mathematics in the early 16th century help us understand the role of the market in promoting new work.

A vernacular tradition grew up, especially in algebra and arithmetic. In algebraic problems in which the unknown is a number, the first power of this unknown quantity was called the *cosa* (meaning 'the thing') in Italian, and the *chose* in French. German writers took over the word from Italians with whom they traded, and called the unknown quantity

[4]See (Smith 2008) on double-entry bookkeeping; for an extract from Pacioli's work, see F&G 7.B3.

Figure 10.3. Luca Pacioli as a geometry teacher

coss in German, so algebra become known as the 'cossic art' in Germany; indeed, the most successful algebra of note to appear in Germany in the early 16th century was Christoff Rudolff's *Die Coss* (1525).

Cossist algebra was a feature of schools in Germany that taught reading and writing and how to use the abacus; both the Hindu–Arabic numerals and Roman numerals were taught. Those who taught the new Hindu–Arabic numerals also taught how to conduct the basic operations of arithmetic with them, and so it is no surprise that a feature of this vernacular tradition was a profusion of abbreviations for mathematical operations. The authors and their readers were practical people — their books typically solved problems that might arise in trade — and literary style was subordinate to getting the job done. Nonetheless, while there is a use of abbreviations as symbols in, for example, the writings of Diophantus and other authors, the late 15th century did see an increased use of arbitrary signs to symbolize mathematical operations — a habit that by the early 17th century began to disrupt the Renaissance tradition of respect for the ancient authors and their texts.

Notation. The first minus sign (replacing a verbal instruction or the abbreviation m) seems to have occurred in a book called the *Deutsche Algebra* (German Algebra, 1481), and the first plus sign occurs in a marginal note to a copy of that book: it derives from a medieval abbreviation for the Latin 'et' meaning 'and'. But that book also used other

Figure 10.4. A polyhedron by Leonardo da Vinci

signs for plus (including the sign ×, confusingly enough!) and we should be careful not to confuse the first use of a sign with its eventually entering standard usage. Curiously, the modern sign for multiplication seems to have come about quite late, in the book *Clavis Mathematicae* (Key to Mathematics) by the English mathematician William Oughtred[5] in 1632, although it has its origins in a similar (but more complicated) sign of the Welsh mathematician Robert Recorde. Leibniz later objected to the sign on the grounds that it was easily confused with the letter X, and advocated the use of the dot instead. Signs for division in the form of two numbers separated by a horizontal bar go back to Leonardo of Pisa, who may have acquired it from Arabic sources; the modern sign may have first appeared in print in Rahn's *Teutsche Algebra* (German Algebra) of 1659 or in work by John Pell, an English mathematician.

Leonardo of Pisa had already used the symbol R to denote a square root. The more familiar $\surd$ sign was brought in by the cossists Adam Ries, Christoff Rudolff, and Michael Stifel; Descartes, or his printer, later added the horizontal bar. Cube and higher roots were denoted by suffices. Powers of the unknown originally retained their geometrical names: *census* or *quadratus* for a square, *cubus* for cube, with higher powers being denoted ad hoc — we shall see examples below. The equals sign was introduced by the Welsh mathematician Robert Recorde (see Section 11.1) in *The Whetstone of Witte* (1557).

There seem to have been two aspects to the widespread acceptance of mathematical signs: their convenience, and their use in a popular book (which partly requires the willingness of printers to adopt the new signs). As we shall see when we look at the English case in more detail (see Chapter 11), more complicated processes seem to have been at work with

[5]We meet Oughtred's work in Chapter 11.

the introduction of new terms, involving people's feelings about their native languages, the introduction of loan words (words taken more or less directly from a foreign language — 'le weekend' is an example in French), and outright neologisms.

10.2 Renaissance translators: Maurolico and Commandino

Serious attention to recovering the Greek heritage of mathematics is now chiefly associated with two men who were active some fifty years later than Regiomontanus: Francesco Maurolico and Federigo Commandino. Between them they produced many reconstructions and new translations of Greek mathematical works and their writings provide us with good evidence about the motivations that underlay the 16th-century Greek translation programme.

Francesco Maurolico. Francesco Maurolico was born of Greek parents who had fled to Sicily after the fall of Constantinople in 1492. His father, originally trained as a physician, took a special interest in mathematics and astronomy and taught these subjects to his son, while professional tutors instructed him in the humanities (at that time dominated by the rediscovered Greek learning).

We know nothing of Maurolico's career until the year 1521, when he was ordained and also produced his first mathematical work, on optics. His next major work, the *Grammaticorum* published in 1528, is of interest in that it includes his views on the nature of mathematics — the only area of study, he felt, where there is certainty of demonstration. He cited his studies of the ancients, claiming that he had independently demonstrated a number of propositions and theorems, and further announced a programme of publication of mathematical works, reminiscent of a programme that Regiomontanus had proposed in 1474. This included a *Compendium Mathematicae* made up from texts studied by him, together with some contributions of his own (see Box 10.1).

During the 1530s and 1540s, Maurolico worked on editions of Euclid, Heron, Apollonius, and Archimedes, with the intention of producing reliable texts by collating manuscripts, and thereby identifying and removing corruptions and unacceptable insertions introduced by earlier commentators. By 1547 he had completed his version of the *Conics* of Apollonius, including a reconstruction of the lost Books V and VI: it was not published until the following century, however, by which time Commandino's translation (published in Bologna in 1566) had become firmly established as the authoritative text.

Box 10.1. Maurolico's publishing programme.

Works on geometry by Euclid, Apollonius, Archimedes, and Menelaus.

Works on astronomy by Euclid, Ptolemy, Thābit ibn-Qurra, Peurbach, and Regiomontanus.

Works on various topics (mechanics, optics, and instruments) by Aristotle, Heron, Ptolemy, Boethius, and Jordanus.

His own works included *Photismi* and *Diaphano* (on optics), *Grammaticorum*, *Cosmographia*, and *Trochitia* (on clocks).

We can glean something of Maurolico's approach and basic principles from a letter that he wrote to his patron, Juan de Vega, Viceroy of Sicily, in August 1556.[6] Maurolico

[6]Discussed in (Rose 1975); an extract appears in F&G 8.B3.

set great store by the *certitude* of mathematics. He was concerned about the unsatisfactory nature of Greek texts, which he found to be full of textual errors and changes introduced through translators' ignorance of the mathematics involved and which could be eliminated only through the patient comparison of existing texts. He was especially critical of translations compiled by those who did not understand the mathematical content of the works. In the case of Euclid's *Elements*, for example, he rejected not only the additions and clarifications of translators such as Campanus of Novara, but also those of the later Greek commentators. Indeed, Maurolico was not averse to reworking texts, and did not restrict himself to literal translation. He considered it permissible to shorten, add to, or amend proofs, and even to omit them altogether where necessary: the mathematics always had to have priority over textual purity.

The task was urgent. With the exception of Cardano's writings, which we discuss in Section 10.3, Mauricolo considered that existing modern works were of poor quality and that mathematicians should therefore study the ancients. This willingness to enhance the utility of the ancient texts by going beyond them was characteristic of Maurolico's work, and derived from his sense of mission.

After a period in which he deserted mathematics for religion, Maurolico turned particularly to mechanics, and in 1569 completed his *Problemata Mechanica*, a critical treatment of Aristotelian mechanics. By the time of his death in 1575 Maurolico had completed an arithmetic in two books — essentially an extension of Diophantus's *Arithmetica*. In this work we find, significantly, the introduction of letters in place of numbers, and the detailing of algebraic algorithms: this was not Diophantus in his own words, but Diophantus brought up to date, one might say. The intention was to provide numerical operations with the generality that was to be found in geometry.

Maurolico's determination to recover the purity of Greek mathematics is strikingly underlined by his extreme view of Islamic algebra, which he regarded as no more than a corrupt derivative of the work of Diophantus. This reflected his position in Sicily, where he was the Abbot of Santa Maria del Parto, professor of mathematics at the University of Messina, and at various times in charge of the fortifications of the city of Messina on behalf of the Emperor Charles V. He was prominently involved in the moves to create a buoyant Christian Europe capable of driving back the strong forces of Islam to the South and West, and his mathematics offered a European tradition going back to the Greeks which the Arab enemy had at best merely transmitted.

Federigo Commandino. We know more about the life of Federigo Commandino than about Maurolico's. His student Bernardino Baldi wrote his biography, and there are a number of extant manuscript sources against which Baldi's account can be checked and supplemented.

Commandino was Italian, born in Urbino of noble parents in 1506 or 1509 (the sources disagree). One of his tutors subsequently entered the service of Cardinal Ridolfi in Rome, one of the most noted collectors of manuscripts of the day. Through him Commandino had access, while still only in his 20s, to many of the more important Greek, Islamic, Byzantine, and medieval works, both in mathematics and the humanities. Through the Cardinal he was able to obtain the patronage of Pope Clement VII, but on the Pope's death in 1534 he took up studies in medicine and philosophy at the University of Padua, and subsequently at Ferrara.

Like Maurolico, Commandino was a searcher after certainty, and when he lost his wife and son through illness he turned his back on the 'uncertainty' of medicine to devote himself exclusively to mathematics. He worked on manuscripts in Rome until the mid-1550s, subsequently returning to Urbino (see Box 10.2).

Box 10.2. Commandino's publishing programme.

Works by Apollonius, Archimedes, Aristarchus, Euclid, Heron, Jordanus, Pappus, Ptolemy, Serenus, Theodosius and Vitruvius.

His own works included *Liber de Horologiorum Descriptione* (On Sundials), which he added to his edition of *Claudii Ptolemaei Liber de Analemmate, etc.* of 1562, and his *Liber de Centro Grauitatis Solidorum* (On Centres of Gravity) of 1565, which was inspired by his reading of a Latin edition of Archimedes's *On Floating Bodies*.

At this point we take up his story in an extract from a letter from the English mathematician John Dee to Commandino, which we can compare with Baldi's biography.[7] Commandino's fame as a translator spread widely, and Dee's letter urged him to publish a work which he thought was an Arabic version of Euclid's lost book *On the Division of Figures*.[8] He was able to publish Dee's book as well as a number of his own works because, through patronage, he was given special copyright privileges that allowed him to set up his own printing press. This was a considerable help in his mathematical translating and publishing enterprise.[9]

Dee to Commandino.

> Having now for many years set my selfe chiefly (my most learned Frederick) how to preserve from utter ruine, the most famous moniments of our Ancestors (such as I could) in all the more curious or elegant kinds of Philosophy; least that either so worthy men should be robbed of their due renown, or we longer want the most abundant benefit of such books, I say that I so bestowing my pains among other most ancient writings of Philosophers, did at length happen upon this small Book, written indeed, in a very blinde ilfavoured character, and also by reason of its age, hard to read. But that I might the better see, I used helps to my sight: and by often study and exercise therein, I got the knack of reading it. Whereupon being hereby better persuaded of the excellencie and worth of the Book, I earnestly wished that the same might forthwith be communicated to the Society of Philosophers. But while I was pondering this in my minde, I found none more worthy that your self (my Commandine) in this our age, to enjoy these our Labours; who have also your own selfe revived certain most excellent Works of Archimedes and Ptolemee almost lost, and to have brought them forth into publick view in a most magnificent dresse. Therefore this little Book, as a perpetual pledge, even of the affection wherewith I ever embrace you, I commit to you and your trust; and do earnestly beseech you, that you will not suffer this our common labour to go forth into the World destitute of that adornment wherewith you are wont

[7] See F&G 8.B1, 8.B2. We meet John Dee in Chapter 11.

[8] Later historians have shown that Dee's arguments for identifying his Arabic text with the lost work of Euclid do not stand up.

[9] In (Dee 1661, 605–608) and F&G 8.B1.

to send abroad others. Yea, I surely hope (if I well know you and your endeavours) that you will some time other so enrich this subject, as that you will neither permit it to rest in a Pentagonal form; nor suffer the Solids themselves long to want the like sections by Plains. Verily, if you would but a little put forward these things, they will be themselves go on to the other kind of Superficies. But that they many be applyed to Solids they will require your sound knowledge, and more than ordinary pains in the Mathematicks. As for the Authors name, I would have you understand, that to the very old copy from whence I writ it, the name Machomet Bagdedine[10] was put ... rather ... this may be deemed a book of our Euclide, all whose Books were long since turned out of the Greeks into the Syriak and Arabic Tongues. Whereupon, It being found some time or other to want its Title with the Arabians or Syrians, was easily attributed by the transcribers to that most famous Mathematician among them, Machomet: Which I am able to prove by many ancient testimonies, to be often done in Moniments of the Ancients; and certain friends of mine doe know (that I may bring one of many) that we have by this means restored one small Book, incomparable in occult and mysticall Philosophie, of that most ancient and excellent Phylosopher Anaxagoras, everywhere now through many ages enobled with the name of Aristotle to Anaxagoras himselfe, and that by the most sure proofs. Yea further, we could not yet perceive so great acuteness of any Machomet in the Mathematicks, from their moniments, which we enjoy, as everywhere appears in these Problems. Moreover that Euclid himselfe wrote one Book ... that is to say, of Divisions, as may be evidenced from Proclus's Commentaries upon his first of *Elements*: and we know none other extant under this title, nor can we finde any, which for the excellence of its treatment, may more rightfully or worthily be ascribed to Euclid. Finally, I remember that in a certain very ancient piece of Geometry, I have read a place cited out of this little Book in expresse words, even as from a most certain work of Euclid. Therefore we have thus briefly declared our opinions for the present, which we desire may carry with them so much weight as they have truth in them ...

But whatsoever that Book of Euclid was concerning Divsions, certainly this is such an one as may be both very profitable for the studies of many, and also bring much honour and renown to every most noble ancient Mathematician ... I will now direct my discourse to you, who are herein to be greatly intreated by me, that you would with all possible diligence advance your most weighty and usefull labours, which you did yesterday most courteously shew to me in your Study. For so you will make the fairest way to perpetuate your fame, who have in so few years, put forth many Books, so happily, so neatly and so many of your own: who alone in our time dost adorn the most excellent Princes of Mathematicians, Archimedes, Apollonius and Ptolomeus, with their due luster. So you will restore a new and wonderful livelynesse to Mathematicall Learning much decayed ... Now the purpose of Travell calls me away, lest I should be

[10]The identity of this man has become more uncertain with time — an interesting comment on the results of research into the history of mathematics.

put to undergo a greater trouble of this scorching Season round about us, before I can shelter my selfe in the Roman Shades.

Farewell, therefore, the honour and renown of Mathematicians; farewell my most courteous Commandine.

We follow this with Baldi's account of Commandino's life, which brings out the importance of patrons. Commandino put aside a number of translations from Greek on which he was working in order to comply with his patron's request for a translation and commentary on Euclid. Commandino's addition of glosses and commentaries to this edition is characteristic of the way in which 16th-century scholars prepared such editions, but he had less of Maurolico's urge to rewrite the old texts.[11]

Bernardino Baldi on Commandino.

Having suffered such blows, and assured, by experience, of the uncertainty of fortune, he [Commandino] returned to his native land with a mind to find calm and the expectation of quiet and virtuous leisure: which he thought he could do, since he had already seen both his daughters married and had put his family affairs in order. So he expected to complete many works he had already begun; but then Francesco Maria [della Rovere], the son of our Duke Guido Ubaldo, a young man of noble spirit, knowing how well such sciences are suited to one who plays the part of a ruler and turns his hand to the arts of war, would not permit Federico to live cloistered within the walls of his family home; but, making him the most honourable proposals, wished, as his father had done before, to call him [Commandino] to his service: when he had entered [the ruler's household], in reading Euclid's Elements to the prince he gave him much satisfaction by his interpretation [of the text]. Hence, the prince, thinking it unjust to deprive the world of the things he had heard in private, persuaded Federico to translate the work and write a commentary on it. So Commandino, desirous of acting for the common good, and in part obeying his lord's commands, putting aside the translations of Pappus, of Theodosius, of Heron, of Autolycus and of Aristarchus, gave himself wholeheartedly to translating and writing a commentary on Euclid: nor did he labour in vain; for in a short time they [the translation and commentary] were printed at Pesaro and it became clear how great a benefit it was to the world that he had turned his hand to them. Of which, among many others, Christopher Clavius bears witness, affirming that, of all those who up until our time had devoted themselves to studies of this author's [Euclid's] Elements, Commandino alone had restored them to their original clarity, in accordance with the meaning and the tradition of ancient commentators; and had not fallen into those errors he had discovered and noted in [the works of] many others. Federico adorned this book with most perceptive glosses and commentaries, partly original to himself and partly taken from the best books by these [earlier] scholars. Similarly, he added some prefaces so eloquent that anyone who reads them must recognize how outstanding he [Commandino] was in more highly-regarded arts, particularly in the other branches of philosophy. So they were printed; and since the works were

[11] In (Baldi 1859, 528–531) and F&G 8.B2 (tr. J.V. Field). The 'blows' in the first sentence of the extract refer to the death of Commandino's wife and son.

written at the instance of Francesco Maria, and on account of his persuasion, so Commandino dedicated and consecrated them to his name.

At about the same time, an English nobleman from London, called John Dee, a most learned man, a student of the antique and a lover of these studies [i.e. mathematics], being on his way to Rome, attracted by Federico's reputation came to Urbino, with the sole purpose of making his acquaintance and paying him a visit; when he was received there by him with great courtesy he found the reality much greater than what he had learned by repute. The said John brought with him a little book that had not been printed, inscribed with the name Macometto Bagdedino, which deals with the division of areas; and which he [Dee] had rescued, with much patience, from the shadows of antiquity and the barbarities of the Arabs. Hence, wishing it to see the light, he judged it the best opportunity of carrying out his intention to leave the book in the hands of Commandino: which he did, accompanying it with a most elaborate [i.e. complimentary] letter; in which, among many other things, he wrote these words: 'I found none more worthy than your self (my Commandino) in this our age, to enjoy these our Labours; who have also your own self revived certain most excellent Works of Archimedes and Ptolemeus almost lost, and have brought them forth into publick view'. This short work extended only so far as the division of the pentagon; so Federico, not content, as he himself says, that this author's treatise should close with no more than the division of that [polygon], reduced to two very short problems all that the author had gathered into many, and demonstrated the manner in which one might divide all other areas ad infinitum: which done, judging the book worthy of a prince, he printed it, and dedicated it to the name of Francesco Maria, in the year 1570. Later this little book was translated into the vernacular [Italian], and published, by Fulvio Vianni de'Malatesti da Montefiore a young man of very noble spirit.

The heirs saw to it that the printing of Heron's work was completed, and it was dedicated to the cardinal of Urbino, since that had been Commandino's intention while he lived. The works which death had forced him to leave unfinished, or which he had not been able to publish were these: the six books of Pappus's Collections; all the other works of Euclid; two books of Theodosius, one on habitations and the other on days and nights; two books by Autolycus, on rising and setting and another on the moving sphere; the work of Leonardo Pisano [Fibonacci] and that of Fra Luca [Pacioli], which he intended to correct and modernize.

Commandino died in 1575, the same year as Maurolico. They never met, despite sharing a common patron for a time, but were well aware of each other's work and sometimes corresponded by letter. Their principles were virtually identical: both saw in the recovery of Greek mathematical texts a significant resolution of the quest for 'certainty'; both were concerned about the purity of the restored texts, though not averse to adding or amending essential proofs; and both had an extensive mathematical programme that included works of their own in addition to the collation and translation of texts of the ancients.

Commandino is best known for his Latin editions of Euclid, Archimedes, Apollonius, Heron, Ptolemy, and Pappus; Latin was chosen because, being the language of the Church, it was also the language of educated Europe. This list of authors on its own is sufficient to attest not only to his importance for the history of mathematics in the 16th century, but also to the importance that he attached to such work.

He was quite explicit about this in his Latin edition of the *Elements*, published in 1572. The book is dedicated to the heir of Urbino, Duke Francesco Maria (mentioned earlier by Baldi), who had commissioned it and who employed Commandino as his tutor. The dedication of the work spells out his programme for the renaissance of mathematics, which was now in a low state and almost excluded from the universities, by comparison with its dignity and fame in antiquity. Then he divided philosophical study into three parts: divine, natural, and mathematical, of which only the last of these is 'certain' — Euclid, especially, was seen as laying down the certain basis of mathematics. Finally, not only was mathematical knowledge certain, it was also useful because, although it does not have to rely directly on the phenomena experienced through the senses, mathematics was nevertheless directly applicable to the physical world.

In keeping with the emphasis on certainty combined with utility, Commandino regarded Archimedes as the greatest mathematician of all time, and his editions of Archimedes were to remain the basis of most translations until the 19th century. He also wrote original works, including a tract on sundials published in 1575, which was stimulated by current research on conic sections and by Ptolemy's account of the motion of the Sun, which Commandino felt was too narrowly theoretical.

The results of Commandino's programme were to exert great influence on subsequent generations of mathematicians. With the support of powerful patrons, and despite the poor status of mathematics in the universities, Commandino and others were able to develop a considerable renaissance of Greek mathematics. Although primarily interested in geometry, he also knew some of the most innovative algebraists of his time, whose work we study in the next section. His contact with Cardano and Tartaglia at Bologna moved him to take an interest in algebra, and in a letter of 1574 he praised Bombelli's *Algebra* and encouraged Bombelli and a friend to complete the translation of Diophantus. To appreciate their work, we now step backwards chronologically and look at the tradition of vernacular algebra that was developing at that time.

10.3 Cubics and quartics in 16th-century Italy

The impression is sometimes given that European thought slumbered until the rediscovery of ancient Greek mathematics awoke it. This over-simplification contains a considerable degree of truth, but one thing it neglects is the advances in algebra, which owe more to the vernacular than the classical aspects of the time.

The state of European algebra at the close of the 15th century is fairly represented by the vernacular works of Pacioli and Chuquet. Both of these writers had recognised that any general solution for cubic equations would require new mathematical ideas. Pacioli remarked that no algebraic treatment of cubics was possible using the methods then in use (which handled quadratic equations). Chuquet was a little more hopeful, saying that such things 'are left for those who would wish to proceed more deeply'.[12]

Cubic equations. The classification of problems as algebraic, and specifically cubic, emerged slowly but steadily in Arabic and later Western writings. Problems that give rise to what could now be regarded as cubic equations had been investigated from ancient times. Certain types of such problems are to be found on Mesopotamian tablets, and examples of cubic problems can be found in Greek mathematics — for example, two of the three

[12]See (Flegg, Hay, and Moss 1985, 196).

'classical problems' (see Section 4.3) give rise to cubic equations: the Greeks had regarded these as geometrical problems solvable by *neusis* constructions or by the intersection of conics and other curves.

In addition, some cubic problems are dealt with in Diophantus's work, and some were also discussed in Hindu and Arabic works. Bhāskara II explored the possibility of solving cubics by finding appropriate substitutions that would reduce them to problems involving only quadratic and linear relations, but he could find no general method for doing this. The high point in the Islamic study of cubics was reached earlier in that century, with the work of Omar Khayyām that we mentioned briefly in Chapter 8. This provided a thorough analysis of the geometrical solution of different types of cubic, somewhat in the classical Greek style although not something that any Greek mathematician seems to have done. So a *solution* of a cubic problem was not a number, but the point of intersection of two curves. It was not known whether arithmetical solutions to cubic equations were possible in general, though Omar Khayyām had spoken with guarded optimism on the subject. In any case, his work on this was not known in the West until the 17th century, after the developments in Italy that we are about to pursue.

In the West, individual cubic equations continued to be solved by guesswork and checking the answer, and consideration of some general solutions appeared in Italian manuscripts. For example, we can find an unavailing discussion in an anonymous 13th-century manuscript of methods for solving cubics. Such efforts to find solution methods, coupled with Pacioli's later claim that they had been fruitless, provided a challenge that 16th-century Italian mathematicians chose not to ignore. The story of their eventual success is a remarkable tale of intrigue and secrecy which is worth telling for its own sake. It also throws light on the general conditions under which the academics of that day had to work — conditions that we would like to think are very different from those in our universities today!

The 16th century was a period of considerable political unrest in Northern Italy, then a collection of comparatively small states. Charles V, who had been crowned Holy Roman Emperor in Bologna, ruled over Sardinia, Sicily, Naples, Milan, and lands to the north-east of the Adriatic and to the immediate North and West of Italy (see Figure 10.5). He faced hostility both from the increasing power of the Turks in the East and from France in the West. Wars with France made life in the Italian universities highly uncertain. From time to time, individual universities had to be closed and the academics and students had to seek refuge elsewhere.

There was also further uncertainty for the academic staff: they had no appointments with 'tenure'. Appointments, even of senior professors, were temporary, often for only one year at a time. In order to increase the likelihood of a continued appointment to a chair in mathematics, it was advisable for the occupant to take part in the many public problem-solving contests that were arranged, and it was a great advantage in such contests to have a secret method unknown to the other participants for solving specific problems. It was in this context of secrecy that the story of the solution of cubic equations was played out.

Scipione del Ferro. It appears that the first general algebraic method for solving a specific form of cubic equation was due to Scipione del Ferro, a professor of mathematics at the University of Bologna. Early in the 16th century he discovered how to solve equations of the form *'a cube and things equal to numbers'* (which we might now write as $x^3 + cx = d$). His method involved introducing two new unknowns to replace the unknown *cosa* or 'thing' (our x). As we shall see, this was done in such a way that the cubic equation was

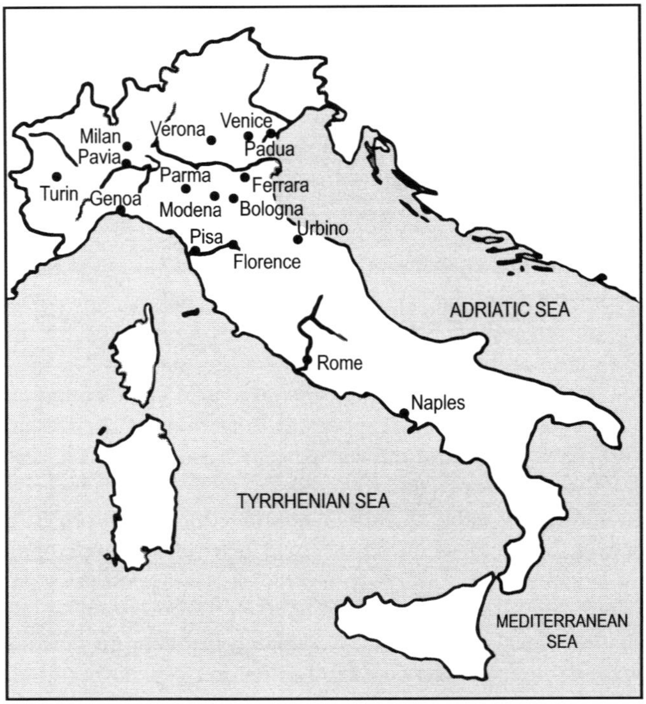

Figure 10.5. Cities in Renaissance Italy

replaced by a pair of simultaneous equations of a form whose method of solution had been known since Mesopotamian times.

Some years after his discovery, del Ferro revealed his method for solving cubic equations to his pupil, Antonio Fior, and also to his son-in-law, Annibale della Nave, his successor as professor at Bologna. It was impossible to conceal from the mathematical community the knowledge that some cubics were solvable, but for a time del Ferro's method was known only to those to whom he had revealed it. Rumours as to the method abounded, and attempts by other mathematicians to find it were redoubled.

Niccolò Tartaglia. In 1535 a certain Niccolò of Brescia was independently and in highly charged circumstances to discover how to solve del Ferro's cubic equations. As a boy, Niccolò had been cut in the face by the sabre of a soldier during the sack of Brescia (between Verona and Milan) in 1512, and as a result had developed a significant stammer: he was known as 'Tartaglia' (the stammerer), by which name he is now chiefly remembered. Although brought up in a poor family, he managed to teach himself Latin, Greek, mathematics, and science. He then earned a living as a travelling science teacher: he taught mathematics and science successively in Verona, Plaisance, Venice, Brescia, and again in Venice, where he eventually settled until his death in 1557. Most of his life was spent under conditions of financial difficulty — often he was unable to obtain payment for the lessons he gave. He was especially interested in military science, and spent what little money he had on developing his inventions in the field of artillery.

After the death of del Ferro in 1526, Fior felt himself free to exploit the secret that had been entrusted to him. In 1535 he heard a rumour that Tartaglia had solved a cubic equation and, believing him to be an imposter, eventually challenged him to a public contest. Each

LA PRIMA PARTE DEL GENERAL TRATTATO DI NVMERI, ET MISVRE DI NICOLO TARTAGLIA,

NELLAQVALE IN DIECISETTE LIBRI SI DICHIARA TVTTI GLI ATTI OPERATIVI, PRATICHE, ET REGOLE NECESSARIE NON SOLAMENTE in tutta l'arte negotiaria, & mercantile, ma anchor in ogni altra arte, scientia, ouer disciplina, doue interuenghi il calculo.

MALIGNITA'

NOIAR NON PVO

A' FORTEZZA

CON LI SVOI PRIVILEGII.

In Vinegia per Curtio Troiano dei Navò.

M D LVI.

Figure 10.6. Tartaglia's *General Treatise on Number and Measure* (Venice, 1556)

contestant had to set thirty questions for the other by 22 February 1535, two months were allowed for solving the problems, and the loser was to pay for thirty dinners, to be enjoyed by the winner and his friends — a substantial outlay.

A few days before the 22nd, Tartaglia heard that Fior had been entrusted 'by a great master ... thirty years ago' with the solution to at least one type of cubic equation, and he began to suspect that Fior, who was not otherwise known to be a good mathematician, would perhaps set thirty questions involving cubic equations. Alarmed, and by now highly motivated, Tartaglia independently rediscovered the method of solution, apparently during a sleepless night, 12–13 February 1535, just eight days before the contest deadline. He was now free to set cubic equations as well as other types of problem to Fior.[13]

It turned out that Tartaglia had guessed correctly. Fior relied entirely on del Ferro's secret method for solving cubic equations of the form *a cube and things equal to numbers*, and all of the questions he posed to Tartaglia were of this form. He was unable to find solutions to the more general mathematical problems set by his opponent. Tartaglia, however,

[13]See (Ore 1953), which our account generally follows, and (Nordgaard 2004, 156).

was able to solve all of Fior's problems and win the contest; apparently the honour was enough and he did not insist on the thirty dinners.

The problems involved have come down to us through Tartaglia's account of the contest, and we give some of them here:[14]

Fior's challenge to Niccolò Tartaglia.

> These are the thirty problems proposed by me Antonio Maria Fior to you Master Niccolò Tartaglia.
>
> (1) Find me a number such that when its cube root is added to it, the result is six, that is 6. [This is equivalent to the equation $x^3 + x = 6$.]
>
> (2) Find me two numbers in double proportion such that when the square of the larger number is multiplied by the smaller, and this product is added to the two original numbers, the result is forty, that is 40. [This is equivalent to the equation $((2x)^2 \times x) + x + 2x = 40$, that is, $4x^3 + 3x = 40$.]
>
> (3) Find me a number such that when it is cubed, and the said number is added to this cube, the result is five. [$x^3 + x = 5$.]
>
> ...
>
> (15) A man sells a sapphire for 500 ducats, making a profit of the cube root of his capital. How much is this profit? [$x^3 + x = 500$.]
>
> ...
>
> (17) There is a tree, 12 *braccia* high, which was broken into two parts at such a point that the height of the part which was left standing was the cube root of the length of the part that was cut away. What was the height of the part that was left standing? [$x^3 + x = 12$.]
>
> ...
>
> (30) There are two bodies of 20 triangular faces [icosahedra] whose corporeal areas added together make 700 *braccia*, and the area of the smaller is the cube root of the larger. What is the smaller area? [$x^3 + x = 700$.]

It seems that only four of Tartaglia's questions to Fior survive:[15] they are of different types. One asks:

> Find an irrational quantity such that when it is multiplied by its square root augmented by 4, the result is a given rational number.

This is $x^3 + 4x^2 = d$ and therefore of the form $x^3 + bx^2 = d$, so it is not of Fior's type, whereas Tartaglia's question:

> Find an irrational quantity such that when it is added to four times its cube root, the result is thirteen.

is $x^3 + 4x = 13$, which is of the form $x^3 + cx = d$.

---

[14]See (Tartaglia 1546) in (Masotti 1959, 106–108, 120–133) and F&G 8.A1 (transl. F.R. Smith).

[15]Tartaglia's questions are quoted from (Nordgaard 2004, 156–157).

Gerolamo Cardano. A dominant new figure now entered the stage, Gerolamo Cardano, one of the most fascinating of Renaissance scholars. He was born in 1501, and after an unhappy childhood he eventually graduated in medicine at the University of Pavia, having also acquired a taste for physics, mathematics, and gambling. Cardano achieved European fame as a physician, being summoned as far afield as Scotland to diagnose an illness of the Archbishop of St Andrews. At various times he held chairs in medicine or mathematics at a number of Italian universities, and was a prolific writer of books. His written works embrace medicine, mathematics, physics, astronomy, and games of chance (including advice on cheating). His medical writings include outstanding sections on anatomy, but he was also in considerable demand for the casting of horoscopes. At one time he was imprisoned by the Church authorities for casting the horoscope of Christ, but he was subsequently pardoned and in later years enjoyed a pension granted by the Pope. In 1576 he predicted his own death and (it has been claimed) starved himself to ensure that his prediction came about exactly as he had foretold.

Cardano's chief scientific work was *De Subtilitate* (On Subtlety), in which he discussed at great length the Aristotelian physics handed down through the scholastic tradition. He was interested in the history of science and mathematics, and in *De Subtilitate* he gave an account of twelve individuals whom he considered great: his list includes Archimedes ('the first of the twelve'), Ptolemy, Aristotle, Euclid, and Apollonius — and also al-Khwārizmī, whom he called 'Mahomet, son of Moses the Arab'. He seems to have had none of Maurolico's religious agenda.

Cardano's two most notable mathematical works are his *Practica Arithmeticae* (The Practices of Arithmetic) (1539) and the *Ars Magna* (The Great Art) (1545); the latter established him as a major figure in the history of algebra, and we shall return to it shortly. He also wrote a surprisingly frank account of his life in *De Vita Propria Liber* (The Book of my Life).[16]

Cardano heard of the contest between Fior and Tartaglia, and attempted to coax the solution of cubic equations from the latter. For some time he was unsuccessful in this, but eventually he invited Tartaglia to stay at his house in Milan on the pretext that he would provide him with an introduction to the Spanish Governor of the city in the expectation that the latter would finance Tartaglia's researches in military science. In return, the latter revealed the secret of solving cubic equations to Cardano, but only after extracting a solemn oath that it would never be published. Our explanation of Tartaglia's method immediately follows the extract.

Tartaglia's account of the agreement he reached with Cardano begins:[17]

Cardano and Tartaglia.

> CARDANO: I hold it very dear that you have come now, when his Excellency the Signor Marchese has ridden as far as Vigevano, because we will have the opportunity to talk, and to discuss our affairs together until he returns. Certainly you have, alas, been unkind in not wishing to give me the rule that you discovered, on the case of the thing and the cube equal to a number, even after my greatest entreaties for it.
>
> TARTAGLIA: I tell you, I am not so unforthcoming merely on account of the solution, nor of the things discovered through it, but on account of those things which it is possible to discover through the knowledge of it, for it

[16]See (Cardano 1962); extracts from this book appear in F&G 8.A4(c).

[17]Tartaglia (1546) in (Masotti 1959, 120–122), transl. F.R. Smith, and F&G 8.A2.

is a key which opens the way to the ability to investigate boundless other cases. And if it were not that at present I am busy with the translation of Euclid into Italian (and at the moment I have translated as far as his thirteenth book), I would already have found a general rule for many other cases. But as soon as I have completed this work on Euclid that I have already begun, I intend to compose a book on the practice [of arithmetic], and together with it a new algebra, in which I have resolved not only to publish to every man all my discoveries of new cases already mentioned, but many others which I hope to find; and, more, I want to demonstrate the rule that enables one to investigate boundless other cases, which I hope will be a useful and beautiful thing. And this is the reason which makes me refuse them to everyone, because at present I am not working on them (being, as I said, busy with Euclid), and if I teach them to any speculative person (as is your Excellency), he could easily with such clear information find other solutions (it being easy to combine it with the things already discovered), and publish it, as inventor. And to do that would spoil all my plans. Thus this is the principal reason that has made me so unkind to your Excellency, so much more as you are at present having your book printed on a similar subject, and even though you wrote to me that you want to give out these discoveries of mine under my name, acknowledging me as the inventor. Which in effect does not please me on any account, because I want to publish these discoveries of mine in my books, and not in another person's books.

CARDANO: And I also wrote to you that if you did not consent to my publishing them, I would keep them secret.

TARTAGLIA: It is enough that I did not choose to believe that.

CARDANO: I swear to you, by God's holy Gospels, and as a true man of honour, not only never to publish your discoveries, if you teach me them, but I also promise you, and I pledge my faith as a true Christian, to note them down in code, so that after my death no one will be able to understand them. If you want to believe me now, then believe me, if not, leave it be.

TARTAGLIA: If I did not give credit to all your oaths, I would certainly deserve to be judged a faithless man, but since I have decided to ride to Vigevano to call upon his Excellency the Signor Marchese, because it is now three days that I have been here, and I am sorry to have waited for him so long, when I have returned I promise to demonstrate everything to you.

CARDANO: Since you have decided anyway to ride as far as Vigevano after the Signor Marchese, I want to give you a letter to give to his Excellency, so that he should know who you are. But before you go, I want you to show me the rule for these solutions of yours, as you have promised me.

TARTAGLIA: I am satisfied.

Tartaglia then continued by setting out the method he had discovered. Before we look at that, we pass to the end of his account where the two men re-affirmed their agreement.

> TARTAGLIA: Now, remember your Excellency, and just do not forget your faithful promise, because if by unhappy chance it is broken, that is if you publish these solutions, whether in this book that you are having printed at the moment; or even, in another one different from this one, you publish them giving my name and acknowledging me as the real inventor, I promise you and I swear to publish immediately another book which will not be very agreeable for you.
>
> CARDANO: Do not doubt that I will keep my promise. Go, and be sure that you give this my letter to the Signor Marchese on my behalf.
>
> TARTAGLIA: Now please do not forget.
>
> CARDANO: Go, straight away.
>
> TARTAGLIA: By my faith, but for that, I do not want to go to Vigevano. I would much rather turn in the direction of Venice, let the matter go as it will.

Tartaglia revealed his method in these terms — our comments in modern notation are interpolated to help you follow it:

> TARTAGLIA: But I want you to know, that, to enable me to remember the method in any unforeseen circumstance, I have arranged it as a verse in rhyme, because if I had not taken this precaution, I would frequently have forgotten it, and although my telling it in rhyme is not very concise, it has not bothered me, because it is enough that it serves to bring the rule to mind every time that I recite it. And I want to write down this verse for you in my own hand, so that you can be sure that I am giving you the invention accurately and well.
>
> When the cube and the things together
> Are equal to some discrete number,
>
> [To solve $x^3 + cx = d$,]
>
> Find two other numbers differing in this one.
> Then you will keep this as a habit
> That their product should always be equal
> Exactly to the cube of a third of the things.
>
> [Find u, v such that $u - v = d$ and $uv = (c/3)^3$.]
>
> The remainder then as a general rule
> Of their cube roots subtracted
> Will be equal to your principal thing.
>
> [Then $x = \sqrt[3]{u} - \sqrt[3]{v}$.]
>
> In the second of these acts,
> When the cube remains alone,
>
> [In the second case, to solve $x^3 = cx + d$,]
>
> You will observe these other agreements:
> You will at once divide the number into two parts
> So that the one times the other produces clearly
> The cube of a third of the things exactly.
>
> [Find u, v such that $u + v = d$ and $uv = (c/3)^3$.]

Then of these two parts, as a habitual rule,
You will take the cube roots added together,
And this sum will be your thought.

[Then $x = \sqrt[3]{u} + \sqrt[3]{v}$.]

The third of these calculations of ours
Is solved with the second if you take good care,
As in their nature they are almost matched.

[The third case, to solve $x^3 + d = cx$, is similar to the second.]

These things I found, and not with sluggish steps,
In the year one thousand five hundred, four and thirty.
With foundations strong and sturdy
In the city girdled by the sea.

This verse speaks so clearly that, without any other example, I believe that your Excellency will understand everything.

CARDANO: How well I will understand it, and I have almost understood it at the present. Go if you wish, and when you have returned, I will show you then if I have understood it.

His method is perhaps not as clear to us as Cardano found it, and it is worth a closer look. The first nine lines of Tartaglia's verse explain his solution to equations of the form *a cube and things equal to numbers*:

$$x^3 + cx = d.$$

Here we have to find two numbers, u and v (say), whose product uv is equal to $(c/3)^3$ and whose difference $u - v$ equals d. The solution to the cubic equation is then stated to be

$$x = \sqrt[3]{u} - \sqrt[3]{v}.$$

We can follow this working by taking a specific example of Cardano's:

a cube and six things equal to twenty $(x^3 + 6x = 20)$

from Chapter XI of his *Ars Magna*. We have to find u and v such that

$$uv = (6/3)^3 = 8 \quad \text{and} \quad u - v = 20.$$

First, what is u? From the second equation, we can see that $v = u - 20$, and substituting this expression for v into the first equation gives

$$uv = u(u - 20) = u^2 - 20u = 8\,.$$

On solving this quadratic equation, we obtain $u = 10 + \sqrt{108}$, and since $v = u - 20$, we have $v = -10 + \sqrt{108}$. The solution obtained by Tartaglia's method for the cubic equation is thus claimed to be

$$x = \sqrt[3]{u} - \sqrt[3]{v} = \sqrt[3]{(10 + \sqrt{108})} - \sqrt[3]{(-10 + \sqrt{108})}.$$

Remarkably, this forbidding expression is equal to 2, as we explain in Box 10.3.

Box 10.3. Simplifying Tartaglia's solution.
To see why

$$x = \sqrt[3]{u} - \sqrt[3]{v} = \sqrt[3]{(10 + \sqrt{108})} - \sqrt[3]{(-10 + \sqrt{108})} = 2,$$

we exploit the fact that $\pm 10 + \sqrt{108} = \pm 10 + 6\sqrt{3}$ and look for numbers of the form $a + b\sqrt{3}$ whose cubes are $10 + 6\sqrt{3}$ and $-10 + 6\sqrt{3}$.
A little algebra shows that

$$(1 + \sqrt{3})^3 = 10 + 6\sqrt{3} \quad \text{and} \quad (-1 + \sqrt{3})^3 = -10 + 6\sqrt{3}.$$

Putting all this together, we deduce that, as claimed,

$$x = \sqrt[3]{(10 + \sqrt{108})} - \sqrt[3]{(-10 + \sqrt{108})} = (1 + \sqrt{3}) - (-1 + \sqrt{3}) = 2.$$

Cardano's *Ars Magna*. In 1536, at the age of 14, Ludovico Ferrari became Cardano's secretary and pupil. He worked with his master on writing the *Practica Arithmeticae* and turned out to be an independent mathematician in his own right, becoming a public lecturer on the subject in Milan before he was 20 and going on to become a tax assessor to Ferrante Gonzaga, Governor of Milan. Little is known about Ferrari, except for stories of his hot-blooded character; it is said that even Cardano was afraid of him at times, but Ferrari was always grateful to Cardano for giving him his chance to succeed in life, and happily called himself 'Cardano's creation'. It was Ferrari who saw how to reduce the solution of quartic equations (those of degree 4) to cubics, and who thereby became the first person to find a general method of solving quartic equations (it is given as Rule Two of Chapter 39 in the *Ars Magna*).

Armed with the first methods for solving both cubic and quartic equations, Cardano prepared to publish a major work on algebra, but his oath of secrecy initially prevented him from including Tartaglia's solution of cubic equations. Cardano also included the work by Ferrari on biquadratic (quartic) equations in the *Ars Magna*, presenting several separate cases of such equations. This was undoubtedly original, and Cardano must have chafed even more under his oath.

In 1542, however, della Nave made del Ferro's papers available to Cardano and Ferrari, and from these they learned that the earliest discovery was due not to Tartaglia, but to del Ferro. Cardano now felt free to ignore his oath and include the secret method for cubics in his *Ars Magna*. Here Cardano stated quite clearly in the *Ars Magna* the correct attributions of various solutions of equations, and he offered this defence of his action in making Tartaglia's secret public.[18]

> Scipio del Ferro of Bologna about thirty years ago invented [the method set forth in] this chapter, [and] communicated it to Antonio Maria Florido of Venice, who when he once engaged in a contest with Nicolo Tartalea of Brescia announced that Nicolo also invented it; and he [Nicolo] communicated it to us when we asked for it, but suppressed the demonstration. With this aid we sought the demonstration, and found it, though with great difficulty, in the manner which we set out in the following.

[18]Cardano, *Ars Magna*, Chapter XI, and F&G 8.A4(a),(b).

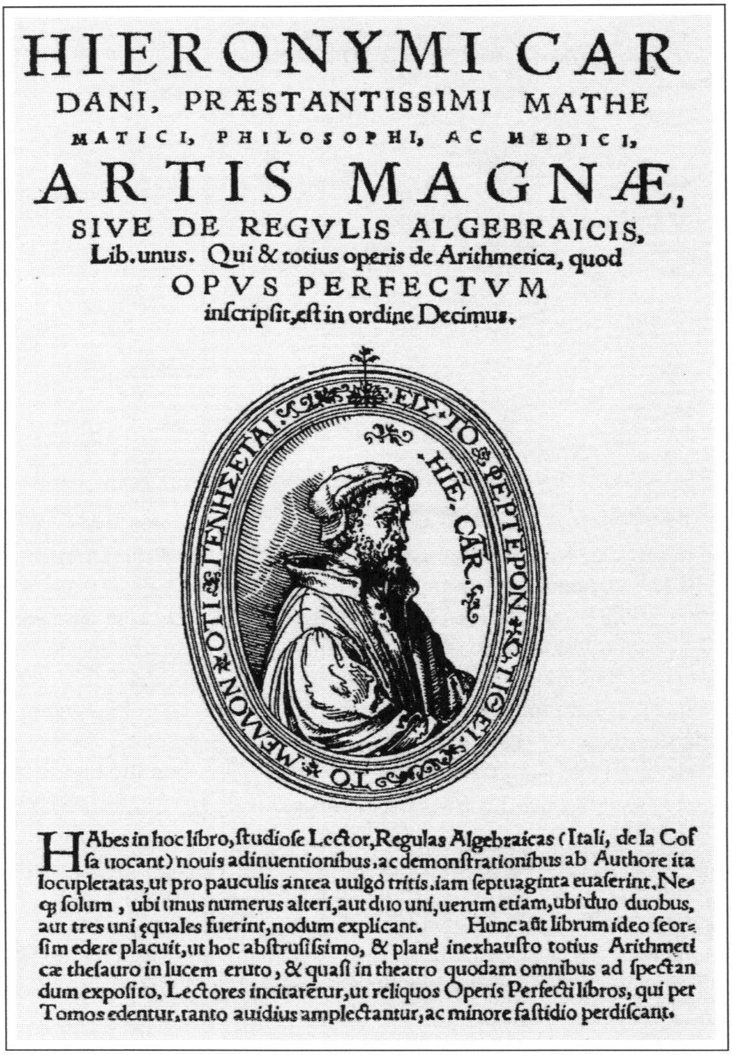

HIERONYMI CAR
DANI, PRÆSTANTISSIMI MATHE
MATICI, PHILOSOPHI, AC MEDICI,
ARTIS MAGNÆ,
SIVE DE REGVLIS ALGEBRAICIS,
Lib. unus. Qui & totius operis de Arithmetica, quod
OPVS PERFECTVM
inſcripſit, eſt in ordine Decimus.

HAbes in hoc libro, ſtudioſe Lector, Regulas Algebraicas (Itali, de la Coſſa uocant) nouis adinuentionibus, ac demonſtrationibus ab Authore ita locupletatas, ut pro pauculis antea uulgò tritis, iam ſeptuaginta euaſerint. Neq; ſolum, ubi unus numerus alteri, aut duo uni, uerum etiam, ubi duo duobus, aut tres uni ęquales fuerint, nodum explicant. Hunc aũt librum ideo ſeorſim edere placuit, ut hoc abſtruſiſſimo, & planè inexhauſto totius Arithmeticæ theſauro in lucem eruto, & quaſi in theatro quodam omnibus ad ſpectandum expoſito, Lectores incitarẽtur, ut reliquos Operis Perfecti libros, qui per Tomos edentur, tanto auidius amplectantur, ac minore faſtidio perdiſcant.

Figure 10.7. Cardano's *Ars Magna* (Nurnberg, 1545)

Tartaglia was outraged, and in 1546 he published his protest in a pamphlet entitled *Quesiti e Invenzioni Divers* (Questions and Diverse Inventions). A long exchange of pamphlets and letters ensued between Tartaglia and Ferrari, each more vitriolic than its predecessor, and these continued until the former's death in 1557. Eight years later Ferrari, by then a professor of mathematics in Bologna, died — poisoned in mysterious circumstances, possibly by his sister. Thus, Cardano survived Tartaglia by nineteen years and Ferrari by eleven. Writing of Ferrari in his autobiography (Chapter 35), Cardano quoted the Roman poet Martial:

> Life is exceedingly short and old age rare;
> whoever you love, do not desire them to please too much.

The *Ars Magna* influenced work on algebra throughout the next century. In the opening words of its first chapter Cardano offered this account of the nature of the work:[19]

Cardano on the history of algebra.

This art originated with Mahomet, the son of Moses the Arab. Leonardo of Pisa is a trustworthy source for this statement. There remain, moreover, four propositions of his with their demonstrations, which we will ascribe to him in their proper places. After a long time, three derivative propositions were added to these. They are of uncertain authorship, though they were placed with the principal ones by Luca Paccioli. I have also seen another three, likewise derived from the first, which were discovered by some unknown person. Notwithstanding the latter are much less well known than the others, they are really more useful, since they deal with the solution of [equations containing] a cube, a constant, and the cube of a square.

In our own days Scipione del Ferro of Bologna has solved the case of the cube and first power equal to a constant, a very elegant and admirable accomplishment. Since this art surpasses all human subtlety and the perspicuity of mortal talent and is a truly celestial gift and a very clear test of the capacity of men's minds, whoever applies himself to it will believe that there is nothing that he cannot understand. In emulation of him, my friend Niccolo Tartaglia of Brescia, wanting not to be outdone, solved the same case when he got into a contest with his [Scipione's] pupil, Antonio Maria Fior, and, moved by my many entreaties, gave it to me. For I had been deceived by the words of Luca Paccioli, who denied that any more general rule could be discovered than his own. Notwithstanding the many things which I had already discovered, as is well known, I had despaired and had not attempted to look any further. Then, however, having received Tartaglia's solution and seeking for the proof of it, I came to understand that there were a great many other things that could also be had. Pursuing this thought and with increased confidence, I discovered these others, partly by myself and partly through Ludovico Ferrari, formerly my pupil. Hereinafter those things which have been discovered by others have their names attached to them; those to which no name is attached are mine. The demonstrations, except for the three by Mahomet and the two by Ludovico, are all mine. Each is individually set out under a proper heading and, following the rule, an illustration is added.

Although a long series of rules might be added and a long discourse given about them, we conclude our detailed consideration with the cubic, others being merely mentioned, even if generally, in passing. For as positio [the first power] refers to a line, quadratum [the square] to a surface, and cubum [the cube] to a solid body, it would be very foolish for us to go beyond this point. Nature does not permit it. Thus, it will be seen, all those matters up to and including the cubic are fully demonstrated, but the others which we will add, either by necessity or out of curiosity, we do not go beyond barely setting out. In everything, however, the worth of the preceding books [of this work], especially the third and fourth books,

[19] See Cardano in (Smith 1959, 203–206), and F&G 8 A4(a).

should be kept in mind, lest I be thought trifling when I repeat or obscure when I skip over something.

We shall shortly examine in more detail Cardano's discussion of his solution method for the first case of the cubic equation — *a cube and things equal to numbers* (see Figure 10.8). To pursue Cardano's precise argument in its geometrical language is quite hard and time-consuming, so you might prefer to read Box 10.4 (on page 307), which explains, in our algebraic language, the justification for the Rule that Cardano adopted.

The heavily geometrical aspect of the Demonstration is nonetheless interesting. Cardano's language of justification is deeply imbued with classical Greek geometry. Cardano referred explicitly to Euclidean propositions, using the technical terms 'binomial' and 'apotome' from *Elements*, Book X. (These are the names for what are called u^3 and v^3 in Box 10.3, arising from the way that they are magnitudes involving square roots — the details are unimportant, but the fact that Cardano uses this language is significant.) No-one would dispute that Cardano's achievement is part of the history of algebra, but at this point Cardano seems to be much more involved with geometry. However, as with the Islamic examples we looked at in the previous chapter, geometrical language is being used in quite a flexible way, without any special concern for dimensional constraints. This is apparent in the Demonstration's opening sentence, where Cardano unabashedly added a cube to six times a side to produce a number. As with al-Khwārizmī, it seems as if algebraic solution methods have to be supported by geometrical vindications.

Here is Cardano's 'Rule' which is a recipe method or algorithm for solving cubic equations of the stated type, preceded by his 'Demonstration' of the rule, in which he gives a proof of the validity of his general method, with the particular example we discussed earlier: $x^3 + 6x = 20$.[20]

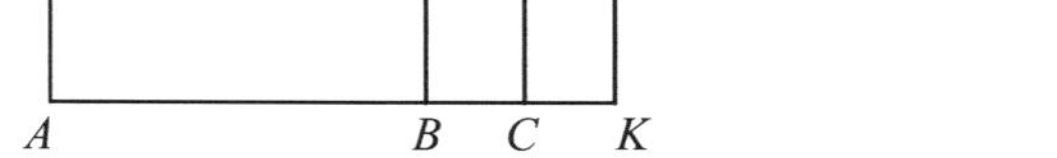

Figure 10.8. Cardano: cube and things equal to a number

Cardano on the solution to a cubic equation.

Demonstration: For example, let the cube of GH and six times the side GH be equal to 20. I take two cubes AE and CL whose difference shall be 20, so that the product of the side AC by the side CK shall be 2, — i.e. a third of the number of 'things'; and I lay off CB equal to CK, then I say that if it is done thus, the remaining line AB is equal to GH and therefore to the value of the 'thing', for it was supposed of GH that it was so [i.e. equal to x], therefore I complete, after the manner of the first theorem of the 6th

[20]See Cardano in (Smith 1959, 203–206), and F&G 8 A4(b).

chapter of this book, the solids DA, DC, DE, DF, so that we understand by DC the cube of BC, by DF the cube of AB, by DA three times CB times the square of AB, by DE three times AB times the square of BC. Since therefore from AC times CK the result is 2, from 3 times AC times CK will result 6, the number of 'things': and therefore from AB times 3 AC times CK there results 6 'things' AB, or 6 times AB, so that 3 times the product of AB, BC, and AC is 6 times AB. But the difference of the cube AC from the cube CK, and likewise from the cube BC, equal to it by hypothesis, is 20; and from the first theorem of the 6th chapter, this is the sum of the solids DA, DE, and DF, so that these three solids make 20.

But taking BC minus, the cube of AB is equal to the cube of AC and 3 times AC into the square of CB and minus the cube of BC and minus 3 times BC into the square of AC. By the demonstration, the difference between 3 times CB times the square of AC, and 3 times AC times the square of BC, is [3 times] the product of AB, BC, and AC. Therefore since this, as has been shown, is equal to 6 times AB, adding 6 times AB to that which results from AC into 3 times the square of BC there results 3 times BC times the square of AC, since BC is minus. Now it has been shown that the product of CB into 3 times the square of AC is minus; and the remainder which is equal to that is plus, hence 3 times CB into the square of AC and 3 times AC into the square of CB and 6 times AB make nothing. Accordingly, by common sense, the difference between the cubes AC and BC is as much as the totality of the cube of AC, and 3 times AC into the square of CB, and 3 times CB into the square of AC (minus), and the cube of BC (minus), and 6 times AB. This therefore is 20, since the difference of the cubes AC and CB was 20.

Moreover, by the second theorem of the 6th chapter, putting BC minus, the cube of AB will be equal to the cube of AC and 3 times AC into the square of BC minus the cube of BC and minus 3 times BC into the square of AC. Therefore the cube of AB, with 6 times AB, by common sense, since it is equal to the cube of AC and 3 times AC into the square of CB, and minus 3 times CB into the square of AC, and minus the cube of CB and 6 times AB, which is now equal to 20, as has been shown, will also be equal to 20. Since therefore the cube of AB and 6 times AB will equal 20, and the cube of GH, together with 6 times GH, will equal 20, by common sense and from what has been said in the 35th [of Book I] and 31st of the 11th Book of [Euclid's] *Elements*, GH will be equal to AB, therefore GH is the difference of AC and CB. But AC and CB, or AC and CK, are numbers or lines containing an area equal to a third part of the number of 'things' whose cubes differ by the number in the equation, wherefore we have the Rule:

Rule: Cube the third part of the number of 'things', to which you add the square of half the number of the equation, and take the root of the whole, that is, the square root, which you will use, in the one case adding the half of the number which you just multiplied by itself, in the other case subtracting the same half, and you will have a 'binomial' and 'apotome' respectively; then subtract the cube root of the apotome from the cube root of the binomial, and the remainder from this is the value of the 'thing'.

> In the example, the cube and 6 'things' equals 20; raise 2, the 3rd part of 6, to the cube, that makes 8; multiply 10, half the number, by itself, that makes 100; add 100 and 8, that makes 108; take the root, which is $\sqrt{108}$, and use this, in the first place adding 10, half the number, and in the second place subtracting the same amount, and you will have the binomial $\sqrt{108}+10$, and the apotome $\sqrt{108}-10$, take the cube root of these and subtract that of the apotome from that of the binomial, and you will have the value of the 'thing', $\sqrt[3]{(\sqrt{108}+10)} - \sqrt[3]{(\sqrt{108}-10)}$.

But if we grant that the Demonstration is an adequate theoretical defence of the Rule, we can still ask about the utility of the method — for example, does it provide intelligible answers? This is not a trivial question. Even in the case of quadratic equations, the square root in the formula throws up the need to understand numbers like $\sqrt{5}$, which are not rational numbers. We saw with al-Khwārizmī's work that the classification of equations is built around the tacit idea that the solutions must be positive numbers; thus, for him, the equation $x^2 + 10x = 39$ had the unique solution $x = 3$; the other solution that we would find ($x = -13$) was simply invisible to him. Cardano's division of cubic equations into types ('a cube equal to things and numbers', and so on) was clearly intended to serve something of the same purpose, but he was to find that matters were no longer so tidy. The problem that this highlighted was to prove of considerable significance for the history of mathematics.

Let us pause to observe that one influential aspect of Cardano's treatment of the different types of cubic was that he showed how to reduce equations in such a way that the 'square' term could be made to disappear. He did not take equations at their face value, so to speak, but showed how to reduce them to standard forms to which his methods applied. This was a great step forward in the recognition that equations have an internal structure. It implied, for example, that the case we examined above is more general than it looks, because it also serves to solve more complicated-looking equations, once the equation has been transformed in a way that Cardano indicated. So, considering that his equations were written down in such a way that all terms are positive (which accounts for the large number of cases that he needed to consider), we see that Cardano's treatment was remarkably thorough.

As for the answers that his method produces, Cardano accepted negative numbers if that is what his rules led to. He described such solutions as 'false' or 'fictitious', which is a reasonable view to hold if numbers are taken to be counting numbers — there is a difference between plus-three sheep in a field and minus-three sheep.

Square roots of negative numbers. But Cardano's procedure for solving cubics could also lead to results that are even stranger than negative numbers. Take the case of $x^3 = 14x + 8$: this evidently has $x = 4$ as one solution, and also has two further solutions that are also real numbers, $x = -2 + \sqrt{2}$ and $x = -2 - \sqrt{2}$. But if we apply Cardano's method to the equation (as he, prudently, did not) we find

$$x = \sqrt[3]{2 + \sqrt{-\tfrac{2636}{27}}} + \sqrt[3]{2 - \sqrt{-\tfrac{2636}{27}}},$$

a result that Cardano was unable to make sense of, because it involves the square root of a negative quantity. He observed that the problem could be solved geometrically, and also by some ad hoc methods if one was lucky, but his general method seemed to lead to absurdities on occasions like this. He even specified that the method breaks down for equations of the form $x^3 = bx + c$ when $(b/3)^3 > (c/2)^2$. This is a remarkable achievement, made as a

Box 10.4. Why does Cardano's rule work?
Consider the similarity in form between these two equations:

$$\begin{aligned} x^3 \quad &+ \quad cx \quad = \quad d, \\ (u-v)^3 &+ 3uv(u-v) = u^3 - v^3. \end{aligned}$$

The first of these is the cubic equation whose solution we seek; the second is an algebraic identity, true for all values of u and v (as can be seen by multiplying the terms out).

If it is possible to arrange matters so that c is $3uv$ and d is $u^3 - v^3$, then x must be $u - v$, because of the formal similarity between the equations.[a] So it remains only to discover those values of u and v that satisfy $3uv = c$ and $u^3 - v^3 = d$. We write $v = c/3u$ and deduce that

$$u^3 - \left(\frac{c}{3u}\right)^3 = d,$$

from which it follows that

$$27(u^3)^2 - 27d(u^3) - c^3 = 0,$$

which is a quadratic equation in u^3. We deduce that

$$u = \sqrt[3]{(d/2) + \sqrt{[(d/2)^2 + (c/3)^3]}},$$

the cube root of the 'binomial', and

$$v = \sqrt[3]{(-d/2) + \sqrt{[(d/2)^2 + (c/3)^3]}},$$

the cube root of the 'apotome'.

The solution to the original cubic is then $x = u - v$.

Cardano's rule is effectively a set of instructions for inserting into this formula the particular values of c and d that specify the given cubic equation.

---

[a]This is the argument that Cardano covers in his Demonstration with the words 'by common sense and from what has been said in the 35th of Book I and 31st of the XIth book of Euclid's *Elements*'; Book I, Prop. 35, is Euclid's proof that parallelograms on the same base and between the same parallels are equal (in area), and Book XI, Prop. 31, is the analogous result in three dimensions.

throwaway remark in the opening chapter of the book, and again in Chapter XII, and shows how extensive was his mastery of the theory.

Nonetheless, the unfortunate feature of the method is that here there are three real solutions, but the method fails to find any of them. This turned out to be unavoidable: whenever the equation has three real solutions, the method churned out at best a mysterious answer that involves the square roots of negative numbers. As such, it was to prove a spur to later writers on the subject — notably Rafael Bombelli, whose work we consider below.

It is easy to believe that the whole idea of square roots of negative numbers is a complete nonsense: any number, other than zero, gives a positive number when multiplied by itself. A justifiable reaction to such objects as the square root of a negative number occurring in an argument is that some basic rule has been broken. But whereas they can easily be avoided in quadratic equations — and we saw Diophantus head them off in Problem I.7, (see p. 181) — they arise with Cardano's method exactly where they could be expected not to.

In fact, Cardano showed a greater readiness than most of his contemporaries to contemplate the square roots of negative numbers. As he put it in the *Ars Magna*:[21]

> I will give an example: If it should be said, Divide 10 into two parts the product of which is 40, it is clear that this case is impossible. Nevertheless, we will work thus: We divide 10 into two equal parts, making each 5. These we square, making 25. Subtract 40, if you will, from the 25 thus produced ... leaving a remainder of 15, the square root of which added to or subtracted from 5 gives parts the product of which is 40. These will be
>
> $$5+\sqrt{-15} \text{ and } 5-\sqrt{-15}.$$
>
> Putting aside the mental tortures involved, multiply $5+\sqrt{15}$ by $5-\sqrt{-15}$, making $25-(-15)$ which is $25+15$. Hence this product is 40 ... So progresses arithmetic subtlety the end of which, as is said, is as refined as it is useless.

Cardano's words were well chosen. It is certainly subtle and refined. It is progress, because answers have been obtained where before there were none. But what use could there be for numbers that are neither counting numbers nor measuring numbers?

As we noted above, Cardano's procedure, when applied to cubic equations with real and distinct roots, gives rise to expressions involving the square roots of negative numbers. Tartaglia had already described such cases as 'not reducible', and this type of equation became known under the Latin name of the 'casus irreducibilis' or 'irreducible case'. At the end of the century Viète introduced a solution method for such cubics that drew on trigonometric identities, thereby by-passing the procedure that had given rise to such paradoxical or 'impossible' answers. This finessed the problem for cubic equations, but it left unexplained the mysterious square roots of negative numbers.

Cardano also included in his *Ars Magna* a method devised by Ferrari for quartic equations. This method involves introducing a second unknown, then solving a cubic, then feeding the solution back into the original quartic, and all this without any notation to distinguish between one unknown and another, or to assist in the mental manipulation of lengthy equations. Little wonder that Cardano said of it: 'Carrying out such operations as these is about the greatest thing to which the perfection of human intellect or, rather, of human imagination, can come'.[22]

The *Ars Magna* became well known, but it was not easy reading. Cardano's insights into the structure and transformation of equations were without precedent in European (or any) mathematics, and it was some time before they were taken up. Rafael Bombelli in Italy and Simon Stevin in the Netherlands, both engineers, wrote books that clarified and systematised Cardano's work, and gradually his ideas percolated into mainstream mathematics and set the agenda for the study of equations well into the 17th century.[23] If a single book can be said to mark the beginning of European (as opposed to classical or Islamic) mathematics, it must be the publication of the *Ars Magna* in 1545.

10.4 Bombelli and Viète

Rather dramatically, two of the next major contributions to the study of algebra were made by mathematicians who were interested in studying Greek mathematical texts residing unpublished in the stores of the Vatican. Their names were Rafael Bombelli and François Viète, and the work that particularly drew their attention was by Diophantus.

[21] See (Cardano 1968, 219–220).

[22] See (Cardano 1968, 246).

[23] Stevin, *L'Arithmétique* (1585).

As we noted above, Regiomontanus in the 1460s was the first mathematician to seek to interest mathematicians of the West in the existence of a text by Diophantus in the original Greek, but this had no immediate effect. In the 1550s Bombelli was alerted by a friend to a passing reference to a Greek manuscript of Diophantus in the Vatican library.[24] He planned to translate and publish the work, but such an edition has never wholly come to light, although a partial Italian translation exists and portions of the *Arithmetica* were incorporated into a revised version of Bombelli's *Algebra* (Figure 10.9). The sheer oddness of Diophantus's work, and perhaps the difficult nature of some of its problems, seem to have brought it a steady following.

After Bombelli, Wilhelm Holzmann (writing under the Greek name of Xylander) published a Latin translation of Diophantus's *Arithmetica* at Basel in 1575, which in turn was translated by Stevin into French ten years later. Xylander's translation was severely (and somewhat unfairly) criticised by Stevin, and also by Claude Bachet who published another Latin translation (together with the Greek original) in Paris in 1621. Bachet's edition formed the basis for subsequent studies, until the researches of Paul Tannery in the 19th century established a new definitive text.

Bombelli's *Algebra*. As for Bombelli himself, we know very little of his life, except that he lived and worked at Bologna as engineer to the Bishop of Melfi and was an expert in hydraulics. It would be quite natural that he would turn also to mathematics. In his thirties he prepared an *Algebra* that was essentially a systematisation of Cardano's *Ars Magna*. However, once he was aware of the Diophantus manuscript in the Vatican library, he carried out a thorough revision of his *Algebra*, which was eventually published in 1572, the year of his death; a second edition appeared seven years later with only the title page changed. In both editions only Books I–III were published; two further books, which included discussion of geometrical problems to be solved by algebra, remained in manuscript form only and were not published until the 20th century.

A clue to Bombelli's originality, or at least his novel source, is given in the preface to his book, which indicates at least some of the earlier work still known in his day.[25]

Bombelli on the history of algebra.

> I have decided first to consider the majority of the authors who up to now have written about [algebra], so that I can fill in what they have missed out. They are very many, and among them certainly Mohammed ibn Musa, an Arab, is believed to be the first, and there is a little book of his, but of very small value. I believe that the world 'algebra' came from him, because some years ago, Brother Luca [Pacioli] of Borgo San Sepolcro of the Minorite order, having set himself the task of writing on this science, as much in Latin as in Italian, said that the word 'algebra' was Arabic, and means in our language 'position', and that the science came from the Arabs. Many who have written after him have believed and said likewise, but in recent years, a Greek work on this discipline has been discovered in the Library of our Lord in the Vatican, composed by a certain Diophantus of Alexandria, a Greek author, who lived at the time of Antoninus Pius. When it had been shown to me by Master Antonio Maria Pazzi, from Reggio, public lecturer in mathematics in Rome, and we had judged him an author very intelligent in numbers (although he did not treat irrational numbers, but in

[24]See (Bombelli 1966, 8–9); see also F&G 8.A5(a) for Bombelli's account of this discovery.

[25]From the Preface to Bombelli's *Algebra*, 8–9, this translation by F.R. Smith appears in F&G 8.A5(a).

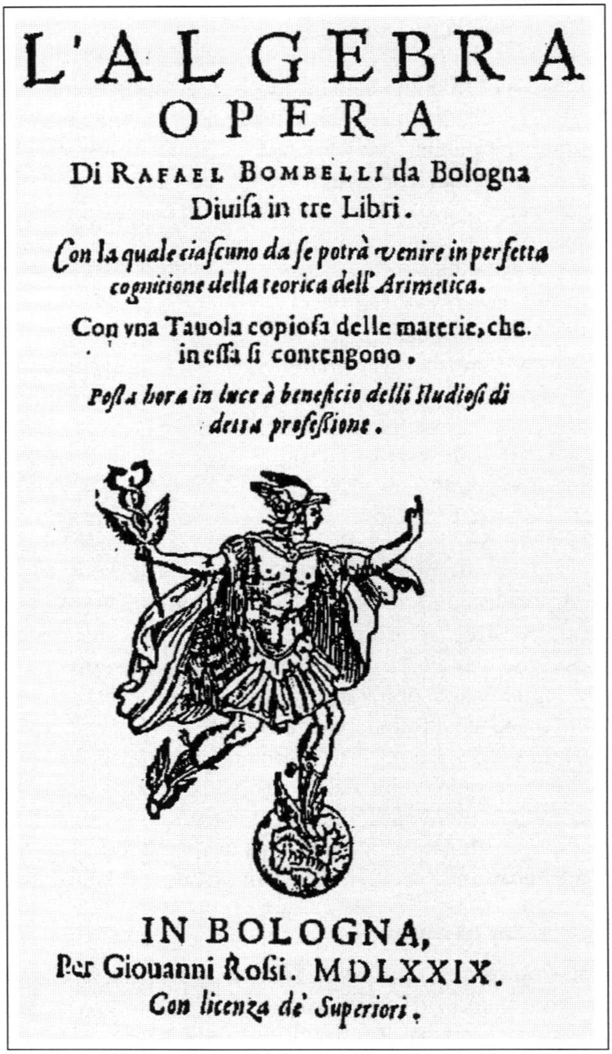

L'ALGEBRA
OPERA
Di RAFAEL BOMBELLI da Bologna
Diuiſa in tre Libri.
Con la quale ciaſcuno da ſe potrà venire in perfetta cognitione della teorica dell'Arimetica.
Con vna Tauola copioſa delle materie, che in eſſa ſi contengono.
Poſta hora in luce à beneficio delli ſtudioſi di detta profeſſione.

IN BOLOGNA,
Per Giouanni Roſſi. MDLXXIX.
Con licenza de' Superiori.

Figure 10.9. Bombelli's *Algebra*, 2nd edition (1579)

> him alone is evident a perfect order of working), he and I, in order to enrich the world with such a complete work, set ourselves to translate it, and we have translated five books (of the seven). The remainder we have not been able to finish because of the troubles that have occurred to us, the one and the other. In this work we have found that he cites the Indian authors many times.

After this indication of the importance he attached to a previously unknown Greek author, Bombelli turned to writers much nearer his own time.

> After him (after a long interval of time) Leonardo Pisano wrote, in Latin, but after him there was no one who said anything useful until the above-mentioned Brother Luca, who indeed (even though he was a careless writer and for that reason made some mistakes) was nonetheless the first

> to shed light on this science, even though there are some who set themselves up and take to themselves all the credit, wickedly finding fault with the few errors of the Brother and saying nothing of his good work. Then in our times both Italians and foreigners have written, such as Oroncio, Scribelio, and Boglione, all French; young Stifel, a German, and a certain Spagnard, who wrote copiously in his own language, but truly there has been no one who has opened the secret of algebra other than Cardano of Milan in his *Ars Magna*, where he spoke a lot about this science, but in the telling was obscure. He treated it likewise in certain of his broadsheets, which with Ludovico Ferrari, from our Bologna, he wrote against Niccolò Tartaglia of Brescia. In these there are extremely fine and ingenious problems in this science, but with such little modesty from Tartaglia (who of his nature was so accustomed to speaking evil, that when he thought he had made an honoured sage of himself, he had actually been talking slander) that he offended almost all honourable minds, seeing clearly how he talked nonsense about both Cardano and Ferrari, geniuses of these our times more divine than human. There are others still who have written on the subject, but if I wanted to name them all I would have enough to do. But because their books have been of little benefit I will say nothing of them, and only (as before) say that having thus seen how much of the subject has been dealt with by the above-mentioned authors, I too then in my turn have put together the present work for the common good, dividing it into three books.

We see that Bombelli had a high opinion of Diophantus. He was also aware of the Indian roots of algebra although he gave no details; his opinion of al-Khwārizmī was low; he mentioned, but gave no assessment of, Leonardo of Pisa; Pacioli's work was 'good' and 'useful', despite its mistakes; he thought that Cardano's *Ars Magna* was at times obscure, but that Cardano himself was a genius, and he was on the side of Cardano and Ferrari in their disputes with Tartaglia.

The most significant changes that Bombelli made to his *Algebra* arising from his study of Diophantus occur in Book III, where problems in a practical vein in the original version were replaced by 143 abstract problems taken from the *Arithmetica*. Bombelli also revised his terminology throughout, changing cossist terms into terms derived from Diophantus: for example, *cosa* (for the unknown) became *tanto* ('so many'). Bombelli saw his revisions in terms similar to those expressed by Maurolico and Commandino — that is, as achieving a restoration of the 'effectiveness' of the ancients.

Bombelli's *Algebra* is notable in several ways, in addition to its role in further disseminating the methods of Cardano and an awareness of Diophantus. In particular, it includes the first extensive discussion of what we today call *complex numbers* and the rules for calculating with them.[26]

Bombelli on imaginary numbers.

> I have found another kind of cube root of a compound expression very different from the other kinds, which results from the case of the cube equal to so many [things] and a number, when the cube of the third of the things is greater than the square of the half of the number, as will happen

[26]Bombelli, *Algebra*, 133–134, and F&G 8.A5(b) (transl. F.R. Smith). In our bracketed additions the symbol i stands for $\sqrt{-1}$, the supposed square root of -1.

in this case. This kind of square root has different arithmetical operations from the others and a different denomination, because when the cube of the third of the things is greater than the square of the half of the number, the excess can be called neither plus nor minus. But I shall call it 'plus of minus' when it is to be added, and when it is to be subtracted I shall call it 'minus of minus', and this operation is most necessary, more than the other cube roots of compound expressions, in regard to the cases of powers of powers [fourth powers], accompanied by cubes, or things, or both together, which are the equations where this sort of root appears much more than the other. This will seem to many to be more artificial than real, and I held the same opinion myself, until I found the geometrical demonstration (as will be shown in the proof of the above case on a plane surface). And first I shall deal with multiplication, setting down the rule of plus and minus.

Plus times plus of minus, makes plus of minus. $[(+1)(i) = +i]$
Minus times plus of minus, makes minus of minus. $[(-1)(i) = -i]$
Plus times minus of minus, makes minus of minus. $[(+1)(-i) = -i]$
Minus times minus of minus, makes plus of minus. $[(-1)(-i) = +i]$
Plus of minus times plus of minus, makes minus. $[(+i)(+i) = -1]$
Plus of minus times minus of minus, makes plus. $[(+i)(-i) = 1]$
Minus of minus times minus of minus, makes minus. $[(-i)(-i) = -1]$

As we saw above, Cardano had become aware that square roots of negative numbers can arise in the solution of equations, even those that are known to have real numbers as answers (we discuss when imaginary *roots* arise in Box 10.5). Bombelli made the important discovery that these complex expressions always appear in 'conjugate pairs' — pairs of the form $a + b\sqrt{-1}$ and $a - b\sqrt{-1}$. In one of his examples, derived from the cubic equation $x^3 - 15x - 4 = 0$ whose roots are the real numbers $x = 4$ and $x = 2 \pm \sqrt{3}$, Cardano's method expressed the root $x = 4$ in the form

$$\sqrt[3]{2 + \sqrt{-121}} + \sqrt[3]{2 - \sqrt{-121}},$$

which does not look anything like the number 4.

However, Bombelli observed that

$$2 + \sqrt{-121} = 2 + 11\sqrt{-1} \quad \text{and} \quad 2 - \sqrt{-121} = 2 - 11\sqrt{-1},$$

and that

$$(2 + \sqrt{-1})^3 = 2 + 11\sqrt{-1} \quad \text{and} \quad (2 - \sqrt{-1})^3 = 2 - 11\sqrt{-1}.$$

It follows that

$$\sqrt[3]{2 + \sqrt{-121}} + \sqrt[3]{2 - \sqrt{-121}} = \left(2 + 11\sqrt{-1}\right) + \left(2 - 11\sqrt{-1}\right) = 4.$$

This showed that appropriate manipulation of the pairs according to his rules could eliminate the 'imaginary' parts and arrive at numbers that do not involve roots of negative quantities. He formulated the four basic operations with complex numbers, although he continued to regard such numbers as 'sophistic' and 'useless'. They appeared useless because he was unable to use them in the solution of cubics, since it was necessary first to know one of the roots, or, alternatively, to know how to find a cube root of a number in this form — a non-trivial task. Although Bombelli is often praised by historians of mathematics for his insight, *complex numbers* (as numbers of the form $a + b\sqrt{-1}$ are usually called) continued to cause perplexity for several centuries to come.

Bombelli's symbolism is interesting. He used the by-then familiar symbols $p.$ and $m.$ for plus and minus, and powers of the unknown were represented by $\overset{1}{\smile}$, $\overset{2}{\smile}$, etc.; thus, our expression $4x^3 - 16x + 16$ appeared as $4\overset{3}{\smile}m.16\overset{1}{\smile}p.16$. Square roots were denoted by $R.q.$, cube roots by $R.c.$, fourth roots by $R.R.q.$, and so on. For our round brackets he used $\lfloor \ldots \rfloor$, so that our $\sqrt[3]{(72 - \sqrt{1088})}$ appeared as $R.c.\lfloor 72.m.R.q.1088 \rfloor$, and 'imaginary' numbers, such as $\sqrt{-121}$, were written as $R.q.\lfloor 0m.121 \rfloor$. Such notation for exponents was new in Italian works, although it had appeared in France in Chuquet's manuscript of 1484.

Bombelli's *Algebra* was certainly the best treatment of the subject by an Italian writer to date. It was to exert its influence over two centuries at least: we find it quoted by Leibniz in the 17th century and by Euler in the 18th. It is an interesting confluence of the tradition of the Italian algebraists of the 16th century and the Renaissance return to the Greeks. His *Algebra* arose naturally from that of Cardano.

Box 10.5. When does *'a cube and things equal to numbers'* have complex roots?

We have already seen that in solving the equation $x^3 + cx = d$, both the 'binomial' and the 'apotome' include the square root of the expression $(d/2)^2 + (c/3)^3$.

Replacing c by $-c$ gives $x^3 = cx + d$, so for this general cubic problem we need the square root of $(d/2)^2 + (-c/3)^3$ – that is, of $(d/2)^2 - (c/3)^3$. This is negative, and therefore its square root is complex, whenever

$$(c/3)^3 > (d/2)^2.$$

However, following his discovery of Diophantus, he wished to restore the 'effectiveness' of the ancients and was responsible for disseminating much of the content of the *Arithmetica*. Thus, he reflected the diverse visions of the 16th-century mathematical renaissance where, in particular, we find that writers who were trained in the sciences turned to Greek mathematics in order to obtain a secure foundation in mathematics and its applications. Bombelli may not have been able to provide secure foundations for the square roots of negative numbers, but his exposure to Diophantus did lead him to seek a more general theory and to diminish the status of problems. This was not a paradoxical response so much as a natural attempt to impose order on the jumble of examples that Diophantus offered.

François Viète. The Frenchman François Viète, often known by the Latin form of his name, Franciscus Vieta, was both an original mathematician and an important transitional figure in the recovery of Greek mathematical texts. He was born at Fontenay-le-Comte in 1540, and followed in the footsteps of his father, a lawyer, by graduating in law at the University of Poitiers in 1559.[27]

After practising at the bar for some four years, he entered the household of a Huguenot nobleman as tutor to his daughter and subsequently became secretary to his widow.[28] The daughter had a special interest in astrology and astronomy, and Viète was obliged to study these subjects, along with trigonometry, in order to fulfil his role as a tutor. From 1571 he held various political appointments in Paris and Rennes, becoming a privy counsellor in

[27]His engraved portrait, Figure 10.10, appeared in Van Schooten's 1646 edition of Viète's collected works.

[28]Huguenots were Protestants at the time of the religious wars in France. Many were killed on the orders of King Charles IX in the St. Bartholomew's Day massacre of 23 August 1572 which led to murderous assaults on Protestants across France that lasted for several weeks.

Figure 10.10. François Viète (1540–1603)

1580. Political intrigues led to his banishment from Paris in 1584, although he was recalled to court office at Tours in 1589. He became a specialist in deciphering coded messages — so successfully that his enemies accused him of sorcery.

Apart from his period as a tutor, Viète turned to astronomy and mathematics largely as a recreation for two periods of his life: 1564–1571, when he wrote primarily on astronomy and trigonometry, and 1584–1589, when he wrote primarily on algebra and geometry. Viète also began to write a *Canon Mathematicus* in the 1560s, in which he was to systematise and extend the Greek knowledge of plane and spherical trigonometry. This work, published in the 1570s, is the first extensive work on trigonometry in the West and is remarkably 'modern' in its approach. Expressions for all six trigonometric ratios are included, and tables are given for every minute of arc — previously, Rheticus in 1551 had produced tables only for every ten minutes of arc.

His computations were made by an iterative process that corresponded to considering inscribed and circumscribed polygons with very many sides — for the sine of 1 minute of arc the inscribed polygon had over 6000 sides, and the circumscribed polygon had over twice that number. He was, however, conscious of errors and at one time attempted to withdraw the *Canon* from circulation.

In these tables, Viète introduced decimal fractions that were later to be made popular by Simon Stevin. Initially, these were written with the fraction bar underneath and the denominator understood — thus, $\underline{205}$ was read as if it had a denominator of 1000 (and stood for 0.205). Later, he separated integral and decimal parts of numbers by means of a vertical stroke — for example 112246|205 represented our $112,246.205$.

Another area of interest was the reform of the Julian calendar being undertaken towards the close of the century by Pope Gregory XIII, with the advice of the Jesuit astronomer-mathematician Christoph Clavius and others. Viète bitterly attacked the accuracy of

Clavius's calculations, thereby earning for himself considerable animosity. Acrimonious correspondence ensued, which ended only with Viète's death in 1603. The root cause of his attacks on Clavius seems, however, to have been less a matter of disinterested accuracy of calculation (and not a religious matter) but rather that Clavius had ignored Viète's principles of the 'analytic art', to which we now turn.

The first of Viète's works on algebra, *In Artem Analyticem Isagoge* (Introduction to the Analytic Art), was published at Tours in 1591. In its first chapter he gave his interpretation of 'analysis' and introduced a new kind of analysis which he called 'rhetics' or 'exegetics'. The term 'analysis' reminds us of the Greek interest in that approach to geometry, and indeed Viète is another figure who tried to raise the standards of mathematics by a self-conscious return to the ancients.

Chapter I begins with the Greek notion of analysis, which Viète contrasted with synthesis.[29] The former, he said, assumes the required result and argues from this to a known truth; the latter assumes this known truth and argues so as to reach the required truth. The ancients, he now informed his readers, had two forms of analysis: *zetetics* (setting up equations and proportions) and *poristics* (testing theorems by equations or proportions). Viète's contribution was to add a third kind of analysis: *rhetics* or *exegetics* (solving equations or proportions). This threefold 'analytic art' is 'the science of discovery in mathematics'. He then commented on a shortcoming of 'the old analysis', which was limited to numbers, and referred to a 'newly discovered symbolic logistic'.

Viète broke with the carefree tradition of adding squares to sides, and in Chapter III returned to the Greek requirement of homogeneity: 'Homogeneous terms must be compared with homogeneous terms'.[30] There was the usual scale of magnitudes: side or root, square, cube, square–square, square–cube, and so on, their kinds being length or breadth, plane, solid, plane–plane, plane–solid, and so on. Since 'plane–plane' is four-dimensional, we see that Viète observed no restriction to three dimensions.

Chapter IV begins with the important statement: 'Numeral logistic is [a logistic] that employs numbers, symbolic logistic one that employs symbols or signs for things as, say, the letters of the alphabet'. The important point here is that Viète was making a conscious effort to generalise from numbers to symbols that can represent numbers or magnitudes. This is made explicit in Chapter VII, where Viète argued (none too clearly) that rhetics can be applied to both numbers and magnitudes, and is therefore a more general method superior to each.[31] In fact, Viète hoped that that his method was actually universal. He then went on to advocate that the preferred form of solution is the one that obtains the equation from synthesis, thus allowing the synthesis to be its own proof, and not the reverse. So, for Viète, the skilful geometer conceals the fact that he is doing analysis and presents his work synthetically, setting out the equation or proportion merely as an aid to the arithmetician.

Viète's overall aim was hardly a modest one. We find it at the end of the last chapter of the *Isagoge*: 'The analytic art ... claims for itself the greatest problem of all, which is TO SOLVE EVERY PROBLEM'. It was not possible for Viète to solve every equation, let alone every problem. He did, however, solve equations up to the fourth degree, either by algebraic methods or by *neusis* constructions. He was also a committed European, so much so that he never used the word 'algebra' because it was non-European; all the old vocabulary (by which he meant terms of Arabic origin) had, he said, to be purged, as he

---

[29]Part of Chapter I (translated into English) can be found in F&G 8.C2(b).

[30]Homogeneity means that every term in an equation must be of the same degree.

[31]See (Smith 1959, 205–206) and F&G 8.C2(c).

put it in the preface to the *Isagoge*, lest the subject 'retain its filth and continue to stink in the old way'.[32]

In looking back on Viète's achievements, modern historians tend to make special mention of the advances in algebraic notation. He introduced letters of the alphabet to represent numbers or magnitudes, with consonants for those known and vowels for the unknowns. The signs $+$ and $-$ appear in his works, but $\times$ and $\div$ do not: multiplication was denoted by the word 'in', and division by a line. In writing out his algebraic expressions, he indicated the type of each quantity so that homogeneity was preserved. For example:[33]

> A cubum, + A quadrato in B ter, + A in B quadratum ter, + B cubo aequari $A+B$ cubo

represents, in modern notation,

$$A^3 + 3A^2B + 3AB^2 + B^3 = (A + B)^3.$$

The 1646 edition of his collected works (edited by Frans van Schooten) shortened the Latin terms to single letters: for example, 'A quadrato' becomes AQ, but the word *aequatur* for equality was retained in full. Thus, in the van Schooten edition we find the numerical equation:[34]

$$1QC - 15QQ + 85C - 225Q + 274N \text{ aequatur } 120 \ ,$$

which we would write as

$$x^5 - 15x^4 + 85x^3 - 225x^2 + 274x = 120 \ .$$

The point of commenting on Viète's notation is not merely to enable us to read what he wrote. Still less is it to place it in some hierarchy or chronology of notations that leads up to our own system. Rather, it is to observe the major conceptual step involved in introducing the idea of an indeterminate quantity, which nonetheless can be treated as if it were a number.

The concept of an indeterminate quantity is a novelty, and it is a little tricky to grasp at first. As the historian Jacob Klein argued in his book, for Viète an indeterminate quantity is not a variable, which would be something that changed over time, or as the result of the change in some other quantity upon which it depends.[35] Nor is to be treated as if it has a definite value which you happen not to know. That is what would be called an 'unknown quantity', and it is what comes up in problems, where, at the end of the solution its true, definite value is revealed. An indeterminate quantity is a kind of quantity or magnitude, which might be a number, or a length, and which can be manipulated according to stated rules. Viète intended it as a novel kind of quantity which belongs in a universal theory embracing algebra and geometry.

In Viète's work, algebra was being presented as a universal method of reasoning, embracing, or surpassing, both arithmetic and geometry, which when taken up by others was to lead to its replacing geometry as the main language of mathematics. Algebraic manipulation could now replace geometrical construction. The new mode of mathematical thought was thus gradually to become algebraic rather than geometrical.

Furthermore, Viète coupled his intellectual advance with a polemical historical position. His aim was to purify algebra by restoring what he saw as the lost analytical art of the Greeks. To that end, he made a special study of Pappus, Apollonius, and Diophantus.

---

[32] In (Klein 1968, 318) and F&G 8.C2(a). The remark occurs towards the end of a long, rambling panegyric to Princess Mélusine and the apparent glories of the Lords of Rohan and the Bretons.

[33] Viète, *Isagoge*, in (Viète 1646, 17).

[34] See (Vieta 1646, 158), the last page of *De Emandatione*, Caput XIV.

[35] See (Klein 1968, 165).

Pappus's *Collection* had, in fact, included a 'treasury of analysis'[36] in which — once you have sufficient algebra — it is possible to see the principles of algebraic methods being applied to geometry.

This may have provided Viète with the stimulus for his most significant contribution to mathematical thought — the promotion of algebra as the most universal method in mathematics. At all events, it was grist to his Eurocentric mill. He argued that by forming a tradition out of Euclid's geometry, Diophantus's algebra, and the philosophy of Plato and Aristotle, one could recover both Greek rigour and a creative process of discovery that had been known to the Greeks but since lost (indeed, buried). The mathematical details of this claim would take us too far afield, but the claim was made repeatedly by Viète. We quote just one example, where Viète contrasted Regiomontanus's unsuccessful mixture of geometry and algebra with his own brand:[37]

> But the problems which Regiomontanus solved algebraically he confesses he could sometimes not construct geometrically. But was that not because algebra was up to that time practised impurely? Friends of learning, embrace a new algebra; farewell, and look to the just and the good.

As Klein commented:[38]

> This 'new' and 'purified' algebra, which is represented by the 'ars analytice', is for Vieta quite as much 'geometric' as it 'arithmetical'. Thus he conjectures that a generalized procedure which is not confined to figures and numbers lies not only behind the geometric analysis of the ancients but also behind the Diophantine *Arithmetic*.

We have seen enough of Greek and Islamic mathematics to see through Viète's historical claim, but we should note that it was neither the first nor the last of its kind. His mathematics, however, was to inspire subsequent generations of mathematicians. His pupil Marino Ghetaldi developed Viète's work on Apollonius, and used algebra freely in the solution of geometrical problems. Fermat was to adopt Viète's notation, including the principle of homogeneity that accompanied it. In England, Thomas Harriot was to study Viète's work in detail and rewrite much of it in his own much clearer notation. William Oughtred was to include a popular exposition of Viète's ideas in his *Clavis Mathematicae* (1631) and so introduce a whole generation of British mathematicians to algebraic geometry, while Henry Briggs's decimal division of trigonometrical degrees was suggested by Viète's decimal fractions. Indeed, it is possible here only to hint at the extent of Viète's influence on the subsequent development of mathematics in Europe.

With the work of Viète the ground rules for a truly independent branch of mathematics had been well laid. Algebra was no longer only a calculation process for solving problems. Algebraic rules and theorems no longer depended upon geometrical justification, but could be defended algebraically. A whole new way of thinking was thus open to mathematicians — thinking that would be directed towards casting new light on classical geometry and opening up new vistas of mathematical research and activity.

---

[36]See F&G 5.B3.

[37]See (Vieta 1646, 324) in 'Responsum ad problema quod omnibus mathematicis totius orbis construendum proposuit Adrieanus Romanus' (Response to the problems proposed by Adrieanus Romanus to be considered by all mathematicians in the entire world).

[38]See (Klein 1968, 158).

10.5 Further reading

Huff, T.E. 1993. *The Rise of Early Modern Science*, Cambridge University Press. This is a comparative study of intellectual life in the Islamic world, Europe, and China, with the aim of explaining why science arose only in Europe.

Ore, O. 1965. *Cardano, the Gambling Scholar*, Dover. A highly readable account of one of mathematics' more colourful characters.

Rose, P.R. 1975. *The Italian Renaissance of Mathematics*, Droz. A full study of the mathematical sciences and the rediscovery of Greek mathematics in 15th- and 16th-century Italy.

Stedall, J. 2011. *From Cardano's Great Art to Lagrange's Reflections: Filling a Gap in the History of Algebra*, European Mathematical Society: Heritage of European Mathematics. The subtitle says it all, but the filling is rich, clear, and fascinating.

11

The Renaissance of Mathematics in Britain

Introduction

There are several ways to investigate the history of mathematics, depending on what kind of question is being asked. After four chapters that traced some mathematical concerns over the course of a millennium or more, we now come to one that covers little more than seventy years, and after a chapter in which the transmission of mathematical ideas was the prime focus, we now turn to one in which we explore the range of mathematical activity within one culture and period: that of Britain in the 16th and early 17th centuries.

We shall proceed by pursuing various questions and examining the contributions of particular individuals: Robert Recorde, John Dee, Thomas Harriot, John Napier, and Henry Briggs. We consider why they were important, against the backdrop of a period in which the mathematical arts and sciences attained prominence in British culture as never before, and as Britain set about acquiring an Empire and began to colonise North America.

How was mathematics taught, and why? What kind of mathematics was it? What role did different groups in society play in this? What factors influenced the development of mathematical activity? Complete answers to these questions are unattainable, but we reach partial ones and try to make sense of a very complex period by concentrating on some specific themes: the language in which mathematics was communicated and taught, the public perception of mathematics and mathematicians, and the interplay of mathematical tradition and contemporary needs in fostering and directing mathematical practices.

Note that most of the texts with which this chapter is concerned were printed when English spelling was less fixed or consistent than now. We have kept a few of these texts in their original spelling, while the rest have been modernised without further comment.

11.1 Mathematics in the vernacular: Robert Recorde

Robert Recorde lived from about 1510 to 1558. His first book *The Ground of Artes* (see Figure 11.1), published around 1543, was not the earliest arithmetic text published in England: that was Cuthbert Tonstall's *De Arte Supputandi* of 1522. Nor was it the earliest book on arithmetic written in English: such a work had been published in London in 1526 under the title *The Arte and Science of Arismetique*, although only a single leaf survives. The earliest book on arithmetic written in English to survive in its entirety was published at St Albans under the altogether informative title of *An Introduction for to Lerne to Reken with Pen and with the Counters, after the True Cast of Arismetyke or Awgrym in hole Numbers and also in Broken*. But Recorde's book was unprecedentedly popular — there were to be some fifty printings of it over the next century and a half, the last being in 1699. Investigating Recorde's work — *The Ground of Artes* and his other mathematical texts — will help us to understand certain features of the mathematical life of 16th-century England.

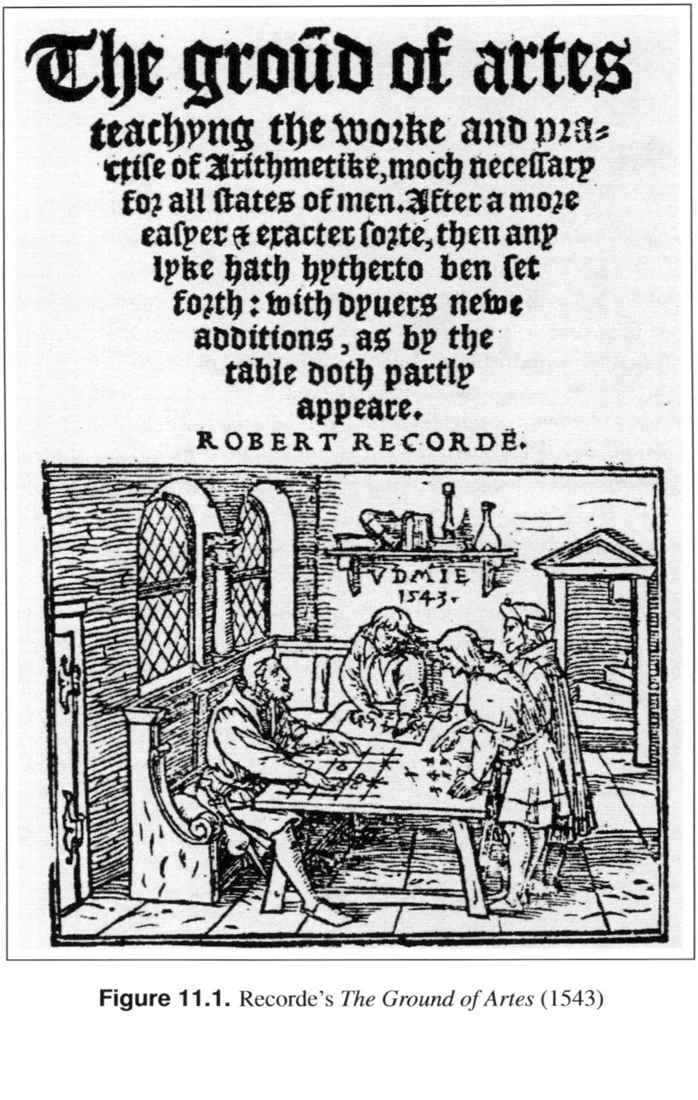
The groūd of artes
teachyng the worke and pra-
ctise of Arithmetike, moch necessary
for all states of men. After a more
easyer & exacter sorte, then any
lyke hath hytherto ben set
forth: with dyuers newe
additions, as by the
table doth partly
appeare.
ROBERT RECORDE.

Figure 11.1. Recorde's *The Ground of Artes* (1543)

Why was *The Ground of Artes* so popular? Both the work itself, and the historical context in which it appeared, are relevant here. Recorde's work was well written and carefully thought through, as well as being original. Of the other books mentioned above, Tonstall's was closely modelled on Italian commercial texts, and the St Albans work was largely a translation of a 1508 Dutch arithmetic, *Die Maniere om te Leeren Cyffren*. Robert Recorde was the first writer to consider afresh what mathematics should be taught, and how to teach it, in the circumstances of the time: increasing literacy, growing trade and prosperity, and developing technological capabilities for peace and war. Writing most of his books in the form of dialogues between master and pupil greatly contributed to their appeal, and the teaching decisions that Recorde took happily coincided with the needs of a rapidly growing readership for such texts.

But before looking in more detail at how and what Recorde taught, we consider the historical context more broadly. Two 16th-century developments are especially helpful perspectives for locating Recorde's work: these are the *humanist* tradition and the *vernacular* tradition.

In the present context, *humanism* is the name given to the movement for the critical recovery of classical writings, emphasising humane thought and culture as distinct from theological studies. Maurolico and Commandino were humanist scholars in this sense, as was the influential Dutch scholar Desiderius Erasmus, whose stay in England in the early years of the 16th century helped to stimulate a new spirit in educational thought and practice. Erasmus particularly emphasised the educational value of studying the writings of the ancients; other humanists sought a broader learning in which experience of the world played a greater part. Some humanists were strongly anti-clerical, seeing nothing but harm arising from the scholastic educational tradition, with its overwhelming emphasis on the writings of Aristotle and a teaching style that valued rote learning and logical disputation above thinking or comparison with the real world.

The humanist critique of scholastic modes of learning was powerfully expressed in the lively satire *Gargantua and Pantagruel* by a French monk and physician, François Rabelais, published in around 1532. In this passage the giant Gargantua is being taught by 'a great Sophister–Doctor':[1]

> To try masteries in School disputes with his condisciples, he would recite by heart backwards and did sometimes prove on his fingers ends to his mother, quod de modis significandi non erat scientia [that there was no knowing about the method of signifying]. Then did he reade to him the compost [computus], for knowing the age of the Moon, the seasons of the year, and tides of the sea, on which he spent sixteen yeares and two moneths ...
>
> At the last his father perceived, that indeed he studied hard, and that although he spent all his time in it, did neverthelesse profit nothing, but which is worse, grew thereby foolish, simple, doted and blockish ... he found that it were better for him to learne nothing at all, than to be taught such like books, under such Schoolmasters, because their knowledge was nothing but brutishnesse, and their wisdome but blunt foppish toyes, serving only to bastardize good and noble spirits, and to corrupt all the flower of youth.

This is not an attack on *computus* (ecclesiastical computation, the calculation of the dates in the church calendar) as such, but on its being badly taught and taking an unconscionable time. Rabelais was in favour of teaching the liberal arts of the quadrivium, provided that it was done in an intelligent way.

[1]See (Urquhart 1653, Chs. xiv, xv).

So Recorde, who was born in Tenby, South Wales, graduated from the University of Oxford in 1531 at a time when the higher educational traditions of the preceding three centuries were under attack, their practice and value were called into question, and there was a liberalising humanistic movement to spread educational benefits more widely. We shall shortly see ways in which Recorde's works partake of this spirit of the new learning.

The vernacular movement provides another dimension for perceiving Recorde's achievement. This is not the same as humanism — we saw in Chapter 10 that the translations of Maurolico and Commandino were into Latin, not Italian, and Erasmus was keener that schoolchildren should be taught Greek than that Greek and Latin works should be translated into English. But there was much contemporary debate, in which many humanist scholars were involved, over the appropriateness and possibility of writing matters of any subtlety or technicality in English. Discussions about translating classical texts into the vernacular — and indeed about writing texts in the vernacular in the first place — are found throughout the century. The increased availability of printed books, a related increase in the proportion of the population who could read, and a new confidence in the expressive power of the English language itself, led some to realise that it was now feasible and desirable to open up learning to people who did not read Latin.

When Recorde came to write his great series of textbooks in the 1540s and 1550s he found, naturally enough, that the English language did not contain enough technical terms. This gave him two choices: he could use the Latin and Greek terms, trying to pass them off as English words and hoping that they would become familiar, or he could coin new words or adapt them from those already in the language.

Recorde chose the latter path at first, as can be seen most clearly in his geometry textbook *The Pathway to Knowledg* (1551), where he set out to introduce Euclidean geometry to English readers. He began with a discursive account of the definitions in Book I of Euclid's *Elements*, where he introduced several terms that we no longer use: *pricks* for points, and *sharp, square* and *blunt corners* (or *angles*) for acute, right, and obtuse angles. So he made some use of synonyms, giving some Latin-derived words (*right, angle*) as well as English ones.

Recorde's discussion of lines is interesting. A *straight line* is about the only one of his English definitions to have lasted, in place of the more literal Latin term *right line* (*linea recta*) which he also gave. And besides the great pains he took to explain what a *crooked line* is — because he enjoyed drawing them, by the looks of things — Recorde had to explain when a sharply crooked line is considered as two lines (namely, at an angle or corner). He was more open than we are, too, to considering angles between non-straight lines.[2]

Recorde went on to create further terms: a *touch line* is a tangent; an equilateral triangle is *threelike*; a parallelogram is a *likejamme* (unless all four sides are equal, in which case it is a *likeside*). These and many more were the result of a thoughtful scholarly endeavour to anglicise mathematical terminology, in order to minimise difficulties for those learning the subject for the first time through the medium of the English language.

Recorde's new terms have generally not lasted.[3] That they did not survive into everyday usage is due to the countervailing pressures to which Recorde himself bowed in the end; not only did a completely fresh set of terms make life harder for the people who spoke Latin, but for English people coming new to mathematical studies, one word seemed to be just as good as another.

[2] Recall the contribution by Geminus in Chapter 5.

[3] In contrast, more such coinages by German authors survive in German.

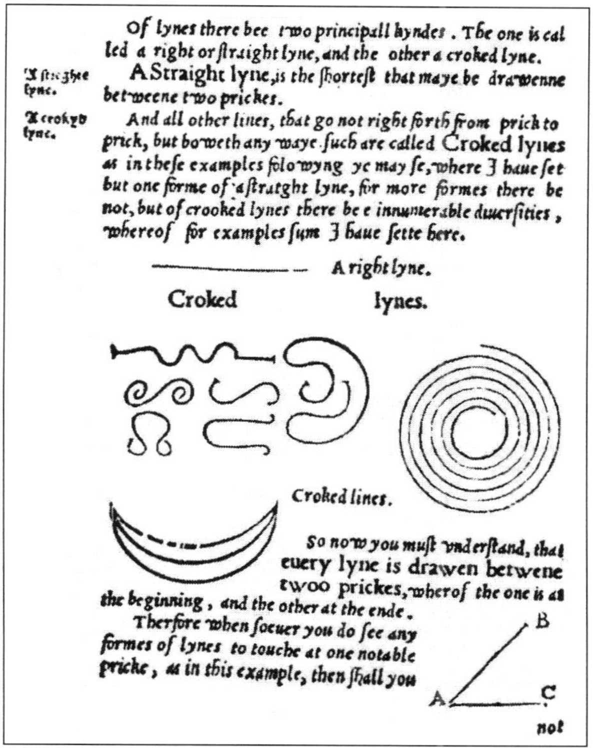

Of lynes there bee two principall kyndes. The one is called a right or straight lyne, and the other a croked lyne.

A straight lyne. — A Straight lyne, is the shortest that maye be drawenne betweene two prickes.

A croked lyne. — And all other lines, that go not right forth from prick to prick, but boweth any waye, such are called Croked lynes as in these examples folowyng ye may se, where I haue set but one forme of a straight lyne, for more formes there be not, but of crooked lynes there be e innumerable diuersities, whereof for examples sum I haue sette here.

A right lyne.

Croked lynes.

Croked lines.

So now you must vnderstand, that euery lyne is drawen betwene twoo prickes, wherof the one is at the beginning, and the other at the ende.

Therfore when soeuer you do see any formes of lynes to touche at one notable pricke, as in this example, then shall you

not

Figure 11.2. Crooked and straight lines, from Recorde's *Pathway to Knowledg* (1551)

His concern to be understood extended to his care about notation. He introduced the equals sign in *The Whetstone of Witte* (Figure 11.3) to avoid 'the tediouse repetition of these woordes: is equalle to', as he put it, writing that 'Gemowe lines of one lengthe . . . because noe. 2. thynges, can be moare equalle'.[4]

Such carefully considered endeavours would be wasted unless there were readers eager to study mathematics. Lest there be any doubt on the matter, Recorde described, in detail that verged on the overwhelming, the benefits accruing from the study of arithmetic (in *The Ground of Artes*) and those brought by geometry (in *The Pathway to Knowledg*).

Here he is extolling arithmetic:[5]

Recorde on The profit of arithmetick.

MASTER: Wherefore as without numbring a man can do almost nothing, so with the help of it, you may attain to all things.

SCHOLAR: Yes, sir, why then it were best to learn the Art of numbring, first of all other learning, and then a man need learn no more, if all other come with it.

MASTER: Nay not so: but if it be first learned, then shall a man be able (I mean) to learn, perceive, and attain to other Sciences; which without it he could never get.

---

[4]'Gemowe' is derived from Gemini, meaning twins.

[5]See (Recorde 1552) in F&G 9.A1(b).

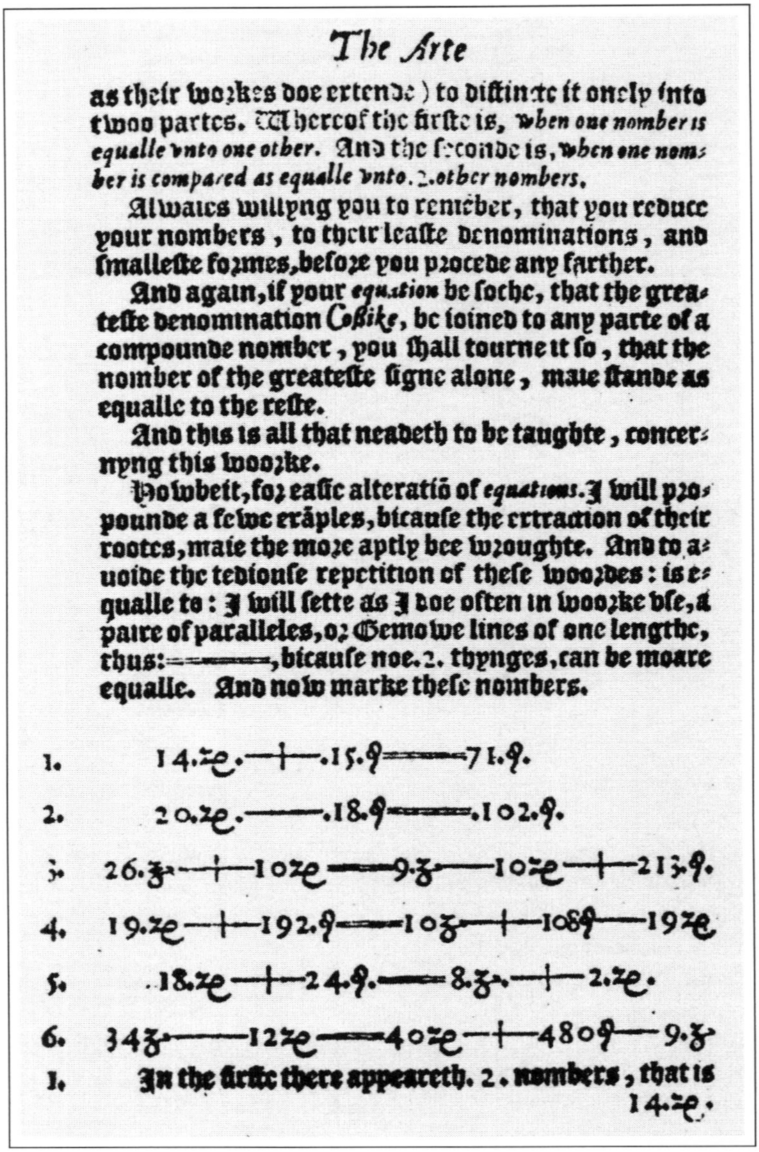

The Arte

as their workes doe extende) to distincte it onely into twoo partes. Whereof the firste is, *when one nomber is equalle unto one other*. And the seconde is, *when one nomber is compared as equalle unto .2. other nombers*.

Alwaies willyng you to remember, that you reduce your nombers, to their leaste denominations, and smalleste formes, before you procede any farther.

And again, if your *equation* be soche, that the greateste denomination *Cosike*, be ioined to any parte of a compounde nomber, you shall tourne it so, that the nomber of the greateste signe alone, maie stande as equalle to the reste.

And this is all that neadeth to be taughte, concernyng this woorke.

Howbeit, for easie alteratiō of *equations*. I will propounde a fewe exāples, bicause the extraction of their rootes, maie the more aptly bee wroughte. And to auoide the tediouse repetition of these woordes: is equalle to: I will sette as I doe often in woorke use, a paire of paralleles, or Gemowe lines of one lengthe, thus: =======, bicause noe .2. thynges, can be moare equalle. And now marke these nombers.

1. 14.ze. —+— .15.φ ======= 71.φ.
2. 20.ze. ——— .18.φ ======= .102.φ.
3. 26.ʒ. —+— 10ze ======= 9.ʒ. ——— 10ze —+— 213.φ.
4. 19.ze —+— 192.φ ======= 10ʒ —+— 108φ ——— 19ze.
5. 18.ze —+— 24.φ. ======= 8.ʒ. —+— 2.ze.
6. 34ʒ ——— 12ze ======= 40ze —+— 480φ ——— 9.ʒ.

1. In the firste there appeareth. 2. nombers, that is 14.ze.

Figure 11.3. The elongated 'equals' sign in Recorde's *The Whetstone of Witte* (1557)

SCHOLAR: I perceive by your former words, that Astronomy and Geometry depend much on the help of numbring: but that other Sciences, as Musick, Physick, Law, Grammer, and such like, have any help of Arithmetick, I perceive not.

MASTER: I may perceive your great Clerk-linesse by the ordering of your Sciences: but I will let that passe now, because it toucheth not the matter that I intend, and I will shew you how Arithmetick doth profit in all these somewhat grosly, according to your small understanding, omitting other reasons more substantiall.

First (as you reckon them) Musick hath not onely great help of Arithmetick, but is made, and hath his perfectnesse of it: for all Musick standeth by number and proportion: And in Physick, beside the calculation of criticall dayes, with other things, which I omit, how can any man judge the pulse rightly, that is ignorant of the proportion of numbers?

And so for the Law, it is plain, that the man that is ignorant of Arithmetick, is neither meet to be a Judge, neither an Advocate, nor yet a Proctor. For how can hee well understand another mans cause, appertaining to distribution of goods, or other debts, or of summes of money, if he be ignorant of Arithmetick? This oftentimes causeth right to bee hindered, when the Judge either delighteth not to hear of a matter that hee perceiveth not, or cannot judge for lack of understanding: this commeth by ignorance of Arithmetick.

Now, as for Grammer, me thinketh you would not doubt in what it needeth number, sith you have learned that Nouns of all sorts, Pronouns, Verbs, and Participles are distinct diversly by numbers: besides the variety of Nouns of Numbers, and Adverbs. And if you take away number from Grammer, then is all the quantity of Syllables lost. And many other ways doth number help Grammer. Whereby were all kindes of Meeters found and made? was it not by number?

But how needfull Arithmetick is to all parts of Philosophy, they may soon see, that do read either Aristotle, Plato, or any other Philosophers writings. For all their examples almost, and their probations, depend of Arithmetick. It is the saying of Aristotle, that hee that is ignorant of Arithmetick, is meet for no Science. And Plato his Master wrote a little sentence over his Schoolhouse door, Let none enter in hither (quoth he) that is ignorant of Geometry. Seeing hee would have all his Scholars expert in Geometry, much rather he would the same in Arithmetick, without which Geometry cannot stand.

And how needfull Arithmetick is to Divinity, it appeareth, seeing so many Doctors gather so great mysteries out of number, and so much do write of it. And if I should go about to write all the commodities of Arithmetick in civill acts, as in governance of Common-weales in time of peace, and in due provision & order of Armies, in time of war, for numbering of the Host, summing of their wages, provision of victuals, viewing of Artillery, with other Armour; beside the cunningest point of all, for casting of ground, for encamping of men, with such other like: And how many wayes also Arithmetick is conducible for all private Weales, of Lords and all Possessioners, of Merchants, and all other occupiers, and generally for all estates of men, besides Auditors, Treasurers, Receivers, Stewards, Bailiffes, and such like, whose Offices without Arithmetick are nothing: If I should (I say) particularly repeat all such commodities of the noble Science of Arithmetick, it were enough to make a very great book.

SCHOLAR: No, no sir, you shall not need: For I doubt not, but this, that you have said, were enough to perswade any man to think this Art to be right excellent and good, and so necessary for man, that (as I think now) so much as man lacketh of it, so much hee lacketh of his sense and wit.

MASTER: What, are you so farre changed since, by hearing these few commodities in generall: by likelihood you would be farre changed if you knew all the particular Commodities.

And elsewhere he breaks into verse to advocate 'The commodities of geometry':[6]

Recorde on The commodities of geometry.

Sith Merchauntes by shippes great riches do winne,
I may with good righte at their seate beginne.
The Shippes on the sea with Saile and with Ore,
Were first founde, and styll made, by Geometries lore.
Their Compas, their Carde, their Pulleis, their Ankers,
Were founde by the skill of witty Geometers.
To sette forth the Capstocke, and eche other part,
Wold make a great showe of Geometries arte.
Carpenters, Carvers, Joiners and Masons,
Painters and Limners with suche occupations,
Broderers, Goldesmithes, if they be cunning,
Must yelde to Geometrye thankes for their learning.
The Carte and the Plowe, who doth them well marke,
Are made by good Geometrye. And so in the warke
Of Tailers and Shoomakers, in all shapes and fashion,
The woorke is not praised, if it wante proportion.
So weavers by Geometrye hade their foundacion,
Their Loome is a frame of straunge imaginacion.
The wheele that doth spinne, the stone that doth grind,
The Myll that is driven by water or winde,
Are workes of Geometrye straunge in their trade,
Fewe could them devise, if they were unmade.
And all that is wrought by waight or by measure,
Without proofe of Geometry can never be sure.
Clockes that be made the times to devide,
The wittiest invencion that ever was spied,
Now that they are common they are not regarded,
The artes man contemned, the woorke unrewarded.
But if they were scarse, and one for a shewe,
Made by Geometrye, then should men know,
That never was arte so wonderfull witty,
So needfull to man, as is good Geometry.

We need not take such passages as these completely at face value; this kind of praising of one's subject was a rhetorical feature of books of the time, particularly of mathematical books written in the vernacular. As the historian Elizabeth Eisenstein has pointed out, in relation to contemporary French works:[7]

French tutors and reckon-masters were just as eager as their counterparts elsewhere to attract wealthy pupils and patrons while achieving the more elevated status conferred by literary fame ... The same literary devices that had earlier

[6] Recorde, *The Pathway to Knowledg*, in F&G, 9.A2(a).
[7] See (Eisenstein 1979, 541–542).

> adorned anthologies of devotional literature began to appear in arithmetic books which dealt with tables of interest and accounting. The importance of cultivating business arts by financiers and statesmen was stressed ... The same authors also found ways of dignifying applied or practical geometry. Its usefulness to gentry with landed estates and to nobles entrusted with military command could be underlined.

To evaluate Recorde's aims, it is helpful to see how well he fits into Eisenstein's pattern of writers. If we look back, for example, at the extract from the preface to *The Ground of Artes* where Recorde addressed what he called 'The profit of Arithmetick', it would seem, if anything, that Recorde seeks to avoid being put in Eisenstein's category. Although he mentioned military applications of arithmetic ('numbering of the Host, summing of their wages, ...'), it was in rather a rushed way and the Scholar soon makes it clear that he wishes to hear no more about it. Again, contrary to the writers described by Eisenstein, Recorde's account of 'The commodities of geometry' is entirely devoted to peaceful benefits. Finally, if someone were really anxious to attract and appease wealthy patrons, writing an open letter to the King inveighing against the 'miserable case' of the realm in which 'the ministers and interpreters of the lawes, are destitute of all good sciences' seems an odd way to go about winning friends and influencing people (depending, perhaps, on the detailed political situation at the time).

From these extracts, Recorde seems to have been independently minded, and concerned (perhaps more altruistically) with the spreading of mathematical knowledge. He was clearly a humanist scholar who had read widely and critically in ancient and modern authorities, but it is not known how much of his income was from teaching or writing about mathematics — he probably practised medicine for a while, and he was also a civil servant (Surveyor of Mines and Monies in Ireland) in the early 1550s.

The attitudes that Recorde expressed are as important as his influence in explaining mathematics in the vernacular. One example is the lead he gave to those trying to balance the implications of the rediscovery of Greek knowledge: just because some Greek had stated something, was it thereby necessarily true?

Recorde gave a clear answer to this pressing question of the time, in his astronomy text *The Castle of Knowledge* (see Figure 11.4). This is one of the firmest statements of something that Recorde emphasised repeatedly: the *reason* for something is important in our deciding whether to accept it, rather than the *authority* of whoever said it. While one might hope that this is a commonplace of education now, in making such strong assertions in the 16th century Recorde was aligning himself with others who were seeking a judicious balance between everyday experience and the unfolding Greek heritage.

There was also a political and religious message in what Recorde said:[8]

> No man can worthely praise Ptolemye, his travell being so great, his diligence so exacte in observations, and conference with all nations, and all ages, and his reasonable examination of all opinions, with demonstrable confirmation of his owne assertion, yet muste you and all men take heed, that both in him and in al mennes workes, you be not abused by their autoritye, but evermore attend to their reasons, and examine them well, ever regarding more what is saide, and how it is proved, then who saieth it: for autoritie often times deceaveth many menne.

The judiciousness and moderation of Recorde's views are worthy of remark, by comparison with the more extreme radical position taken by his near-contemporary in France,

[8]See (Recorde 1556) and F&G 9.A3.

Figure 11.4. Recorde's *The Castle of Knowledge* (1556)

Peter Ramus. Ramus reputedly made his name as a young man by defending in public the daring thesis that 'Everything that Aristotle has said is false'.[9]

In a lively career of noisy controversy, Ramus pressed for the teaching of logic and the quadrivium subjects to be made more relevant, practical, and methodical, before dying the cruel but (to others) glamorous death of a Protestant martyr in the St Bartholomew's Day massacre (23 August 1572). Subsequently, a popular educational movement known as Ramism spread widely, particularly in Protestant northern Europe, committed to clear and

[9]This claim was as significant a symbol of changing scholarly attitudes, we might say, as Luther's celebrated nailing of the 95 theses to the church door at Wittenberg a few years earlier was a symbol of the Reformation.

systematic expositions of mathematics and of every other subject, often in the vernacular. In the words of Ramus's Scottish translator, Roland M'Kilwein in 'The logike of P. Ramus martyr ... per M. R. Makylmenæum Scotum' (1574)[10]

> Shall we then think the Scottische or Englische tongue is not fit to wrote any art into? No in dede.

As the historian of Ramism Walter Ong has pointed out, the ambivalence that still attached to the vernacular can be seen in the way that M'Kilwein voiced these admirably progressive sentiments in a translation issued under the splendid Latinised form of his name!

In Britain, Ramism had quite an impact in the later 16th century, particularly among the young at the universities (Glasgow, Aberdeen, and St Andrews, as well as at Oxford and Cambridge), who were attracted by its anti-hierarchical message and its claim that knowledge was accessible to all people of whatever station. In geometry and arithmetic Ramus stressed, as had Recorde, the importance of a teaching style that arises out of practical situations and usefulness in the first place. But Recorde was the more solid mathematical scholar, and his works had established themselves as the prime English textbooks long before translations of Ramus's books began to be made in the 1570s and 1580s.

We may see something of the style of teaching pioneered by Recorde, in an extract from *The Pathway to Knowledg*. After the opening definitions that we looked at earlier, there is a section called 'The practike workinge of sondry conclusions Geometrical', where Recorde shows how to bisect an angle. The differences between Recorde's and Euclid's accounts are interesting. The most substantial difference between them is that in Book I, Prop. 9 Euclid proved this result[11] — that is, he gave a construction and then proved that the line thus constructed is indeed the bisector required — whereas Recorde did not even describe a construction in any very helpful way, let alone show why it worked. His construction would work — on the assumption that halving an arc of a circle is a more basic or easier construction than halving an angle — but he did not show how to bisect an arc.

> To divide an angle of right lines into ii equal partes. First open your compasse as largely as you can, so that it do not excede the length of the shortest line that incloseth the angle. Then set one foote of the compasse in the verye point of the angle and with the other fote draw a compassed arch from the one lyne of the angle to the other, that arch shall you devide in halfe, and then draw a line from the angle to the middle of the arch, and so the angle is divided into ii equall partes.
>
> *Example*. Let the triangle be A, B, C, then set I one foot of the compasse in B, and with the other I draw the arch D, E, which I part into ii equall parts in F, and then draw a line from B, to F, & so I have mine intent.[12]

So in terms of the standards of classical geometrical propositions, Recorde's discussion is woeful. But this should alert us to the inference that he must have been trying to do something different here — Recorde was perfectly able to translate Euclid quite precisely, had he wanted to, and he had clearly studied the *Elements* carefully before reworking Euclid's ideas in this way. A reasonable guess about what Recorde's intentions might be was that he was simply trying to familiarise his readers with the notion of bisecting an angle, and accustom them to using compasses; for example, his very first instruction is to 'open

[10]Cited in (Ong 1958, 14).

[11]See F&G 3.B3(b) and our account in Chapter 3.

[12]See (Recorde 1551) and F&G 9.A2(b).

your compasse as largely as you can' — a remark of no theoretical import, but practically very sensible.

Our inference that Recorde was pursuing a carefully considered pedagogical strategy is confirmed by a discussion later in the book, where he explains his teaching decisions:[13]

> It is not easy for a man that shall travail in a strange art, to understand at the beginning both the thing that is taught and also the just reason why it is so. And by experience of teaching I have tried it to be true, for when I have taught the proposition, as it imported in meaning, and annexed the demonstration with all, I did perceive that it was a great trouble and a painful vexation of mind to the learner, to comprehend both those things at once ... This thing caused me in both these books to omit the demonstrations, and to use only a plain form of declaration, which might best serve for the first introduction for so shall men best understand things: First to learn that such things are to be wrought, and secondarily what they are, and what they do import, and then thirdly what is the cause thereof.

Not only was Recorde's teaching strategy within each book carefully considered, all his mathematical texts form an ordered progression, as their titles make clear:[14]

The Ground of Artes — The Pathway to Knowledg — The Gate of Knowledge — The Castle of Knowledge — The Treasure of Knowledge — The Whetstone of Witte

Everyone who studied some aspect of mathematics in Elizabethan England would have done so by joining Recorde for some part of this metaphorical journey, or by learning from someone who had done so. For example, a book published at the start of Elizabeth's reign, William Cuningham's *Cosmographical Glasse* (1559), clearly expected its readers to have made a preparatory study of Recorde's works: in it a pupil explains his mathematical training by saying, 'Yes sir I have redde the ground of Artes, The whetstone of wytte, and the path way'.[15]

Who, though, would the readers have been for such works? That there was a great need for the books, if even modest levels of numeracy were to be attained, becomes clear from a salutary passage from an educational writer of 1612, well after Recorde's death. John Brinsley, a school headmaster, wrote a dialogue advocating reform between a traditional teacher (Spoudeus) and a progressive teacher (Philoponus). When they came to discuss the level of arithmetical understanding Brindley had them agree 'about the ordinarie numbers or numbring':[16]

> *Spoudeus*: ... For I am much troubled about this, that my readers and others above them, are much to seeke in all matters of numbers, whether in figures or in letters [that is, in Hindu–Arabic or in Roman numerals]. Insomuch, as when they heare the Chapters named in the Church, many of them cannot turne to them, much lesse to the verse.
>
> *Philoponus*: This likewise is a verie ordinary defect, & yet might easily be helped by common meanes, in an hour or two. I call it ordinarie, because you shal have schollers, almost readie to go to the Universitie, who yet can hardly tell you the number of Pages, Sections, Chapters, or other divisions in their bookes.

[13]See (Recorde 1551, a.iiv–a.iiir).

[14]*The Gate of Knowledge* and *The Treasure of Knowledge* have not survived; it is not certain whether Recorde completed them.

[15]Quoted in (Bennett 1982, 8).

[16]See (Brinsley 1627, 25–26).

Then, after Philoponus explained how the Roman and Hindu–Arabic numerals work, he continued:

> In a word, to tell what any of these numbers stand for, or how to set downe any of them; will performe fully so much as is needfull for your ordinarie Grammar scholler. If you do require more for any, you must seeke Records The Arithmetique, or other like Authors, and set them to the Cyphering schoole.

From this we learn that even grammar school pupils about to go to university might be unable to use or distinguish numerals. It seems safe to infer that the distribution of mathematical understanding, even at the most basic level, was somewhat patchy. Nor was mathematics the aspiration of the brightest and the best. The Charterhouse[17] orders of 1627 instructed the Master that:[18]

> It shall be his care and the Usher's charge, to teach the scholars to cypher and cast an account, especially those that are less capable of learning, and fittest to be put to trades.

Charterhouse may have been exceptional in making even this much provision for teaching arithmetic.

So the question of who were the readers of Recorde's books, and indeed of works such as Cuningham's, which covered a wide range of mathematical practices such as surveying, navigation, the use of mathematical instruments, determining longitude, and so on, is a complicated one. In a rather banal sense, the answer is the surveyors and navigators. So perhaps we can make more progress by drawing out two further questions. How did the surveyors and navigators come to feel that mathematical knowledge would aid them? Was there an increasing number of people who felt that mathematics was useful to them?

Some answers to these questions will become evident as this chapter progresses. For now, let us register that the last two extant texts by Recorde were connected with the interests of the Muscovy Company, a company chartered in 1555 for trading with Russia:[19] *The Castle of Knowledge*, on the elements of astronomy, was written for the Muscovy Company's navigators, and Recorde's elementary algebra text, *The Whetstone of Witte*, was dedicated to the Governors of the Company. But before probing to discover what use these mathematics texts might have been to a trading company (if indeed they were), we try to gain a broader picture of the role and perception of mathematical activity in Elizabethan society, and consider how this may be seen in the life and work of the most influential mathematician of the generation after Recorde, John Dee. Dee carried on Recorde's work, both in editing new editions of *The Grounde of Artes* and in being technical adviser to the Muscovy Company; like Recorde, he was a Welshman, but in other respects he forms rather a contrast.[20]

The pages that Recorde devoted to the benefits of arithmetic and geometry were no empty rhetoric. He rightly judged that most potential readers would have somewhat vague ideas of what mathematics was about, or was good for. He was concerned, not so much to boost his books above those by rival authors, as to explain why anyone should read mathematics texts at all, in order to advance the claims of his subject in the eyes of the newly literate book-reading public.

---

[17] Charterhouse, now an English independent school, had been founded in 1611.

[18] See (O'Day 1982, 62).

[19] The company had rights to explore and trade to the north, north-east, and north-west, a much greater scope than its name would suggest.

[20] As we shall see below, in Section 11.2.

Despite Recorde's efforts there was — and remained throughout this period — some ambivalence in the public mind over what kind of an enterprise mathematical activity represented. In the next section we briefly consider the public image of mathematical sciences, before turning to the most famous Elizabethan discussion of them, by John Dee.

In 1551, around the time that Recorde would have been penning his preface to a later edition of *The Ground of Artes*, which spoke of arithmetic being 'so many waies needful unto the fyrst planting of a common welthe', Royal commissioners visited the University of Oxford to reform it along Protestant lines. According to John Aubrey writing over a century later:[21]

> My old cosen Parson Whitney told me that in the Visitation of Oxford in Edward VI's time they burned Mathematical bookes for Conjuring bookes, and, if the Greeke Professor had not accidentally come along, the Greeke Testament had been thrown into the fire for a Conjuring booke too.

This account was supplemented by Aubrey's friend, the antiquary Anthony à Wood, who commented,[22]

> Sure I am that such books wherein appeared Angles or Mathematical Diagrams, were thought sufficient to be destroyed because accounted Popish, or diabolical, or both.

Historians now consider that these accounts are somewhat exaggerated; although medieval manuscripts *were* lost over this period, the causes were general neglect as much as ideological, and there was a general desire to update from manuscripts to printed texts. Nonetheless, there is other testimony to the ambivalent status of mathematics in earlier generations. Francis Osborne wrote in the 1650s:[23]

> My Memory reacheth the time, when the Generality of People thought her [mathematics] most useful Branches, Spels and her Professors, Limbs of the Devil; converting the Honour of Oxford, due for her (though at that time slender) Proficiency in this study, to her Shame: Not a few of our then foolish Gentry, refusing to send their Sons thither, lest they should be smutted with the Black-Art.

So from these 17th-century memoirists we learn of a strange and unexpected perception of mathematics: 'conjuring books', 'Popish', 'diabolical', 'Limbs of the Devil', the 'Black-Art'. This is not at all how Robert Recorde comes across to us! So there are two questions here: Were the mid-17th-century perceptions of earlier times correct? And if so why was there this connection between mathematics and occult studies?

There had long been a clear overlap between the skills of astrologers and those of other mathematical practitioners. It accounts for St Augustine's hostility to mathematics, and in later times the Latin term *mathematicus* meant what we would call an astrologer. That this overlap was indeed a live issue in the early 17th century may be seen, for example, from the statutes drawn up in 1619 by Sir Henry Savile for the Chair of Astronomy he founded at Oxford: the professor 'is utterly debarred from professing the doctrine of nativities and all judicial astrology without exception' — there would be no point in such an injunction unless it were otherwise likely that the professor might profess astrology. It is understandable, in the case of astrology, that only someone with mathematical training would have the appropriate calculation skills.

In his preface to *The Pathway to Knowledg*, Recorde observed that:[24]

---

[21] Cited in (Dick 1962, 21), see F&G 9.A1(a).

[22] See (Anthony à Wood 1792, 107), quoted in (Yates 1939, 229).

[23] Quoted in (Feingold 1984, 79).

[24] See (Recorde 1551). In medieval times, astronomy was taken to include some astrology.

> When [inventions] be wrought, and the reason thereof not understande, then say the vulgar people, that those things are done by negromancy. And hereof it came that fryer Bakon was accounted so great a necromancier, which never used that art (by any conjecture that I can find) but was in geometry and other mathematical sciences so expert, that he could do by them such things as were wonderful in the sight of most people.
>
> Great talk there is of a glass that he made at Oxford, in which men might see things that were done in other places, and that was judged to be done by power of evil spirits. But I know the reason of it to be good and natural, and to be wrought by geometry (since perspective is a part of it) and to stand as well with reason as to see your face in common glass.

We can thus identify different strands of perceptions about mathematics in this period. Some people, such as Recorde and Savile, seem to have had a straightforwardly no-nonsense view, distinguishing mathematics from other activities in a way that is quite comprehensible to most of us today. Yet such cool detachment was very much the exception. Belief in the applied astronomical science of astrology — at some level or other, from the credulous to the cynical — was more common than not, as it would continue to be for a further century or more. To understand the role and functioning of mathematics in Renaissance Britain, we may need to open ourselves to awareness of a different way of conceiving the world. Even in Recorde, there are hints of quite a different sense of the relationship between mathematics and the world.

In his preface[25] to *The Whetstone of Witte*, Recorde's argument for claiming that certain infallible knowledge comes only through mathematics is to do with the status of *number*, and how, by the precedent of Plato, Aristotle and others, it is used to 'searche all secrete knowledge and hid misteries', going as far as 'the composition of manne, yea and the verie substaunce of the soule'. While the precise meaning of this is not really spelled out, the general drift is clear: Recorde seems to have been aligning himself in a detached but sympathetic way with the long philosophical tradition, stemming at least from Plato and through Boethius permeating the quadrivium, of number as a fundamental constituent of the world. To understand number was to understand the world, and conversely no certain knowledge was possible except through mathematics.

11.2 Mathematics for the Commonwealth: John Dee

This theme was taken up fifteen or so years later, in the most profound, far-reaching, and influential discussion of the mathematical sciences to appear in the later 16th century. This was the *Mathematicall Praeface*, an extended essay by John Dee (see Figure 11.5). Let us start our consideration of Dee's work by looking at how he consolidated contemporary views, such as had already appeared in Recorde, of the place of mathematics and the certainty of mathematical argument. Dee's view is that mathematics *mediates* between the material and the spiritual worlds:[26]

Dee on the neutrality of mathematics.

> All things which are and have being are found under a triple diversity general. For either they are deemed supernatural, natural, or of a third being.

[25] In F&G 9.A4(a).

[26] See (Dee 1570, 44–45); in the original the pages are unnumbered.

Figure 11.5. John Dee (1527–1608)

Things supernatural, are immaterial, simple, indivisible, incorruptible and unchangeable. Things natural, are material, compounded, divisible, corruptible and changeable. Things supernatural are of the mind only comprehended; things natural, of the sense exterior are able to be perceived. In things natural, probability and conjecture hath place; but in things supernatural, chief demonstration and most sure science is to be had. By which properties and comparisons of these two, more easily may be described the state, condition, nature and property of those things which we before termed of a third being; which, by a peculiar name also, are called things mathematical ...

A marvellous neutrality have these things mathematical, and also a strange participation between things supernatural, immortal, intellectual, simple and indivisible: and things natural, mortal, sensible, compounded

> and divisible. Probability and sensible proof may well serve in things natural, and is commendable. In mathematical reasonings, a probable argument is nothing regarded, nor yet the testimony of sense any whit credited; but only a perfect demonstration, of truths certain, necessary and invincible, universally and necessarily concluded, is allowed as sufficient for an argument exactly and purely mathematical.

Opinions can legitimately differ about Dee's classically balanced prose. After the simple and unaffected style of Recorde we seem to have gone from popular writing to a much more erudite approach. Nonetheless, his overall concern should have come over clearly. The main contrast that Dee emphasised was between the material, corruptible world around us, and the immaterial, incorruptible, spiritual world, from which the following question then arose: Where does mathematics fit into this? In particular, where does the certainty — 'truths certain, necessary and invincible, universally and necessarily concluded' — of mathematical results arise? Dee's answers to these two questions are connected: the certainty of mathematics arises from its position, intermediate between the material and the spiritual worlds, partaking of attributes of both.

Plato is the predecessor who most strongly comes to mind in Dee's discussion. The kind of concerns that Dee voiced, and the analytical distinctions he made, go back further — indeed, they can be traced to the distinctions we met in Parmenides (see Chapter 4) between experience and reason, and between probable and certain truth. But this whole way of looking at things had been passed on most strongly through the words and ideas of Plato.

Dee indeed began his essay by speaking of 'Divine Plato, the great Master of many worthy Philosophers'. This discussion, and much more, appeared as the preface to an English translation of Euclid's *Elements*, published in 1570 (see Figure 11.6). The translator was Henry Billingsley, a wealthy haberdasher who later became Lord Mayor of London. The book was the first full English translation — in contrast with the adaptation of Books I–IV that Recorde presented in *The Pathway to Knowledg* — and was a most lavish affair, a handsome folio volume a little over twelve inches tall. It carries notes extracted from all the most important commentaries from Proclus onwards, and at 928 pages long (excluding Dee's preface) was clearly meant for patrons of substance (in contrast to Recorde's works, which were designed for the educated general public).

The book also has 'pop-up' diagrams to show the real form of solid figures. The figures are made of paper and their edges are pasted on to the page in the book so that they can be opened up to make actual models of the solid figures represented.[27]

Just how broad was Dee's vision of mathematics may be gathered from his *Groundplat* (see Figure 11.7), the outline plan in which he summarised his view of the classification of mathematical arts and sciences. It has rather more components than we can study in depth, but some aspects stand out and are worth noticing.

In one respect his classification follows earlier practice, in recognising two different kinds of arithmetic and of geometry: each occurs both as 'Principall' and as 'derivative from the Principalls', or 'vulgar'. This contrast goes back at least to Plato — you may recall a similar distinction in the *Philebus* passage in Section 3.2. The difference, as Dee saw it, is described for arithmetic as follows:[28]

---

[27]Two examples of these 'pop-up diagrams' are reproduced at *http://www.math.ubc.ca/ cass/Euclid/dee/dee.html* .

[28]In F&G 9.B1(b),(c).

Figure 11.6. Euclid's *Elements*, translated by Henry Billingsley, 1570

Dee on the certainty of mathematics.

(b) To the unfained lovers of truthe, and constant Studentes of Noble Sciences, John Dee of London, hartily wisheth grace from heaven, and most prosperous successe in all their honest attemptes and exercises ... Of *Mathematicall* thinges, are two principall kindes: namely, *Number*, and *Magnitude*. ... Neither *Number*, nor *Magnitude*, have any Materialitie. First, we will consider of *Number*, and of the Science *Mathematicall*, to it appropriate, called *Arithmetike*: and afterward of *Magnitude*, and his Science, called *Geometrie*. But that name contenteth me not: whereof a

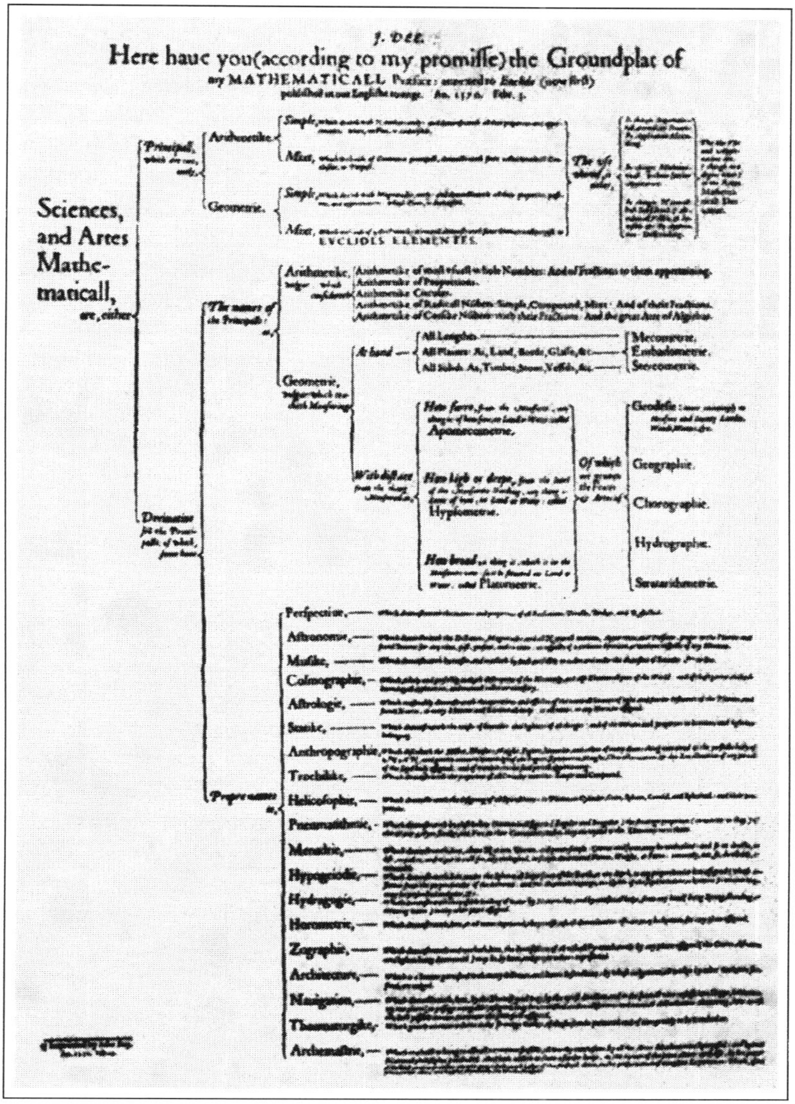

Figure 11.7. Dee's *Groundplat*

> word or two hereafter shall be sayd. How Immateriall and free from all matter, *Number* is, who doth not perceave? yea, who doth not wonderfully wonder at it? . . . And therefore the great & godly Philosopher *Anitius Boetius*, sayd: . . . All thinges *(which from the very first originall being of thinges, have bene framed and made) do appeare to be Formed by the reason of Numbers. For this was the principall example or patterne in the minde of the Creator.*

At this point Dee has made his claim for the elevated, philosophical, value of arithmetic. Now he turned to the practical question of saying what arithmetic involves, and it is interesting to see how far he went.

> But farder understand, that vulgar Practisers, have Numbers, otherwise, in sundry Considerations: and extend their name farder, than to Numbers, whose least part is an *Unit*. For the common Logist, Reckenmaster, or Arithmeticien, in hys using of Numbers: of an Unit, imagineth lesse partes: and calleth them *Fractions*. As of an *Unit*, he maketh an halfe, and thus noteth it, $\frac{1}{2}$. And so of other, (infinitely diverse) partes of an *Unit*. Yea and farder, hath, *Fractions of Fractions*. &c. And, forasmuch, as, *Addition, Substraction, Multiplication, Division* and *Extraction of Rotes*, are the chief, and sufficient partes of *Arithmetike*: which is, the *Science that demonstrateth the properties, of Numbers, and all operations, in numbers to be performed*: How often, therfore, these five operations, do, for the most part, of their execution, differre from the five operations of like generall property and name, in our Whole numbers practisable, So often, (for a more distinct doctrine) we, vulgarly account and name it, an other kynde of Arithmetike.

Dee has now indicated that there are five basic operations in arithmetic, and that they extend the reach of mathematicians well beyond what can be done with whole numbers. But he did not stop there.

> Practise hath led *Numbers* farder, and hath framed them, to take upon them, the shew of *Magnitudes* propertie: Which is *Incommensurabilitie* and *Irrationalitie*. (For in pure *Arithmetike*, an *Unit*, is the common measure of all Numbers.) And, here, Numbers are become, as Lynes, Playnes and Solides: some tymes *Rationall*, some tymes *Irrationall*. And have propre and peculier characters, (as PROBLEM and so of other. Which is to signifie *Rote Square, Rote Cubik: and so forth*:) & propre and peculier fashions in the five principall partes: Wherfore the practiser, estemeth this, a diverse *Arithmetike* from the other ... And yet (beside all this) Consider: the infinite desire of knowledge, and incredible power of mans Search and Capacitye: how, they, joyntly have waded farder (by mixtyng of speculation and practise) and have found out, and atteyned to the very chief perfection (almost) of *Numbers* Practicall use. Which thing, is well to be perceived in that great Arithmeticall Arte of AEquation: commonly called the *Rule of Coss.* or *Algebra*. The Latines termed it, *Regulam Rei* & *Census*, that is, the *Rule of the thyng and his value*. With an apt name: comprehendyng the first and last pointes of the worke.

So it seems that arithmetic includes all of contemporary algebra, including the solution of equations. The preface traces a journey that begins 'Arithmetic Principall', the science of number, which is the study that had come down through Nicomachus and Boethius and the 'unit arithmetic' of Euclid's *Elements*, Books VII–IX. Everything thereafter is derivative, from the daring break into fractions needed by 'vulgar Practisers' and 'the common Logist, Reckenmaster, or Arithmeticien', to the arithmetic of irrational numbers and algebra of the sort we studied in Chapter 10. There was thus in Dee's work a profound mixture of aspects of Greek thought, most strongly the ideas of Plato, with a radical and up-to-date perception of everything in which mathematics might conceivably play any part.[29]

[29]Dee's distinction between sciences and arts lies somewhere between true and certain theoretical knowledge and practical skills with a mathematical component.

In this blend of ancient and modern, Dee was entirely typical of his age, although he went further than many of his contemporaries in striving to apply the essentially Platonic insight that mathematics really is the key to understanding the Universe, to such an extent that his vision of the cosmos is hard, if not impossible, for us to recapture now. John Aubrey's account,[30] records that Dee 'was accounted a conjuror', and did little to discourage the popular suspicion that mathematics and occult activities intermeshed: indeed, at least three well-known plays of the period contain portraits reflecting aspects of John Dee: the title-characters of Marlowe's *Dr Faustus* (1593) and Ben Jonson's *The Alchemist* (1610) and the character of Prospero in Shakespeare's *The Tempest* (1611). This point is well illustrated by two modern biographical accounts of Dee.[31]

Throughout Dee's life the reputation of magician and necromancer attended him, and may be what led to a mob ransacking his house at Mortlake near London in 1583, destroying astronomical and navigational instruments, two globes made by his friend Gerardus Mercator, quantities of chemical apparatus from his three laboratories, and documents, books, and manuscripts. Dee had a library of some 4000 volumes, the greatest in England (and possibly in Europe) in its quantity of contemporary scientific works, as well as medieval manuscripts salvaged from the dissolution of the monasteries half a century earlier.

It is clear that even earlier Dee had experienced public antagonism, for his 1570 *Mathematicall Praeface* broke into an affirmation of the purity of his endeavours and indicated the kinds of misunderstanding to which he felt he was subject:[32]

> Ought any honest Student, and Modest Christian Philosopher, be counted, & called a Conjurer? Shall the folly of Idiots, and the Malice of the Scornful, so much prevail, that He, who seeketh no worldly gain or glory at their hands: But only, of God, the treasure of heavenly wisdom, & knowledge of pure verity: Shall he (I say) in the mean space, be robbed and spoiled of his honest name and fame? He that seeketh (by St Paul's advertisement) in the Creatures Properties, and wonderful virtues, to find just cause, to glorify the Æternal, and Almighty Creator by: Shall that man, be (in hugger mugger) condemned, as a Companion of the Hellhounds, and a Caller, and Conjurer of wicked and damned Spirits? ... Well: Well. O (you such) my unkind Country men. O unnatural Country men. O brainsick, Rash, Spiteful, and Disdainful Country men. Why oppress you me, thus violently, with your slandering of me: ... Have I, so long, so dearly, so far, so carefully, so painfully, so dangerously fought & travailed for the learning of Wisdom, & attaining of virtue: And in the end (in your judgement) am I become worse, than when I began? Worse, than a Mad man? A dangerous Member in the Common Wealth: and no member of the Church of Christ?

We shall not probe further what occasioned this outburst, but notice Dee's total conviction that his endeavours were for 'the learning of Wisdom, & attaining of virtue', and his distress that he could be judged 'a dangerous Member in the Common Wealth'. For there lay behind Dee's activities a deep political as well as religious impulse, concerned with Britain's role in the world, with the expansion of the British Empire, and the central place of mathematical sciences and arts in achieving this.[33] For instance, his *General and Rare Memorials Pertayning to the Perfect Arte of Navigation* (1577) was a mixture of navigational theory, practical navigational tables, and learned and impassioned argument on the

[30] F&G 9.B3(a).

[31] F&G 9.B3(b),(c).

[32] See (Dee 1570, 44–45).

[33] Dee was the first to use the term 'British Empire', in his *Brytanici Imperij Limites* (The limits of the British Empire), written late in the 1570s.

importance of the navy and its role in defending and expanding British interests. Thus the title page (see Figure 11.8) shows Queen Elizabeth at the helm of a ship labelled Europa (in Greek), amidst much further symbolism conveying an imperial message: for example, the Latin inscription around the title means 'more things are hidden (or unknown) than are evident', and the Greek letters around the picture signify that it is a 'British hieroglyphic', contributing to the imperial theme.

Figure 11.8. Dee's *General and Rare Memorials*

We come back to the importance of navigation in the next section. But Dee's overall concerns were expressed in the final couple of pages of his essay.[34] Here he defended

[34]See F&G 9.B1(g).

his use of the vernacular as part of his defence of publishing Euclid's *Elements* in English translation. He seems to have anticipated objections from 'the Universities' (Oxford and Cambridge), which he answered by pointing out that no harm came to Italian universities from translations of Greek mathematical works into Italian. Dee, rarely content to let a dying horse lie unflogged, went on to give a similar argument in respect of the universities of Germany, Spain, Portugal, and France. He then gave two further arguments for why the English universities should welcome the new translation: that students could now more easily study mathematics which is the foundation of so many other subjects, and that it was in the universities' interests to have more alert students such as the opportunity of studying mathematics provides.

The other theme that Dee addressed was the value and use of the subject matter of Euclid's *Elements*: that is, a brief summary, as we have seen from other authors, of the benefits of geometry. Dee remarked that people who already use arithmetic and geometry could practically make further progress through studying the theoretical subjects for private pleasure and gain and also 'for sundry purposes in the Common Wealth'.

We asserted earlier that Dee's *Mathematicall Praeface* was influential. What are our grounds for this claim? The book it appeared in was evidently a very expensive one, well outside the means of most members of the book-buying public, so Dee's noble vision of 'Common Artificers' studying Billingsley's Euclid seems somewhat unrealistic on financial grounds alone. And Dee's essay was not to be reprinted for over eighty years (1651): unlike the works of Recorde, it was not on every book stall. Coupled with Dee's elevated prose style, turned out in great haste (he twice remarked on the printer's impatience for his copy), it would not be surprising to learn that this work sank without trace.

Yet there are many tributes to its importance, and to Dee's stature as friend and adviser to a generation of mathematical practitioners. Many subsequent writings show the influence of Dee's analysis, with its overwhelming emphasis (as may be seen in his *Groundplat*) on the role of mathematics in practical applications. Even people who cared nothing for Dee's brand of high-minded Platonism, or who may have been positively suspicious of some of the authorities that Dee cited, valued the practical message that the way to achieve things in investigating nature, controlling it, and contributing to the common good, is through the mathematical sciences and arts. The influential philosopher Francis Bacon, for example, made essentially this point some thirty years later.[35]

There were also other writers and thinkers concerned with advancing the use of mathematics, developing Britain's place in the world, investigating and controlling nature, and applying the insights of mystical-sacred knowledge from the distant past to further understanding of the world and human interaction with it. Dee was not the only person to work through any of these ideas, but he was unique in combining all of these concerns in the way he did. He was too rare a person to be 'typical' in any helpful sense: we study him not because there were fifty other Dee's wandering the country, but because nearly every facet of mathematical activity and belief during the period seems to have intersected his life in some way or other. If there seems a strangeness at the heart of his concerns, we are seeing the strangeness in the heart of the Renaissance.

11.3 The mathematical practitioners

What *is* a mathematician? When we use the term in any particular context, what does that signify? Generally we have treated the word, and cognate words like 'mathematics', as

[35] See F&G 9.C5.

relatively unproblematic. To explain what is meant by describing someone in any particular period as 'a mathematician' would be so lengthy as to crowd out other topics of legitimate interest. But we shall attempt to do so here, in the context of late Elizabethan England, as a way of understanding better the variety of modes of mathematical practice at that time. For what became increasingly apparent as the 16th century progressed was the extraordinary number of different kinds of ways in which people participated in mathematical activity of some sort, at some level or other, for diverse reasons, occupying greater or lesser portions of their lives. Extracting something called 'the history of mathematics' from this in a succinct and illuminating fashion is an exercise in simplifying matters for clarity, without trivialising them or importing our own preconceptions.

It helps to move from these generalities to see the nature of the problem in a more specific way. In 1593 the Cambridge Ramist Gabriel Harvey described some of his contemporaries in a variety of ways:[36]

> He that remembreth Humfrey Cole, a Mathematicall Mechanician, Matthew Baker a Ship Wright, John Shute an Architect, Robert Norman a Navigatour, William Bourne a Gunner, John Hester a Chimist, or any like cunning, and subtile Empirique, (Cole, Baker, Shute, Norman, Bourne, Hester will be remembered, when greater Clarkes shal be forgotten) is a prowd man, if he contemne expert artisans, or any sensible industrious practitioner, howsoever unlectured in Schooles, or Unlettered in bookes . . . and what profounde Mathematician, like Digges, Hariot, or Dee, esteemeth not the pregnant Mechanician? Let every man in his degree enjoy his due: and let the brave engineer, fine Daedalist, skilfull Neptunist, marvellous Vulcanist, and every Mercuriall occupationer, that is, every Master of his craft, and every Doctour of his mystery, be respected according to the uttermost extent of his publique service, or private industry.

As well as straightforward terms, such as 'Ship Wright', 'Architect', 'Navigatour', and 'Gunner', there are words here that carry a rough gist such as 'Mathematical Mechanician' and 'Empirique', a string of classical-sounding terms such as 'Daedalist' and 'Neptunist', and various general words such as 'Clarkes', 'artisans', and 'practitioner'. It is not clear just how many of these mark real distinctions, nor is it clear when Harvey uses different words for the same thing in the cause of literary style. But there does seem to be a major distinction between 'Mathematician' and 'Mechanician', even though some of the latter are clearly practitioners of the mathematical arts, as Dee described them.

We shall follow the historian E.G.R. Taylor in labelling most of the people with whom we are concerned as *mathematical practitioners* — that is, people who practised mathematics in some form, for some purpose. In order to indicate the range of activities that this overall label covers, we give details of some of the people mentioned by Harvey, mostly drawn from Taylor's book *The Mathematical Practitioners of Tudor & Stuart England*, in which the known biographical details of 582 practitioners are given. These spotlights give some indication of the variety of mathematical practices of this period. Calling everyone in sight a 'mathematical practitioner' focuses our attention on the range of interest and approach under the one label. But there were indeed tensions at the time between the 'mechanician' and the 'mathematician' ends of the practitioner spectrum.[37]

Thus the well-educated mathematician Thomas Digges actually took to sea (for all of fifteen weeks) in order to prove to mariners that his navigational rules were better than theirs. Thomas Digges was a pupil of his father Leonard Digges, another mathematical

[36] See (Harvey 1593, 190) and F&G 9.C3.

[37] See Gabriel Harvey's discussion in F&G 9.C3, which seems to attempt to soothe such a dispute.

practitioner, and of John Dee. The foremost English Copernican, he wrote on astronomy, ballistics, military organisation, and navigation. He was a member of the Parliaments of 1572 and 1584. Robert Norman, the 'navigatour', on the other hand, made a vigorous defence of his end of the spectrum:[38]

> But I doe verily thinke, that notwithstanding the learned in those Sciences being in their studies amongst their books, can imagine greate matters, and set down their farre fetcht conceits ... yet there are in this land diuers Mechanicians, that in their seuerall faculties and professions, have the vse of those Artes at their fingers endes, and can apply them to their seuerall purposes, as effectually and more readily, than those that would most condemne them.

Figure 11.9. Part of an astrolabe made by Humfrey Cole

Among other examples, we find Humfrey Cole, 'a Mathematicall Mechanician', described in these terms:[39] an engraver and die-sinker, an expert in mining, metallurgy, and metalwork, and the most famous instrument maker of Elizabethan times. A remarkable range of instruments made by him survive: astrolabes (see Figure 11.9), an armillary sphere, ring-dials, a sector, a nocturnal, gunners' scales, a geometrical table, part of two theodolites, a garden dial, and [pocket] compendiums.

Robert Norman was apparently an experienced seaman, who settled down after almost twenty years at sea to make nautical instruments, particularly sea-compasses and marine

[38] Preface to *The Newe Attractive*, cited in (Bennett 1986, 13).

[39] See (Taylor 1954, 178).

charts. His book *The Newe Attractive* (1581) was the first to reveal the phenomenon of magnetic dip (compass needles do not lie horizontally), as well as analysing magnetic variation (the amount by which compass needles deviate from pointing true north varies from place to place). Norman described himself as an 'unlearned Mechanician', and his book is important for its early advocacy of an approach to science that became more popular in the following century: arguments should be based, he said, 'onely upon experience, reason, and demonstration, which are the groundes of Artes'.

Finally, we mention William Bourne, 'a gunner', an innkeeper of Gravesend who served as a gunner before turning to write textbooks on surveying, gunnery, optics, the design and use of instruments, and the first English manual on navigation (*A Regiment for the Sea*, 1574). This was an amplification and popularisation of the discussion of navigation in John Dee's *Mathematicall Praeface*.

Through this ferment of activity more and more people became persuaded that mathematics was a crucial underpinning of a wide range of practical interactions with the world. In this respect, school education lagged behind. The purpose of grammar schools remained overwhelmingly the teaching of Latin grammar, as their name suggests. For just as the educated Byzantine thought in Attic Greek (see Chapter 6), so educated Europeans continued to write and think in Latin. Throughout this period, and well into the 18th century, it was Latin, the language of Virgil and Cicero, that formed the intellectuals' framework of sensibility — a point to remember when we evaluate the progress of the vernacular. Brinsley's school would have been typical of most grammar schools in making no provision for even basic mathematical training: as his final remark indicated, pupils might have to attend special ciphering schools to learn any skills above basic numeration.

Evidence on the teaching of mathematics at the two universities is somewhat ambivalent. Mathematics seems to have been available for students who were interested, depending perhaps on the enthusiasms of particular tutors, but it was possible to graduate with barely more arithmetical knowledge than one of Brinsley's pupils.

Certainly all the leading mathematicians and scientific intellectuals of the period were university graduates, and Mordechai Feingold, the historian who has most recently studied the matter, has argued that the stereotyped view of Oxford and Cambridge as 'institutions devoid of mathematical instruction and inimical to new scientific modes of thought' is unfounded.[40]

Outside the formal educational institutions of schools and universities, there was the world of the mathematical practitioner, mostly centred on what a contemporary writer called the 'third university of England': the capital city of London.[41] Teaching was often an aspect of the practitioners' role: teaching the use of instruments they were making and selling, instructing people who bought their books, being paid to teach people the skills that they needed, or were encouraged to think they did. It was in London, too, that various educational initiatives were taken in the later 16th century. A group of London merchants joined with the city authorities in 1588, the year of the Spanish Armada, to found a public Mathematical Lectureship. This was given for some four years by Thomas Hood, a follower of the educational views of Ramus, who lectured on subjects such as astronomy and navigation, arithmetic and geometry, and sold the textbooks that he wrote on the subjects of his lectures.

---

[40] See (Feingold 1984, 21). For a fuller discussion, see *Oxford Figures*.

[41] *The Third Vniuersitie of England* was the title of a book on the colleges in London by Sir George Buck, cited in (Stow 1615).

Gresham College, London. A more lasting public educational venture was the founding in 1596 of Gresham College in the City of London, under the will of a wealthy financier, Sir Thomas Gresham, founder of the Royal Exchange (see Figure 11.10).

Seven professors were appointed, to deliver lectures in Astronomy, Divinity, Geometry, Law, Music, Physic (Medicine), and Rhetoric. The particular responsibilities of each professor were spelled out: the Geometry professor was to teach arithmetic as well as theoretical and practical geometry, while the Astronomy professor was to teach geography and navigation as well as 'the principles of the sphere and the theoriques of the planets, and the use of the astrolabe and staff, and other common instruments for the capacity of mariners'.[42]

Figure 11.10. The original Gresham College

One interesting feature of the instructions that the astronomy and geometry professors were given, in the light of our exploration of the vernacular, is that each lecture was to be read twice: first in Latin for visiting scholars, then in English for the London audience. There is unfortunately little evidence on the content, level, or attendances at the lectures themselves, so the actual contribution of Gresham College to the education of London's mathematical practitioners is hard to pin down. But the mathematical professors were themselves generally energetic and influential practitioners, none more so than the first Professor of Geometry, the Halifax-born Henry Briggs. Briggs seems to have had a wide capacity for friendship, keeping in touch with people and getting things done — we shall examine his great contribution to the dissemination of logarithms in the next section — and in 1619 he was chosen by Sir Henry Savile as the first occupant of the Chair of Geometry which Savile founded at Oxford. Arising from his interest in geography and navigation, Briggs became involved with the Virginia Company and advised them on the quest for the North-West Passage, as well as investing in the company itself: he held two shares of £12. 10s., each one quarter of his annual salary as a Gresham professor.

---

[42]Cited in (Adamson 1980, 17); Gresham College still exists for the purpose of giving free public lectures.

Navigation. To investigate the needs of mathematical practitioners, and where mathematics came in, we now concentrate on navigation, a mathematical art of great importance. Recorde's last two books were connected with the activities of the Muscovy Company, while John Dee was mathematical adviser to the company for some thirty years, teaching its navigators. The increased 16th-century need for navigational theory and practice is obvious enough in general terms, but the problems may usefully be spelled out, in order to see how mathematics was crucial for their solution.

The old methods of coastal navigation were inadequate for the increasing oceanic voyages of trade and discovery. If you were on a ship in the middle of the ocean, how could you tell where you were, and how could you get to where you wanted to be? There were two main methods. One was to work out the course you had travelled, by estimating speed and distance along successive directions, making allowance for wind, current, and so forth: this should tell you your present position in relation to where you started. This method, called 'dead-reckoning', was somewhat rough and ready and particularly liable to error, because errors in judging each day's course could be cumulative. The other method was to try to find your position directly, by using astronomical instruments to measure the positions of Sun and stars: this should at least make it possible to determine your latitude north or south. Longitude was a much more difficult problem, to be satisfactorily solved only in the 18th century. Although this second method of determining position was theoretically more satisfactory, celestial observations that were accurate on dry land were harder to take from the deck of a rolling ship; they also called for more mathematical understanding and skill.

A deeper contrast between these two methods is that they make quite different presuppositions about the world. Whereas a dead-reckoning navigator treated the oceans as an extension of familiar coastal waters, relating positions dynamically to ports and other positions, a user of the second method was working within the tradition of scientific geography and cosmography going back even before Ptolemy in the 2nd century AD. In this the Earth was regarded as spherical, and positions on it were described by reference to a coordinate grid of latitude and longitude, which could be determined instrumentally by reference to the celestial sphere above. As his prime meridian (or northerly axis) Ptolemy took a line of longitude through the Canaries, the most westerly place he knew, as Figure 11.11, from the 1562 edition of his *Geography*, shows.

In practice, both methods might be used. However, for practical purposes, the problem of longitude could not be overcome by steering to the desired latitude (as determined by measuring the angle of the pole-star or the midday sun) and then heading due east or west until land was reached. This was for two reasons. First, it was eventually necessary to know how far you were from the coast where you intended to land. Second, the desired latitude was that of the prevailing winds, once they were reliably known, which may have differed from that of the port you were aiming for.

That said, there remained problems in trying to fit together these two traditions, the Earth-based and the astronomical. The magnetic compass, for instance, rightly hailed by Cardano as one of the formative discoveries of the era,[43] had somewhat quirky behaviour, which mariners took until the end of the 16th century to make sense of (see the above discussion of Robert Norman's *The Newe Attractive*); for a long time it was not clear whether the phenomenon of compasses deviating erratically from pointing true astronomical north was due to badly-made instruments, or to the general nature of things.

[43] See F&G 8.A4(c).

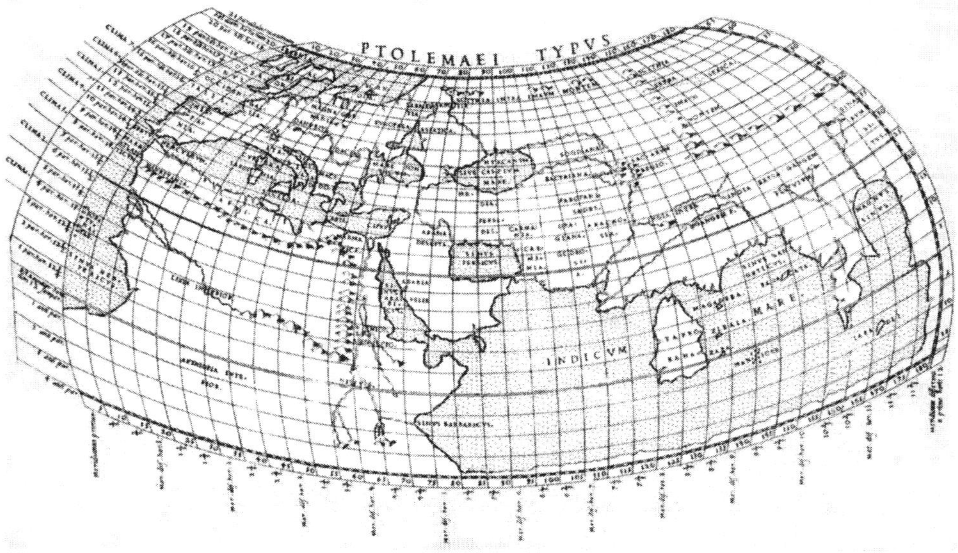

Figure 11.11. A map projection by Ptolemy

Another problem lay in the relation between navigating on a spherical Earth and the flat charts on which one's position or course were marked: the problem of *map projection*. Where small familiar areas such as the Mediterranean were concerned, empirical charts had been developed over generations and served their local purposes well. But the obvious extrapolation of this to oceanic charts, where lines of latitude and longitude were represented as straight lines at right angles to one another, led to difficulties in higher latitudes — in the north Atlantic, for instance, where the prevailing easterly winds carried British ships to the new colonies in North America — due to the fact that the meridians (the lines of longitude) converge towards the poles. The distance between successive meridians is not constant (as is that between successive lines of latitude), but depends on one's latitude.

For navigators steering by compass course, it was even more important that a constant compass direction (and therefore a steady setting of the helm) should correspond to a straight line on the chart. But a constant compass course cannot correspond to the shortest distance between two points on the Earth's surface unless the two points lie on a common meridian. It was the Portuguese mathematician Pedro Nunes who pointed out, in the middle of the 16th century, that a course of constant compass bearing (a *rhumb line*) corresponds to a spiral on the sphere; this is not the same thing as a *great circle*, the shortest path between any two points on a sphere (see Figure 11.12).

Mercator's projection. The most important feature of the 1569 world map drawn by Dee's friend Mercator was that, with the projection he devised, a rhumb line maps onto a straight line, crossing all meridians at the same angle. This made it possible to answer the question; given that I know the longitude and latitude of the point I wish to reach, what course do I steer to get there? Mercator's projection involved a variable stretching along the meridian lines, spacing them out towards the poles; thus, countries near the equator are accurately represented, while a country like Greenland appears much larger than its actual area might suggest.

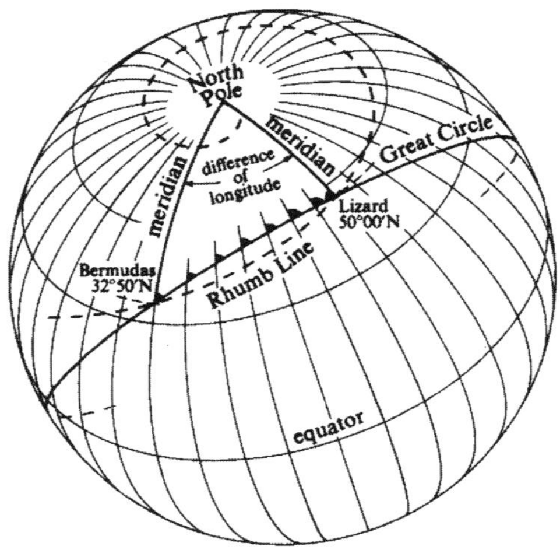

Figure 11.12. The rhumb line constantly changes direction

But Mercator did not explain how he had achieved this, and it may be that he did it by judgement and guesswork. It was not until the end of the century that the English mathematical practitioner Edward Wright, in his influential work *Certaine Errors in Nauigation* (1599), first explained the mathematical basis for Mercator's projection in print. The clarity of Wright's explanation is a tribute, with his appealing image of a swelling bladder, to the tradition of vernacular exposition founded by Recorde:[44]

> Suppose a spherical superficies [surface] with meridians, parallels and the whole hydrographical description drawn thereupon to be inscribed into a concave cylinder, their axes agreeing in one.
>
> Let this spherical superficies swell like a bladder (whilst it is in blowing) equally always in every part thereof (that is as much in longitude as in latitude) till it apply, and join itself (round about and all along also towards either pole) unto the concave superficies of the cylinder: each parallel upon this spherical superficies increasing successively from the equinoctial [equator] towards either pole, until it comes to be of equal diameter with the cylinder, and consequently the meridians still widening themselves, till they come to be so far distant every where each from the other as they are at the equinoctial. Thus it may most easily be understood, how a spherical superficies may (by extension) be made a cylindrical, and consequently a plain parallelogram superficies.

In this way Wright explained that Mercator's projection is basically a cylindrical projection, in which positions on the sphere are considered as being projected outwards onto a cylinder touching round the equator, which is then unrolled and stretched by the right amount to straighten the rhumb lines (see Figure 11.13).[45] All this should be fairly clear in a qualitative way, but the question of concern was: what is 'the right amount'? This was the mathematical problem, solved both by Wright and by the third 'profounde mathematician' named by Gabriel Harvey. This was Thomas Harriot, and it is to Harriot that we now turn.

---

[44]Quoted in (Boas 1970, 1945).

[45]As pictured by the historian David Waters: meridional segments are shown in Figure 11.13 in different stages for explanatory purposes — on Wright's bladder they all swell simultaneously; notice how the spiral rhumb line becomes straight.

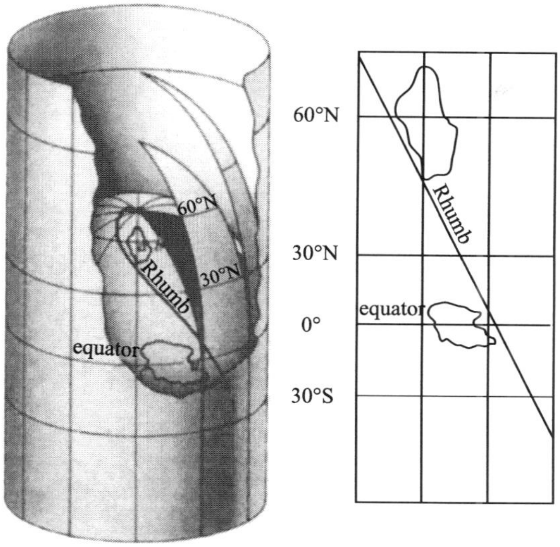

Figure 11.13. Wright's explanation of the Mercator projection

11.4 Thomas Harriot

Harriot published so little, and his range was so great, that it has always been difficult to gauge his influence — scholars are still working through the 8000 surviving manuscript pages that he left behind. He was greatly respected and admired by his friends: a poem of 1618, written towards the end of his life when he was 58 and already very ill, speaks of 'deep Harriots Minde In which there is noe drosse but all's refin'd'.[46]

The first North American colonies. By the 1580s it was apparent, to some of those concerned to further England's imperial ambitions, just how pressing was the need for improvements in theoretical and practical navigating skills. For instance, one officer on a 1582 voyage made the somewhat sobering observation that in the mid-Atlantic the navigators could not agree where their ships were, to within a hundred miles east-west and forty miles in latitude. This kind of observation added weight to John Dee's plea that 'in navigation, none ought to have greater care to be skilful than our English pilots'.[47]

Walter Ralegh, a vigorously rising star in the Elizabethan firmament, was one who realised the need for navigational training, in fulfilment of his Letters Patent from the Queen, granted on 25 March 1584, 'to discover search fynde out and view such remote heathen a barbarous landes Contries and territories not actually possessed of any Christian Prynce'.[48] Ralegh engaged the young Oxford graduate Thomas Harriot to become a member of his household and assist in the preparations for his colonising venture to America, to what were eventually to be called Virginia and North Carolina. Harriot's success is testified to by the geographer Richard Hakluyt, who wrote (addressing Ralegh) in 1587:[49]

---

[46]Corbet, in (Feingold 1984, 137).

[47]F&G 9.B1(f)).

[48]Online in *From Revolution to Reconstruction*, `http://www.let.rug.nl/usa/D/1501-1600/hakluyt/raleigh.htm`.

[49]Quoted in (Shirley 1983, 80).

> Ever since you perceived that skill in the navigator's art, the chief ornament of an island kingdom, might attain its splendour amongst us if the aid of the mathematical sciences were enlisted, you have maintained in your household Thomas Hariot, a man pre-eminent in those studies, at a most liberal salary, in order that by his aid you might acquire those noble sciences in your leisure hours, and that your own sea-captains, of whom there are not a few, might link theory and practice, not without almost incredible results.

In fulfilling his remit, Harriot busied himself with every aspect of navigational theory and practice: the design of instruments for taking observations, assessing and allowing for observational errors, the construction of tables for interpreting one's observations, and of tables that enabled Mercator's projection to become a helpful navigational aid. In producing the latter tables, an enterprise that continued on and off in later periods of his life, Harriot developed new and significant mathematical techniques and results that we shall consider shortly.

Harriot also sailed with Ralegh in 1585–86 on the ill-fated journey to Roanoke, and on arrival learned the language of the Algonquin Indians, the native Americans who lived there.[50] He described the land, its produce, and its people in his account of the voyage, the *Briefe and True Report of the New Found Land of Virginia* (see Figure 11.14), but his positive remarks were largely ignored by the British colonists who sailed to Virginia (named after Elizabeth, the 'Virgin Queen') and later Jamestown (named for her successor, James I of England and VI of Scotland), and who hoped for easy access to wealth. In a passage that mixes respect, condescension, and a desire for conquest, Harriot wrote:[51]

> In respect of vs they are a people poore, and for want of skill and iudgement in the knowledge and vse of our things, doe esteeme our trifles before thinges of greater value: Notwithstanding in their proper manner considering the want of such meanes as we haue, they seeme very ingenious; For although they haue no such tooles, nor any such craftes, sciences and artes as wee; yet in those thinges they doe, they shewe excellencie of wit. And by howe much they vpon due consideration shall finde our manner of knowledge? and craftes to exceede theirs in perfection, and speed for doing or execution, by so much the more is it probable that they shoulde desire our friendships & loue, and haue the greater respect for pleasing and obeying vs. Whereby may bee hoped if meanes of good gouernment bee vsed, that they may in short time be brought to ciuilitie, and the imbracing of true religion.

This book, accompanied by John White's fine drawings of Algonquin people and their villages, made Harriot the authority on North America.

On the next page of his *Report*, Harriot wrote of how amazed the Algonquins were by

> Mathematicall instruments, sea compasses, the vertue of the loadstone in drawing yron, a perspectiue glasse whereby was shewed manie strange sightes, burning glasses, wildefire woorkes, gunnes, bookes, writing and reading, spring clocks that seeme to goe of themselues.

---

[50]The settlement at Roanoke was begun in 1585, and more settlers joined in 1587, but it could not be supported during the Spanish–English War of 1587–90, and it had disappeared entirely by 1590 when it was revisited. The reasons for the disappearance remain famously unresolved.

[51]*Report*, p. 26. *The brief and true report of the New Found Land of Virginia* is available online at `http://docsouth.unc.edu/nc/hariot/hariot.html`.

Figure 11.14. Harriot's *Briefe and True Report* (1590)

This confirms that he must have travelled with a magnetic compass and a telescope (the perspective glass), the latter a few months before Galileo made his name with it, as we shall see in Chapter 12.

Harriot's work on maps. Harriot showed that another map projection, known as *stereographic projection* (see Figure 11.15) has the property of preserving angles — that is, the angle between two lines on the globe is always the same as the angle between the image of those lines under this projection. The stereographic projection was an old one, described by Ptolemy for the celestial sphere in the 2nd century and later studied by Commandino. The image of any point (other than the South Pole) is obtained by drawing a straight line from the South Pole to the point in question; where this line cuts the plane of the equator is the image of the point.

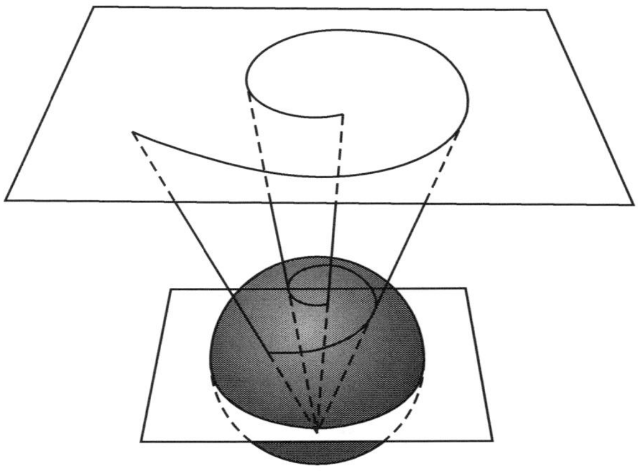

Figure 11.15. Stereographic projection takes a rhumb line on the globe to a spiral on the plane

What happens to curves on the globe under this projection? The meridians (the lines of longitude all converging on the North Pole) become a set of straight lines radiating from a central point (the image of the pole), lines of latitude become concentric circles, and a rhumb line becomes a spiral, spiralling towards the pole in such a way as to make equal angles with all the meridian radius lines. The curve came to be called an *equiangular* or *logarithmic spiral*, the latter term being introduced by Varignon[52] in 1704.

Harriot proceeded to establish some properties of the equiangular spiral; these were of independent interest, as well as making possible further inferences about the calculations needed for rhumb lines under the Mercator projection. In particular, he achieved a remarkable result, the rectification and quadrature of the equiangular spiral — rectification is the measure of the length of a curve; quadrature is the measure of the area bound by the curve.

Figure 11.16 illustrates the kind of ideas that Harriot was using: it shows a polygonal spiral cut up and reassembled into a triangle. It surely has something to do with the equiangular spiral, but what exactly? How might Harriot have found the length and area of the equiangular spiral?

The polygonal spiral of Figure 11.16 is an approximation to an equiangular spiral, in the sense that it spirals inwards by following a succession of straight lines, each of which cuts a radial line at a constant angle α. So if we imagine constructing many diagrams like this, with shorter and shorter straight-line segments, we will approach the equiangular spiral more and more closely. It is clear from the way that the polygonal spiral is divided into small triangles that they are then reassembled into one large triangle, and that the areas inside the spiral and the triangle are the same. The perimeter of the spiral has been rearranged as two sides of the triangle, which therefore becomes isosceles in the limiting case as the straight-line segments have shorter and shorter lengths. It follows that the area

[52]This curve is different from the Archimedean spiral that we studied in Chapter 5; in modern terms, the equation of the Archimedean spiral is $r = a\theta$, whereas that of the equiangular spiral is $r = ae^{\theta}$.

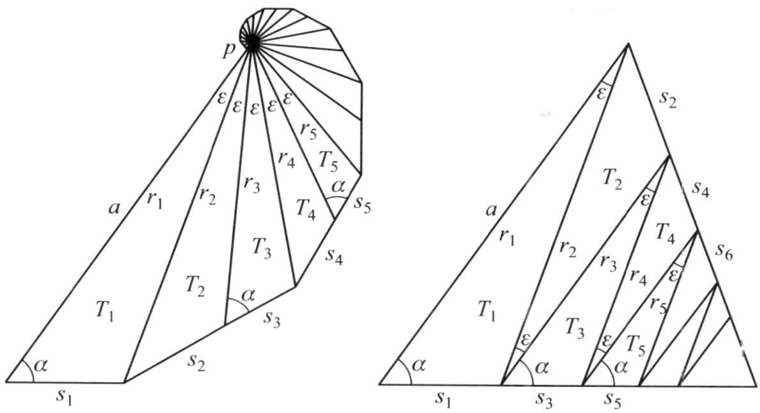

Figure 11.16. Triangles and spirals

and perimeter of the triangle can be found: the area is $\frac{1}{4}a^2 \tan\alpha$, and the perimeter is $a \sec\alpha$.[53]

What is especially interesting about this rectification of the equiangular spiral is that it had generally been believed that the length of such a curve could not be obtained. Indeed, René Descartes wrote forty years later that 'the ratios between straight and curved lines are not known, and I believe cannot be discovered by human minds'.[54] So an enquiry that began as serving the needs of navigational calculation blossomed in Harriot's hands into an advance of independent theoretical interest.

From his work on the stereographic projection, Harriot derived a fundamental formula relating two latitudes, for which the *meridional part* of one (its stretching factor for the Mercator projection) is a whole-number multiple of that of the other: namely, if the stretching required for latitude θ is n times that for latitude φ, then

$$\tan(45^\circ - \theta/2) = (\tan(45^\circ - \varphi/2))^n .$$

One important thing to notice about this formula is that it introduces an nth power; the procedure is related to the principle of what would shortly be called logarithms. It enabled Harriot to make much progress, which his work on the spiral enabled him to complete, in constructing tables for the Mercator projection.

Harriot and others devoted much energy to the construction of tables for these meridional parts. Why were they so important, and what was their mathematical significance?

Suppose that a ship steers on a constant course for some way, and in doing so changes its latitude by a certain amount, which is comparatively straightforward to measure. Then, by constructing a 'nautical triangle', as in Figure 11.17, we can easily find the distance travelled and the *departure* (the distance from the starting longitude) by elementary trigonometry. However, the departure is not a measure of longitude change, except on the equator, because the longitude lines converge as one gets closer to the pole. At any particular latitude φ the departure is connected to the difference in longitude by a factor of $\sec\varphi$, while the distance you think you have 'departed' east or west of your original longitude is related to the actual longitude change by the equation accompanying Figure 11.17.

---

[53]To prove this, note that the altitude of the isosceles triangle with base a and base angles α is $\frac{1}{2}a\tan\alpha$, and the two equal sides are each $\frac{1}{2}a\sec\alpha$.

[54]Descartes, *La Géométrie*, p. 91, in F&G 11.A8.

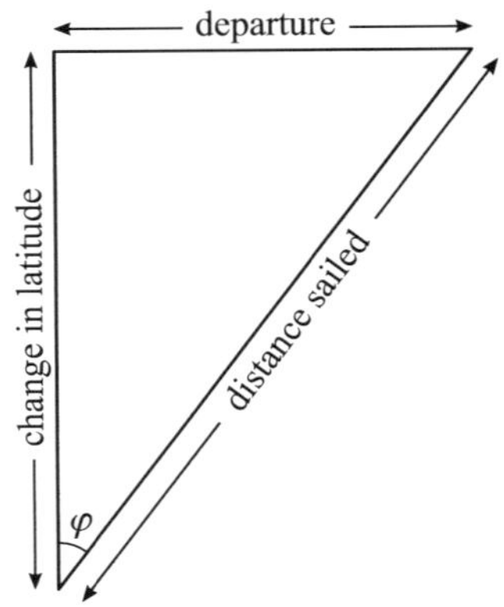

Figure 11.17. The difference in longitude = departure $\times$ sec ϕ

This stretching factor of $\sec\varphi$ is what is required for the Mercator projection, but the problem is that it varies with latitude. To gauge the effect is not a matter of noting one's starting and ending latitude and multiplying by the secant of the difference in latitude (or some similarly simple calculation), but of adding up the contributions of all the latitudes in between — that is, calculating the sum of each $\sec\varphi$ for every latitude φ between the starting and ending points. This can be done *approximately* by adding up secants for very small intervals of latitude, which is how Harriot and his contemporaries, Edward Wright and John Dee some years previously, originally calculated the tables. But by 1614 Harriot had produced tables of meridional parts by a direct procedure not unlike that used by Napier in his independent construction of logarithm tables.

We shall not pursue the details of Harriot's method, for a naturally complicated subject is made even more so by Harriot's having been one of the most taciturn of mathematicians. The historian Jon Pepper, who has studied Harriot's work on navigation, remarked ruefully during his exegesis of Harriot's papers:[55]

> It should be mentioned that nowhere in the nearly 675 sheets relevant to these meridional-parts tables does Harriot explain any step he takes. A consequence of this is that it is not always possible to be sure that when Harriot appears to be carrying out a particular process, he is consciously doing exactly what one might like to suppose he is doing.

This difficulty is a useful reminder of a theme that has dogged us sporadically — Harriot's many pages of figures and calculations offer no overt clue to the thought processes that gave rise to them. All the same, Harriot was evidently working his own way through issues that began to dominate mathematics three-quarters of a century later. One of the intriguing and difficult aspects of studying the history of mathematics is trying to judge what statements can helpfully be made about past mathematical work in the light of subsequent developments.

Harriot's algebra. Harriot worked not only on navigation, but on ballistics, optics, astronomy, and algebra. His algebra is of particular interest because in it we see the first signs of Viète's influence in England.[56]

[55] See (Pepper 1968, 383–384); some of Pepper's views on Harriot's achievement can be found in F&G 9.D7(a).

[56] We looked at Viète's work at the end of Chapter 10.

Viète had produced an algebraic formulation for many problems in mathematics, with the virtue of being able to express arbitrary products of magnitudes. In order to deal with the prevalent idea that the product of two lengths is an area, of three lengths a volume, and of four lengths a nonsense (since space has only three dimensions), Viète had devised a language to express the novel quantities and a supporting notation; for example, he wrote A *in* B for the product of A and B, A *quadratus* for the square of A, and A *cubus* for the cube of A.[57] Harriot knew Viète's work thoroughly, and made considerable advances in notation, replacing A *in* B by ab, A *quadratus* by aa, A *cubus* by aaa, and so on. This is strikingly like the conventions we use today, down to the preference for lower case over upper case letters, all re-introduced later by Descartes, from whom their use derives. Harriot was also the first to write the inequality signs $<$ and $>$, although his signs were longer and more curved than those we use today.

Harriot wrote a *Treatise on Equations* that was his response to Viète's *De Potestatum Resolutione* (1600), in which he was the first algebraist to assert that polynomials can always be written as products of linear or quadratic factors. This insight, combined with the clarity of his notation, enabled Harriot to see relationships between the roots and the coefficients of polynomial equations, and to write a systematic treatise on solving cubic and quartic equations. He was clear that negative solutions of an equation had to be accepted on a par with positive ones, and he accepted complex solutions too.

A few days before he died, Harriot wrote his will, entrusting his unpublished *Treatise* to his friend and fellow mathematician Nathaniel Torporley. Torporley gave up his position as Rector of Salwarpe in Worcestershire and moved to either Syon House or Petworth, both houses of Harriot's patron the Duke of Northumberland, but the project seems to have stalled and Torporley lost the trust of Harriot's executors, who gave the papers to Walter Warner, also named by Harriot in his will. The result was a publishing disaster. Warner published Harriot's algebra as the *Artis Analyticae Praxis* in 1631, but he re-ordered the manuscript pages. As the historian Jacqueline Stedall has observed, 'In so doing he not only destroyed the coherence of Harriot's treatise but made it appear considerably less sophisticated than in fact it was'.[58]

Harriot's original manuscript sheets on equations became dispersed among the several hundred sheets he left behind after his death, and only in recent years has his algebra been reassembled and reassessed by Stedall in *The Greate Invention of Algebra: Thomas Harriot's Treatise on Equations*. One unfortunate result of this dispersal was that it became a matter of some controversy during the 17th century and beyond as to whether Descartes knew of Harriot's work when he wrote *La Géométrie* (1637), which took up some of the same ideas.[59] All that we can say with certainty is that this was not impossible, and that, because his manuscripts circulated amongst his friends and colleagues during his lifetime and for twenty years or so after his death, the material in them may have been better known then than is now realised.

Harriot was typical of the mathematical practitioner of the period in covering a great range of mathematical arts and sciences, but less typical in that he was not a 'freelance' mathematician; he had patrons throughout his working life, first in Sir Walter Ralegh and later in the Earl of Northumberland. He was not in the position of needing to publish his work, either for his reputation or for financial security, and he may have been discouraged from doing so by the labour of preparing it for printing. Nevertheless, the impression

[57]Michael Stifel had published his *Arithmetica Integra* in 1544, in which the product of $3A$ and $9B$ was written as $27AB$, but the practice had not caught on.

[58]See (Stedall 2003, 20).

[59]See F&G 9.D6(a).

remains of someone who was cautious to the point of secretiveness. Not even the interest of Kepler, who wrote to him in 1606, could persuade him to reveal that he already knew the sine law of refraction (now usually called *Snell's law*, after one of its rediscoverers twenty years later). Nor did the admonishments of his friends prevail, despite a pleading letter from one of them, Sir William Lower:[60]

> Doe you not here startle, to see every day some of your inventions taken from you; for I remember longe since you told me as much, that the motions of the planets were not perfect circles. So you taught me the curious way to observe weight in Water, and within a while after Ghetaldi comes out with it, in print a little before Vieta prevented you of the Gharland for the great Invention of Algebra. Al these were your deues and manie others that I could mention; and yet too great reservednesse hath robd you of these glories. but although the inventions be greate, the first and last I meane, yet when I survei your storehouse, I see they are the smallest things, and such as in Comparison of manie others are of smal or no value. Onlie let this remember you, that it is possible by too much procrastination to be prevented in the honor of some of your rarest inventions and speculations. Let your Countrie and friends injoye the comforts they would have in the true and great honor you would purchase your selfe by publishing some of your choise works.

The result of Harriot's 'too much procrastination' was that as recently as 1936 one historian could write that 'Harriot's mathematical ability rarely rose above mediocrity', a judgement that can now be seen to have been perverse.[61]

Harriot's life ended painfully, for he succumbed to cancer of the nose, a form of skin cancer that became apparent in 1615, and it took his life in October 1621. It is likely that Harriot was a smoker, and that this was an early example of the dangerous side effects of tobacco, on which export the wealth of Jamestown was founded.

11.5 *Excellent briefe rules*: Napier and Briggs

We now turn to a mathematician whose influence has never been doubted, on 'pregnant Mechanician' and 'profounde Mathematician' alike.[62] For many years, John Napier spent his leisure time devising means of making arithmetical calculations easier. Just why a Scots laird at the turn of the 17th century should have thus devoted the energies left over from the management of his estates remains a puzzle. Up to the publication of his description of logarithms in 1614, three years before his death, Napier was best known to the world for his Protestant religious treatise *A Plaine Discovery of the Whole Revelation of Saint John* (1594), a well-received work that was translated into several languages. Napier's interest in mathematics, and in computational methods in particular, seems to have started in the 1570s, when he was in his early 20s. It continued, mostly unknown to the wider world, until the flurry of publishing activity forty years later which revealed first his table of logarithms, *Mirifici Logarithmorum Canonis Descriptio* (1614, see Figure 11.18) (*A Description of the Admirable Table of Logarithms*), then three further computation aids, *Rabdologie* (1617), and after his death an account of how his logarithms were calculated, *Mirifici Logarithmorum Canonis Constructio* (1619) (*The Construction of the Admirable Table of Logarithms*).

---

[60] In (Shirley 1983, 400) and F&G 9.D4.

[61] See (Scott 1936, 344).

[62] Harvey (1593, 190).

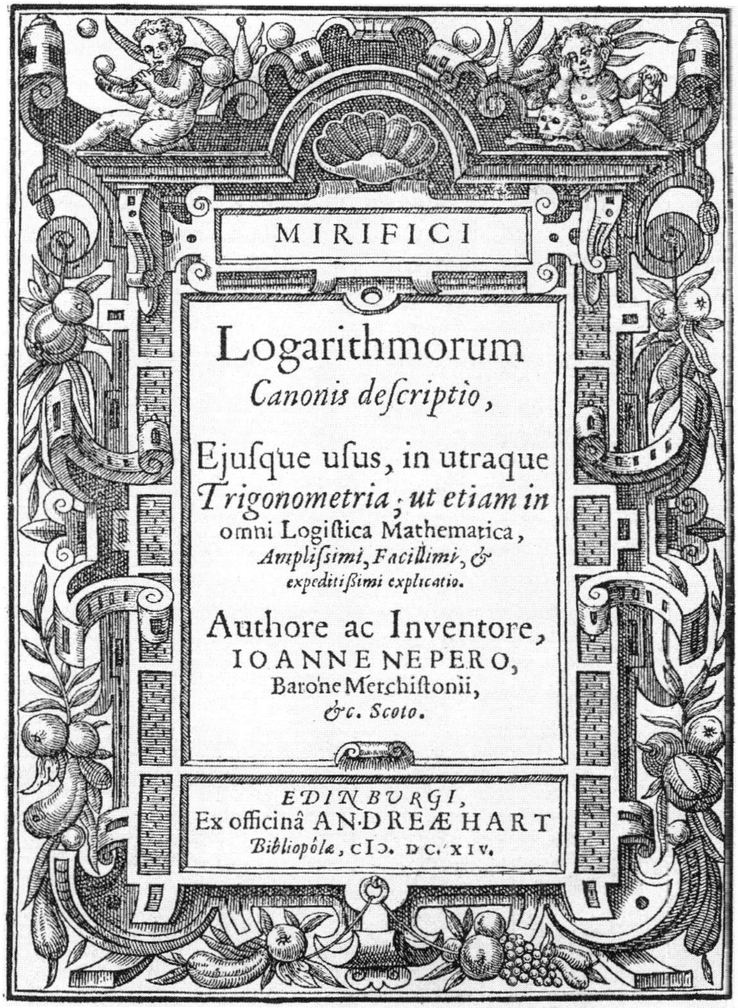
MIRIFICI

Logarithmorum
Canonis deſcriptio,

Ejuſque uſus, in utraque
Trigonometria; ut etiam in
omni Logiſtica Mathematica,
Amplíſsimi, Facillimi, &
expeditiſsimi explicatio.

Authore ac Inventore,
IOANNE NEPERO,
Barone Merchiſtonii,
&c. Scoto.

EDINBURGI,
Ex officinâ ANDREÆ HART
Bibliopolæ, CIↃ. DC. XIV.

Figure 11.18. Napier's *Mirifici Logarithmorum Canonis Descriptio* (1614)

The titles of his theological and mathematical works form a revealing contrast: the mathematical titles are in Latin whereas that of his theological tract is in English, and thus we might conjecture that Napier had different readers in mind. Given the substantial vernacular textbook tradition by this time, we may infer that Napier was not aiming at the home-grown practitioner, but at a more learned, and possibly international, audience.

Our hypothesis, that Napier had the international mathematical community in mind as his primary audience, is confirmed from the preface to his *Descriptio*, translated by Edward Wright into English as *A Description of the Admirable Table of Logarithms*:[63]

> Seeing there is nothing (right well beloved students in the Mathematics) that is so troublesome to Mathematicall practise, nor that doth more molest and hinder Calculators, than the Multiplications, Divisions, square and cubical Extractions

[63] See F&G 9.E1.

of great numbers, which besides the tedious expence of time, are for the most part subject to many slippery errors. I began therefore to consider in my minde, by what certaine and ready Art I might remove those hindrances. And having thought upon many things to this purpose, I found at length some excellent briefe rules to be treated of (perhaps) hereafter. But amongst all, none more profitable than this, which together with the hard and tedious Multiplications, Divisions, and Extractions of rootes, doth also cast away from the worke it selfe, even the very numbers themselves that are to be multiplied, divided and resolved into rootes, and putteth other numbers in their place, which performe as much as they can do, onely by Addition and Subtraction, Division by two or Division by three; which secret invention, being (as all other good things are) so much the better as it shall be the more common; I thought good heretofore to set forth in Latine for the publique use of Mathematicians.

From this we learn that Napier wrote his book in Latin 'for the publique use of Mathematicians', to ensure that it shall be 'the more common' — that is, more widely spread. The English translation is for the benefit of 'our Countrymen in this Island' and 'the more publique good'. Napier sounds pleased, all the same, that non-Latin speakers in the United Kingdom might find logarithms useful too.

Before pursuing what logarithms are, and who they were designed for, we look briefly at another of Napier's computational aids. For, in the years following his death, it was *Napier's bones* (see Figure 11.19) that were more widely known and used.

Figure 11.19. Napier's bones: a set of rods that display the basic multiplication tables

These consisted of the columns of a multiplication table inscribed on rods, designed to make the multiplying of two numbers easier by setting down the partial products more swiftly. This simple contrivance was derived from an ancient multiplication method called 'lattice multiplication' (see Box 11.1), which involved a special layout on paper of the two numbers to be multiplied so as to display the partial products of the multiplication; this facilitated adding them in the right way.

Box 11.1. Multiplication with Napier's bones.
Consider first various ways of laying out the multiplication of 934 and 314, as given in the *Treviso Arithmetic* of 1478 (the first printed arithmetic book). Note that the result, 293,276, is reached in the middle configuration by adding the figures in the matrix along the diagonal bands, starting from the bottom right, and carrying a digit over to the next band where appropriate.
To do the same multiplication using Napier's bones we align the rods for 9, 3, and 4 and look horizontally along the times 3, times 1, and times 4 rows to form the partial products to be added, as here.

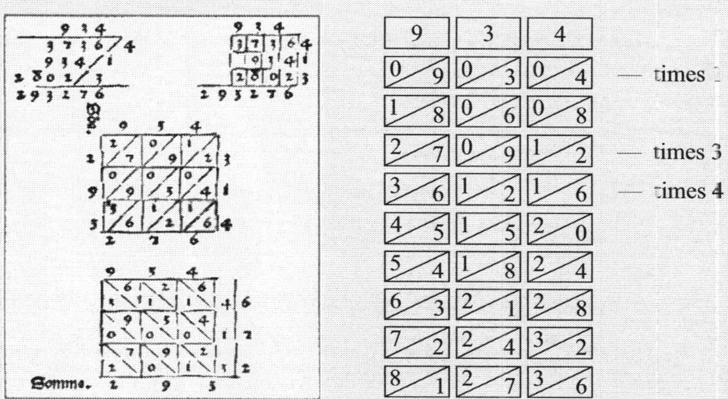

Figure 11.20. Lattice multiplication and Napier's bones

Napier's logarithms.

Napier's major and more lasting invention, that of logarithms, forms an interesting case study in mathematical development. Within a century or so, what started life as merely an aid to calculation, a set of 'excellent briefe rules' as Napier called them, came to occupy a central role within the body of theoretical mathematics.

The basic idea of logarithms is straightforward: to replace the wearisome task of multiplying two numbers by the simpler task of adding together two other numbers, and similarly to replace division with subtraction. To each number is associated another, which Napier first called an 'artificial number' and later a 'logarithm' (a term he coined from two Greek words, 'logos' for word and 'arithmos' for number), with the property that, from the sum of two logarithms the result of multiplying the two original numbers can be recovered.

In a sense, this idea had been around for a long time. Since Greek times or even earlier, it had been known that the multiplication of terms in a geometrical progression correspond to the addition of terms in an arithmetical progression. For instance, consider

$$\begin{array}{cccccc} 2 & 4 & 8 & 16 & 32 & 64 \\ 1 & 2 & 3 & 4 & 5 & 6 \end{array}$$

and notice that the *product* of 4 and 8 in the top line (= 32) lies above the *sum* of 2 and 3 in the bottom line (= 5). Here, the top line is a *geometrical progression*, because each term is twice its predecessor: there is a constant *ratio* between successive terms. The lower line is an *arithmetical progression*, because each term is one more than its predecessor: there is a constant *difference* between successive terms. These two lines appear as parallel columns of numbers on a Mesopotamian tablet, although we do not know the scribe's intention in writing them down, and a continuation of these progressions is the subject of a passage in Chuquet's *Triparty* of 1484, as we saw in Section 10.1.[64]

So the idea that addition in an arithmetical series parallels multiplication in a geometrical one was not unfamiliar. Nor was the notion of obtaining the result of a multiplication by means of an addition, for this was quite explicit in certain trigonometrical formulas well known by 16th century, such as:

$$2\cos A\cos B = \cos(A - B) + \cos(A + B)$$
$$2\sin A\sin B = \cos(A - B) - \cos(A + B).$$

Thus, if you want to multiply two cosines or two sines together — a nasty calculation on fiddly numbers — you can reach the answer through the much simpler operation of adding or subtracting two other numbers. This method was much used by astronomers towards the end of the 16th century, particularly by the great Danish astronomer Tycho Brahe who was visited by a young friend of Napier, John Craig, in 1590. So Napier may have been aware of these techniques around the time that he started serious work on his own idea, although conceptually it was entirely different.

Napier's definition of logarithms is rather interesting. We shall not pursue all its details, but include enough to see its approach and character.[65] Imagine two points, P and L, each moving along its own line (see Figure 11.21). The line PQ is of a fixed finite length, but L's line is endless. L travels along its line at constant speed, but P slows down. P and L start from the points P_0 and L_0 with the same speed, but thereafter P's speed drops in proportion to the distance that it has still to go: so at the half-way point between P_0 and Q, P is travelling at half the speed it started with, at the three-quarter point it is travelling with a quarter of its original speed, and so on. So P will never reach Q, any more than L will reach the end of its line, and at each instant the positions of P and L uniquely correspond. In Napier's definition, the distance L_0L at any instant is the *logarithm* of the distance PQ: thus, the distance that L has travelled at any instant is the logarithm of the distance that P has yet to travel.

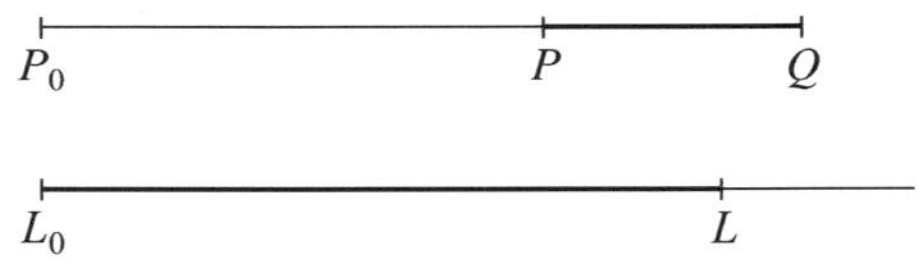

Figure 11.21. Napier's kinematic definition of logarithms

How does this fit in with the ideas we spoke of earlier? The point L moves in an arithmetic progression: there is a *constant difference* between the distance it moves in equal time intervals — that is what 'constant speed' means. The point P, however, is slowing down in a geometrical progression: in its motion it is the *ratio* of successive distances that remains constant in equal time intervals. The major novelty in Napier's concept of logarithm is in his use of the concept of motion, of points moving along lines with speeds defined in various ways. By comparison both the exponents of Chuquet's table, and the trigonometrical formulas, are 'static' objects — there is evidently a deep difference of mathematical style here.

[64] In *The Sand-reckoner*, long before, Archimedes proved a similar result for any geometrical progression; see *The Works of Archimedes*, 229–230, and F&G 4.A2.

[65] A good, brief account will be found in (Edwards 1979, 143–151).

What seems so clever about Napier's approach is that he could cope with any number, and not just those that happened to form part of some particular discrete geometric progression. This was effected by his intuition, springing from the *continuous* nature of the straight line and of the motion.

The idea of *motion* found in Napier's work seems to have come from a concept used by Archimedes in his study of spirals (as you saw in Chapter 5), so there was a classical precedent for propositions about points moving along lines.[66] Furthermore, although much of the Western mathematical tradition had hitherto been nervous of the concept of motion, there had been exceptions to this: three centuries or so earlier, both the Merton School in 14th-century Oxford, and Nicole Oresme at the University of Paris, had made prolonged studies of issues involving this concept. The details of Napier's education are obscure – we know that he spent a year at the University of St Andrews in his early teens, but not what he did or learned thereafter — but it is not implausible that he became aware of such medieval studies of motion at some stage.

In Napier's description, it is the length P_0Q whose logarithm is 0, and that the shorter the length the larger is its logarithm. Napier chose P_0Q to be $10^7 = 10{,}000{,}000$, because he intended the table of logarithms to be used trigonometrically — it was the logarithms of sines and tangents that he calculated and tabulated in his 1614 *Descriptio*, and not the logarithms of numbers in general — and it was usual in that context to multiply the value of sines and tangents by a conventional number such as 10,000,000 in order to be able to work with whole numbers.

As for the calculation of the logarithms, as Edwards (1979) explains, Napier effectively called the logarithm of a number x the number y such that

$$x = 10^7(1 - 10^{-7})^y.$$

Thus the steps in the arithmetic progression are each 10^{-7}, and Napier regarded that amount as sufficiently small. It follows that if $x' = 10^7(1 - 10^{-7})^{y'}$ then

$$\frac{x}{x'} = (1 - 10^{-7})^{y-y'},$$

so the differences in the logarithms y and y' depends only on the ratio of x to x'. Accordingly, as the xs diminish geometrically, their logarithms y increase arithmetically.

The problem Napier then faced was that it would seem to require 10^7 calculations to compute a table of these logarithms, and this is an impossible amount of work. However, Napier was able to come up with a further series of clever arguments that enabled him to compute only relatively few values and accurately interpolate the rest. However, on occasion even seven significant figures were not sufficient for the accuracy he desired, and he used four more, beyond the 'decimal point' — one of the earliest occurrences in print of our decimal point symbol, which helped to stabilise this notation in its now-familiar form.

We should also notice one defect in Napier's approach. If the intention is that addition of logarithms mimics multiplication of numbers, then the logarithm of 1 should be zero, because $1 \times x = x$ implies that

$$\text{logarithm of } (1 \times x) = (\text{logarithm of } 1) + (\text{logarithm of } x).$$

However, writing Naplog x for Napier's logarithm of x, we have

$$\text{Naplog } xy = \text{Naplog } x + \text{Naplog } y - \text{Naplog } 1,$$

[66] See, for example, Proposition 1 of *On Spirals* in F&G 4.A7.

and Naplog $1 = 10^7 \times 16.11809565$. For the details, see our account in Box 11.2, which, for the sake of simplicity, draws on some mathematics not available in Napier's time. It is also easy to see that Naplog $10^7 = 0$.

Box 11.2. Finding logarithms.
Napier supposed that the line PQ was of length 10^7 and that the moving point P was at a distance x from P_0 and therefore the distance PQ was $10^7 - x$. At this moment, its velocity is proportional to $10^7 - x$.
The calculus tells us that we can write

$$\frac{dx}{dt} = 10^7 - x,$$

$$\text{so} \quad \frac{dx}{10^7 - x} = dt,$$

and so, integrating both sides, we find that

$$\log(10^7 - x) = -t + c,$$

where c is a constant of integration.

But $x = 0$ when $t = 0$, so $c = \log 10^7$, and we conclude that

$$t = \log 10^7 - \log(10^7 - x).$$

Here log stands for the modern logarithm, taken to the base e, and defined by the relationship that if $x = e^y$ then $y = \log x$.

Napier's logarithm of t was Naplog $(x) = 10^7(\log 10^7 - \log x)$, which expresses $t = LL_0$ in terms not of P_0P but of $PQ = 10^7 - x$.

Notice that Naplog $1 = 10^7(\log 10^7 - \log 1) = 10^7(\log 10^7) = 10^7 \times 16.11809565$.

The following short calculation shows that

$$\text{Naplog}\, xy = \text{Naplog}\, x + \text{Naplog}\, y - \text{Naplog}\, 1.$$

In fact

$$\text{Naplog}\, xy = 10^7(\log 10^7 - \log xy) = 10^7(\log 10^7 - \log x - \log y),$$

and

$$\begin{aligned} \text{Naplog}\, x + \text{Naplog}\, y &= 10^7(\log 10^7 - \log x) + 10^7(\log 10^7 - \log y) \\ &= 10^7(\log 10^7 - \log x - \log y) + 10^7(\log 10^7) \\ &= \text{Naplog}\, xy + \text{Naplog}\, 1, \end{aligned}$$

as required.

Henry Briggs. Few mathematical inventions have burst on the world so unexpectedly as Napier's logarithms. Although various disparate strands — the idea of doing multiplication via addition, the idea of comparing arithmetical and geometrical progressions, the use of the concept of motion — had all been floated at some stage, the enthusiasm with which Napier's work was received makes it clear that this was perceived as a novel invention and that it fulfilled a pressing need.

Foremost among those who welcomed the invention was Henry Briggs, the first Professor of Geometry at Gresham College, who wrote to the Biblical scholar James Ussher in 1615:[67]

> Naper, lord of Markinston, hath set my Head and Hands a Work with his new and admirable Logarithms. I hope to see him this Summer if it please God, for I never saw Book which pleased me better or made me more wonder.

Briggs did indeed 'see him this Summer' in a visit recorded some time later by the astrologer William Lilly.[68]

During this visit in 1615, and a further one in the following year, Briggs and Napier discussed some simplifications to the idea and presentation of logarithms. It is fortunate that they profited from each other's company in this way, for their world-views were very different. Napier not only considered the Pope to be the Antichrist and expected the Day of Judgement quite shortly (probably between 1688 and 1700), but was also 'a great lover of astrology' (according to Lilly), and may even have practised witchcraft, as contemporary rumour had it. Briggs, on the other hand, had no time for such astrological practices.

It was these two very different people who worked together on simplifying logarithms. They agreed that it would generally be more useful if the logarithm of 1 were to be 0, and the logarithm of 10 were to be 1 (see Box 11.3), and Briggs spent two years recalculating the tables on this basis. Thus the early history of logarithms exemplifies well a remark that the historian Clifford Truesdell has made in another context: 'the simple ideas are the hardest to achieve; simplicity does not come of itself but must be created'.[69] Napier and Briggs had to work hard to create the basic property of logarithms:

$$\log(x \times y) = \log x + \log y.$$

As a result of Briggs's discussions with Napier, log 1 was redefined to be 0, thus simplifying the formula and making logarithms easier to use.

Briggs, too, developed the calculation of logarithms of ordinary numbers, using the correlation log 10 = 1, log 100 = 2, log 1000 = 3, etc. These logarithms are called *ordinary logarithms*, or *logarithms to the base 10* (see Figure 11.22). 30,000 of these numbers were calculated to fourteen places of decimals, with a gap in the tables, between 20,000 and 90,000, that Briggs hoped others would fill.

One way to compute the logarithm of every number from 1 to 100,000 would be to take each number in turn and apply the method given in the box. For each number, we would compute sixteen successive square roots, a subtraction, and the multiplication by $2^{16} = 65{,}536$. But notice the amount of work involved. There is a routine method for finding the square root of a number, which involves little more than breaking a decimal into successive pairs, halving and carrying, but if we suppose that this takes 2 minutes then it would take three-quarters of an hour to find the logarithm of each number. We would be lucky to do 60 in a week at that rate, so the logarithm table would take 1800 weeks at this rate, or about 36 years! Of course, as we compute $\log 2, \log 3, \log 4, \ldots$, there is almost no work in computing the logarithm of a product: $\log 4 = 2 \times \log 2$, $\log 6 = \log 2 + \log 3$, etc. But there are nearly 8600 primes less than 100,000. That is 120 weeks or more on the logarithms of the prime numbers, and a certain amount of time on the rest, which then have to be factorised (not an easy business — for example, which of 23453 and 23459 is

---

[67]Quoted in (Hallowes 1961–1962, 80–81). Ussher was the scholar whose studies of biblical chronology revealed that the Creation took place at 9 A.M. on 23 October in 4004 BC.

[68]Quoted in F&G 9.E3.

[69]Quoted in F&G 14.C1(b).

Box 11.3. Finding logarithms.

The fundamental property of logarithms is that they convert multiplication to addition. In symbols:

$$\log(x \times y) = \log x + \log y \ .$$

It follows from this that $\log 1 = 0$. But how do we do find the logarithm of an arbitrary number?

Suppose we want to find $\log 5$. We know that $x \times y$ is very nearly y when x is very nearly 1. This suggests that the logarithm of $1 + z$ is very close to zero, and could perhaps be bz when z is very small. (We shall give Briggs's defence of this argument shortly, and we have a use for this mysterious b which we will come to later.) Now, how do we get from 5 to a very small number? Notice that the fundamental formula above implies that $\log x^k = k \log x$ for all numbers x and all k. If we let $x = 5$ and raise $\frac{1}{2}$ to a high power k, such as $k = 16$, we can write that

$$\log(5^{1/2^{16}}) = (1/2^{16}) \log 5 \ ,$$

which implies that

$$2^{16} \log(5^{1/2^{16}}) = \log 5 \ .$$

We now write $5^{1/2^{16}} = 1 + z$, where z is a number that depends on all our choices so far (5, 2, and 16). This turns our last equation into this equation for $\log 5$:

$$2^{16}(bz) = \log 5 \ .$$

We now have to find $5^{1/2^{16}} = 1 + z$, and deduce the value of z. This involves finding the square root of 5 (which is $5^{\frac{1}{2}}$), the square root of that (which is $5^{\frac{1}{4}}$), the square root of that (which is $5^{\frac{1}{8}}$), then the square root of that (which is $5^{\frac{1}{16}} = 5^{1/2^4}$) until we have done this successively sixteen times (to find $5^{1/2^{16}}$). We find that $z = 0.000024558375$ to 12 decimal places, so

$$\log 5 = 2^{16} \times 0.00024558375 \times b.$$

This works out to be $1.60946 \times b$, which agrees with the natural logarithm of 5 to four decimal places (which is 1.609437912), if we make the natural choice of $b = 1$.

We can find the natural logarithm of 2 in the same way. It works out to be $0.69315 \times b$. This gives us $\log 10 = \log 2 + \log 5 = 2.30261 \times b$. Since we have already chosen $b = 1$, this gives us $\log 10 = 2.30261$, which compares very well with the accurate value of $\log 10 = 2.302585$.

It is worth noticing that the value of $\log x$ can be found in this way for every value of x simply by using the fundamental formula and the observation that $\log 1 = 0$.

prime?). Plainly, good insightful shortcuts are required if this is to be done in a reasonable length of time, and if the method is to be accurate (which the above method is not, even after 16 extractions of a square root).

As for what Briggs actually did, he advocated the method in Box 11.3 for its accuracy, and used it to compute the logarithms of prime numbers. To compute the logarithm of a number to 16 decimal places he argued that one should proceed to take square roots until the first 32 decimal places are correct and the first 16 are all zeros. He gave as his reason that the non-zero decimals are then directly proportional to the logarithm — in other words, that $\log(1+x) = bx$ for some b. This is true, although he did not explicitly say so, because when x is very small the square root of $1 + x$ is known to be very close to $1 + \frac{1}{2}x$; in fact,

2		Logarithmi.			Logarithmi.
1	0	0000,00000,00000	34	1	5314,78917,04226
2	0	3010,29995,66398	35	1	5440,68044,35028
3	0	4771,21254,71966	36	1	5563,02500,76729
4	0	6020,59991,32796	37	1	5682,01724,06700
5	0	6989,70004,33602	38	1	5797,83596,61681
6	0	7781,51250,38364	39	1	5910,64607,02650
7	0	8450,98040,01426	40	1	6020,59991,32796
8	0	9030,89986,99194	41	1	6127,83856,71974
9	0	9542,42509,43932	2	1	6232,49290,39790
10	1	0000,00000,00000	43	1	6334,68455,57959

Figure 11.22. Some of Henry Briggs's logarithms

if $x = 10^{-6}$ then the error in that approximation is about 10^{-12}. But, on finding that it could be necessary to take the square root 53 times, he looked for simplifications. One of these was his observation that (in base 10) $\log 10 = 1$, $\log 10^{1/2} = 0.5$, $\log 10^{1/4} = 0.25$, and so, repeating this process a further 50 times he found that

$$\log(1.000\,000\,000\,000\,000\,1) = 0.000\,000\,000\,000\,000\,043\,429\,448\,190\,325\,1804.$$

From this value he could find an excellent approximation to the value for the logarithm of numbers near to $1 + 10^{-15}$ simply by adding. From the values of $\log 5$ and $\log 10$ $(= 1)$ he could then find the value of $\log 2$. He could now find the logarithms of many other numbers directly, and of numbers very close to these by interpolation. Simplifications and sophisticated interpolations were something that Briggs proved to be very insightful about. But no wonder he appealed to others to finish the job.

Another person besides Briggs who immediately recognised the importance of Napier's concept was the navigational practitioner Edward Wright, and knowledge of logarithms spread rapidly in various ways. Wright's English translation of Napier's *Mirifici Logarithmorum Canonis Descriptio* was one source of knowledge; when Wright died before its publication it was dedicated by his son to the East India Company, another of the major trading and exploration companies of the time. This dedication suggests an audience for whom the knowledge was thought to be useful — Wright had been navigational consultant to the Company for the last year or two of his life.

Knowledge of logarithms spread by word of mouth too; Briggs lectured on them at Gresham College, as did Edmund Gunter who was appointed Professor of Astronomy there in 1620. Within a decade, editions or similar tables had been published in France, Germany, and the Netherlands. The gap in Briggs's table was filled by Adrian Vlacq, whose tables were calculated to ten places of decimals, and were published in Holland in 1627 as *Het Tweede Deel van de Nieuwe Telkonst* (or, The Second Part of the New Art of Counting). The calculations of Briggs and Vlacq were the essential basis of all subsequent tables of 'common logarithms'.

Johannes Kepler. As an example of the impact of logarithms abroad, we consider the response of the astronomer Johannes Kepler.[70] Kepler was precisely the kind of practitioner for whom logarithms were of greatest benefit: a professional astronomer driven at times to distraction, he tells us, by the magnitude and complexity of the calculations he needed to do. So when he was able to study a copy of Napier's *Descriptio* around 1619, he welcomed it warmly, dedicating his next book to Napier (not realising that Napier had been dead for two years).

Kepler went further. Napier's book contained the definition of logarithms and a set of logarithm tables, but not how to get from one to the other. So Kepler set to work recalculating their construction from first principles, basing his argument upon the classical theory of proportion in Euclid's *Elements*, Book V. This approach was consciously quite different from Napier's geometrical and kinematical considerations; Kepler wrote that, for him, logarithms were not associated[71]

> inherently with categories of trajectories, or lines of flow, or any other perceptible qualities, but (if one may say so) with categories of relationship and qualities of thought.

It is interesting to note that Kepler felt that the appeal to kinematical intuition in Napier's work lacked rigour. He emphatically asserted that what he was supplying was *rigorous proof*, which was, for him, something cast in Euclidean mould.[72] In the event, just as Napier had done, he had to approximate to reduce potentially endless numbers to a suitable finite approximation, but Kepler's overall approach was influential, nonetheless.

Kepler's *Rudolphine Tables* of 1627 were planetary tables on which he had been working for twenty-six years, latterly with the aid of logarithms. The Frontispiece (see Figure 11.23), designed by Kepler himself, includes amongst a host of symbolic detail the muse of arithmetic on the roof, her halo displaying the number 6,931,472, which is 10^7 times the logarithm of 2 (which reflects the fact that $\log\sin 30^\circ = \log\frac{1}{2} = -\log 2 = -0.6\,931\,472$).

We should not leave this initial phase of logarithms without drawing attention to one of the few expressions of dissent. Kepler's former astronomy teacher, Michael Mästlin, Dean of Arts at Tübingen University in Germany, was most distrustful of logarithms; he felt that mathematicians should not use numerical tables whose construction they did not know about or understand, and pointed out that even if their calculations seemed to work in specific instances, there was no guarantee that this would always be the case. He even went so far as to tell Kepler that 'it is not seemly for a professor of mathematics to be childishly pleased about any shortening of the calculations'.[73]

However, Kepler's approach can now be seen as responding successfully to the challenge of explaining how logarithms could be rigorously grounded, so that mathematicians could be confident that the tables were mathematically sound and reliable. And Kepler would have emphatically disagreed with Mästlin's last point. Mästlin's apparently masochistic approach to calculation is perhaps to be understood as coming from a practitioner of an older generation, no longer actively engaged in the computations that Kepler found so wearisome.[74]

---

[70] We consider Kepler's astronomical work in Section 12.2.

[71] Quoted in (Belyi 1975, 656).

[72] For his style of argument, see F&G 9.E4.

[73] Quoted in (Caspar 1959, 309).

[74] For a time Kepler had an assistant in Prague called Joost Bürgi, a globe maker, who came to the idea of logarithms independently of Napier, but he published his tables only in 1620, too late for them to have any influence.

Figure 11.23. Frontispiece of Kepler's *Rudolphine Tables* (1627)

In England, one of the most influential promulgators of logarithms was Edmund Gunter. Apart from discussing them in his astronomy lectures at Gresham College, Gunter incorporated a logarithmic scale on a ruler (that is, with marks spaced in geometrical progression but numbered in arithmetical progression); with the aid of a pair of dividers for transferring distances along the ruler, this enabled rough-and-ready multiplications to be carried out swiftly. This conversion of the idea behind abstract mathematical tables into a physical object was an ingenious and fruitful step. As William Oughtred wrote:[75]

[75]Oughtred, *Just Apologie* (1634, 11).

> The honour of the invention of logarithms, next to the Lord of Merchiston and our Mr Briggs, belongeth to Master Gunter, who exposed their numbers upon a straight line.

It was probably Oughtred, or possibly Delamain, who created the slide rule. He placed two such logarithmic scales side by side, so that, when one is slid against the other, the result of a multiplication (that is, an addition of lengths) could be read straight off. Gunter invented or improved several instruments — Gunter's staff, Gunter's sector, Gunter's quadrant, Gunter's chain, and Gunter's line all became familiar to the mathematical practitioner.

An assessment of Gunter's influence in a biographical note written by the antiquary John Aubrey in 1690 may not, however, be entirely reliable.

> Captain Ralph Gretorex, Mathematical-Instrument Maker in London, sayd that he was the first that brought Mathematicall Instruments to perfection. His Booke of the Quadrant, Sector, and Crosse-staffe did open men's understandings and made young men in love with that Studie. Before, the Mathematicall Sciences were lock't-up in the Greeke and Latin tongues; and so lay untoucht, kept safe in some Libraries. After Mr Gunter published his Booke, these Sciences sprang up amain, more and more to that height it is at now.[76]

Aubrey's view of the mathematical sciences as 'lock't-up in the Greeke and Latin tongues' until liberated by Gunter seems rather an exaggeration, and at least half a century out, when one considers that Euclid's *Elements* were 'unlocked' by Billingsley in 1570. Aubrey seems to have conflated two different things: the publication of mathematical works in English, and an increasing use and popularity of practitioners' instruments. Gunter may have contributed largely in the latter respect, and this did indeed help to spread awareness of mathematical practice. Nonetheless, Aubrey's remarks usefully remind us that the pursuit of mathematics and of religion were not then embodied so separately as now.

Figure 11.24 shows the 1673 edition of Gunter's *Works*. The upper right-hand figure is using Gunter's cross-staff; at bottom right is Gunter's quadrant; at bottom left is Gunter's cross-bow; at top left is Gunter's sector. The original 1623 edition has the same design, but with different words: *The Description and Use of the Sector, the Crosse-staffe and other Instruments, for Such as are Studious of Mathematicall Practise*.

Although Gunter died at the age of 45, his collected writings went through several editions in the course of the century: Isaac Newton's well-thumbed copy of the second edition is now in the library of Trinity College, Cambridge — he bought it in 1667 for five shillings, not a small sum. The title page of the fifth edition gives us a good representative impression of the world of the mathematical practitioner in the 17th century. From the title page, it seems that many people contributed to the book. Apart from Gunter himself, Mr Bond added some Questions in Navigation, Mr Foster invented some instruments (with more lines than Mr Gunter's), and "diligent Corrections and so forth" were done by William Leybourn, Philomath. So this title page provides evidence of a book-writing instrument-inventing mathematics-teaching tradition, extending through several hands over half a century.

The content sounds fairly mixed, too. The book seems to have been designed for people interested in arithmetic, geometry, astronomy, navigation, dialling (the design of

[76] In (Aubrey 1949) and F&G 9.G4.

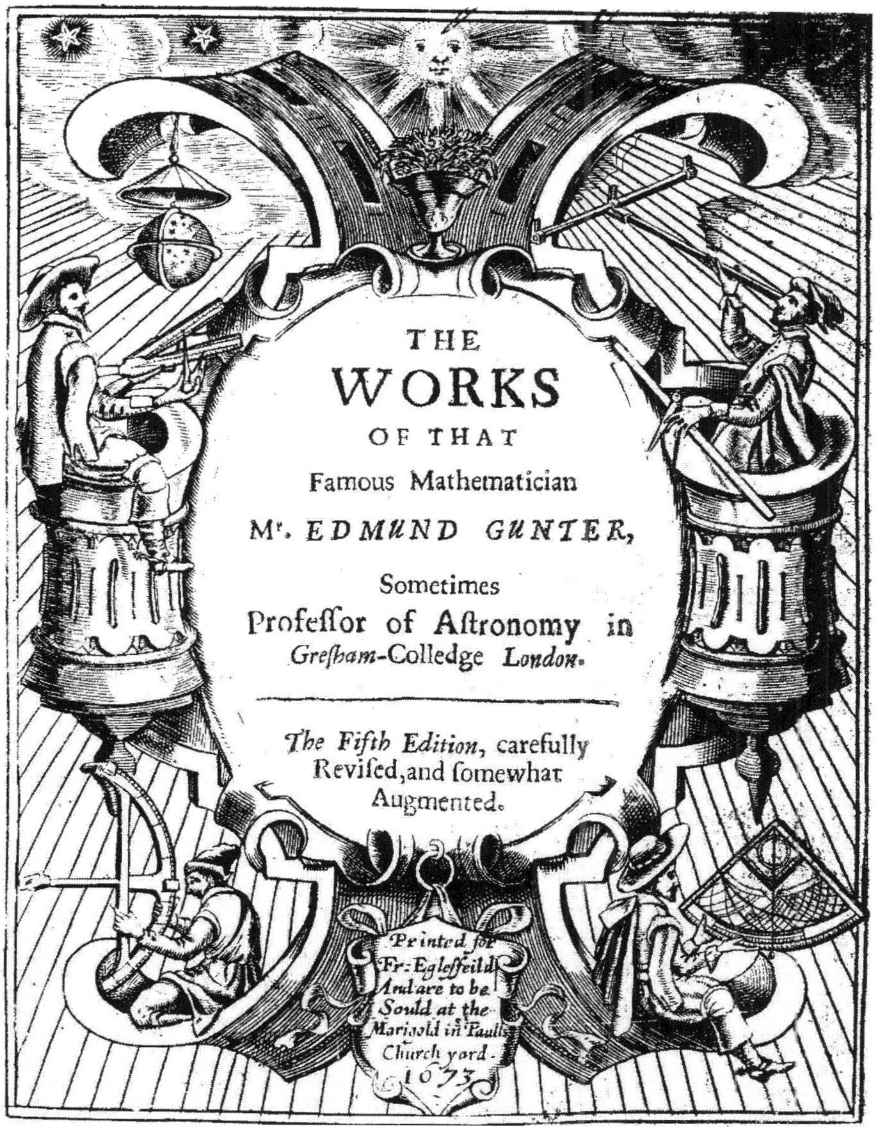

Figure 11.24. Frontispiece of Gunter's *Works*

sundials), and fortification, to which end various instruments are described and logarithmic tables are given. This corresponds with the range of *Arts and Sciences Mathematical Professed and Taught* by William Leybourn.[77]

In such ways the use of logarithms as a practical calculating aid became more widely known, and grew to become one of the most valuable components of the practitioner's skills. Over the course of the 17th century further theoretical investigations involving logarithms took place; the device that started life as a set of 'excellent briefe rules' for

[77]Napier's bones is one of the arithmetic instruments whose use he taught.

removing the hindrances that, as Napier put it, 'molest and hinder Calculators' developed into a powerful concern at the heart of mathematical science, as you will see in Chapter 12.

Meanwhile, it was these kinds of instrumental concerns that mediated between mathematics and the world. A century later, the mathematician Edmund Stone summarised the development of mathematics as a useful practice in this way:[78]

> Mathematicks are now become a popular study, and make a part of the Education of almost every Gentleman ... Mathematical Instruments are the means by which those Sciences are rendered useful in the Affairs of Life. By their assistance it is that subtile and abstract Speculation is reduced into Act. They connect as it were, the Theory to the Practice, and turn what was bare Contemplation, to the most substantial Uses. The Knowledge of these is the Knowledge of Practical Mathematicks So that the Descriptions and Uses of Mathematical Instruments make, perhaps, one of the most serviceable Branches of Learning in the World.

11.6 Further reading

Chartres, R. and Vermont, D. 1997. *A Brief History of Gresham College 1597–1997*, Gresham College, London. A well-illustrated account of the rise, decline, and recovery of this important institution.

Clucas, S. (ed.) 2006. *John Dee: Interdisciplinary Studies in English Renaissance Thought*, Springer. A set of essays examining Dee's activities from a wide range of perspectives.

Eisenstein, E.L. 1979. *The Printing Press as an Agent of Change: Communications and Cultural Transformations in early-modern Europe*, Cambridge University Press. A thorough historical guide to one of the most important influences on the subject matter of this chapter.

Fauvel, J., Flood, R., and Wilson, R. (eds.) 2013. *Oxford Figures: Eight Centuries of The Mathematical Sciences*, second edition, Oxford University Press. A valuable source of new information for many chapters in this book, and exceptionally well illustrated.

Feingold, M. 1984. *The Mathematicians' Apprenticeship: Science, Universities and Society in England, 1560–1640*, Cambridge University Press. Full of informative detail and illuminating quotations.

Havil, J. 2014. *John Napier: Life, Logarithms and Legacy*, Princeton University Press. As the title suggests, this not only covers Napier's route to his discovery of logarithms but his theological work, including his prediction that the apocalypse would occur in the year 1700.

Hill, C. 1982. *Intellectual Origins of the English Revolution*, Oxford University Press. A broad sweep by one of the most distinguished historians of the period, in which the role of mathematical practitioners is given unusually full weight.

Howson, G. 1982. *A History of Mathematics Education in England*, Cambridge University Press. Only the first two chapters concern our period, but this is a readable and interesting book that looks at aspects of this subject up to the last century.

Roberts, G. 2016. *Robert Recorde: Tudor Scholar and Mathematician*, Scientists of Wales, University of Wales Press. An engaging account of Recorde's difficult career in a time of religious and political dangers.

[78]See (Stone 1973, 51).

Roberts, G. and Smith, F. 2012. *Robert Recorde: the Life and Times of a Tudor Mathematician*, University of Wales Press. This is both a biography and a thorough overview of his work, the influences upon it and its contributions to mathematical and scientific life.

Shirley, J.W. 1983. *Thomas Harriot: A Biography*, Oxford University Press. A good guide to the breadth of Harriot's activities, as well as to his relations with friends and colleagues, though short on Harriot's mathematics.

Stedall, J.A. 2003. *The Greate Invention of Algebra: Thomas Harriot's Treatise on Equations*, Oxford University Press., Oxford University Press. A masterly reconstruction of Harriot's study of equations, with a fine short introduction.

Taylor, E.G.R. 1954. *The Mathematical Practitioners of Tudor & Stuart England*, Cambridge University Press. A historical survey (1485–1714) with 582 biographies of practitioners and 628 mathematical works listed briefly — a most useful work.

Website: *The Manuscripts of Thomas Harriot* (1560–1621)
`www.echo.mpiwg-berlin.mpg.de/content/scientific_revolution/harriot`.
As the website says: 'The project aims not only to publish Harriot's surviving papers but to organize them in such a way that readers can find their way more easily through the disordered raw material.'

12

The Astronomical Revolution

Introduction

One of the most significant applications of mathematics has been to astronomy. We begin this chapter by looking at the work of Nicolaus Copernicus, whose book *De Revolutionibus Orbium Coelestium* (On the Revolutions of the Celestial Spheres) famously put the Sun at the centre of the universe and thereby, controversially, displaced the Earth.

Then we consider the attitude to mathematics of an outstanding figure of the early 17th century, the astronomer and cosmologist Johannes Kepler. Kepler's views were marked by a strong Platonism: for him, God was the supreme geometer. So Kepler was at once an intellectual conservative and a profound innovator, a fascinating if not uncommon mixture; indeed, it was this combination of qualities that led to his most famous success (ascertaining the nature of the true orbit of Mars) but also held him back from a radical breakthrough in other fields (notably, the theory of musical harmony). So Kepler presents us with a varied picture of a mathematical scientist at work, his mind stocked with an array of both original and received ideas.

Finally, we introduce his great Italian contemporary Galileo Galilei (usually simply called Galileo), in order to show you more of the different ways in which the relationship between mathematics and the physical world was seen at the time. While Galileo too contributed to an acceptance of the new cosmology, he came from a different philosophical background. His approach had weightier consequences for the developing tradition of experimental science, with its subtle yet crucial invocation of mathematics.

12.1 The Copernican revolution

Ptolemy's Earth-centred system of epicycles and equants lasted without essential alteration until the 16th century. By that time, a number of different events were creating questions for the study of astronomy, a subject that had been allowed to lapse in the West for centuries from the sheer difficulty of the work, before being jolted back into life by the rediscovery of Ptolemy's *Almagest* and some knowledge of Islamic astronomy.

A major issue at this time was calendar reform, a topic much debated by the Roman Catholic Church, which needed to calculate the date of Easter and certain other events in the Church calendar. New editions of Greek works cast doubt on the accuracy of some texts, including Ptolemy's, and the exploration of the Atlantic and expeditions that sailed beyond Africa to the east showed that Ptolemy's geography was badly wrong about the size of the Earth and the world beyond the Mediterranean, further suggesting that he might also be wrong in his astronomy.

We can get a measure of the problem by looking at events in the years after Copernicus's death in 1543. By the 16th century, use of the 19-year Metonic cycle (which we discussed in Section 6.2) had accumulated enough errors for the phases of the Moon to depart from their predicted values by four days, a fact obvious to any informed observer, and one that had implications for the date of Easter. The Council of Trent in 1563 therefore proposed that the calendar be corrected, and they did this by adjusting the estimate of the length of the year and by altering the rules for defining leap years.

The final result, after many years of discussion, was the *Gregorian calendar*, named for its proponent Pope Gregory XIII. It was proclaimed in 1582 and adopted after many struggles at various dates. While Catholic countries took it up almost at once, many Protestant ones delayed until 1700, Britain and America waited until 1752, Russia did not adopt it until 1918, and Greece only in 1923. In each case the calendar leapt forward and days were 'lost': early countries moved from 4 to 15 October 1582, losing ten days, while the British lost eleven days and the Russians lost fifteen. Although the resolution of the calendar crisis came after Copernicus's death, it made use of much of his work, as we now see.

Nicolaus Copernicus. Nicolaus Copernicus was born in Torun in Poland in 1473 and educated in Krakow (where he became fascinated by astronomy), before going to Bologna (where he studied medicine) and Ferrara (where he studied law). On his return to Poland he worked for the church in Frombork; remarkably, he was never a professional astronomer.

The heliocentric hypothesis, that the Sun is at the true centre of the universe, did not originate with him. It had been held by the Greek astronomer Aristarchus, could be found in the astronomical writings of Nasīr al-Dīn, and was revived by Oresme and by Nicolaus Cusanus, a 15th-century bishop. Copernicus himself mentioned Aristarchus, but without giving a source, and also referred to related comments by ibn Sina and by ibn Battuta, a 14th-century Islamic geographer and traveller whose journeys took him from his birthplace in Morocco to China.

In fact, the whole complicated issue of influences from Islamic astronomers on Copernicus has been discussed by historians for many years. It is clear that Islamic astronomers had long disliked Ptolemy's use of equants — al-Haytham polemicised against them in the early 11th century, for example — but the names that Copernicus mentioned are of Islamic writers in al-Andalus (later, Spanish Andalucia), and there is good evidence that they represented an unsuccessful group with little influence on astronomers in the Eastern Islamic world. Copernicus did not mention members of the Eastern group by name, but one of his ingenious mathematical devices for producing straight-line motion by means of epicycles had been discovered by Nasīr al-Dīn, and is described in Copernicus's book in a way very close to al-Ṭusī's; this device allows epicycles to perform tasks that Ptolemy had described with equants. More strikingly, Copernicus's lunar theory is identical to that of ibn al-Shāṭir, who died around 1375. The problem facing all these historical investigations is the absence of evidence that would connect this Islamic work with Copernicus: there are no documentary sources linking the two, and yet it is hard to balance the relative dearth of top-class astronomical work in the West with the extensive analyses of the East.

It has recently been suggested that the heliocentric hypothesis might itself be part of the link.[1] Ptolemy gave reasons for his belief that the Earth was genuinely at rest at the centre of the universe, and Aristotelian physics made this idea fundamental to both astronomy and physics. But Ptolemy had also expressed the view that mathematics has a higher claim to certainty than physics does. If astronomical evidence could be interpreted to suggest that the Earth moves, then a new non-Aristotelian physics would be required.

As it happens, Islamic theologians had never embraced Aristotle's ideas about causes with the enthusiasm of Christian theologians, so it was possible for Islamic astronomers to consider the heliocentric hypothesis more easily than could later Europeans. What seems to have happened is that al-Ṭusī was the first to argue that Ptolemy's arguments for a stationary Earth were wrong — he thought that the matter could not be decided either way — and this spurred three centuries of debate on the issue. The historian F.J. Ragep has observed that when Copernicus came to rebut Ptolemy's arguments for a stationary Earth he did so in terms very close to al-Ṭusī's, suggesting, but not establishing, a link.[2] He has also found a late Islamic astronomer, 'Alī Qūshhjī, who lived at the start of the 15th century, and whose work seems to have been known and used by Regiomontanus. These are delicate issues that should be further illuminated as investigations into the history of Islamic science deepen.

Copernicus first discussed the heliocentric hypothesis in a short work, the *Commentariolus* (Little Commentary) of 1514, but he delayed publishing the full account of his ideas until 1543, the year of his death. A famous, but unverifiable, story has him being presented with the first copy of his major work on his deathbed. The reasons for the long delay of twenty-nine years have been much discussed by historians. It does not seem that the Church was hostile, but rather that Copernicus feared the disagreement of professional astronomers. Like the best Islamic astronomers before them, they had become aware of many inadequacies in Ptolemy's treatise, the *Almagest*, as they struggled to produce and to master better editions of his work. Regiomontanus had written an important commentary on it in 1463, in which he pointed out that according to Ptolemy's lunar theory the Moon would sometimes be twice as far from the Earth as it is on other occasions, but its apparent size does not vary anything like that much. Professional astronomers would be a difficult audience to convince of a new theory, not so much from any allegiance to the ancient work as from a deep awareness of the many problems that would have to be solved by any successful replacement theory.

Copernicus seems to have begun his reworking of planetary astronomy by trying to eliminate Ptolemy's equants. He expressed his dislike of them in the opening page of his *Commentariolus*:[3]

> Yet the planetary theories of Ptolemy and most other astronomers, although consistent with the numerical data, seemed likewise to present no small difficulty. For these theories were not adequate unless certain equants were also conceived; it then appeared that a planet moved with uniform velocity neither on its deferent nor about the center of its epicycle. Hence a system of this sort seemed neither sufficiently absolute nor sufficiently pleasing to the mind.
>
> Having become aware of these defects, I often considered whether there could perhaps be found a more reasonable arrangement of circles, from which every

[1] See (Ragep 2007).
[2] See (Rageq 2007, 73).
[3] In (Rosen 1971, 57–58).

> apparent irregularity would be derived and in which everything would move uniformly about its proper center, as the rule of absolute motion requires.

This led Copernicus to a complicated theory with more epicycles than Ptolemy's; this was possibly another reason for his fear that professional astronomers might criticise him. He produced a heliocentric system, with the Moon orbiting the Earth and all the planets (including the Earth) orbiting the Sun. He explained how the steady motion of the planets round the Sun can produce apparently retrograde motion when seen from the (now moving) Earth. But he did not set out the mathematics that supported his theory.

Nor did he publish the work: he circulated it privately. But when the comments he received were mostly encouraging, he set to work (amid his ecclesiastical duties) to write the much larger work that would be necessary — with its detailed mathematics — and to set about finding a printer. It seems that the manuscript was largely written when the mathematician Georg Joachim Rheticus arrived in Frombork from the University of Wittenberg to assist him.

Rheticus, who was to be elected Dean of the liberal arts faculty on 18 October 1541 when he returned to Wittenberg, had produced a set of trigonometrical tables to aid in the work, and the scale of his achievement is worth noting. He published the tables in 1542 under the title *De Lateribus et Angulis Triangulorum* (On the Sides and Angles of Triangles) and added a table of half-chords subtended in a circle. As Rosen commented:[4]

> Such a half-chord is actually a sine, although both Copernicus and Rheticus studiously avoided the use of that term. The table of sines in the *On the sides and angles of triangles* differs from the corresponding table in *De revolutionibus orbis coelestium* by increasing the length of the radius from one hundred thousand to ten million and by diminishing the interval of the central angle from $10'$ to $1'$. Furthermore, by indicating the complementary angle at the foot of the columns and at the right-hand side of the page, the 1542 table became the first to give the cosine directly, although that term is not mentioned. Rheticus did not ascribe the authorship of this table to Copernicus nor, presumably out of modesty, to himself. Nevertheless, the table was undoubtedly his doing. His independent place in the history of mathematics is due precisely to his computation of innovative and monumental trigonometrical tables.

After working with Copernicus for two years, Rheticus wrote and published his *Narratio Prima* (First Report), which is an introduction to Copernicus's major work. Rheticus then obtained the permission of Duke Albrecht of Prussia for Copernicus's work to be published, and started to see it through the press before handing over to Andrew Osiander, a Lutheran minister who was also interested in mathematics and astronomy. In 1543 the job was completed and *De Revolutionibus Orbium Coelestium* was published (see Figure 12.1). The book itself was dedicated to Pope Paul III, doubtless for political reasons, and Rheticus's name is not mentioned, in order not to detract from the dedication.

De Revolutionibus Orbium Coelestium is divided into six Books. The first Book puts the planets in their order from the Sun, calculates their mean distances from the Sun, and gives the times of their orbits; by comparison, Ptolemy's system gave the Sun, Mercury, and Venus the same annual period. Book II goes into more mathematical detail, to make the theory quantitatively accurate. Book III deals with the motion of the Sun and the precession of the equinoxes — it turns out that the Sun itself is placed not at the exact centre of the universe, but slightly off centre. Book IV deals with the motion of the Moon.

[4]See (Rosen 2008, 396).

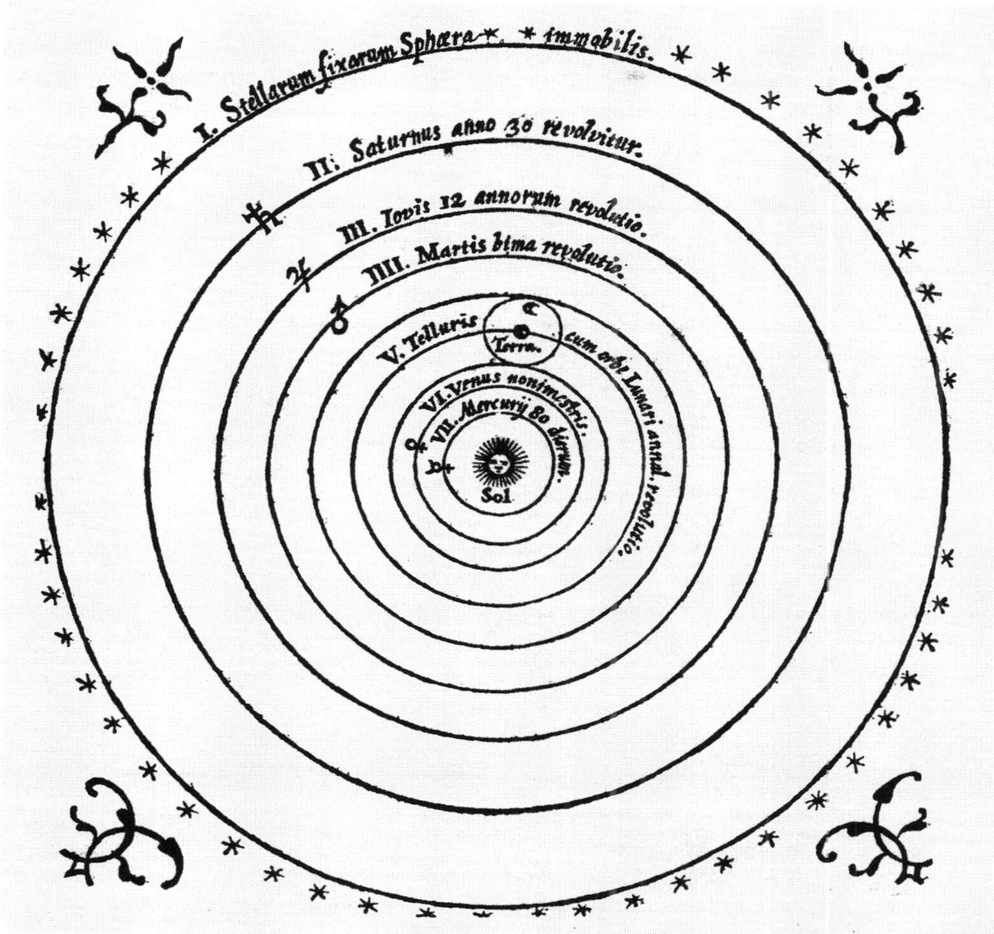

Figure 12.1. Copernicus's heliocentric system, with Sol (the Sun) at the centre and Terra (the Earth) as one of the planets orbiting it.

Books V and VI deal with the latitude and longitude of the planets on the celestial sphere; the treatment of their longitude was widely acclaimed by astronomers and did much to win acceptance for the work.

Its path was smoothed by Osiander's preface, which asserted that *De Revolutionibus* offered an account of the motions of the heavens, but did not provide a true cause or explanation. As he put it, it is the job of the astronomer:[5]

> since he cannot by any line of reasoning reach the true causes of these movements — to think up or construct whatever causes or hypotheses he pleases such that, by the assumption of these causes, those same movements can be calculated correctly from the principles of geometry for the past and for the future too . . . it is not necessary that these hypotheses be true, or even probably [so].

[5]See (Copernicus 1952, 505).

Rheticus objected — he even crossed out the words in his own copy – but the remark helped to separate the question of the accuracy of Copernicus's theory from the controversial question of whether the Earth or the heavens really move. The University of Wittenberg became associated with promoting Copernicus's ideas, and Erasmus Reinhold, an astronomer there who later became Dean and Rector of the University, set about producing a set of planetary tables based on them. These became known as the Prutenic tables — 'Prutenic' being the Latin for 'Prussian' — and their accurate predictions helped to spread the Copernican system, even though Reinhold himself did not accept the heliocentric hypothesis.

Copernicus's *De Revolutionibus Orbium Coelestium* is often hailed as one of the major intellectual events of human history, and as one of the key works that ushered in modern science. The philosopher Immanuel Kant considered it so in the 18th century, and the historian Thomas Kuhn called his 1957 book *The Copernican Revolution: Planetary Astronomy in the Development of Western Thought*. In keeping with the presumed momentous consequences of *De Revolutionibus*, some historians like to find weighty antecedents for it, and many have argued that the book had a difficult reception.

Kuhn suggested that no such great step could have been taken without Copernicus subscribing to the heliocentric hypothesis on other grounds, to do with what is called Renaissance neo-Platonism. He observed that veneration for the Sun fits poorly with the Ptolemaic system, but well with the Copernican one: this might be a reason for formulating a heliocentric theory. Moreover, after putting the Sun (almost) at the centre, Copernicus immediately wrote of it:[6]

> He is rightly called the Lamp, the Mind, the Ruler of the Universe; Hermes Trismegistus names him the Visible God, Sophocles' Electra calls him the All-seeing. So the Sun sits as upon a royal throne ruling his children the planets which circle around him.

Hermes Trismegistus was a Greek god, often referred to by neo-Platonists, who supposed that he and his followers had written numerous surviving texts that might even allow for a synthesis of Platonic and Christian writings. Only in 1614 were these texts shown to be forgeries from AD 300.

Kuhn's thesis is attractive at first sight. However, this one remark is the only reference that Copernicus made to neo-Platonic ideas: in view of his lengthy struggle against equants, most recent historians consider it too flimsy a basis for a life's work.

As for the reception of the book, it is often said that the Catholic Church was particularly hostile. The sources of this belief have to do with many different currents in modern life, including various forms of hostility to the Catholic Church, and a much debated 'war' between science and religion. The current consensus among historians is that there was indeed to be hostility from the Church, but not until some decades after Copernicus's death. In fact, it was the Protestant leader Martin Luther who inveighed against Copernicus's work, on the grounds that it contradicts the Bible, and specifically Joshua's command (in Joshua X, xii) that the Sun stand still:

> On the day when the LORD gave the Amorites over to the Israelites, Joshua spoke to the LORD; and he said in the sight of Israel, "Sun, stand still at Gibeon, and Moon, in the valley of Aijalon." And the Sun stood still, and the Moon stopped.

As late as 1600 there was no official Church position on *De Revolutionibus*, and the leading Jesuit astronomer Clavius used it in his teaching without endorsing the heliocentric

[6]Quoted in (Kuhn 1957, 131).

hypothesis. But if there is no evidence of a difficult reception, there is also no doubt of its ultimate impact. How that came about, and how it depended on later work by others, will occupy us for the rest of this chapter and beyond.

12.2 Kepler

We now turn from the reception of Copernicus's ideas and look at the next major development in the study of astronomy that also belongs in any history of mathematics: the work of Johannes Kepler. It started in Denmark, on the island of Hven (now in Sweden), where the astronomer Tycho Brahe ran an observatory on money paid by the King of Denmark (see Figure 12.2).

Figure 12.2. Tycho Brahe in his observatory

Brahe was the leading astronomer of the late 16th century, and was famed for the accuracy of his observations, but he did not accept the Copernican system. He found it difficult to accept that the Earth might move, and move as swiftly as Copernicus required it to do, and he argued that if it did then there would be an observable stellar parallax, whereas there was none.[7] By this he meant that if the Earth is really in different places on a night in January and exactly six months later, then we should have to look for the stars in slightly different directions — but we do not. (In fact his argument was correct, but the stars were much further away than he imagined and the relevant observations were too delicate for his instruments to detect.)

Accordingly, Brahe devised a system in which all the planets go round a point near the Sun, but the Sun itself orbits the Earth. This is observationally equivalent to the Copernican heliocentric system, but keeps the Earth at the centre of the universe where tradition and the Church (after 1616) required it to be. Copernicus's conviction that the Sun should lie at the centre of all planetary orbits was overturned in Brahe's theory.

At this point a question in celestial mechanics becomes important: What makes the planets move? In Ptolemy's geocentric theory each planet is confined to a sphere and moves about on its system of epicycles and equants. All these spheres are nested with no gaps between one and the next, so one may say that each planet moves in its spherical 'shell', and the size of each shell is broadly determined by the observational data. The motion of the planets is then driven by the motion of the nested spherical shells. One might want more detail in this description, but it is not obviously unworkable. On the models of Copernicus and Brahe, however, the order of the shells is altered, their sizes are very different, gaps appear between the shells, and all of them penetrate the shell of the Moon. There is no way that the shells can form any physical mechanism for driving the planets in their orbits. So by 1601 a fairly accurate description of how the planets move was available, but the associated explanation of how this motion occurs was inadequate. The discovery of a more convincing explanation was to lead to the abandoning of the entire system of epicycles.

The architect of the new theory was Johannes Kepler, who succeeded Brahe as the Hapsburg Emperor Rudolph II's Imperial Mathematician. One consequence of the new theory was that revised astronomical tables were prepared by Brahe and Kepler and published in 1627; these Rudolfine tables superseded the Prutenic ones, and were to remain in use for the rest of the 17th century. The endless hours of work involved in making such tables is one reason why Kepler was so pleased to hear of Napier's logarithms, which promised to shorten the time spent in astronomical calculations.[8]

Kepler's philosophy. Kepler first studied astronomy with Michael Mästlin at Tübingen, where he learned the rudiments of Copernican theory, and soon wrote a book, the *Mysterium Cosmographicum* (The Mystery of the Universe) (1596), in which he ardently embraced Copernicanism as correct. His passion on the subject reflected the equal strength with which he believed that a divine Creator would have created a perfect, and hence a mathematical, universe.

In 1596 Kepler attempted to solve the problem of why there are just six planets. The solution he proposed was a striking geometrical one involving the five regular solids nested inside six spheres — one sphere for each planet (see Figure 12.3). As Kepler knew well, Euclid's *Elements*, Book XIII, had established that there are precisely five regular solids.

[7] *Stellar parallax* is the apparent variation in position of a star seen from different positions in the Earth's orbit, allowance being made for the Earth's daily rotation.

[8] We mentioned Kepler's involvement with logarithms in Chapter 11.

Even before the *Mysterium Cosmographicum* was published, Kepler had encountered opposition from the theologians in Tübingen. They forced the removal of a chapter in which Kepler sought to reconcile heliocentrism with Scripture, and urged Kepler[9]

> to proceed in the presentation of such hypotheses clearly only as a mathematician, who does not have to bother himself about the question whether these theories correspond to existing things or not.

[9] Quoted in (Caspar 1959, 68).

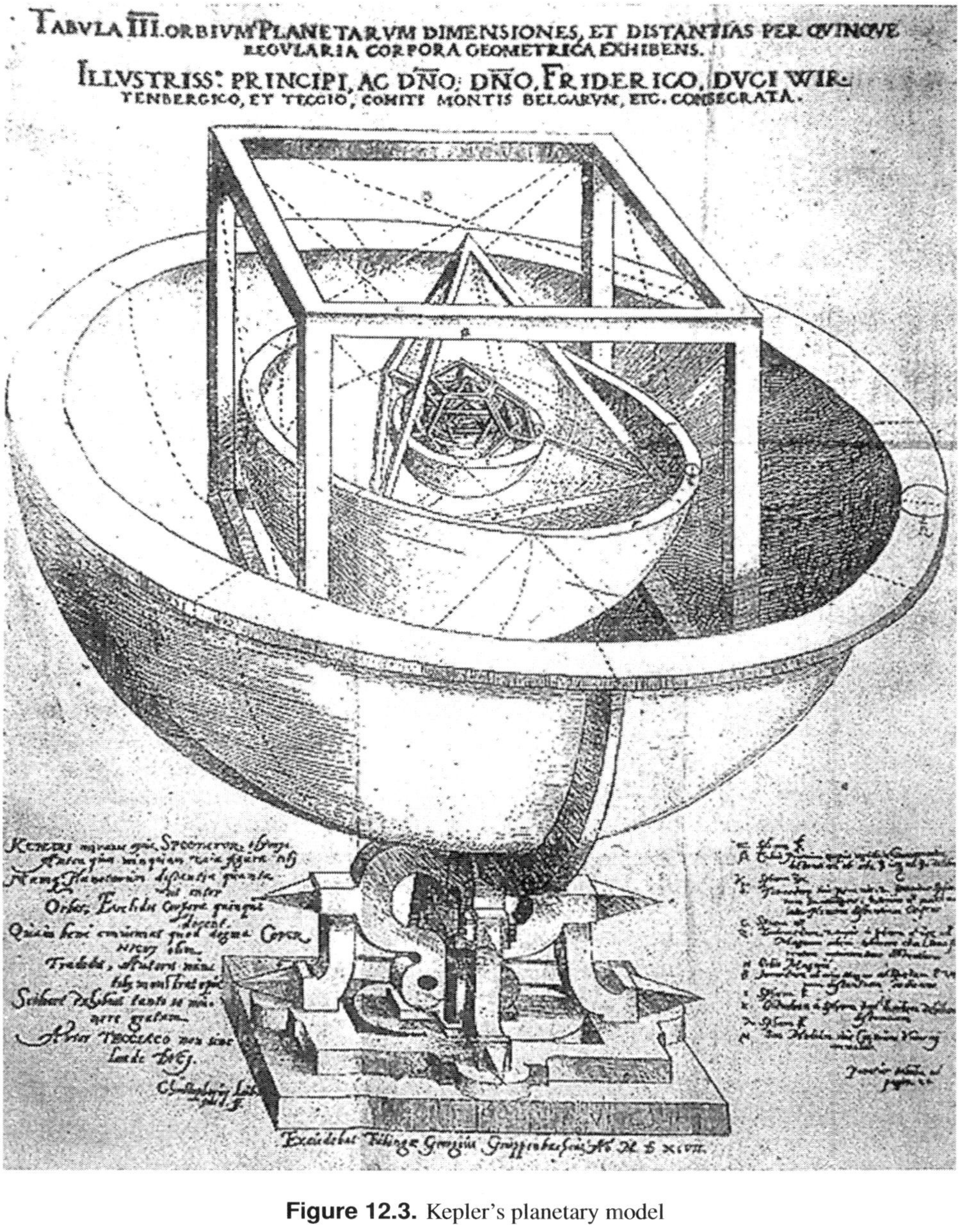

Figure 12.3. Kepler's planetary model

This was not something Kepler was willing to accept: one of the driving features of his work was a need to ground astronomy in a convincing physics, and this is a significant departure from Copernicus's approach.

Kepler regarded it as his humble duty to divine the mind of the Creator, but he soon discovered that the Creator must have made a plan other than the polyhedral one, because it did not quite fit the data, and in 1609 he rejected his description as being insufficiently accurate. It is notable that Kepler adhered to a view of God as a geometer, but also had a scrupulous respect for observational evidence: either belief alone is quite common, but to find them united in one person is rare — this is one reason why Kepler is so important and marks such a break with previous work.

His devotion was to carry him through years of laborious calculations. Moreover, he brought two mathematical skills to the task of understanding the heavens: a capacity for three-dimensional visualisation, and a prodigious ability for numerical computation (his many mistakes notwithstanding). He was also thoroughly familiar with Apollonius's *Conics*. His geometrical training and skill decisively affected his ideas about the role of geometry in physical theory, as did his ardent neo-Platonism, which we now briefly discuss.

Neo-Platonism. We recall from Chapter 6 that neo-Platonism derived ultimately from Plato's ideas, but that these ideas were reformulated many times by successive schools. One key belief that was related to mathematics, as emphasised by Proclus in his 5th-century commentary on Euclid's *Elements*, was that of the purity of mathematical ideas. In discussing whether mathematical knowledge derives from sense objects or exists independently of the physical world, Proclus took the latter (Platonic) view: we know mathematical truths, not by abstracting from objects in the world, but because they exist inherently in us:[10]

> We must therefore posit the soul as the generatrix of mathematical forms and ideas.

Proclus defended this opinion for several pages in a way that so excited Kepler that he incorporated a translation of them into his work *Harmonice Mundi* (Harmony of the World). His reference to soul was not to the Christian concept, for Proclus was not a Christian. There was, however, a strand of Christianity in neo-Platonism, deriving partly from St Augustine and partly from a prudent desire not to offend Christian orthodoxy.

Within neo-Platonism one could be an enthusiastic mystic, a passionate mathematician or (with Augustine) hostile to mathematics and its uses. Neo-Platonism was not a school, nor even a coherent philosophy, although it certainly involved a disposition to take abstract thought seriously. Kepler's brand of neo-Platonism was focused intently on the idea of the universe as a harmonious whole, which helps explain the title of his work, *Harmonice Mundi*. In a further passage from this work his neo-Platonism was manifested in his enthusiasm for harmony, his deep reverence for God, and his speculation that 'the nature of things ... was the finger of God', coupled with his devotion to abstract mathematical thought.[11]

Kepler's neo-Platonicism.

> There, beyond my expectations and with the greatest wonder, I found approximately the whole third book given over to the same consideration of celestial harmony, fifteen hundred years ago. But indeed astronomy

[10]See (Proclus 1970, 11).

[11]See F&G 10.A2.

> was far from being of age as yet; and Ptolemy, in an unfortunate attempt, could make others subject to despair, as being one who, like Scipio in Cicero, seemed to have recited a pleasant Pythagorean dream rather than to have aided philosophy. But both the crudeness of the ancient philosophy and this exact agreement in our meditations, down to the last hair, over an interval of fifteen centuries, greatly strengthened me in getting on with the job. For what need is there of many men? The very nature of things, in order to reveal herself to mankind, was at work in the different interpreters of different ages, and was the finger of God — to use the Hebrew expression; and here, in the minds of two men, who had wholly given themselves up to the contemplation of nature, there was the same conception as to the configuration of the world, although neither had been the other's guide in taking this route. But now since the first light eight months ago, since broad day three months ago, and since the sun of my wonderful speculation has shone fully a very few days ago: nothing holds me back. I am free to give myself up to the sacred madness, I am free to taunt mortals with the frank confession that I am stealing the golden vessels of the Egyptians, in order to build of them a temple for my God, far from the territory of Egypt. If you pardon me, I shall rejoice; if you are enraged, I shall bear up. The die is cast, and I am writing the book — whether to be read by my contemporaries or by posterity matters not. Let it await its reader for a hundred years, if God Himself has been ready for His contemplator for six thousand years.

His ideas extended to geometry and music, and in his *Mysterium Cosmographicum* he first attempted to explain musical harmonies by interpreting the ratios of the pitches involved as one or another aspect of the five regular solids. Later he became disenchanted with this theory and put forward another one, based on divisions of the Zodiac. In his *Harmonices Mundi* he based his theory of harmony on a distinction between 'knowable' and 'unknowable' polygons — those that are constructible by straight-edge and compasses alone, and those that are not. We need not enter into the details, but note that here too Kepler saw geometry as the means whereby God expressed his perfection.

The purity of mathematical ideas, and their origin in the soul, seem to have suggested to Kepler that they are sufficiently elevated to be appropriate to a divine purpose. However, Kepler's strongly-held view that God is a geometer had the atypical corollary that astronomical observations are to be taken very seriously indeed, since they are observations of what God truly made.

Kepler and Brahe. The *Mysterium Cosmographicum* impressed Kepler's peers, but it failed to convince them. Kepler decided to seek Tycho Brahe's help in getting access to the best possible observations of the motions of the planets, while Brahe, for his part, was happy to recruit another bright young astronomer to his team, which was already the largest of its kind in Europe. So Kepler came to Hven in February 1600 to work as one of Brahe's assistants. But Brahe did not entirely trust the younger man and gave him only the observations of Mars to work on, making him swear to keep them secret until he (Brahe) was ready to publish: it was to be a most fortunate restriction.

During his short time in Hven, which lasted less than a year, and with redoubled efforts after the death of Brahe in October 1601, Kepler struggled to fit various curves through the observed positions of Mars. Brahe had opposed Kepler's physical speculations, but now Kepler was unconstrained. None of the curves that Kepler devised quite satisfied him, in spite of all their ingenious constructions. Initially, solely as a means of simplifying his calculations, Kepler tried out the assumption that Mars moves in such a way that the line joining it to the Sun sweeps out equal areas in equal intervals of time. This idea is now known as *Kepler's second law* (the area law), so it is notable that he thought of it before he came to what we term his first law, and that for a while he entertained it as one of two conflicting simplifications of his problem.

At a later stage Kepler made a further simplification, and tried to fit the data to an elliptical orbit, with the Sun placed off-centre. He did not actually say that it was at one focus of the ellipse, although that is where his calculations put it. This is odd because Kepler was very familiar with the geometry of conics: indeed, he introduced the word 'focus' into mathematics. To his surprise, these two 'simplifications' fitted extremely well with Brahe's observations, so Kepler decided that they had to be correct: Mars, he proclaimed, moves in an elliptical orbit around the Sun, with the line joining it to the Sun sweeping out equal areas in equal times. These are now known as *Kepler's first and second laws*, respectively (see Figure 12.4). To sum up, Kepler stated two laws:

> *Kepler's first law*: all planets go round the Sun in elliptical orbits with the Sun at a focus of the ellipse.
> *Kepler's second law*: all planets sweep out equal areas in equal times.

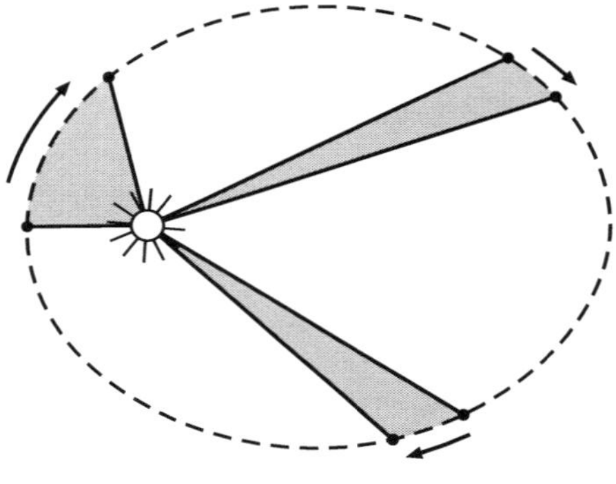

Figure 12.4. Kepler's first and second laws

Kepler's ideas about how the Sun drives the planets in their orbits now led him, via his derivation of a novel result about conic sections, to the happy realisation that the area law, when coupled with the observations, could be made to yield quite specific predictions. Moreover, this combination actually implies that the orbit has to be elliptical. It was this realisation – which Kepler supported by an argument involving approximations to the true orbit, and not by a rigorous proof — that gave him the satisfying harmony he required of a theory. It put the Sun, which Kepler wanted to be important in astronomy, at a geometrically and a physically important place with respect to the planets.

Finally, in 1609, Kepler published the first modern theory of planetary motion, in his *Astronomia Nova* (the full title of which, in English, is nothing less than 'New Astronomy, Based upon Causes, or Celestial Physics, Treated by means of Commentaries on the Motions of the Star Mars, from the Observations of Tycho Brahe, Gent'). This swept away the Ptolemaic and Copernican machinery, while Kepler's ellipses are still with us. We used the word 'modern', not because Kepler's theory was attentive to observations, for astronomy traditionally was, and not because the motions were correctly explained, because they were not, but because his explanation for planetary motion was based on the concept of a force that drove the planets in their orbits.[12] He also went on to list other regularities in planetary motion that surpassed the Copernican machinery by being simple and mathematically elegant.

If we read the preface to Kepler's *Astronomia Nova* we can get a good impression of the mathematical knowledge that he brought to astronomy and of the chief difficulties he encountered — one can imagine Kepler's contemporaries doing much the same.[13] The passage is quite long — 29 paragraphs — but it follows a remarkable chain of thought and documents one of the decisive moments in the emergence of modern mathematical science.

Kepler began by painting an engaging picture of how he himself approached the difficulties in understanding astronomical theory.[14]

Kepler on the motion of the planets.

[1] It is extremely hard these days to write mathematical books, especially astronomical ones. For unless one maintains the truly rigorous sequence of proposition, construction, demonstration, and conclusion, the book will not be mathematical; but maintaining that sequence makes the reading most tiresome, especially in Latin, which lacks the articles and gracefulness possessed by Greek when it is expressed in written symbols. Moreover, there are very few suitably prepared readers these days: the rest generally reject such works. How many mathematicians are there who put up with the trouble of working through the *Conics* of Apollonius of Perga? And yet that subject matter is the sort of thing that can be expressed more easily in diagrams and lines than can astronomy.

[2] I myself, who am known as a mathematician, find my mental forces wearying when, upon rereading my own work, I recall from the diagrams the sense of the proofs, which I myself had originally introduced from my own mind into the diagrams and text. But then when I remedy the obscurity of the subject matter by inserting explanations, it seems to me that I commit the opposite fault, of waxing verbose in a mathematical context.

[3] Furthermore, prolixity of phrases has its own obscurity, no less than terse brevity. The latter evades the mind's eye while the former distracts it; the one lacks light while the other overwhelms with superfluous glitter; the latter does not arouse the sight while the former quite dazzles it.

Kepler next described the table of contents, which he had designed as a careful guide to the whole book, before he continued by contrasting the Ptolemaic and Copernican theories, and set out his reasons for preferring the latter.

[4] The reader should be aware that there are two schools of thought among astronomers, one distinguished by its chief, Ptolemy, and by the assent of the large majority of the ancients, and the other attributed to more recent proponents, although it is the most ancient. The former treats the individual planets separately and assigns causes to the motions of each in its own orb, while the latter relates the planets to one another, and

[12] We discuss his ideas about this force below.

[13] See F&G 10.A1.

[14] See (Kepler 2015, 17–19, 20–22, 24–25, 34–35). For another translation see F&G 10.A1.

deduces from a single common cause those characteristics that are found to be common to their motions. This latter school is again subdivided. Copernicus, with Aristarchus of remotest antiquity, ascribes to the translational motion of our home the earth the cause of the planets' appearing stationary and retrograde, and I too subscribe to this opinion. Tycho Brahe, on the other hand, ascribes this cause to the sun, in whose vicinity he says the eccentric circles of all five planets are connected as if by a kind of knot (not physical, of course, but only quantitative). Further, he says this knot, as it were, revolves about the motionless earth, along with the solar body . . .

[5] My aim in the present work is chiefly to reform astronomical theory (especially of the motion of Mars) in all three forms of hypotheses, so that what we compute from the tables may correspond to the celestial phenomena. Hitherto, it has not been possible to do this with sufficient certainty. In fact, in August of 1608, Mars was a little less than four degrees beyond the position given by calculation from the *Prutenic Tables*. In August and September of 1593 this error was a little less than five degrees, while in my new calculation the error is entirely removed.

[6] Meanwhile, although I place this goal first and pursue it cheerfully, I also make an excursion into Aristotle's *Metaphysics*, or rather, I inquire into celestial physics and the natural causes of the motions [of the planets]. The eventual result of this consideration is the formulation of very clear arguments showing that only Copernicus's opinion concerning the world (with a few small changes) is true, that the other two are proved false, and so on . . .

[7] Now my first step in investigating the physical causes of the motions [of the planets] was to demonstrate that that coming together of the eccentrics occurs in no other place than the very center of the solar body (not some nearby point), contrary to what Copernicus and Brahe thought . . .

[8] With these things thus demonstrated by an infallible method, the previous step towards the physical causes is now confirmed, and a new step is taken towards them, most clearly in the theories of Copernicus and Brahe, and more obscurely but at least plausibly in the Ptolemaic theory.

[9] For whether it is the earth or the sun that is moved, it has certainly been demonstrated that the body that is moved is moved in a nonuniform manner, that is, slowly when it is farther from the body at rest, and more swiftly when it has approached this body.

[10] Thus the physical difference of these three opinions is now immediately apparent — by way of conjecture, it is true, but yielding nothing in certainty to conjectures of doctors on physiology or to any other natural philosophers whatever.

[11] Ptolemy is certainly hooted off the stage first. For who would believe that there are as many theories of the sun (so closely resembling one another that they are in fact equal) as there are planets, when he sees that for Brahe a single solar theory suffices for the same task, and it is the most widely accepted axiom in the natural sciences that Nature makes use of the fewest possible means?

> [12] That Copernicus is better able than Brahe to deal with celestial physics is proven in many ways ... For Brahe no less than for Ptolemy, besides that motion which is proper to it, each planet is still actually moved with the sun's motion, mixing the two into one, the result being a spiral. That it results from this that there are no solid orbs, Brahe has demonstrated most firmly. Copernicus, on the other hand, entirely removed the five planets from this extrinsic motion, the cause of the deception arising from the circumstances of observation. Thus the motions are still multiplied to no purpose by Brahe, as they were before by Ptolemy.

His aim, as set out in paragraph [5], is traditional: to get an accurate description of the motions of the planets. But then, in paragraph [6], he 'made an excursion' into the underlying physics, and claimed that the Copernican theory offers the best, indeed the only, chance of a systematic causal theory of the motion. Next, Kepler started to draw out the implications for a physical theory of planetary astronomy that would be based on some sort of action of the Sun.

> [13] Second, if there are no orbs, the conditions under which the intelligences and moving souls must operate will be made very difficult, since they have to attend to so many things to endow the planet with two intermingled motions. They will at least have to attend at one and the same time to the principles, centers, and periods of the two motions. But if the earth is moved, I show that most of this can be done with physical rather than animate faculties, namely, magnetic ones ...
>
> [14] For if the earth is moved, it has been demonstrated that it receives the laws of speed and slowness from the measure of its approach towards the sun and of its receding from the same. And in fact the same happens with the rest of the planets: they are urged on or held back according to the approach towards or recession from the sun. So far the demonstration is geometrical.
>
> [15] And now, from this very reliable demonstration, the conclusion is drawn, using a physical conjecture, that the source of the five planets' motion is in the sun itself. It is therefore likely that the earth is moved, since a likely cause of its motion is apparent.
>
> [16] That, on the other hand, the sun remains in place in the center of the world, is most probably shown by (among other things) its being the source of motion for at least five of the planets ... And it is more likely that the source of all motion should remain in place rather than move.

Some pages later Kepler went on to discuss how this motion might be caused, as follows, starting with evidence that the Earth attracts everything close to it.

> [17] The true theory of gravity rests upon the following axioms. All corporeal substance, to the extent that it is corporeal, has been so made as to be suited to rest in every place in which it is put by itself, outside the orb of power of a kindred body.
>
> [18] Gravity is a mutual corporeal disposition among kindred bodies to unite or join together; thus, the earth attracts a stone much more than the stone seeks the earth. (The magnetic faculty belongs to this order of things.) ...

> [19] If the earth should cease to attract its waters to itself, all the sea water would be lifted up, and would flow onto the body of the moon.
>
> [20] The orb of the attractive power in the moon is extended all the way to the earth, and calls the waters forth beneath the torrid zone, in that it calls them forth into its path wherever the path is directly above a place. This is imperceptible in enclosed seas, but noticeable where the beds of the oceans are widest and there is much free space for the waters' reciprocation ...
>
> [21] For it follows that if the moon's power of attraction extends to the earth, the earth's power of attraction all the more extends to the moon and far beyond, and accordingly, that nothing that consists to any extent whatever of terrestrial material, carried up on high, ever escapes the mightiest grasp of this power of attraction ...

If the Earth's attraction reaches at least as far as the Moon, argued Kepler, then it is plausible that the attraction of the Sun reaches to all the planets.

> [22] I had begun to say that in this work I treat all of astronomy by means of physical causes rather than fictitious hypotheses, and that I had taken two steps in my efforts to reach this central goal: first, that I had discovered that the planetary eccentrics all intersect in the body of the sun, and second, that I had understood that in the theory of the earth there is an equant circle, and that its eccentricity is to be bisected.
>
> [23] Now we come to the third step, namely, that it has been demonstrated with certainty ... that the eccentricity of Mars's equant is also to be precisely bisected, a fact long held in doubt by Brahe and Copernicus.
>
> [24] Therefore, by induction extending to all the planets ... it has been demonstrated that, since there are (of course) no solid orbs, as Brahe demonstrated from the paths of comets, the body of the sun is therefore the source of the power that drives all the planets around. Moreover, I have specified the manner [in which this occurs] as follows: that the sun, although it stays in one place, rotates as if on a lathe, and out of itself sends into the space of the world and immaterial *species* of its body, analogous to the immaterial *species* of its light. This *species* itself, as a consequence of the rotation of the solar body, also rotates like a very rapid whirlpool throughout the whole breadth of the world, and carries the bodies of the planets along with itself in a gyre, its grasp stronger or weaker according to the greater density or rarity it acquires through the law governing its diffusion.
>
> [25] Once this common power was proposed, by which all the planets, each in its own circle, are driven around the sun, the next step in my argument was to give each of the planets its own mover, seated in the planet's globe (you will recall that, following Brahe's opinion, I had already rejected solid orbs) ...
>
> [26] By this train of argument, the existence of the movers was established. The amount of work they occasioned me in Part 4 [of the book] is incredible to speak of, when in producing the planet-sun distances and the eccentric equations that were required, the results came out full of flaws and in disagreement with the observations. This is not because they were falsely

introduced, but because I had bound them to the millstones (as it were) of circles, under the spell of common opinion. Restrained by such fetters, the movers could not do their work.

[27] But my exhausting task was not complete until I had made a fourth step towards the physical hypotheses. By most laborious proofs and by computations on a very large number of observations, I discovered that the course of a planet in the heavens is not a circle, but an oval path, perfectly elliptical.

[28] Geometry stepped in, and taught that such a path results if we assign to the planet's own movers the task of making the planet's body reciprocate along a straight line extended towards the sun. Not only this, but also the correct eccentric equations, agreeing with the observations, resulted from such a reciprocation.

[29] Finally, the pediment was added to the structure, and proven geometrically: that such a reciprocation is apt to be the result of a magnetic corporeal faculty. Consequently, these movers belonging to the planets individually are shown with great probability to be nothing but properties of the planetary bodies themselves, like the magnet's property of seeking the pole and catching up iron. As a result, every detail of the celestial motions is caused and regulated by faculties of a purely corporeal nature, that is, magnetic, with the sole exception of the whirling of the solar body as it remains fixed in its space. For this a vital faculty seems required ...

Kepler was evidently familiar with the theory of conic sections (see Box 12.1) trailed at the beginning with a reference to Apollonius and consummated at the end by a reference to the elliptical orbit of Mars. But he found mathematics difficult to write — Kepler's description here is obviously heartfelt — and the calculations were very burdensome (in [27]), especially when the hypothesis turns out to be false (in [26]).

The same passage makes it clear that he claimed to have introduced three specifically new ideas that to astronomy: the reality of the Earth's motion ([13], [15]), the elliptical orbit of Mars ([27]) which replaces the tramping round in circles of [26], and the physical explanation in terms of forces ([21], [24]) that he provided for the motion of the planets.

Kepler's book contained not only the first accurate qualitative description of the motion of a planet, but also the first attempt to explain how they were moved by forces. In this connection, the historian of science R.S. Westfall made an interesting point. When Kepler published a revised edition of his *Mysterium Cosmographicum* in 1621, he said that his theories were originally based on the concept of an *anima motrix* (soul of motion), but were now based on that of a *vis motrix* (motive force). Westfall commented:[15]

> From anima motrix to vis, from the animistic to the mechanistic — the development of Kepler's thought foreshadowed the course of 17th century science.

This observation is worth keeping in mind as we proceed. By 1621 Kepler was thinking of it as a 'magnetic force', by analogy with the then-mysterious force of terrestrial magnetism, and he argued that such a force emanates from the Sun and pushes Mars around in its orbit. This dynamical theory was soon to be discarded, but it is significant because it extends a force found on the Earth to the heavens, and because it is the prototype of modern scientific theories in which a physical cause is sought for the motions.

[15] See (Westfall 1977, 10).

Box 12.1. Kepler on conics.

Kepler's interest in optics led him to consider how the different sections of a cone can be considered as members of a single family. He drew a fresh lesson from Apollonius's construction of conic sections as the intersection of a plane at various angles to one conic surface, and he noticed that each conic section changes smoothly into the next as the plane is continuously rotated. Thus,

> a straight line is transformed through infinitely many hyperbolas into a parabola, and then through infinitely many ellipses into a circle ... The bluntest hyperbola is a straight line, and the sharpest one, a parabola; the sharpest ellipse is a parabola, and the bluntest one, a circle.

To see what Kepler meant, point a torch at a ceiling. Directly upwards, the beam produces a circular patch of light, but if you tilt the torch, you produce a series of ellipses, then a parabola, and then a series of hyperbolas which become more and more like a straight line. In this experiment the ceiling acts as the plane section through the cone of the light beam.

Kepler argued that this principle of continuity ought also to apply to properties of the individual conic sections. He asked himself why the ellipse and the hyperbola have two focuses, whereas the parabola seems to have only one. His answer was that the parabola does have two focuses, but that one of them is infinitely distant, for as an ellipse becomes more elongated, its focuses move further apart so an ellipse becomes a parabola when its second focus becomes infinitely far away.

Both of these features, the recognition that the family of conic sections is continuous, and the introduction of the concept of a point at infinity, were important contributions to later developments in geometry, as well as standing Kepler in good stead for his use of conic sections in optics and in his 'war' with Mars.

The reception of Kepler's book is also interesting. Kepler's former teacher Mästlin, a lifelong Copernican, disliked it precisely because it invoked physical causes, and he wrote to Kepler in 1616 advocating a return to the traditional geometry and arithmetic associated with the Ptolemaic and Copernican theories. Younger men like Descartes and some of the next generation of astronomers, however, accepted the idea of a physical cause for the motion, even if not one of Kepler's type. By 1670 the elliptical orbits themselves, however caused, were widely accepted in books on astronomy in circulation, and any controversy was now concerned with how the motion was produced. The crucial discussion of that topic was to be provided by Isaac Newton in his *Principia Mathematica* of 1687.

In 1619, just as he was finishing his *Harmonice Mundi*, Kepler found one more law of motion for the planets, now known as *Kepler's third law*. He noticed that if a denotes the average distance of any planet from the Sun, and T is its period of revolution around the Sun, then a^3/T^2 is the same for every planet in the solar system. This law (often also called 'the three-halves power law') is easy to confirm and was generally accepted.

Kepler presents us with a fascinating picture of a man at work, drawing on deep philosophical and aesthetic convictions, as well as on a considerable technical mastery of mathematics. Had we the space, we could enrich this picture by considering his attitude to astrology, or by examining his discovery of some of the stellated regular solids.[16] But here

[16] See *Harmonices Mundi* (1619) in Kepler, *Gesammelte Werke* VI (1940), facing p. 79, and F&G 10.A3(c).

we can present him only as an exemplar of the mathematical scientist, and show how his mathematics enabled him to change the science of astronomy.

There were similar scholars all over Europe. Simon Stevin, who was born in Bruges in Flanders (now part of Belgium) in 1548 or 1549, and the powerful figure of Galileo are perhaps the two most directly comparable, but many engineers and architects were finding that their practice led them to extend their mathematical knowledge considerably. This is the context in which the journey towards the calculus began, and is also the background to the discovery of projective geometry. There were sometimes also philosophical undercurrents at work (as we saw with Kepler). But our next task is to consider Kepler's Italian contemporary, Galileo.

12.3 The language of nature: Galileo

Figure 12.5. Galileo Galilei (1564–1642)

Galileo was an almost exact contemporary of William Shakespeare, and his world was the northern Italian world of bustling independent cities, competitive merchants, princes and patrons, not unlike the setting of some of Shakespeare's plays. Although he became a professor of mathematics in his native Pisa in 1589, at the age of 25, the isolated life of scholarship was not for him. In 1609, by which time he had become professor of mathematics at Padua, some 25 miles from Venice, he heard of a Dutch invention of the telescope. He quickly learned how to make the lenses, and was soon the best lens-grinder and telescope-maker around. Apart from its novelty value, the telescope had a sound commercial use for spotting ships at sea, and for this reason Galileo was rewarded by the Venetian senate with a salary of 1000 florins a year for life, virtually doubling his previous salary.

In January 1610, out of sheer curiosity, he turned a telescope on Jupiter, then clearly visible, and watched it for some two weeks. Despite the distortions caused by his lenses, he could see many unfamiliar objects, most of them stars, but as Jupiter moved, at first three and later four objects could be seen moving with it. They lay in a plane, and moved gradually with respect to Jupiter. After carefully plotting their positions nightly, Galileo leapt to the answer — they were satellites of Jupiter. He was the first to discover new bodies in the domain of the planets.

With equal agility he realised that such a discovery needed to be properly marked. Already in contact with the rich and powerful Medici family of Florence, he wrote to assure them that their glory could be truly measured only if it were matched in the heavens — as now it could be, because he, Galileo, had discovered the four Medician satellites; there were, conveniently, four prominent Medici. They took up the offer, paid Galileo for it, and he transferred his affections from Venice to Florence. Careful observation, and the intellectual quality of mind capable of seeing four spots travelling with and around Jupiter as a miniature solar system, had led Galileo to his first great discovery.

Soon after, a former pupil named Benedetto Castelli wrote to Galileo from Prague to suggest a crucial experiment that would distinguish between the Copernican and Ptolemaic systems. Castelli pointed out that in the Ptolemaic system we should always see more or less the same amount of Venus, because it is constrained to move in an orbit that keeps it between the Earth and the Sun. Therefore the Earth–Venus–Sun angle is always large, and so Venus must always appear more or less crescent-shaped, and its apparent size cannot vary much. However, in the Copernican system the Sun is now at the centre of the orbit of Venus, the Earth–Venus–Sun angle can vary greatly, and so Venus would show much more varied phases and its apparent size would vary greatly (see Figure 12.6).

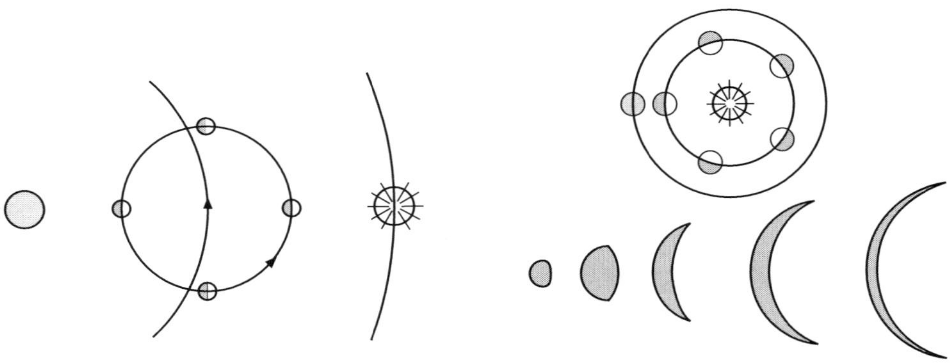

Figure 12.6. The phases of Venus: Ptolemaic system (left) and Copernican system (right)

Had Galileo, asked Castelli, detected these phases with his telescope? To his consternation, he had not, but on that very night he did. His luck held, for now Venus was prominent in the sky, and in a few nights' observations he had discovered the phases of Venus, and with them observational evidence in favour of the Copernican system over the Ptolemaic one. So now the crucial question was: did his telescope allow him to show that the Earth really moves, or was the Earth at rest (as the Church and astronomers such as Tycho Brahe believed) with its 'motion' in the Copernican system being purely a way of simplifying the calculations? Upon this question was to hang Galileo's rise to fame and his eventual punishment by the Catholic Church in 1633.

Galileo's observations of the planets could not discriminate between the geocentric theory advocated by Brahe and a heliocentric one. This is because in both theories every planet orbits the Sun, and so no planetary observation can distinguish between them. Nor did his best theoretical argument in that direction get observational support. Like Tycho Brahe before him, Galileo argued that the existence of stellar parallax would confirm the reality of the Earth's motion and refute Brahe's ideas. Alas for Galileo, its detection required much more accurate telescopes than he had available.

This inability to detect a parallax meant that the simple geometrical arguments that underpinned Galileo's astronomical work were successful only when their conclusions served to enrich mankind's view of the universe: Jupiter has satellites and Venus has phases. But when his theory that the Earth moves and the heavens are at rest conflicted with everyday experience and the established doctrine of the Church, it fared less well. This was to make Galileo more and more certain that it would be necessary to elucidate the concept of motion itself, but in the short term his stock rose as it was recognised that his arguments about the revelations of the telescope could be correct.

During a trip to Rome in 1611, undertaken to spread the news of his discoveries, he was elected a member of Rome's scientific society, the Accademia dei Lincei, founded in 1603. On this occasion the Lincei chose to raise their numbers from five to eleven, although they continued to exclude university professors who tended to adhere to the old Aristotelian interpretations — with scandalously little interest in rigour at times, as Galileo enjoyed pointing out. Moreover, in 1611 the Jesuits in Rome under Clavius became convinced about the reality of what one saw through a telescope. But philosophers in universities and the Jesuits of the Collegio Romano still disagreed with the new scientists in academies and elsewhere on what astronomical observations implied. Specifically, did the Earth move, or was it really at rest?

Galileo's naive quantifications and elementary geometry (which we look at in more detail below) had led him to the same intellectual position as Kepler: Copernicus's view that the Earth and all the other planets go round the Sun was physically correct. Both Kepler and Galileo looked for simple physical explanations of phenomena; both could say that God was a geometer and that geometry was the key to the Universe, albeit with strikingly different interpretations of that claim. But beneath these important points of agreement lay interesting disagreements. Westfall has made the shrewd observation that:[17]

> Kepler, who treated the solar system in mechanical terms and sought to comprehend the physical forces that govern its motions, employed a system of mechanics based on principles overthrown by Galileo. Galileo, who formulated the basic concepts of the new mechanics, ignored the problems to which Kepler's celestial mechanics addressed itself and held that the planets move naturally in circular orbits.

Those who denied that the Earth moves had recourse to the prevailing and well-developed Aristotelian theory of motion, which begins very seductively. According to that theory, ordinary objects are kept moving by forces — remove the force and they stop — whereas celestial objects move in circles, this being their 'natural' motion, as the theory termed it. This agrees with our day-to-day experience. Locally, objects stay where you put them, and to move them requires an effort; the heavens move all the time without apparent effort.

The dispute about the motion of the Earth therefore led directly to investigations into the theory of motion. The modern theory of dynamics, begun by Galileo and given its

[17]See (Westfall 1971, 18).

lasting form by Newton, is very different from the Aristotelian one. It asserts that terrestrial or celestial objects will always be in uniform motion in a straight line when left alone, and that it takes a force to accelerate them, slow them down, or change their direction: forces like friction or air resistance slow objects down when they are moving, while gravity keeps them in orbit. This theory is not immediately plausible — everyone has to be taught it — and to see it being discovered, and thus to see new mechanical explanations being proposed for natural phenomena, we must go back to the earlier work of Galileo.

In order to analyse motion, Galileo tried to quantify it. One problem that he confronted was that falling bodies move too fast for accurate measurement. He ingeniously solved this problem by rolling balls down gently sloping planes and measuring the distances rolled in given times, apparently using his musical skills to do the timing. Even so, his route to success was slow and painful. In 1604 he had been able to state the *distance law* for falling bodies:

the distance fallen is proportional to the square of the time of the descent.

By 1609, he had discovered the *time law*:

at any instant the velocity is proportional to the elapsed time.[18]

From 1608 onward, Galileo's preferred world was an idealised one of unencumbered motion amenable to simple mathematical laws, and not the cluttered world of daily life. The sole reason for accepting an idealised world is that mathematics applies in it: you pay the price of artificial assumptions, but you reap the reward of simple physical laws. If your conclusions turn out to be empirically acceptable, then your original assumptions may not have been artificial after all but full of insight. If so, a further stage of mathematics is suggested in which the physical concepts you have isolated are made to explain why the phenomena develop as they do. In this way, science, although empirical, was to grow out of the certain knowledge of the mathematician. Science is correct (so its practitioners would argue), because what is deduced is empirically sound, and because its means of deduction, being mathematical, are logically sound.

An example may help to explain what this means. One of Galileo's earliest observations was his discovery in 1602 that the period of oscillation of a pendulum is independent of the length of arc through which it swings: this observation is popularly supposed to have been made while he was contemplating the chandeliers in Pisa Cathedral. Upon what, then, does the period of swing depend? The answer, Galileo discovered, is that the period is proportional to the square root of the length of the pendulum, but independent of the weight of the pendulum bob (the chandelier).

The idealisation here is to neglect air resistance, which in any real situation may either slow down a pendulum until it stops, or move it in a variety of ways by means of air currents. Notice, too, that the elaborate chandeliers have been idealised as points, and that their heavy supporting chains are taken to be weightless. Once the idealisation is made, the reward is the simple correlation of period and length. This reward then invites a mathematical proof, which would elucidate suitable physical concepts and articulate them mathematically. When that stage had been reached, one could hope to invert the whole argument and prove that the time taken for a pendulum to complete a full swing is independent of the size of the arc through which it swings. This confers the status of certain

[18]Compare this with the mean-speed law that we mentioned when discussing the work of Oresme; Galileo seems to have known of some of these medieval discussions.

knowledge on an empirical fact. But in 1602 Galileo contented himself with the numerically tidy observational law, and did not also derive it mathematically.

Galileo's geometry. Galileo believed that geometry is crucial to any attempt to understand nature, and that the way forward was to idealise and simplify the natural processes that he was trying to explain. These might seem to be markedly distinct kinds of activity. A useful way to approach them is to look what Galileo himself said later in his life, in *The Assayer* (1623), the *Dialogue Concerning the Two Chief World Systems* (1632), and the *Two New Sciences* (1638). (See Figure 12.7.)

In *The Assayer* Galileo defended his use of mathematics in terms that have become famous (if not always quoted correctly):[19]

> Philosophy is written in this grand book, the Universe — which stands continually open to our gaze. But the book cannot be understood unless one first learns to comprehend the language and read the letters in which it is composed. It is written in the language of mathematics, and its characters are triangles, circles, and other geometric figures without which it is humanly impossible to understand a single word of it; without these, one wanders about in a dark labyrinth.

In the second of these, the *Dialogue*, which concerned debates about the motion of the Earth, he indicated why he valued geometry so highly. Since this book risked, and eventually incurred, the wrath of the Church, we may assume that he was picking his words with care. This is even more true of the final book, written towards the end of his life, when he was under house arrest and knew very well who his enemies were, and he took the opportunity to spell out his arguments most carefully.

The universe, he said, is a book written in the language of mathematics, and such theorems as we possess are objectively and certainly true. Furthermore, we can understand the universe only if we read it mathematically, forsaking all human authorities, for mathematics is divinely true.

How, then, did Galileo connect geometry with observation? How does this help us to understand the universe? To answer these questions, we may look at a passage from his *Two New Sciences*. The sciences that Galileo had in mind were the strength of materials and the motion of objects. To expound them he continued the dialogue from his earlier book, the *Dialogue Concerning the Two Chief World Systems*. There are the same three characters in each book: Salviati, who represents Galileo himself; Sagredo, an intelligent layman (named after one of Galileo's friends); and Simplicio who, as his name suggests, has generally misunderstood the issues hitherto and is the spokesman for the outdated Aristotelian views that Galileo wished to discredit, and which, he thought, had too much support among the Jesuits. By the time of the *Two New Sciences*, however, Simplicio has at least caught up with the ideas that Galileo had been promoting in his youth.

In the *Two New Sciences* Galileo's characters discuss several topics over several days. One was Galileo's theory of pendulums, and Galileo put his old observations in Salviati's mouth.[20]

Galileo on the motion of a pendulum.

> SALVIATI: We come now to the other questions, relating to pendulums, a subject which may appear to many exceedingly arid, especially to those

---

[19]In (Drake 1957, 237–238) and F&G 10.B1.

[20]Galileo, *Two New Sciences*, 94–97, in F&G 10.B2.

Figure 12.7. Frontispiece to Galileo's *Dialogue Concerning the Two Chief World Systems*

philosophers who are continually occupied with the more profound questions of nature. Nevertheless, the problem is one which I do not scorn. I am encouraged by the example of Aristotle whom I admire especially because he did not fail to discuss every subject which he thought in any degree worthy of consideration.

Impelled by your queries I may give you some of my ideas concerning certain problems in music, a splendid subject, upon which so many eminent men have written: among these is Aristotle himself who has discussed numerous interesting acoustical questions. Accordingly, if on the basis of some easy and tangible experiments, I shall explain some striking phenomena in the domain of sound, I trust my explanations will meet your approval.

SAGREDO: I shall receive them not only gratefully but eagerly. For, although I take pleasure in every kind of musical instrument and have paid considerable attention to harmony, I have never been able to fully understand why some combinations of tones are more pleasing than others, or why certain combinations not only fail to please but are even highly offensive. Then there is the old problem of two stretched strings in unison; when one of them is sounded, the other begins to vibrate and to emit its note; nor do I understand the different ratios of harmony and some other details.

SALVIATI: Let us see whether we cannot derive from the pendulum a satisfactory solution of all these difficulties. And first, as to the question whether one and the same pendulum really performs its vibrations, large, medium, and small, all in exactly the same time, I shall rely upon what I have already heard from our Academician. He has clearly shown that the time of descent is the same along all chords, whatever the arcs which subtend them, as well along an arc of $180°$ (i.e., the whole diameter) as along one of $100°, 60°, 10°, 2°, \frac{1}{2}°$, or $4'$. It is understood, of course, that these arcs all terminate at the lowest point of the circle, where it touches the horizontal plane.

If now we consider descent along arcs instead of their chords then, provided these do not exceed $90°$, experiment shows that they are all traversed in equal times; but these times are greater for the chord than for the arc, an effect which is all the more remarkable because at first glance one would think just the opposite to be true. For since the terminal points of the two motions are the same and since the straight line included between these two points is the shortest distance between them, it would seem reasonable that the motion along this line should be executed in the shortest time; but this is not the case, for the shortest time — and therefore the most rapid motion — is that employed along the arc of which this straight line is the chord.

As to the times of vibration of bodies suspended by threads of different lengths, they bear to each other the same proportion as the square roots of the lengths of the thread; or one might say the lengths are to each other as the squares of the times; so that if one wishes to make the vibration-time of one pendulum twice that of another, he must make its suspension four times as long. In like manner, if one pendulum has a suspension nine times as long as another, this second pendulum will execute three vibrations during each one of the first; from which it follows that the lengths of the suspending cords bear to each other the [inverse] ratio of the squares of the number of vibrations performed in the same time.

SAGREDO: You give me frequent occasion to admire the wealth and profusion of nature when, from such common and even trivial phenomena, you derive facts which are not only striking and new but which are often far removed from what we would have imagined. Thousands of times I have observed vibrations especially in churches where lamps, suspended by long cords, had been inadvertently set into motion; but the most which I could infer from these observations was that the view of those who think that such vibrations are maintained by the medium is highly improbable: for, in that case, the air must needs have considerable judgement and little else to do but kill time by pushing to and fro a pendant weight with perfect regularity. But I never dreamed of learning that one and the same body, when suspended from a string a hundred cubits long and pulled aside through an arc of $90°$, or even $1°$, or $\frac{1}{2}^{\circ}$, would employ the same time in passing through the least as through the largest of these arcs; and, indeed, it still strikes me as somewhat unlikely. Now I am waiting to hear how these same simple phenomena can furnish solutions for those acoustical problems — solutions which will be at least partly satisfactory.

SALVIATI: First of all one must observe that each pendulum has its own time of vibration so definite and determinate that it is not possible to make it move with any other period than that which nature has given it. For let any one take in his hand the cord to which the weight is attached and try, as much as he pleases, to increase or diminish the frequency of its vibrations; it will be time wasted.

To understand how Galileo has yoked observation and geometry here, it helps to rearrange the dialogue a little. First, Sagredo observed that chandeliers often move regularly, but claimed never to have seen the isochrony (the time of the swing of a pendulum depends only on the length of the pendulum, not on the angle through which it swings). Then Salviati observed the isochrony: 'each pendulum has its own time of vibration', adding that 'our Academician [Galileo] has clearly shown' the same thing about the time of descent of objects rolling along chords, while experiment shows isochrony for descent along arcs. The word 'shown' carries a mathematical sense here, which is spelled out at length later in the book, but the structure of the dialogue suggests this, for Salviati is certain of isochrony while Sagredo still finds it 'somewhat unlikely'. Salviati's extra degree of conviction derives from the mathematical proofs that Galileo has found for his claims.

The historian Stillman Drake has observed that:[21]

> *The Assayer* marked a crucial point in the history of Galileo's thought. Before, he had spoken as the experimental scientist; later he was to speak as a theoretical scientist. In this work he speaks as the philosopher of science.

Thus Galileo's work became progressively more mathematical, and the harvest he reaped was a set of conclusions that surpassed the observational skills of the day. To take another example from the *Two New Sciences*, we note that Galileo claimed that the path of a projectile is a parabola, which no-one knew before, and said that he could prove it.

The historian Maurice Clavelin concluded a detailed study of Galileo's science with these words, which further underline Galileo's real significance to the history of mathematics:[22]

---

[21] See (Drake 1957, 227).

[22] See (Clavelin 1974, 486).

> ... the manner in which he employed the resources of traditional geometry was altogether remarkable. Projectile motion was turned into an object of mathematical reasoning ... It was therefore not simply in his definition but throughout his analysis that Galileo asserted the dominance of reason over motion. There is perhaps no better example of the powerful impetus he gave to the science of mechanics.

Mathematised motion. In the *Two New Sciences* mathematics intervenes in a decisive way to enable scientists to organise their observations into a deductive hierarchy. The third day of the discussions is given over to the science of local motion, and the dialogue is interrupted for pages at a time as Salviati expounds Galileo's ideas, somewhat in the pattern of Euclid's *Elements*. First, consequences are drawn from the definition of uniform motion; then Galileo defined what he meant by uniformly or naturally accelerated motion (the definition encapsulates his time law for falling bodies); then he began to deduce consequences of this definition.[23]

Galileo's account of falling bodies.

> For I think no one believes that swimming or flying can be accomplished in a manner simpler or easier than that instinctively employed by fishes and birds.
>
> When, therefore, I observe a stone initially at rest falling from an elevated position and continually acquiring new increments of speed, why should I not believe that such increases take place in a manner which is exceedingly simple and rather obvious to everybody? If now we examine the matter carefully we find no addition or increment more simple than that which repeats itself always in the same manner. This we readily understand when we consider the intimate relationship between time and motion; for just as uniformity of motion is defined by and conceived through equal times and equal spaces (thus we call a motion uniform when equal distances are traversed during equal time-intervals), so also we may, in a similar manner, through equal time-intervals, conceive additions of speed as taking place without complication; thus we may picture to our mind a motion as uniformly and continuously accelerated when, during any equal intervals of time whatever, equal increments of speed are given to it. Thus if any equal intervals of time whatever have elapsed, counting from the time at which the moving body left its position of rest and began to descend, the amount of speed acquired during the first two time-intervals will be double that acquired during the first time-interval alone; so the amount added during three of these time-intervals will be treble; and that in four, quadruple that of the first time-interval. To put the matter more clearly, if a body were to continue its motion with the same speed which it had acquired during the first time-interval and were to retain this same uniform speed, then its motion would be twice as slow as that which it would have if its velocity had been acquired during two time-intervals.
>
> And thus, it seems, we shall not be far wrong if we put the increment of speed as proportional to the increment of time; hence the definition of motion which we are about to discuss may be stated as follows: A motion is said to be uniformly accelerated, when starting from rest, it acquires, during equal time-intervals, equal increments of speed.

[23]Galileo *Two New Sciences*, 161–162, in F&G 10.B3.

Galileo then turned to derive his time law, which he obtained as a consequence of a preliminary result.[24]

The time and distance laws for a falling body.

> *Theorem I, Proposition I The time in which any space is traversed by a body starting from rest and uniformly accelerated is equal to the time in which that same space would be traversed by the same body moving at a uniform speed whose value is the mean of the highest speed and the speed just before acceleration began.*

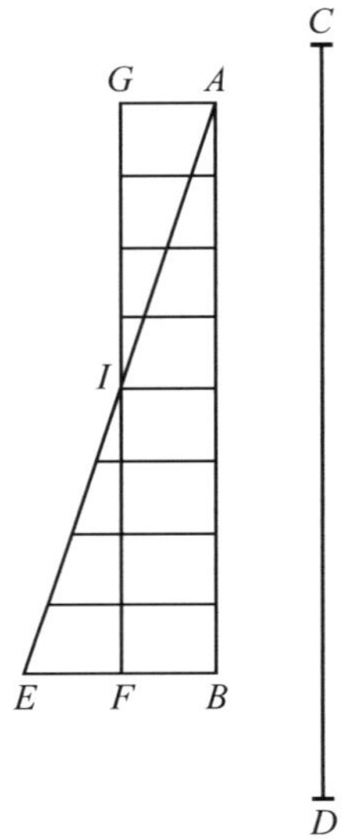

Figure 12.8. On falling bodies

> Let us represent by the line AB the time in which the space CD is traversed by a body which starts from rest at C and is uniformly accelerated [see Figure 12.8]; let the final and highest value of the speed gained during the interval AB be represented by the line EB drawn at right angles to AB; draw the line AE, then all lines drawn from equidistant points on AB and parallel to BE will represent the increasing values of the speed, beginning with the instant A. Let the point F bisect the line EB; draw FG parallel to BA, and GA parallel to FB, thus forming a parallelogram AGB which will be equal in area to the triangle AEB, since the side GF bisects the side AE at the point I; for if the parallel lines in the triangle AEB are extended to GI, then the sum of all the parallels contained in the quadrilateral is equal to the sum of those contained in the triangle AEB; for those in the triangle IEF are equal to those contained in the triangle GIA, while those included in the trapezium $AIFB$ are common. Since each and every instant of time in the time-interval AB has its corresponding point on the line AB, from which points parallels drawn in and limited by the triangle AEB represent the increasing values of the growing velocity, and since parallels

[24] Galileo *Two New Sciences*, 173–176, in F&G 10.B4.

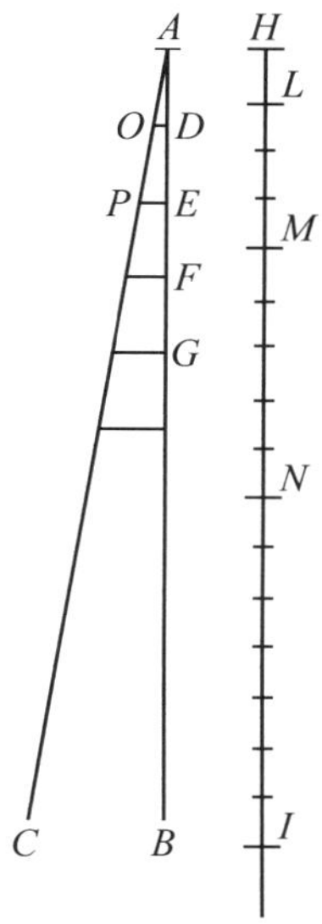

Figure 12.9. Accelerating bodies

contained within the rectangle represent the values of a speed which is not increasing, but constant, it appears, in like manner, that the momenta assumed by the moving body may also be represented, in the case of the accelerated motion, by the increasing parallels of the triangle AEB, and, in the case of the uniform motion, by the parallels of the rectangle GB. For, what the momenta may lack in the first part of the accelerated motion (the deficiency of the momenta being represented by the parallels of the triangle AGI) is made up by the momenta represented by the parallels of the triangle IEF.

Hence it is clear that equal spaces will be traversed in equal times by two bodies, one of which, starting from rest, moves with a uniform acceleration, while the momentum of the other, moving with uniform speed, is one-half its maximum momentum under accelerated motion. Q.E.D.

Galileo now turned to the distance law.

Theorem II, Proposition II The spaces described by a body falling from rest with a uniformly accelerated motion are to each other as the squares of the time-intervals employed in traversing these distances.

Let the time beginning with any instant A be represented by the straight line AB in which are taken any two time-intervals AD and AE [see Figure 12.9]. Let HM represent the distance through which the body, starting from rest at H, falls with uniform acceleration. If HL represents the space traversed during the time-interval AD, and HM that covered during the interval AE, then the space MH stands to the space LH in a ratio which is the square of the ratio of the time AE to the time AD; or we may say simply that the distances HM and HL are related as the squares of AE and AD.

Draw the line AC making any angle whatever with the line AB; and from the points D and E, draw the parallel lines DO and EP; of these two lines, DO represents the greatest velocity attained during the interval AD, while EP represents the maximum velocity acquired during the interval AE. But it has just been proved that so far as distances traversed are concerned it is precisely the same whether a body falls from rest with a uniform acceleration or whether it falls during an equal time-interval with a constant speed which is one-half the maximum speed attained during the accelerated motion. It follows therefore that the distances HM and HL are the same as would be traversed, during the time-intervals AE and AD, by uniform velocities equal to one-half those represented by DO and EP respectively. If, therefore, one can show that the distances HM and HL are in the same ratio as the squares of the time-intervals AE and AD, our proposition will be proven.

But in the fourth proposition of the first book it has been shown that the spaces traversed by two particles in uniform motion bear to one another a ratio which is equal to the product of the ratio of the velocities by the ratio of the times. But in this case the ratio of the velocities is the same as the ratio of the time-intervals (for the ratio of AE to AD is the same as that of $\frac{1}{2}EP$ to $\frac{1}{2}DO$ or of EP to DO). Hence the ratio of the spaces traversed is the same as the squared ratio of the time-intervals. Q.E.D.

Evidently then the ratio of the distances is the square of the ratio of the final velocities, that is, of the lines EP and DO, since these are to each other as AE to AD .

Corollary I Hence it is clear that if we take any equal intervals of time whatever, counting from the beginning of the motion, such as AD, DE, EF, FG, in which the spaces HL, LM, MN, NI are traversed, these spaces will bear to one another the same ratio as the series of odd numbers, $1, 3, 5, 7$; for this is the ratio of the differences of the squares of the lines [which represent time], differences which exceed one another by equal amounts, this excess being equal to the smallest line [viz. the one representing a single time-interval]: or we may say [that this is the ratio] of the differences of the squares of the natural numbers beginning with unity.

While, therefore, during equal intervals of time the velocities increase as the natural numbers, the increments in the distances traversed during these equal time-intervals are to one another as the odd numbers beginning with unity.

How might we think of this problem in the light of Galileo's analysis? Let us suppose that a stone is dropped from rest and (to make it simpler to understand) let us suppose that the acceleration acts every millisecond — think of the stone as getting a tap every millisecond, and so 1000 taps every second; later we will replace this assumption by an argument that allows the acceleration to act all the time. Suppose moreover that the amount of acceleration is independent of the velocity of the stone, and, on some suitable definition of a 'centimetre', this acceleration adds 1 cm per second to the velocity. Then:

- In the first millisecond the stone is given a velocity of 1 cm per millisecond, and so falls 1 cm.
- In the 2nd millisecond its velocity is 2 cm per millisecond, so in that millisecond it falls 2 cm and has now fallen a total of $1 + 2 = 3$ cm.

- In the 3rd millisecond its velocity is 3 cm per millisecond, so in that millisecond it falls 3 cm and has now fallen $1 + 2 + 3 = 6$ cm.

If we continue this argument, we see that at the start of the nth millisecond it acquires a velocity of n centimetres per millisecond, and at the end of the nth millisecond it has fallen a total of

$$1 + 2 + 3 + \cdots + (n-1) + n = \tfrac{1}{2}n(n+1)\ \text{cm}.$$

Its velocity is now n cm per millisecond, so the average of its highest and lowest velocities is $\frac{1}{2}n$ cm per millisecond, as Galileo claimed.

How do we obtain the square of the time, when we have expressions of the form $\frac{1}{2}n(n+1)$? The difference between the distances $\frac{1}{2}n(n+1)$ cm and $\frac{1}{2}n^2$ cm is $\frac{1}{2}n$ cm, but if the bodies fall for as little as one second then n is 1000, the numbers are

$$\tfrac{1}{2}n(n+1) = 500,500 \ \text{and}\ \tfrac{1}{2}n^2 = 500,000\ \text{cm},$$

and the difference between them is 500, or one part in a thousand. And if we imagine that we make one million taps per second, so

$$n = 1{,}000{,}000, \ \text{then}\ \tfrac{1}{2}n(n+1) = 500{,}000{,}500{,}000, \ \text{and}\ \tfrac{1}{2}n^2 = 500{,}000{,}000{,}000.$$

The difference between these numbers is one part in a million, which is negligible even if you imagine that you can measure time to one part on one thousandth of a second (an impossible feat in Galileo's day). So we can say that the distances of $\frac{1}{2}n(n+1)$ cms and $\frac{1}{2}n^2$ cms are nearly the same, and that our assumption that the acceleration is added at discrete intervals is the cause of the discrepancy. We can approximate the acceleration better and better by tapping the stone more and more often: we then make the unit of time smaller and smaller, thus making the number n larger and larger for each second of time. This means that the distances $\frac{1}{2}n(n+1)$ are relatively better approximated by $\frac{1}{2}n^2$.

In this way, we can believe that when we suppose the acceleration to act continually the following is true: the time that a constantly accelerating stone S dropped from rest takes to cover a given distance is equal to the time that another stone U moving at constant velocity would take to cover the same distance, if that velocity is one-half the final velocity of the stone S. In more modern notation, this theorem concerns a body released from rest and accelerating uniformly. Suppose that it covers a distance s in t seconds, and at that moment has velocity v. Then, says the theorem,

> the same distance s is covered by a body moving for the same time with a constant velocity of $\frac{1}{2}v$: in symbols, $s = \frac{1}{2}vt$.

It is easy to feel a little let down – after all, the conclusion is rather simple. But one's next response is one of excitement: this result has been *proved*.

Note, moreover, that in n milliseconds a stone moving with constant speed $\frac{1}{2}n$ cm per millisecond travels $\frac{1}{2}n^2$ cm; it follows by the same argument that the distance that a stone falls is proportional to the square of the length of the time that it has been falling. In modern terms, if s_1 and s_2 are the distances traversed in times t_1 and t_2, respectively, then $s_1 : s_2 = t_1^2 : t_2^2$; this is Galileo's distance law.

Mathematics does not just appear in the statements of the results. Even a glance at Galileo's proofs shows that he was using mathematics in a substantial way to prove things.

It is perhaps most interesting to see how Galileo proved that the path of a projectile must be a parabola, which Salviati had outlined on the fourth day; most of the third day had been taken up with a geometrical discussion of balls rolling on inclined planes. Sagredo is rather nervous, remarking that 'I have not gone very far in my study of Apollonius' and poor Simplicio despairs that the little Euclid he had learned since their previous discussion would not enable him to understand anything.[25] Salviati neatly extracts from Apollonius's *Conics* precisely the theorems that Galileo needs, remarking with what may be charitable intent that

> all real mathematicians assume on the part of the reader perfect familiarity with at least the elements of Euclid.[26]

Galileo then proceeded by recalling his theorems on motion with constant velocity and on motion with constant acceleration that we have just looked at. He then argued that these motions can be thought of as proceeding simultaneously without interfering with one another: no acceleration is produced sideways, for example, although the body is moving sideways and accelerating downwards. Finally, he assumed a modest knowledge of the theory of conic sections, and in particular, the parabola (see Chapter 5).

Modern readers need not feel as nervous as Sagredo or as defensive as Simplicio! The point to feel sure of is Galileo's reliance on mathematics as a tool. We shall not verify his proofs.

So, towards the end of his life, Galileo presented in one package a theory of acceleration, velocity, position, and their variations with time, all held together by mathematical deductions. This theory united the distance law and the time law for falling bodies, which even Galileo had originally not connected, the one being discovered (as we saw) in 1604, and the other five years later. The internal consistency of the package commended it over its rival, the old Aristotelian theory of motion, and this consistency derived in part from its mathematical form. Contrariwise, because this theory was productive as a piece of physics, it invited subsequent mathematicians to improve it, and in so doing to create a mathematics capable of articulating the advancing science.

The hardest thing for us to understand, however, is that Galileo's theory is an accurate description of motion, but it is not a causal theory — you can see this from his adherence to circular orbits in astronomy. For, although Galileo took the decisive step of breaking with Aristotelian ideas, he was unable to give a convincing analysis of why bodies accelerate under their own weight in a particular way. His erosion of the distinction between terrestrial and celestial bodies was in this respect incomplete, and his physics applied best to objects near the surface of the Earth, although conceptually it was consistent with the idea of a genuinely moving Earth.

We have seen the boldness and clarity with which Galileo presented his views, and specifically his opinion that the motion of the Earth is a physical reality. Tragically for him, the Roman Catholic Church saw fit to make an example of him. Historians still discuss the reasons behind the Church's decision to act as it did, which seem to have been complex. Galileo, after all, had become famous as 'the star of Italy' and his most influential patron was Cardinal Barberini, who had become Pope Urban VII in 1623. But the fact remains that the Church, after deliberating for a year, first approved the publication of Galileo's *Dialogue Concerning the Two Chief World Systems* in 1630, and then regretted doing so and summoned Galileo before the Inquisition.

Could the Church leaders have failed to realise that the book contained an argument for the reality of the Earth's motion, based on a theory of the tides (which we shall not discuss)? Perhaps they had been divided on the question of what to do with Galileo, and between 1630 and 1632 the balance of power swung towards the conservatives. Threatened with torture, the aged Galileo withdrew his claim and yet, according to a story that cannot

[25] *Two New Sciences*, p. 245.

[26] *Two New Sciences*, p. 248.

be confirmed, as he left for house arrest in Florence he muttered the famous phrase 'eppur si muove' ('but yet it moves').

The sentence of house arrest meant that Galileo had to smuggle a copy of his *Two New Sciences* out of Italy; it was published in Holland, where the Dutch were expelling the Spanish Inquisition and setting up an independent Protestant state. As we have seen, this book, which must surely be read as Galileo's last reply to his critics, goes much further in the direction of a theoretical mathematical analysis of motion. But it does not even try to deal with the reality of Copernican ideas. No one person could have accomplished such a task as creating a new earthly and heavenly physics. It was Galileo's achievement to present the postulates on which the new physics and astronomy could be based, and to cast them in mathematical form. To weld them into coherence and a new orthodoxy was to be the achievement of others.

12.4 Further reading

Banville, J. 1976. *Doctor Copernicus*, Secker and Warburg. This novel, and the next, give vivid accounts of the title figures, their work and their times through the medium of fiction.

———. 1981. *Kepler*, Secker and Warburg.

Beer, A. & P., 1975. *Kepler: Four Hundred Years*, Pergamon Press. An extensive collection of scholarly but readable articles on all aspects of Kepler's work.

Brecht, B. 1980. *The Life of Galileo* (transl. H. Brenton), Eyre Methuen. Many of the historical assertions in the play have been disputed, but it captures the drama and what can be at stake when science and power collide.

Field, J.V. 1988. *Kepler's Geometrical Cosmology*, Chicago University Press. A clear, scholarly discussion of Kepler's cosmological theories, which embrace music and astrology as well as astronomy, and their grounding in geometry and theology.

Gingerich, O. 2004. *The Book Nobody Read: Chasing the Revolutions of Nicolaus Copernicus*, Walker. An enthralling account both of the impact of Copernicus's great work and of how this impact was documented by many years of tracking down every surviving copy.

Heilbron, J. 2010. *Galileo*, Oxford University Press. A fresh view of Galileo that sees him as a mathematician, a musician, an artist, a writer, a philosopher, and an instrument maker.

Kuhn, T.S. 1957. *The Copernican Revolution: Planetary Astronomy in the Development of Western Thought*, Harvard University Press. A full and readable study of astronomy before Kepler.

Mosley, A. 2007. *Tycho Brahe and the Astronomical Community of the Late Sixteenth Century*, Cambridge University Press. A biography of Brahe.

Rabin, S. 2005. Nicolaus Copernicus, *The Stanford Encyclopedia of Philosophy* (Summer 2005 Edition), Zalta, E.N. (ed.) `http://plato.stanford.edu/archives/sum2005/entries/copernicus/`.

Westfall, R.S. 1971. *The Construction of Modern Science: Mechanisms and Mechanics*, Wiley. A compact and helpful account of 17th-century science from Kepler to Newton.

13

European Mathematics in the Early 17th Century

Introduction

In this chapter we look at some aspects of mathematics on the European continent in the first half of the 17th century. We will see that both the subject itself and the uses to which it was put were changing rapidly in rich and chaotic ways. We cannot attempt to be comprehensive, but will be selective in order to highlight some particular developments that will help us to illuminate certain themes that characterise the period.

The tensions between the rediscovered Greek mathematical tradition and the needs and interests of the 'new science' were still present. But another interesting feature is the range of mathematical activities that had no immediate consequences and established no research tradition, but were in their turn to be 'rediscovered' and further developed a century or two later.

In Section 13.1 we consider the evolving relationships between geometry and algebra and between analysis and synthesis, as they were discussed by such writers as Descartes, Oughtred, Desargues, and Fermat. We also look at some aspects of the working situations and social contexts of mathematicians, by considering the friends and correspondents of the French friar Marin Mersenne. Mersenne was a propagandist for the new science and its mathematics, who speedily established a European network of correspondents that embraced nearly every scientist of note. This has left the historian with a vast collection of letters that provide an unrivalled source of information about the period. Prominent in his achive are letters from nearly all of the French mathematicians of the day — Descartes, Fermat, Desargues, and Pascal, in particular.

In Section 13.2 we look in more detail at Fermat's work in number theory, and in Section 13.3 we compare the ideas of Viète and Fermat with those of René Descartes. Descartes was particularly pleased with his solution to, and considerable generalisation of, a locus problem of Pappus. We examine this in Section 13.4 and discuss why it is so important. Finally, in Section 13.5, we engage with Descartes's ambitious programme for mathematics, with its successes and failures, and with various claims that have since been made for its significance.

By looking at some individual mathematicians and their work, we hope to convey an impression of the excitement and diversity of the early 17th century. During that century an enterprise came into full swing which, for all its habit of reverence to the classical past, took the measure of the Greek achievement and began to surpass it. The world of modern mathematics was born.

13.1 Algebra and analysis

In the course of the 17th century, mathematics acquired a great many new objects to study, problems to deal with, and methods for tackling them, and mathematicians were emancipated from their Greek inheritance. Whereas Viète, and even Fermat, still thought of themselves as rediscoverers of lost Greek knowledge, by the end of the century no-one felt obliged to make that comparison. They were by then their own masters, and one way they demonstrated this mastery was through the growing use of algebra. The central work of this movement, Descartes's *La Géométrie*, published as an appendix to a philosophical treatise in 1637, was a new and decisive way to introduce algebra into geometry that was to transform ways of thinking about geometry altogether.

From geometry to algebra. The historian Michael Mahoney has called the transition from the geometrical to the algebraic mode of thought

> the most important and basic achievement of mathematics at the time.[1]

For the century that saw the invention of the calculus this is a bold claim. We can begin to understand it better by asking: Why did it happen? What made possible so fertile a transformation of mathematical practice? The answer is tied up with a transition from the method of synthesis, as exemplified by Euclid's *Elements*, to a new method, that of analysis, which was often exemplified by algebraic reasoning.

We shall see that it was the power of algebraic analysis as a discovery method that made the introduction of algebra important. What was it about algebra that made it seem so attractive? Mahoney gives a three-point summary of what came to characterise algebra during the 17th century:

- It is symbolic, and it symbolises not only objects (such as lines) but also operations (such as addition) and relations (such as equality).
- What the objects are becomes less important: algebra is ontologically uncommitted, in the sense that you do not need to know what symbols such as x and y mean, as long as you know the rules for handling such symbols.
- The operations and relations become the focus of attention, rather than the objects, and the rules for handling symbols become of prime significance.

In Section 10.4 we argued that Viète's importance rests on his giving algebra the generality of geometry. In his hands the symbols, denoting magnitudes, acquired a logic of their own, the logic of algebra rather than geometry. These symbols were intended to be completely general, and to operate with what Viète called the 'forms of things', not only because they might denote an unknown quantity, but because they might be indeterminate, and might, depending on context, denote either geometrical or arithmetical quantities.[2]

---

[1] Mahoney, in (Gaukroger 1980, 141).

[2] Viète, *Introduction to the Analytical Art*, 1603, in (Klein 1969, 328).

This universal character required novel methods of argument, because a geometrical argument is valid only because its statements accord with the nature of the magnitudes under discussion. Similarly, an arithmetical argument is valid if it proceeds according to the agreed rules for dealing with numbers. But an algebraic or universal argument is valid because its almost meaningless symbols have been validly manipulated, and these rules of manipulation permit either a geometrical or arithmetical interpretation as the occasion demands.

This distinction can be hard to make, because we usually find that an arithmetical or geometrical interpretation is never far away in the sources. We may feel that the generality of algebra is little more than a thin layer of clothing over a more concrete formulation, but there is a sense that algebra, being more formal, is more general.

Analysis and synthesis. One way, then, to explore the question of what made algebra so attractive is to invoke the power of the new methods, and to argue that it was the availability of new techniques that led directly to the new conquests. On this account, the universality of algebra gives it power as a method of analysis that surpasses what can be done with the synthetic methods of geometry. This sense of power can be measured by the following passage from Descartes, in which he compared the methods of analysis and synthesis as part of his reply to critics of his philosophical work, the *Meditations*. It gives some of his reasons for preferring the method of analysis:[3]

Descartes's *Meditations*.

> Analysis reveals the true way in which a thing was found methodically and, as it were, a priori, so that, if the reader wishes to follow it and pay sufficient attention to everything, he will understand the matter no less perfectly and make it no less his own than if he himself had found it. But it has nothing by which to incite belief in the less attentive or hostile reader. For if he should not perceive the very least thing brought forward, the necessity of its conclusions will not be clear; often it scarcely touches on many things which should be especially noted, because they are clear to the sufficiently attentive reader.
>
> Conversely, synthesis clearly demonstrates, in a way opposite to analysis and, as it were, a posteriori (even though the proof itself is often more a priori in the former than in the latter), what has been concluded, and it uses a long series of definitions, postulates, axioms, theorems, and problems, so that, if one of the consequents is denied, it may at once be shown to be contained in the antecedents. Thus it forces assent from the reader, however hostile or stubborn. But it is not as satisfying as analysis; it does not content the minds of those wanting to learn, because it does not teach the manner in which the thing was found.
>
> The ancient mathematicians used to employ only synthesis in their writings, not because they were simply ignorant of the other, but, as I see it, because they made so much of it that they reserved it as a secret for themselves alone.
>
> In fact, I have followed in my *Meditations* only analysis, which is the true and the best way of teaching.

[3]Second Reply to *Objections against the Meditations* (1641), quoted by Mahoney in (Gaukroger 1980, 148).

The idea that synthesis 'forces assent' reminds us of the effect that Euclid's *Elements* was to have on Hobbes,[4] about a decade before Descartes published his *Discourse on Method*.

In a mathematical culture intent on finding new results, new techniques might well drive things a long way, so where did they come from? Partly they derived from the analytic art of Viète. Viète's idea was to use algebra to analyse a problem, in the sense that Pappus used the term *analysis* — that is, you assume that the solution to your problem is known, and then argue until you reach a known truth. This process is to be followed by a *synthesis*, using rigorous deduction from the known truth until the desired solution is validly demonstrated. Viète thought of his work as supplementing an essentially geometrical and Greek way of proceeding; his original contribution, in his own opinion, was in 'exegetics', the solution of the equations produced by the analysis. As the 17th century proceeded, the tendency was to grant algebraic analysis a rigour of its own and to drop the synthesis. Thus a radically new technique and mode of mathematical justification evolved out of interpreting the Greek tradition, a tradition whose emphasis on synthetic proof ran in the opposite direction.

Now, it can be argued that analysis did not often lead to a proof by the method of synthesis, even in classical times. Instead, when successful it generally produced constructions. These are geometrical procedures for finding the solution to a problem, such as finding the centre of a circle through three given points. Recall that the distinction between theorems (which require proofs) and problems (which require solutions, often in the form of constructions) can be found even in Euclid's *Elements*, where Heath, in his edition, added the letters Q.E.D. (*quod erat demonstrandum*, that which was to be proved) at the end of proofs, and Q.E.F. (*quod erat faciendum*, that which was to be done) at the end of constructions; one might then ask for a proof that the stated construction does solve the problem. Thus we have a spectrum of possibilities and pedagogical styles: at one end, the construction is so unintuitive that a proof is essential; at the other end, it is self-explanatory and a proof would be superfluous. We shall see that this spectrum of possibilities, and the tensions that it generates, run through much of the material in this chapter.

If analysis moves to centre stage, then synthesis becomes marginalised and, as Mahoney has argued, this is too major a shift to be accounted for by the efficacy of the new methods alone; other factors must have disposed people to accept the new algebraic approach. Mahoney draws attention to two features of the broader intellectual concerns of the period that may have conditioned the new conception of mathematics: the search for a universal language, and the pedagogical tradition. First, many scholars throughout the 17th century were pursuing the quest for a natural and universal symbolism to reveal truths about the world. Algebraic language came to be seen as a paradigm of successful symbolising that could have wider application. The philosopher John Locke certainly saw a connection. In *An Essay Concerning Human Understanding*, published in 1690, he wrote:[5]

> They that are ignorant of *Algebra* cannot imagine the Wonders in this kind are to be done by it: and what farther Improvements and Helps, advantageous to other parts of Knowledge, the sagacious Mind of Man may yet find out, 'tis not easy to determine . . .
>
> Mathematicians abstracting their Thoughts from Names, and accustoming themselves to set before their Minds, the *Ideas* themselves . . . have avoided thereby

[4]See Chapter 3.

[5]Book IV, Chaps III, XII.

> a great part of that perplexity, puddering, and confusion, which has so much hindred Mens progress in other parts of Knowledge. And who knows what Methods, to enlarge our Knowledge in other parts of Science, may hereafter be invented, answering that of *Algebra* in Mathematicks, which so readily finds out *Ideas* of Quantities to measure others by, whose Equality or Proportion we could otherwise very hardly, or perhaps, never come to know?

So there were wider intellectual currents of the period encouraging the development of symbolic algebra. Locke was writing late in the century, but earlier writers, such as Descartes, also discussed explicitly the power and value of symbolic expression.

Pedagogy. The other intellectual concern singled out by Mahoney was a changing conception of mathematical pedagogy. This stemmed in particular from the influence of the 16th-century teacher and textbook writer Peter Ramus, known also by his French name Pierre de la Ramée. As we discussed briefly in Section 11.1, Ramus had castigated Euclid on the very grounds that others had praised him, for the *Elements*' rigour and deductive logical structure. In Ramus's view this style was a fault, for rigorous proofs conveyed neither clarity nor insight.

This is a debate about the teaching of mathematics that is still with us. But it is more than that, for a constant source of complaint in the early 17th century was that the classical texts were so opaque: the mathematical results were hard to learn and impossible to extend. There was a pressing need for another mathematical way to be found, and by the force of his example Ramus helped to legitimise arguments that made up in intelligibility what they lacked in rigour.

A powerful demonstration of the new algebra arising in the service of a new teaching approach was Oughtred's influential textbook *Clavis Mathematicae* (1631).[6] In the preface to the first English translation (1647), largely made by Robert Wood, Oughtred explained his intentions, and we can judge how emancipated he had become from the Greek tradition.[7]

Oughtred's *Clavis Mathematicae*, Preface to the new edition.

> Many yeares since, I being imployed by the late illustrious Earle of Arundel, to instruct one of his sons in the Mathematicks, penned for his use in Latine, a method of precepts for the more ready attaining, not a superficial notion, but a well-grounded understanding of those mysterious sciences, and of the ancient Writers thereof. This afterward at the request of divers learned and judicious men, I published under the name of *Clavis Mathematicae*. Which Treatise being not written in the usuall syntheticall manner, nor with verbous expressions, but in the inventive way of Analitice, and with symboles or notes of things instead of words, seemed unto many very hard; though indeed it was but their owne diffidence, being scared by the newnesse of the delivery; and not any difficulty in the thing it selfe. For this specious and symbolicall manner, neither racketh the memory with multiplicity of words, nor chargeth the phantasie with comparing and laying things together; but plainly presenteth to the eye the whole course and processe of every operation and argumentation.

[6] The full Latin title is *Arithmeticae in Numeris et Speciebus Institutio: quae tum Logisticae, tum Analyticae, atque Adeo Totius Mathematicae, quasi Clavis est* (*The Method of Calculating in Numbers and Letters: which was the Key first to Arithmetic, then to Analysis and now to the whole of Mathematics*).

[7] See (Oughtred 1647) and F&G 9.F1(a).

> Now my scope and intent in the first Edition of that my *Key* was, and in this New Filing, or rather forging of it, is, to reach out to the ingenious lovers of these Sciences, as it were Ariadne's thread, to guide them through the intricate Labyrinth of these studies, and to direct them for the more easie and full understanding of the best and antientest Authors: such as are Euclides, Archimedes, Apollonius Pergaeus that Great Geometer, Diophantus, Ptolomaeus, and the rest: That they may not only learn their propositions, which is the highest point of Art that most students aime at; but also may perceive with what solertiousnesse, by what engines of Aequations, Interpretations, Comparations, Reductions, and Disquisitions, those antient Worthies have beautified, enlarged, and first found out this most excellent Science.
>
> Truly when I was conversant in reading their bookes, and with wonder observed their most witty demonstrations, so skilfully framed out of principles, as one would little expect or thinke, but laid together with divine Artifice: I was even amazed, whence possibly any power of imagination should be able to sustaine so immense a pile of consequences, and cause that so many things, so far asunder distant, could be at once present to the minde, and as with one consent joyne and lay themselves together for the structure of one argument. Wherefore that I might more cleerly behold the things themselves, I uncasing the Propositions and Demonstrations out of their covert of words, designed them in notes and Species appearing to the very eye. After that by comparing the divers affections of Theoremes in equality, proportion, affinity and dependence, I tryed to educe new out of them. Lastly, by framing like questions problematically, and in way of Analysis, as if they were already done, resolving them into their principles, I sought out reasons and means whereby they might be effected. And by this course of practice, not without long time, and much industry, I found out this way for the helpe and facilitation of Art.

So in some ways Oughtred departed radically from classical practice: his book was not written synthetically and with words, but analytically and with symbols. He also abandoned the classical proof structure as a way of presenting results, because he felt (as Ramus had done earlier) that this did not lead to understanding. But the purpose of his book ('my scope and intent') was to lead to an understanding of the classical authors — Euclid, Archimedes, Apollonius, etc. — it was not to investigate contemporary mathematics, nor to aid current researches. So Oughtred can be seen to have broken with the Greek mathematical tradition, both in proof structure and in his wholesale symbolisation, in order to understand that tradition better. This paradoxical approach of seeming to surpass a traditional way of doing something only to deepen one's grasp of it is a feature of the 17th century. Some writers could not shake off a reverence for a 'golden age', while others were proud to rush beyond it.

Debates about the relative virtues of analysis and synthesis continued, both in mathematics and in the broader intellectual context, mentioned earlier, of what method should be followed for the discovery and teaching of truth more generally. Viète claimed that, while synthesis forces you to acknowledge that a result is correct by showing you that it must be true, analysis is the best way to teach results because it shows you how they are found and so leads to understanding in those who want to learn. As we shall see, Descartes was particularly concerned with such matters.

Mersenne and his circle. It is a striking fact that a friar who spent almost all of the last twenty-nine years of his life in a Minimite monastery in Paris should be a central figure in the rise of mathematics; it tells us how unlike our world was that of Marin Mersenne.[8] But Mersenne made up for his sheltered life by corresponding energetically with nearly every scientist and mathematician of his day, and acting as a clearing house for letters that disseminated knowledge between Holland and Italy, London and Toulouse. His close friend Gassendi, in whose arms he died, described him as:[9]

> A man of simple, innocent, pure heart, without guile ... A man whom all the arts and sciences, to whose advance he tirelessly devoted himself, by investigating or by deliberating or by stimulating others, will justly mourn.

Indeed, it is hardly likely that the competitive, secretive, and controversy-ridden world of the 1600s could have entrusted so much news to anyone whom it was not possible to trust more or less completely.

Mersenne was no stranger to scientific investigation himself, which doubtless helped to earn him the respect of others. Moreover, he was courageous. When in 1616 the Church passed a decree against Copernicanism, Mersenne affirmed the plausibility of the Earth's motion, and when in 1635 he heard that the Church had condemned Galileo on the grounds that the latter proclaimed that the Earth really does move, Mersenne contemplated writing in his defence. He finally abandoned that idea because, like many of Galileo's contemporaries, he found Galileo's reasons for the suggestion unconvincing; but he openly proposed that others try to resolve the question one way or another.

Mersenne's own philosophy of science was pragmatic. He disagreed with those, such as Kepler, Galileo, and Descartes, who thought that scientific conclusions (even if arrived at by mathematical arguments) could be absolutely true. But he allowed that scientific statements could be sufficiently precise to be reliable, finding that the sciences came in a logical order that enabled them to succeed where philosophy had failed, in building up a cumulative picture of the world. So he denied the claims of sceptics, such as the 16th-century essayist Michel de Montaigne, who argued that the only certain knowledge was that of the divine, and disputed the possibility of scientific knowledge at all. Such scepticism was an influential doctrine at the time and Mersenne and the Minimites argued most firmly against it.

Mersenne's own scientific interests were diverse, but his chief love was the study of sound. What makes a stringed instrument sound as it does? Why are some combinations of sounds harmonious and others not? How fast does sound travel? He investigated many questions like these.

Even more importantly, he organised regular meetings, somewhat similar to those of the world's first scientific society, the Accademia dei Lincei in Rome, to which Galileo belonged. Among those who regularly attended Mersenne's Académie Parisienne was Claude Mydorge, a geometer whose book on the conic sections, *Coniques* (1639), was thoroughly classical, though on occasion more elegant than Apollonius's. Others who attended were Gilles de Roberval (a professor of mathematics in Paris), the two Pascals (first Étienne, and later his young son Blaise), and Girard Desargues.

Girard Desargues. Of these, Girard Desargues was an architect and a mathematician who in 1639 circulated a short and densely written essay entitled *Brouillon project d'une*

[8]The Minimites were close to the Franciscans, and in addition to the traditional vows of poverty, chastity, and obedience, they vowed to live as vegetarians. The order was prominent in the Catholic Church's campaigns during the Counter-Reformation against those it described as atheists and free-thinkers.

[9]Quoted in (Crombie 1975, 320).

atteinte aux evenemens des rencontres du Cone avec un Plan (*Rough Draft of an Essay on the Results of Taking Plane Sections of a Cone*).[10] This essay can be taken as the founding document of what later became projective geometry, the study of those properties that a figure shares with its shadows under projection from a point source of light.

Desargues exploited the fact that every conic section can be obtained in this way from a circle. It follows that any question about properties of conic sections that do not alter under a projection can be established simply by proving them for a circle. Unfortunately, there are few such properties, because such projections change lengths and angles in figures, and even the ratios of lengths. However, the fact that three or more points lie on a line is a projective property, and Desargues found a property involving four points on a line that is not altered by any projection (Figure 13.1):[11]

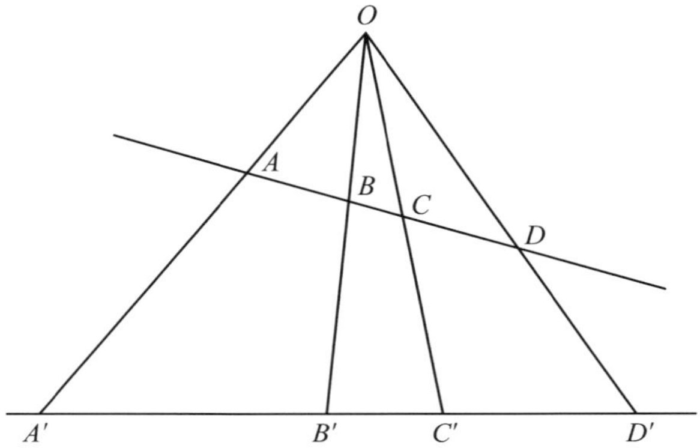

Figure 13.1. Ratios of ratios of lengths are preserved under projection

> if A, B, C and D are any four points on a line, and if A', B', C' and D' are their images under a projection, then A', B', C', D' lie on a line and the ratios $AB.CD/AD.CB$ and $A'B'.C'D'/A'D'.C'B'$ are always equal.

On this slender basis, and on another even more complicated formula involving six points on a line, Desargues built up a remarkably rich theory of conic sections.

Some years later he added another result which he let his friend the artist and engraver Abraham Bosse publish in one of his own works. This theorem still bears the name, *Desargues's theorem* (see Figure 13.2):[12]

> if two triangles ABC and $A'B'C'$ are in perspective from a point O (so O, A, A' lie on a line, as do O, B, B' and O, C, C'), and if the lines BC and $B'C'$ meet at the point P, the lines CA and $C'A'$ meet at the point Q, and the lines AB and $A'B'$ meet at the point R, then the points P, Q and R lie on a line.

Note that everything about this theorem is projective: it is a basic theorem in any account of projective geometry. Non-projective concepts such as distance, length, and angle do not feature here.

---

[10]English transl. in (Desargues 1986); French in (Desargues 1988).

[11]The expression $AB.CD/AD.CB$, called a cross-ratio, can be written as $AB/CB \div AD/CD$ and treated as a ratio of ratios.

[12]This is the first of three geometrical propositions that Desargues allowed Abraham Bosse to publish in *La Perspective de Mr Desargues* (1648, 340–343), transl. in (Desargues 1986, 161–162); see F&G 11.D6.

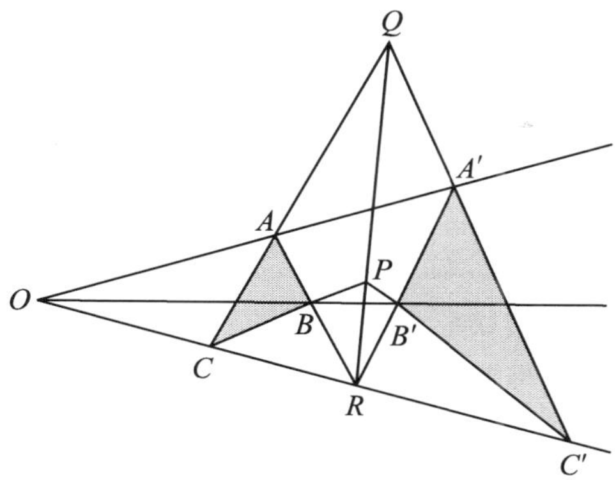

Figure 13.2. Desargues's theorem

Desargues circulated fifty copies of his essay, in the hope that other mathematicians would offer helpful criticism — a naively optimistic view, as it turned out. Mydorge claimed that it had little new in it, while Descartes asserted that it would be much better if it were rewritten in his new style of geometry. Only the 16-year old Blaise Pascal saw the charm of the new subject, and he too came up with a fine new theorem, now called *Pascal's theorem* (see Figure 13.3):

> if six points A, B, C, D, E, F lie on a conic, and if AB and DE meet at the point P, BC and EF meet at the point Q, and CD and FA meet at the point R, then the points P, Q and R lie on a line.

This too is an entirely projective theorem.

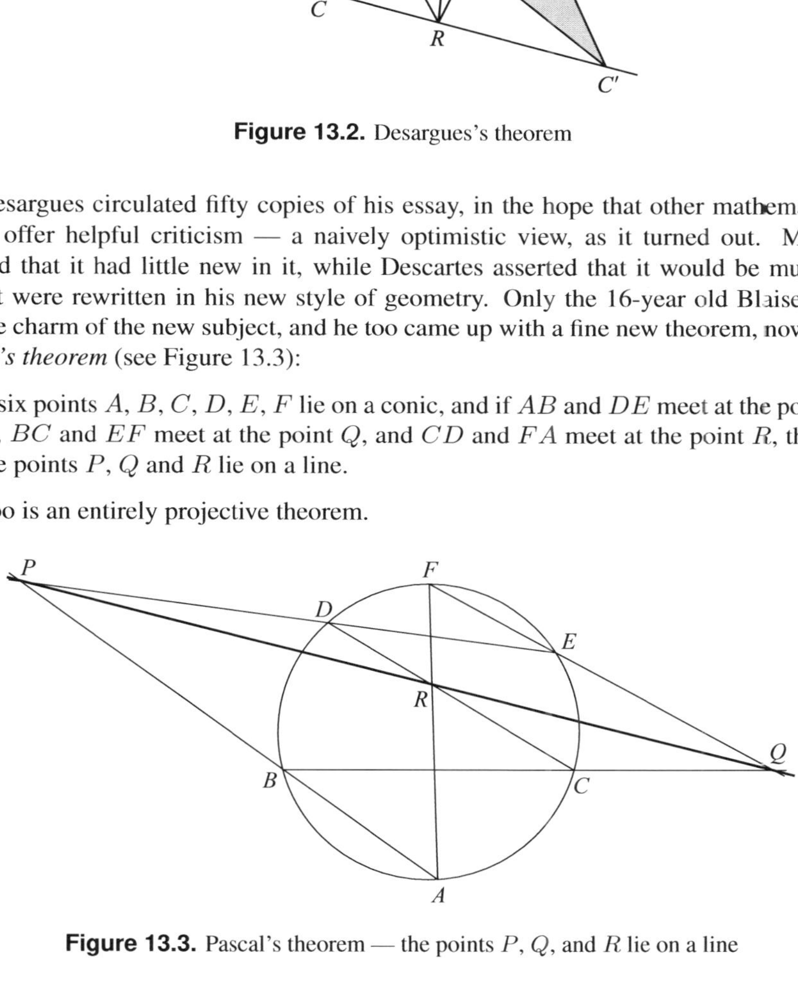

Figure 13.3. Pascal's theorem — the points P, Q, and R lie on a line

Desargues's difficulty was that, apart from the novelty of his work, his ability as a writer was quite limited and he cluttered up his *Essay* with many obscure neologisms. The result is exceedingly difficult to read — one sees the force of Descartes's criticisms — and transfers the study of conic sections, which are figures in the plane, into the more difficult realm of three-dimensional geometry. For these reasons, copies of Desargues's *Essay* became very rare, and the work was known through only one handwritten copy until a single printed original was found in 1950. Pascal never returned to the topic of projective geometry, and it fell out of favour for many years.

Another example of the mixed reception of projective geometry is afforded by the geometer Philippe de la Hire's projective treatment of conics in his book[13] of 1673, but it was poorly received. Six years later he decided that his mistake had been to treat the subject as a three-dimensional one, for not all readers seem to have shared his enthusiasm. The new treatment, *Nouveaux Élémens des Sections Coniques* (1679), fared much better because it was entirely two-dimensional. The historian Julian Lowell Coolidge wrote of it that:[14]

> It would be hard to find a book offering an easier introduction to the conics. Each type of curve is considered separately, starting with some characteristic property ...

and he added that the book would be as usable today as it was on the day it was written.

There is another reason for the failure of projective geometry at this time, and it will lead us to the main theme of this chapter: the tremendous success of Descartes's geometry. But first let us return to Fermat, and investigate his work in a different area, number theory, and his attempts to interest others in it through his correspondence with Mersenne.

Fermat: algebra and geometry. An example of algebraic analysis that exemplifies many of the processes we have been considering is Fermat's study of curves in 1636. A lawyer by profession, Fermat (Figure 13.4) was among the first to extend to curves the approach that Viète had taken towards general magnitudes — namely, a conscious analytical exploration of the connection between problems and their solutions.

We know a little about how Fermat came to work in this way. It seems that his earliest mathematical contact, in the late 1620s, was with one Étienne d'Espagnet who was in possession of some of Viète's unpublished manuscripts. D'Espagnet communicated these to Fermat, who was then in his 20s, and who proceeded to study them avidly.[15] It is therefore not surprising that Fermat's first interest in geometry was an attempt to restore a lost book of Apollonius (*On Plane Loci*), for Viète too had engaged in seeking to restore lost texts. This attempt seems to have suggested to Fermat that the preparatory algebraic analysis he carried out was capable of forming a more general method than the classical one, and in 1636 he wrote a paper describing his own method (*An Introduction to Plane and Solid Loci*). However, he did not publish it, and it first appeared in 1679, after his death. Coming so long after the publication of Descartes's work, its influence on the history of mathematics was to be slight, but its importance as an indication of contemporary thinking about algebra and geometry is considerable.

Fermat used Viète's notation for constants, variables and equations, and set himself the task of analysing equations of the first and second degrees in two unknowns. He found that any such equation can be reduced to one of seven types, which he interpreted geometrically as various kinds of conic section. This was one of the first occasions when algebraic analysis was treated as an end in itself, and not simply as a means to an end. An algebraic object (an equation) was both interpreted geometrically and taken as the main object under investigation. Thus, when Fermat considered (in his Viètan notation) the equation

$$A \text{ in } E \text{ aeq } Z \text{ sq},$$

between variable magnitudes A and E and a constant magnitude Z, he showed that the corresponding locus is a hyperbola. We might write it as $AE = Z^2$, since Fermat wrote

[13]See F&G 11.B2.

[14]See (Coolidge 1945, 41).

[15]The date of Fermat's birth is uncertain, but is often given as 1601 and may be as late as 1607.

Figure 13.4. Pierre de Fermat (1601?–1665)

'in' for 'times', 'aeq' for 'equals', and also 'sq' for 'square'. Switching to our usual letters, we get the more familiar equation $xy = c^2$. What Fermat had done was to show that for the investigation of conics, an analysis by means of simple equations is convenient and productive.

Fermat's investigations exemplify in a number of ways what we have been saying about algebra. His approach was symbolic, its logic that of algebra. Rules of inference were rules about manipulating symbols. It is ontologically uncommitted, in the sense that A and E are perfectly general magnitudes until he chooses to relate them to points on a hyperbola. But it is still subordinate to the Greek legacy: Fermat was offering a way into Apollonius's theory of conics, and not, it would seem, a way beyond it.

13.2 Fermat's number theory

Fermat joined Mersenne's circle (although only as a correspondent) when his friend Pierre de Carcavi went to Paris in 1636. Carcavi took over the role as 'postmaster general' when Mersenne died in 1648, so we have a considerable number of Fermat's letters. These are often our only source for his ideas, since he had an aversion to publishing, and those manuscripts that he allowed to circulate are his only written influence on others in his lifetime.

Fermat seems to have disliked travel, never straying further from his native Toulouse than Bordeaux, and apparently never meeting a fellow mathematician (apart from a three-day visit by Mersenne). His work was published only posthumously, first in a one-volume collection called the *Varia Opera*, edited by his son Samuel in 1679, and later in a four-volume French and Latin edition by P. Tannery and C. Henry between the years 1891 and 1912. Yet the topic we now discuss, which seems to have been his favourite, is one of the oldest in mathematics: the theory of numbers.

As we have seen, number theory, or the *higher arithmetic* as it is sometimes called, deals with the properties of the integers, including the solutions of equations in integers. For example, in Chapter 2, we saw that as early as Mesopotamian times certain whole numbers were known to satisfy the relationship that we now express as $x^2 + y^2 = z^2$; for example, $3^2 + 4^2 = 5^2$ and $5^2 + 12^2 = 13^2$.

Down the centuries, many mathematical cultures have produced mathematicians interested primarily in the properties of numbers; we shall meet a few as this section proceeds. But the same cultures also produced mathematicians who disdained the subject because it appeared to be useless. Its modern history begins with the mathematicians of Mersenne's circle.

One reason why people studied number theory is that the questions are about one of the most fundamental concepts in all mathematics — number. Another is that the questions turn out to be surprisingly delicate. For example, are there solutions to $x^2 - 15y^2 = 2$? The answer is 'yes' if x and y are allowed to be arbitrary numbers, for the equation describes a hyperbola. But if x and y are required to be integers, then the answer is 'no', though this is not obvious. Moreover, the seemingly similar question about $x^2 - 15y^2 = 1$ is answered in the affirmative by infinitely many pairs of integers, such as $x = 4$ and $y = 1$. Subtleties like this give the subject its fascination.[16] It is a paradoxical fact that definite answers can be given to such questions, even though x and y are allowed to range over all positive and negative integers.

Many of the most significant problems in the early study of number theory were asked by Fermat in the 1650s. It is clear from the way that Fermat asked these questions that he could often also solve them and that he had a general theory for dealing with them. But he wrote down singularly few proofs, so his legacy to subsequent generations was mostly the problems themselves. This is tantalising, for he usually claimed only that the problems can be done, but did not show how to do so. He did not even say which of his problems were easy, and which ones were hard.

Of his many contributions that we could discuss, we have selected four; see Box 13.1.

[16] Recall Brahmagupta's treatment of such questions, discussed in Section 7.1.

Box 13.1. Fermat on number theory.

F1. *Fermat's 'last theorem'.* Fermat claimed that there are no solutions in positive integers x, y, z to $x^n + y^n = z^n$, when n is a whole number greater than 2. (The case $n = 2$ is the problem of finding Pythagorean triples, such as $3^2 + 4^2 = 5^2$, which was resolved by the Mesopotamians and again by the Greeks.) Fermat circulated an ingenious argument that showed that there are no positive integers satisfying this equation when $n = 4$, as we shall see.

F2. Characterise those integers that can be written in the form $x^2 + cy^2$, where x and y are integers, for a given positive non-square integer c. In the case $c = 1$, Albert Girard had claimed that the only integers n for which the equation $x^2 + y^2 = n$ has integer solutions are the number 2, the perfect squares $(0, 1, 4, 9, \ldots)$, prime numbers of the form $4k + 1$ (such as $5 = 2^2 + 1^2, 13 = 3^2 + 2^2, 17 = 4^2 + 1^2, \ldots$), and products of these. It is not clear whether Fermat had only empirical evidence for this, or could actually prove it: he claimed to have a proof, but never revealed it.

F3. Find all the solutions in integers to the equation $x^2 - cy^2 = 1$, for a given positive non-square integer c; for example, when $c = 7$, the smallest solution is $x = 8, y = 3$ $(8^2 - 7.3^2 = 1)$. Fermat confronted his contemporaries with the difficult cases $c = 109$ and $c = 149$, where the smallest values for x are 15,140,424,455,100 and 2,113,761,020, respectively. Since these could not have been discovered by trial and error, he must surely have had a general method, but he never disclosed it.

F4. *Fermat's 'little theorem'.* If a prime number p does not divide an integer a, then p must divide $a^{p-1} - 1$; for example, 5 (a prime) does not divide 6 (an integer), but 5 does divide $6^4 - 1 = 1295$.

F1 is the famous 'last theorem' of Fermat. It was not to be proved until 1995, by Andrew Wiles, in a heroic feat that drew on a vast amount of recent work.[17] It was never officially claimed by Fermat; on a date in the 1630s, he merely wrote in the margin of his copy of Bachet's edition of Diophantus's *Arithmetica* opposite Problem 8 in Book II ('Given a square, write it as a sum of two other squares'):

> On the other hand it is impossible for a cube to be written as the sum of two cubes, or a fourth power to be written as a sum of two fourth powers or, in general, for any number which is a power greater than the second to be written as the sum of two like powers. I have a truly marvellous demonstration of this proposition which this margin is too narrow to contain.

Some margin! Alas, Fermat wrote nothing else on this subject that has survived. Since a simple proof has eluded everyone else, it is probable that Fermat had fallen for a fallacious argument, and whether he later found a flaw in his 'truly marvellous demonstration' is an unanswerable question. The marginal note dates from an early stage in Fermat's career, and he never repeated the claim in letters to his various correspondents. The claim was not published until after his death, in the edition of his works made by his son Samuel in 1679 (see Figure 6.11). He did repeatedly claim, however, that he could establish the theorem for cubes and fourth powers, and happily left us a proof in the latter case. We look at it below, after a few preliminary remarks that should make it less obscure.

[17] See (Taylor and Wiles 1995) and (Wiles 1995).

Even the designation 'last theorem' is a misnomer. It was given on the grounds that it was the last result of Fermat to require a proof, but it is not a theorem and it was not even his last unproved claim. Fermat had also raised the question of which integers can be the areas of triangles with sides that are rational numbers. Arabic writers had given an example of a triangle with area 5. Leonardo of Pisa had claimed that a triangle with rational sides cannot have area 1, and Fermat proved this: his argument essentially settles Fermat's last theorem in the case $n = 4$. It is now known that the smallest possible value for the area is 5, but the general problem remains unsolved.

Sums of squares. Fermat claimed that every prime number of the form $4k + 1$ is a sum of two squares, and is so in a unique way (up to the order of the factors), for example, $5 = 2^2 + 1^2$ and $13 = 3^2 + 2^2$. He also claimed that a number is a sum of two squares if and only if it is a product of such primes (and perhaps a power of 2). Thus, $65 = 8^2 + 1^2$, and $65 = 5 \times 13$. He left no proof of this claim, which is, however, correct, and he seems to have been attempting to prove it for a compound number by investigating its divisors, a method that Euler was to make work almost a century later, when Fermat's questions began to generate steady interest.

Mersenne and Frenicle. Fermat had a limited circle of colleagues who shared his interest in number theory. By the 1640s Bachet, the editor of Diophantus's *Arithmetica*, was not among them, for he had died in 1638, but Mersenne himself was interested in perfect numbers. These were discussed in Section 6.1, where we noted that Euclid's *Elements*, Book IX, Prop. 36 shows that $2^{n-1}(2^n - 1)$ is a perfect number whenever $2^n - 1$ is prime.[18] So Mersenne and his correspondents asked 'For which values of n is $2^n - 1$ prime?'. It turns out that for $2^n - 1$ to be prime, n must itself be prime, and such primes are now called *Mersenne primes*.

Let us consider the simplest numbers of this form:

$$2^2 - 1 = 3, \quad 2^3 - 1 = 7, \quad 2^5 - 1 = 31, \quad 2^7 - 1 = 127, \quad \text{and } 2^{11} - 1 = 2047.$$

Of these, the first four are prime, but $2^{11} - 1 = 2047 = 23 \times 89$. Since the sums are already becoming time-consuming, this suggests that it is now right to look for a theorem.

In 1644 Mersenne produced a list of those numbers of the form $2^n - 1$ that he believed to be prime; his list contained a few mistakes and omissions, but is remarkably accurate, given the sizes of the numbers (the largest is $2^{127} - 1$). He also put Fermat in touch with Frenicle du Bessy, a tireless calculator who was very much at home with large numbers and who had asked: 'is $2^{37} - 1 = 137{,}438{,}953{,}471$ a prime number?' In this context Fermat made an interesting discovery in 1640, which he conveyed in a letter to Marin Mersenne on his 'little theorem' in the case $a = 2$.[19]

Some of Fermat's theorems.

> Here are three propositions I have found on which I hope to erect a great building.

[18]Leonhard Euler proved that every *even* perfect number has the form given by Euclid, in a paper presented to the Berlin Academy in 1747. This paper 'De numeris amicibilibus' (E798) is incomplete and was published posthumously in 1849. It is still not known whether odd perfect numbers exist.

[19]Fermat, *Oeuvres* II, 198–199, in F&G 11.C4.

The numbers which proceed from the double progression, reduced by unity, such as

1	2	3	4	5	6	7	8	9	10	11	12	13	
1	3	7	15	31	63	127	255	511	1023	2047	4095	8191	etc.

are called the radicals of perfect numbers because whenever they are prime they produce them. Let us put above them the numbers according to the natural progression 1, 2, 3, 4, 5 etc. which are called their exponents. This done I say that:

(1) When the exponent of a radical number is compound, its radical is also compound. Thus because 6 the exponent of 63 is compound, I say that 63 is also compound.
(2) When the exponent is a prime number, I say that its radical reduced by unity is measured by the double of the exponent. Thus because 7 the exponent of 127 is a prime number, I say that 126 is a multiple of 14.
(3) When the exponent is a prime number, I say that its radical is not measured by any prime number except those which exceed by unity either a multiple of the double of the exponent or the double of the exponent. Thus, because 11, the exponent of 2047, is a prime number, I say that it cannot be measured except by a number which is greater by unity than 22, as 23, or by a number which is greater by unity than a multiple of 22; in fact 2047 is only measured by 23 or 89, from which if you remove unity, 88 remains, a multiple of 22.

Here are three extremely beautiful propositions which I have found and proved, not without difficulty. I could call them the foundations of the invention of perfect numbers. I don't doubt that M. Frenicle got there earlier, but I have only begun and without doubt these propositions will pass as very lovely in the minds of those who have not become sufficiently hypercritical of these matters, and I would be very happy to have the opinion of M. Roberval.

Assertion 2 is his 'little theorem' (F4) in the special case where $a = 2$:

if p is any prime number other than 2, then p divides $2^{p-1} - 1$.

Fermat claimed that when n is prime, then $2^n - 2$, its 'radical less one', is divisible by $2n$ — that is, n divides $2^{n-1} - 1$. He also claimed that the only prime divisors of $2^n - 1$ are $2n+1$ or numbers of the form $2nk+1$, where k is an integer; in particular, he observed that when $n = 11$, the divisors of $2^{11}-1$ are indeed 23 $(= 2.11+1)$ and 89 $(= 2.11.4+1)$.

Fermat's statement of the little theorem for general values of a (not just for $a = 2$) survives in the form of a letter that he wrote to Frenicle, shown in Box 13.2. His proofs do not survive, but plausible reconstructions of them have been given. They are not easy to read, but the examples that Fermat gave are a considerable help; note that Fermat's word 'measures' is the Euclidean word for 'divides'. As you will see, he claims that if we take any number a and any prime number p, and form the progression $a, a^2, \ldots, a^n, \ldots$, then for some exponents n the number $a^n - 1$ is divisible by p, and that when this happens the exponent n divides $p - 1$. Moreover, there is a smallest value of n such that $a^n - 1$ is divisible by p, and if k is another exponent for which $a^k - 1$ is divisible by p, then k is a multiple of n.

Box 13.2. Fermat's letter to Frenicle, 18 October 1640, *Oeuvres* II, 209.
Every prime number infallibly measures one of the powers -1 of any progression whatever, and the exponent of the said power is a sub-multiple of the given prime number -1; and, after one has found the first power which answers this question, all those whose exponents are multiples of the exponent of the first one all answer to the question.

Example: let the given progression be

$$\begin{array}{cccccc} 1 & 2 & 3 & 4 & 5 & 6 \\ 3 & 9 & 27 & 81 & 243 & 729 \end{array}$$

with its exponents above.

Take for example the prime number 13. It measures the third -1, of which 3, the exponent, is a sub-multiple of 12, which is less by unity than the number 13, and because the exponent of 729, which is 6, is a multiple of the first exponent, which is 3, it follows that 13 also measures the said power $729 - 1$.

And this proposition is true generally for all progressions and all prime numbers; concerning which I would send you the demonstration if I did not fear it would be too long.

If you recall our earlier statement of Fermat's little theorem (F4 in Box 13.1), you might have expected to interpret the letter in Box 13.2 as saying that

If p is prime and does not divide a, then p divides $a^{p-1} - 1$.

In fact, such expectations are not quite met. Fermat's statement of the theorem is false when the prime p does divide the number a, and later mathematicians have quietly supplied the necessary piece of small print — perhaps Fermat thought it too obvious to be worth stating explicitly. But Fermat also stated something stronger: he did not merely say that $a^{p-1} - 1$ is divisible by p, but that *all* the numbers $a^k - 1$ that are divisible by p are of the form $a^{nr} - 1$ for some fixed value of n — that is, the exponent is a multiple of n. Later mathematicians may have rescued Fermat from a trivial error, but they also seem to have simplified his theorem.

'Pell's equation'. Fermat's attempts to interest his colleagues seldom met with much success. Only his challenge to the wider community in 1657 to solve problems such as the one described in the next extracts drew any response.[20] This problem (F3 in Box 13.1) can be expressed as follows:[21]

> Given any non-square number c, find an infinite number of squares x^2 such that $cx^2 + 1$ is a square.

In this case, the challenge was met by the English mathematicians John Wallis and William Brouncker, whose approach, sometimes called the 'English method', actually produces all of the infinitely many solutions. It is not clear, however, that it does produce them all, nor indeed that all the numbers that it produces are solutions. Fermat criticised it on those grounds, and so did Huygens in a letter to Wallis in 1658, going on to argue that, in

[20] Fermat, *Oeuvres* II, 335, and F&G 11.C6 .

[21] This question had been discussed by the Indian mathematicians Brahmagupta and Bhāskara II (see Section 7.1), but their work was unknown in the West in the 17th century.

any case, 'There is no lack of better things for us to do'.[22] Once again, although Fermat claimed to have a method that he could prove was satisfactory, it has not come down to us. The matter is not trivial; a rigorous proof eluded even Leonhard Euler, and was first given by Joseph-Louis Lagrange in 1768.

Infinite descent. One of Fermat's greatest achievements in number theory was his resolution of a case of his 'last' theorem: his proof that there are no non-zero integer solutions to the equation

$$x^4 + y^4 = z^4.$$

His interest in this problem is another illustration of how simultaneously 'modern' and 'ancient' he was. As we said earlier, he took the problem from a note that Bachet had put in Book VI of his edition of Diophantus's *Arithmetica.* Diophantus had raised problems about triangles whose areas plus or minus a number are squares, and Bachet added the remark that, if A is the area of a right-angled triangle with integer sides, then there is a number K such that $(2A)^2 + K^4$ is a square. Fermat investigated this in the case when A is also required to be a square, and was led to look for integer solutions to $x^4 + y^4 =$ a square. He then showed that there is not even a solution in integers to the weaker equation $x^4 + y^4 =$ a fourth power; this implies that there are no integer solutions to $x^4 + y^4 =$ a square.[23] To do this, he showed that if there is a solution in positive integers then there is a smaller solution and then a still smaller one, and so on. Since there cannot be an infinite descending sequence of integers the original solution cannot have existed. He argued as follows:

Fermat's last theorem.

> 'Bachet: Find a triangle whose area is a given number.' The area of a triangle in numbers cannot be a square. I am going to give a proof of this theorem, which I have discovered; and I did not find it without painful and laborious thinking about it, but this kind of proof will lead to marvelous progress in the science of numbers.
>
> If the area of a triangle was a square there would be two fourth powers whose difference was a square; it would follow equally that there would be two squares whose sum and difference would be squares. Consequently there would be a square number, the sum of a square and the double of a square, with the property that the sum of the two squares that make it up is likewise a square. But if a square number is the sum of a square and the double of a square its root is likewise the sum of a square and the double of a square, which I can prove without difficulty. One concludes from this that this root is the sum of the two sides of a right angle in a triangle of which one of the square components forms the base and the double of the other square the height.
>
> The triangle will therefore be formed of two square numbers whose sums and differences are squares. But one will prove that the sum of these two squares is smaller than the first two of which one has likewise supposed that the sum and difference are squares. Therefore, if one has two squares whose sum and difference are squares one has at the same

[22] In Fermat, *Oeuvres*, IV, 121, quoted in (Weil 1984, 119).

[23] In Fermat *Oeuvres*, I, 340–341, and F&G 11.C8.

time, in integers, two squares enjoying the same property whose sum is less.

By the same reasoning, one has accordingly another sum smaller than that derived from the first, and continuing indefinitely one will always find smaller and smaller integers satisfying the same condition. But this is impossible, because an integer being given there cannot be an infinity of integers which are smaller.

The margin is too narrow to receive the complete proof with all its developments.

In the same way I have discovered and proved that there cannot be a triangular number, except for one, which is also a fourth power.

Fermat's proof, which is not without its obscurities, survived as a half-page marginal note in his copy of Bachet's edition of Diophantus. The 'modern' thing about it is his method of proof, which is capable of widespread application. Fermat wished to show that there are no solutions, so he argued by contradiction, assuming that there is a solution, finding ever smaller ones, and deducing an absurdity.

It is his method within this *reductio* structure that is so elegant. Specifically, he looked at the value of z, which we can assume to be a positive integer, and showed how to construct from one solution (x_1, y_1, z_1) a second solution (x_2, y_2, z_2) with z_2 strictly less than z_1. By successively repeating this process, he could thus obtain an infinite sequence of decreasing *positive* integers, all less than z — but that is impossible. Consequently, no initial solution can exist and the theorem is proved.

This method, called by Fermat *the method of infinite descent*, is a marvellous contribution to resolving many questions in the negative, and has acquired a life of its own. In a thorough analysis of Fermat's work on number theory, André Weil, a leading number-theorist of the 20th century, argued persuasively that Fermat must have used it on several other occasions, although such proofs have not survived.[24]

Fortunately, just for once, he had found room for this mystery in the margin of the very last proposition of Diophantus; this is how it goes.

Take a Pythagorean triangle whose sides may be assumed mutually prime; then they can be written as $(2pq, p^2 - q^2, p^2 + q^2)$ where p, q are mutually prime, $p > q$, and $p - q$ is odd. Its area is $pq(p+q)(p-q)$, where each factor is prime to the other three; if this is a square, all the factors must be squares. Write $p = x^2, q = y^2, p + q = u^2, p - q = v^2$, where u, v must be odd and mutually prime. Then x, y, and $z = uv$ are a solution of $x^4 - y^4 = z^2$; incidentally, v^2, x^2, u^2 are then three squares in an arithmetic progression whose difference is y^2. We have $u^2 = v^2 + 2y^2$; writing this as $2y^2 = (u+v)(u-v)$, and observing that the g.c.d. of $u + v$ and $u - v$ is 2, we see that one of them must be of the form $2r^2$ and the other of the form $4s^2$, so that we can write $u = r^2 + 2s^2, \pm v = r^2 - 2s^2, y = 2rs$, and consequently

$$x^2 = \tfrac{1}{2}(u^2 + v^2) = r^4 + 4s^4.$$

Thus $r^2, 2s^2$, and x are the sides of a Pythagorean triangle whose area is $(rs)^2$ and whose hypotenuse is smaller than the hypotenuse $x^4 + y^4$ of the original triangle. This completes the proof 'by descent'.

---

[24]In (Weil 1984, 77) and F&G 11.C8. A helpful account is also given in (Edwards 1977, 11–14).

For all that he failed to elicit much of a response, Fermat's career should not be seen as a failure, although we may surely believe that he would have liked more people to share his interests. It represents a typical phenomenon in an exaggerated form: the Janus-like attention of mathematicians in the early 17th century to the progress of their subject. Were they classicists, devotedly rediscovering the legacy of the past? Or were they modernists, boldly creating and discovering new mathematics that surpassed the Greeks?

Later generations were in no doubt about their own originality — Isaac Newton is the last mathematician for whom the comparison with the Greeks matters — and they were to look back at the likes of Fermat and Descartes when they needed to pay homage to their ancestors. But Fermat was deeply committed to his classical heritage, standing deliberately on Diophantine ground. He and others found an abundance of problems about numbers with their origins in the classics of Euclid and Diophantus, and yet he contributed new questions and, above all, new approaches. Even so, at the end of his last (and rather melancholy) letter to Huygens, in 1659, he wrote:[25]

> Maybe posterity will be grateful to me for having shown that the ancients did not know everything ...

and went on to quote Francis Bacon:[26]

> Many will pass away, science will grow.

13.3 Descartes

It is scarcely an exaggeration to say that, as Galileo is to physics and Kepler to astronomy, so is Descartes to mathematics: the modern study of each subject started with these men. At least, this was the opinion of many mathematicians of the following century and later — for example, John Stuart Mill hailed Descartes's work as 'the greatest single step ever made in the history of the exact sciences'.[27]

In fact, much of Descartes's work was in natural science or mathematics, although he is mainly remembered today for his philosophical work. When Descartes was writing, the new experimental science had scarcely established itself as a method of inquiry and it had few triumphs to its name. Opposition to it came largely from philosophical sceptics, who disputed the claim that the use of this novel scientific method could add to knowledge. There was also opposition, to a lesser extent, from the philosophical and theoretical orthodoxy that accepted traditional Aristotelian physics.

The overriding purpose of Descartes's work was to establish that rational inquiry could lead to knowledge, and his mathematical work was partly intended to show how it could be done. Indeed, as we shall see, mathematics was for Descartes the paradigm case of rational inquiry, but he did not disdain science. He went on to investigate a theory of optics, and to outline a theory of planetary motion which Newtonian gravitational theory did not supplant, in France at least, until well into the 18th century.

In his *Discours de la Méthode* (*Discourse on Method*) of 1637, with its celebrated mathematical appendix, *La Géométrie*, Descartes gave an account of how he came to his views about the role of mathematics in the pursuit of true knowledge. On 10 November 1619 he claimed that 'I spent the whole day shut up alone in a stove-heated room', and

[25] See F&G 11.C7.

[26] Quoted in (Weil 1983, 118–119).

[27] See (Mill 1867, 617), cited in (Rée 1974, 28); we consider such grand claims for Descartes later in this chapter.

Figure 13.5. René Descartes (1596–1650)

simply thought, coming to feel that 'buildings conceived and completed by a single architect are usually more beautiful and better planned than those remodelled by several persons using ancient walls ...'.[28] In short, Descartes felt it necessary to carry out a programme for remodelling knowledge completely by himself — an impressive ambition, even for a young man of 23. As he realised, he therefore had to doubt every idea that he might have acquired from childhood.

That night, according to his first biographer Adrien Baillet, he had three dreams that reinforced his sense of mission. Thus began the intellectual odyssey that was to take nearly twenty years before reaching its first published fruits, the *Discours de la Méthode*.

More prosaically, the chief mathematical and scientific influence on Descartes seems to have been Isaac Beeckman, whom Descartes met in Breda in November 1618 while serving as a gentleman volunteer in the army of Prince Maurice of Nassau. Beeckman, who was seven years older than Descartes, was a well-established Dutch scholar, craftsman, and teacher. Descartes wrote to Beeckman on 26 March 1619, outlining his ambitions in mathematics. In this extract from his letter we can see the sort of approach he was advocating to geometrical problems:[29]

---

[28] See (Descartes, 1637, 10).

[29] See (Bos 2001, 232); the original Latin is also there. The Lullius to whom Descartes referred was Ramon Lull (or Llull), a prolific Catalan of the 13th century who produced a combinatorially driven system for combining symbols and ideas, as part of a training whereby Christians could learn to argue against Muslims. It was enjoying a vogue in Descartes's day, and later attracted the attention of Leibniz.

Descartes to Beeckman.

> And to tell you quite openly what I intend to undertake, I do not want to propound a Short art as that of Lullius, but a completely new science by which all questions in general may be solved that can be proposed about any kind of quantity, continuous as well as discrete.
>
> But each according to its own nature. In arithmetic, for instance, some questions can be solved by rational numbers, some by surd numbers only, and others can be imagined but not solved. For continuous quantity I hope to prove that, similarly, certain problems can be solved by using only straight or circular lines, that some problems require other curves for their solution, but still curves which arise from one single motion and which therefore can be traced by the new compasses, which I consider to be no less certain and geometrical than the usual compasses by which circles are traced; and, finally, that other problems can only be solved by curved lines generated by separate motions not subordinate to one another; certainly such curves are imaginary only; the well-known quadratrix line is of that kind. And in my opinion it is impossible to imagine anything that cannot at least be solved by such lines; but in due time I hope to prove which questions can or cannot be solved in these several ways: so that hardly anything would remain to be found in geometry.
>
> This is truly an infinite task, not for a single person. Incredibly ambitious; but through the dark confusion of this science I have seen some kind of light, and I believe that by its help I can dispel darkness however dense.

The first paragraph speaks of a 'complete new science', a topic we return to below in Section 13.5. The second one reads like a version of the classification of geometrical problems given by Pappus (see Section 6.1). But the greater instrumental bias of Descartes's views is noticeable, as is the confidence with which he felt that once geometrical problems had been correctly allocated to their appropriate construction, all geometry would have been sorted out. Notice, too, that what are at stake are problems and their solutions, to be given by means of constructions. The final sentence hints that Descartes believed the problems of his third category to lie outside geometry altogether.

We discuss below what problems lurk in Descartes's ideas about curves, but this is certainly an optimistic programme, strikingly beyond his mathematical abilities to carry out at the time — as the third paragraph not too convincingly acknowledges. We shall also see that his view that some problems are unsolvable geometrically was, by contrast, unduly pessimistic. Descartes had in mind that curves were literally to be drawn by some sort of an idealised machine, made up of hinged rods, that would draw curves exactly as a pair of compasses draws a circle.

Descartes studied mathematics intensely after 1619, attempting to devise a clear and simple algebraic symbolism for analysing mathematical problems. Viète had had the same aim, but it seems that Descartes's later claim that he had not studied Viète's work was true. In 1631 Descartes settled in Holland in order to enjoy the benefits of religious tolerance while continuing to ingratiate himself with the Catholic theologians of the Sorbonne in Paris. In Holland he was introduced by a friend to a problem discussed in the *Mathematical Collection* of Pappus, the 'locus to 3 or 4 lines'. All this time he had tried, with little success, to extract from mathematics the key to correct reasoning. His resolution of the

Pappus problem (which we study in the next section) was to be his sign that he was finally achieving his goal.

Discours de la Méthode. Descartes first published his method of rational enquiry in his *Discours de la Méthode* — or, to give it its full title (in English), *Discourse on the Method of Rightly Conducting the Reason in the Search for Truth in the Sciences.* This work scarcely discusses the method itself, but moves on rapidly to outline the philosophy that Descartes had then developed, and to show off its fruits in three further essays printed at the end. The first of these, on optics, contains the first published statement of the sine law of refraction (Snell's law), and an analysis of lenses and the eye. The second appendix, on meteorology, contains an explanation of primary and secondary rainbows. The third, with which we are chiefly concerned, is the justly celebrated *La Géométrie*, which is itself in three parts.

In the *Discours*, Descartes discussed the role of mathematics, and his intended contribution to it, in this way:[30]

Descartes's *Discours.*

Those long chains of reasoning, so simple and easy, which enabled the geometricians to reach the most difficult demonstrations, had made me wonder whether all things knowable to men might not fall into a similar logical sequence. If so, we need only refrain from accepting as true that which is not true, and carefully follow the order necessary to deduce each one from the others, and there cannot be any propositions so abstruse that we cannot prove them, or so recondite that we cannot discover them. It was not very difficult, either, to decide where we should look for a beginning, for I knew already that one begins with the simplest and easiest to know.

Considering that among all those who have previously sought truth in the sciences, mathematicians alone have been able to find some demonstrations, some certain and evident reasons, I had no doubt that I should begin where they did, although I expected no advantage except to accustom my mind to work with truths and not to be satisfied with bad reasoning.

I do not mean that I intended to learn all the particular branches of mathematics; for I saw that although the objects they discuss are different, all these branches are in agreement in limiting their consideration to the relationships or proportions between their various objects. I judged therefore that it would be better to examine these proportions in general, and use particular objects as illustrations only in order to make their principles easier to comprehend, and to be able the more easily to apply them afterwards, without any forcing, to anything for which they would be suitable. I realized that in order to understand the principles of relationships I would sometimes have to consider them singly, and sometimes comprehend and remember them in groups. I thought I could consider them better singly as relationships between lines, because I could find nothing more simple or more easily pictured to my imagination and my senses. But in order to remember and understand them better when taken in groups, I had to express them in numbers, and in the smallest numbers possible. Thus I took the best traits of geometrical analysis and algebra, and corrected the faults of one by the other.

[30] See (Descartes 1964, 15–16) and F&G 11.A1, paragraph 3.

So we see that Descartes drew from mathematics the idea that all knowledge could be presented in deductive chains, starting from simple truths and using only sound reasoning. He would not learn all mathematics, but, starting from conclusions about straight lines, he would bring together geometry and algebra, thereby improving them both.

In keeping with his youthful ambition, Descartes did not consider himself as just a mathematician, but as a philosopher in the broadest sense of the term. In the *Discours* he described his aim as 'seeking the true method of obtaining knowledge of everything which my mind was capable of understanding': the key word here is *method*. It was his search for a method that brought Descartes to mathematics. But when he turned to mathematics, he found a geometry that was fatiguing and an algebra that was obscure, and so he set out to devise his own rules for discovering the truth.

As Descartes had already made clear, he had by now gone beyond what in 1619 he had said to Beeckman.[31]

> The first rule was never to accept anything as true unless I recognized it to be certainly and evidently such: that is, carefully to avoid all precipitation and prejudgement, and to include nothing in my conclusions unless it presented itself so clearly and distinctly to my mind that there was no reason or occasion to doubt it.
>
> The second was to divide each of the difficulties which I encountered into as many parts as possible, and as might be required for an easier solution.
>
> The third was to think in an orderly fashion when concerned with the search for truth, beginning with the things which were simplest and easiest to understand, and gradually and by degrees reaching toward more complex knowledge, even treating, as though ordered, materials which were not necessarily so.
>
> The last was, both in the process of searching and in reviewing when in difficulty, always to make enumerations so complete, and reviews so general, that I would be certain that nothing was omitted.

His second rule (divide each difficulty into as many parts as possible) looks like analysis, in the sense that Pappus and Viète had used the term. He now talked about using 'symbols' ('chiffres') to express relations between lines.[32]

So the analysis of a problem was to be expressed in algebraic language. This singling out of algebra as the way to analyse a problem is the crucial step taking him beyond his earlier position.

Later generations would not find it easy to say when something was 'certainly and evidently' true, but nor were they to propose any better characterisation of statements that could be taken as bed-rock in an intellectual enquiry. One may suppose that Descartes regarded his four rules as brief notes rather than as definitive statements — indeed, he said as much in a note that he sent to his old schoolmaster, to whom he confided that the *Discours* did not teach but only functioned as a notice for the method. They encapsulated a philosophy that he thought worth proceeding with, so they were, perhaps, pieces of 'good advice' — but Descartes took them very seriously.

---

[31] See (Descartes 1964, 16) and F&G 11.A1, paragraph 2. Descartes had broken with Beeckman in 1629 over a priority dispute, and was unmoved by Beeckman's death in 1637.

[32] The Smith and Latham translation (Descartes 1637) is misleading here. Descartes wrote that he intended to explain things 'par quelques chiffres les plus courts qu'il seroit possible' ('by the briefest possible symbols'); unfortunately, the translation translates the word 'chiffres' as numbers, which makes no sense.

La Géométrie. We now turn to *La Géométrie* to see how successfully Descartes fulfilled his promises. We describe how Descartes set up his new approach to geometry; in the next section we shall see it accomplish one of its most spectacular successes, a triumph of which Descartes was naturally proud. We look at three important topics in *La Géométrie*: his treatment of lengths, the way he treated square roots, and his approach to the solution of equations (mainly, quadratic equations). On the way we pick up one of his most lasting innovations, the use of lower-case letters for known and unknown quantities.

Although Descartes would have agreed with his contemporaries that geometry is the study of magnitudes that can conveniently be represented by lines, he wanted his mathematics to be universal. He meant this in the sense that Viète did, as being applicable to geometry, arithmetic, or any other type of mathematics. At the end of this chapter we see that he considered the lack of universality in Euclidean geometry to be one of its major failings. To achieve this universality, Descartes had to confront the dimensional and homogeneous character of geometry. He therefore introduced a crucial simplification almost at once. Whereas others viewed the product of two lengths as an area, and the product of three lengths as a volume, Descartes explained how magnitudes can be manipulated geometrically and the results described algebraically.[33]

> I shall not hesitate to introduce these arithmetical terms into geometry, for the sake of greater clearness.
>
> For example, let AB be taken as unity, and let it be required to multiply BD by BC. I have only to join the points A and C, and draw DE parallel to CA; then BE is the product of BD and BC.
>
> If it be required to divide BE by BD, I join E and D, and draw AC parallel to DE; then BC is the result of the division.

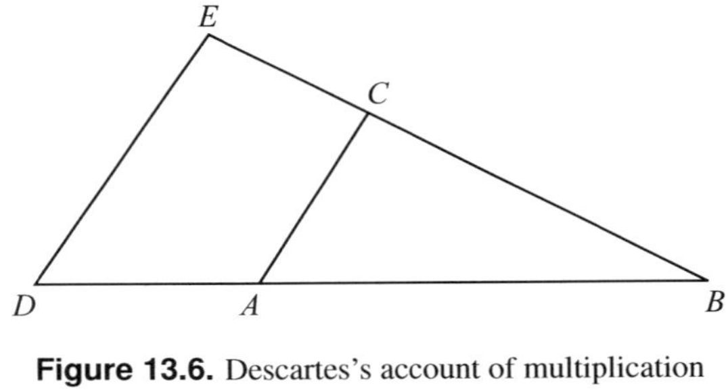

Figure 13.6. Descartes's account of multiplication

> If the square root of GH is desired, I add, along the same line, FG equal to unity; then, bisecting FH at K, I describe the circle FIH about K as a centre, and draw from G a perpendicular and extend it to I, and GI is the required root. I do not speak here of cube root, or any other roots, since I shall speak more conveniently of them later. (See Figure 13.7.)

Taking multiplication as an example, Descartes argued as follows. To multiply BD by BC, he took AB as a unit segment, and drew the segments on different lines through the vertex B (see Figure 13.6). He formed two triangles, ACB and DEB, choosing E so that DE and AC are parallel. Thus these triangles are similar, and so $BC/BA = BE/BD$. It follows that $BD \times BC = BE \times BA$. But BA is of unit length, so BE is equal to the product $BD \times BC$, as claimed. As can be seen from Figures 13.8 and 13.6, Descartes's method is very similar to calculations performed with a sector.

[33] Descartes, *La Géométrie*, p. 5 and F&G 11.A2.

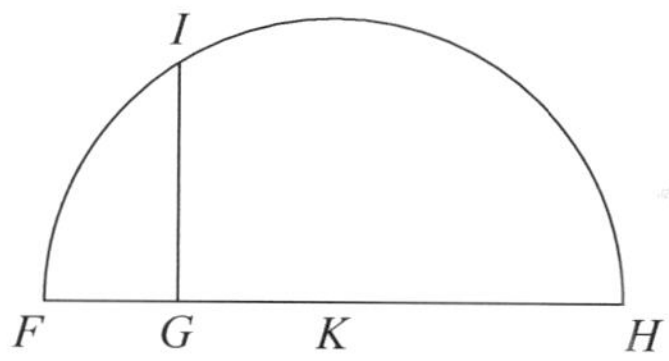

Figure 13.7. Descartes's account of finding a square root

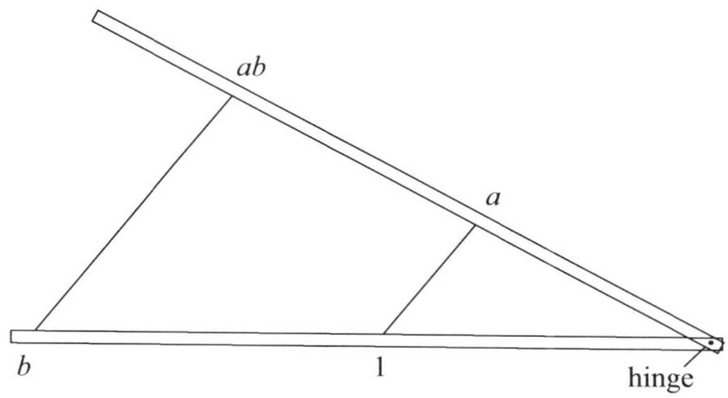

Figure 13.8. A sector used to perform multiplication

The remarkable thing is that Descartes took the length BE to be the answer, whereas the whole of classical antiquity had taken any product of two lengths to be an area. By using the fact that BA is a unit of length, we can express all quantities as line segments (or, one might say in modern terms, one-dimensional) — a decisively new point of view. By deliberately suppressing the fact that BA is a unit of length, Descartes gave the convenient impression that he had introduced dimensionless quantities into mathematics — after all, if all quantities are one-dimensional, what does the dimension concept do for you?

So, where previously geometry had dealt with lengths whose products were areas, and had used the concept of ratio instead of division, it could now deal entirely with lengths whose products and quotients are also lengths. In this respect it is like elementary algebra, which deals with numbers; and the unification of algebra and geometry was a vital part of Descartes's programme. But it is unlike Viète's algebra, which was as conscious of dimension as was Greek geometry. Viète was developing his vision of the true classical mathematics; Descartes was making a fresh start. As Descartes made clear in a letter to Desargues, he thought the classical arguments about ratio and proportion could be complicated for ordinary people to understand, especially in Desargues's hands.[34] Here he presents his streamlined version in which quantities behave like numbers, in that their products and quotients are of the same kind;[35] we know that it was a successful move, because it is the style in which we write today.

Descartes was not the first to propose such a simplification. What was crucial about his proposal was the extent to which he spelled out its implications. For someone who claimed to expect no advantage from his study of mathematics 'except to accustom my mind to work with truths',[36] he went a long way to reformulating the very subject itself.

[34]See (Desargues 1986) and F&G 11.D4.

[35]See F&G 11.D4.

[36](Descartes 1964, 15) and F&G 11.A1.

The same spirit of simplification can also be seen in his construction of a square root. Its significance lies not in the construction itself, which was the standard classical one for finding a mean proportional, given in Euclid's *Elements*, Book VI, Prop. 13,[37] but in the way that Descartes discarded the classical constraints of proportional terminology, treating all lines as representing comparable magnitudes, regardless of the operations producing them.

To this conceptual simplification, Descartes added the notational simplification of representing lines by single lower-case letters. This made possible a much more succinct handling of algebraic formulas, as the underlying geometrical constructions grew more complex. Descartes was now ready to embark on the detailed description of his general programme. It is quite long, and difficult in places — he seems to have been regarded as an arrogantly casual expositor by his successors.[38]

Descartes's general method for solving any problem.

> [1] If, then, we wish to solve any problem, we first suppose the solution already effected, and give names to all the lines that seem needful for its construction — to those that are unknown as well as to those that are known. Then, making no distinction between known and unknown lines, we must unravel the difficulty in any way that shows most naturally the relations between these lines, until we find it possible to express a single quantity in two ways. This will constitute an equation, since the terms of one of these two expressions are together equal to the terms of the other.
>
> [2] We must find as many such equations as there are supposed to be unknown lines; but if, after considering everything involved, so many cannot be found, it is evident that the question is not entirely determined. In such a case we may choose arbitrarily lines of known length for each unknown line to which there corresponds no equation.
>
> [3] If there are several equations, we must use each in order, either considering it alone or comparing it with the others, so as to obtain a value for each of the unknown lines; and so we must combine them until there remains a single unknown line which is equal to some known line, or whose square, cube, fourth power, fifth power, sixth power, etc., is equal to the sum or difference of two or more quantities, one of which is known, while the others consist of mean proportionals between unity and this square, or cube, or fourth power, etc., multiplied by other known lines. I may express this as follows:
>
> $$z = b$$
> $$\text{or} \quad z^2 = -az + b^2,$$
> $$\text{or} \quad z^3 = az^2 + b^2z - c^3,$$
> $$\text{or} \quad z^4 = az^3 - c^3z + d^4, \text{ etc.}$$
>
> That is, z, which I take for the unknown quantity, is equal to b; or, the square of z is equal to the square of b diminished by a multiplied by z; or,

[37] See also F&G 3.C4.

[38] Descartes, *La Géométrie*, p. 13 and F&G 11.A3.

the cube of z is equal to a multiplied by the square of z, plus the square of b multiplied by z, diminished by the cube of c; and similarly for the others.

[4] Thus, all the unknown quantities can be expressed in terms of a single quantity, whenever the problem can be constructed by means of circles and straight lines, or by conic sections, or even by some other curve of degree not greater than the third or fourth.

[5] But I shall not stop to explain this in more detail, because I should deprive you of the pleasure of mastering it yourself, as well as of the advantage of training your mind by working over it, which is in my opinion the principal benefit to be derived from this science. Because, I find nothing here so difficult that it cannot be worked out by one at all familiar with ordinary geometry and with algebra, who will consider carefully all that is set forth in this treatise.

[6] I shall therefore content myself with the statement that if the student, in solving these equations, does not fail to make use of division wherever possible, he will surely reach the simplest terms to which the problem can be reduced.

[7] And if it can be solved by ordinary geometry, that is, by the use of straight lines and circles traced on a plane surface, when the last equation shall have been entirely solved there will remain at most only the square of an unknown quantity, equal to the product of its root by some known quantity, increased or diminished by some other quantity also known. Then this root or unknown line can easily be found. For example, if I have $z^2 = az + b^2$, I construct a right triangle NLM with one side, LM, equal to b, the square root of the known quantity b^2, and the other side, LN, equal to $\frac{1}{2}a$, that is, to half the other known quantity which was multiplied by z, which I supposed to be the unknown line (see Figure 13.9). Then prolonging MN, the hypotenuse of this triangle, to O, so that NO is equal to NL, the whole line OM is the required line z. This is expressed in the following way:

$$z = \tfrac{1}{2}a + \sqrt{\tfrac{1}{4}a^2 + b^2}.$$

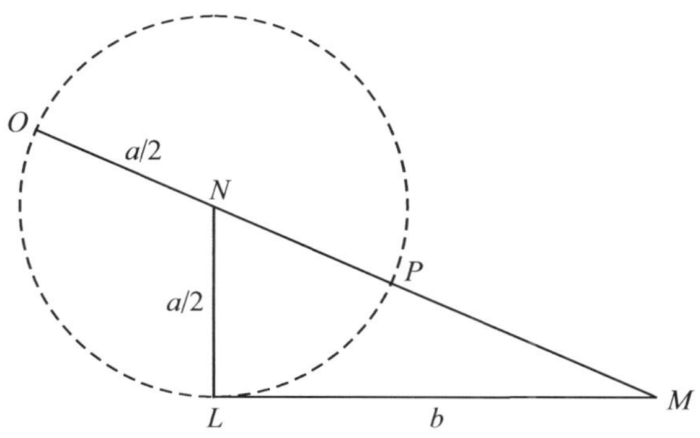

Figure 13.9. Descartes's solution to a quadratic equation

Here Descartes begins with the claim that a method will be developed for solving any problem, and this method is to involve equations. The method is plainly a process of analysis: 'we first suppose the solution already effected', and are to give names to everything in sight. The aim is to obtain a single equation by manipulating the lines until we have obtained equations between them, whence we eliminate all but one of the unknowns. No good reasons are given for believing that this is always possible, but he gave an example, the quadratic equation $z^2 = az + b^2$, as we explain below.

Descartes then extended his method to encompass problems involving many unknowns, and displayed the sort of equation he expected to get on the way to the solution. His conception of what it is to solve a problem is one of the most puzzling features of *La Géométrie*, and is often underestimated; we discuss it in more detail below. Notice here that a problem 'constructed by means of circles and straight lines' does not mean that the answer is a circle or a straight line. Indeed, the answer is most likely to be a number or a line segment representing a magnitude, and not a curve.

Descartes was surely not trying to write an easy book. His generous exercise of self-restraint in the form of an unwillingness to deprive the reader of satisfaction and educational benefit, and a claim that there was 'nothing here so difficult', were surely disingenuous. Indeed, on other occasions he would take the opposite tack. As he wrote to Mersenne in 1637:[39]

> As to the suggestion that what I have written could easily have been gotten from Viète, the very fact that my treatise is hard to understand is due to my attempt to put nothing in it that I believed to be known either by him or by anyone else ... I begin where he left off.

Descartes's example of a quadratic equation is probably the best way in. He put $LM = b$ and $NL = a/2 = NP = NO$, where a and b are given (see Figure 13.9). By the Pythagorean theorem,

$$MN^2 = LN^2 + LM^2 = \tfrac{1}{4}a^2 + b^2 .$$

So

$$MN = \sqrt{\tfrac{1}{4}a^2 + b^2} \quad \text{and} \quad OM = ON + NM = \tfrac{1}{2}a + \sqrt{\tfrac{1}{4}a^2 + b^2} .$$

The marvellous thing is that $OM = z$ is a solution of the equation $z^2 = az + b^2$, and this solution can be obtained entirely algebraically by completing the square.

So we can proceed to solve a quadratic equation with two real roots by a geometrical construction involving circles and lines. By implication, we might infer that Descartes had other constructions for more complicated equations — how else could he claim to be able to solve any problem? — and indeed he went on to outline just such a general method and to explain it in detail in a number of cases.

This makes it clear that Descartes laid claim to a general method for solving problems by algebra, and that the solution of the equations that arise can be effected by geometrical constructions. It remains for us to see the method at work, actually obtain the equations, and solve difficult ones; this forms the subject of our next section.

Descartes also took a significant step forward, from problems that yield an equation in one unknown whose solution is a magnitude, to problems involving two unknowns whose solution is a curve that must then be constructed in some way. We shall see that Descartes took the Pappus problem, which he was so happy to have solved four years earlier, and worked it through — first for three or four lines, and then for any number. He was clearly impressed with how completely he had transcended the classical writers, and so too were his readers. That is the great force of his work.

---

[39] See Descartes, *La Géométrie*, p. 10, n. 17.

13.4 Pappus's locus problem

The *Mathematical Collection* of the late Hellenistic commentator Pappus of Alexandria had been known to the European mathematical community since 1566, when Commandino included extracts from it in his edition of Apollonius's *Conics*. He followed this in 1588 with an edition of the full text, in a Latin translation. It had then been studied intensively for over half a century when Descartes's friend Jacob Golius drew his attention, in 1631, to a problem that Pappus had discussed in connection with Apollonius's *Conics*.[40] Descartes's success in solving and developing his ideas around this problem was to become central to his exposition in *La Géométrie*.

The problem is a *locus problem*, which is a variant on the long-standing Greek preoccupation with defining curves by property. In a locus problem a property is given (expressed in terms of lines, distances, and angles, say) and we want to find the locus of all the points with that property. For example, the locus of all points equidistant from two fixed intersecting lines is a straight line — the line m that bisects the angle made by the fixed lines l_1 and l_2 in Figure 13.10.[41]

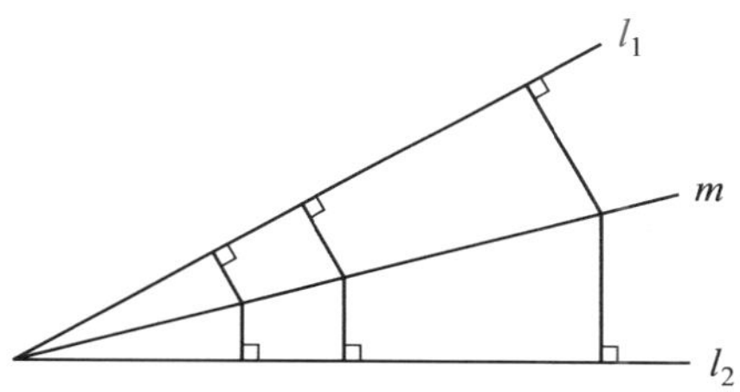

Figure 13.10. The locus of points equidistant from two lines l_1 and l_2

The Pappus problem. Pappus's problem concerns what he called the locus to three or four lines, and its generalisation to any number of lines.[42] It can be summarised as follows in the case of four lines:

Nine ingredients are specified in advance:

four lines (l_1, l_2, l_3, l_4), four angles $(\alpha_1, \alpha_2, \alpha_3, \alpha_4)$, and a fixed ratio k.

Suppose that the lines lie as shown in Figure 13.11. Then we seek the positions of a point P whose distances p_1, p_2, p_3, p_4 from the given lines have the property

$$p_1 \times p_2 = k(p_3 \times p_4),$$

where the distance p_i is measured along the line meeting the line l_i at the angle α_i.

One such point, P, and the locus of all such points are shown in Figure 13.11.

The three-line problem is analogous, and can be thought of as the case where two lines of the four-line problem coincide: that is, there are three lines (l_1, l_2, l_3), three angles $(\alpha_1, \alpha_2, \alpha_3)$, a fixed ratio k, and the condition on P is $p_1 \times p_2 = kp_3^2$.

Pappus had also stated, but did not prove, that the solution to these problems in each case is a conic section. But he did not go into any detail about which conic section arises from which initial configuration of the lines, choice of angles, and the fixed ratio k. It

[40]Golius was a professor of mathematics who later became a brilliant orientalist, based in Leiden.

[41]The distance from a point to a line is measured along the perpendicular from the point to the line.

[42]See F&G 11.A4.

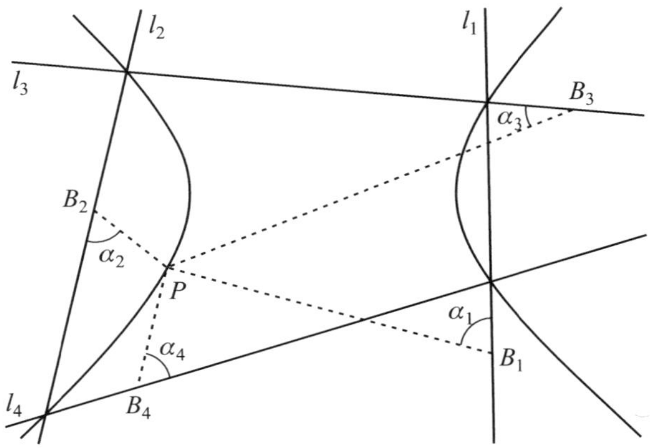

Figure 13.11. A point on the locus to four lines

might be a hyperbola, an ellipse, or a parabola, and Pappus did not distinguish between these cases.

The generalisation to any even number of lines is easy to state, but the problem for an odd number of lines requires a convention. If three lines are given then the condition is

$$p_1 \times p_2 = kp_3^2,$$

but if an odd number of five or more lines are given, then a constant line segment a is introduced to keep the expressions homogeneous, so the problem concerns the property

$$p_1 \times p_2 \times p_3 = k(a \times p_4 \times p_5).$$

But if it was easy to state, it had been almost beyond Pappus's powers to answer. All he could do was to claim that the solution to the problem involving five or more lines would be a curve. That is part of the way towards reaching a solution, for it is not impossible, on considering the matter in advance, that we might find the locus to be a collection of points dotted about the plane. But he went on to say that the curve could not be recognised; he did not know whether it was one of the few curves known to Greek geometers or some new curve altogether.

One difficulty that Pappus had encountered arose in actually specifying the problem. The dimensional restrictions inherent in Greek geometrical language meant that the product of two lines is the rectangle contained by them (that is, an area), and the product of a rectangle and a line is the three-dimensional figure contained by them (and so a volume). So he ran out of dimensions when representing the product of two rectangles; this difficulty would arise in the case of more than six lines. Pappus tried to explain how the difficulty could be got around, by using the language of composition of ratios, but he did not seem very happy about it.

There was therefore a clear challenge to future mathematicians to solve these problems that Greek geometers had been unable to solve: first, to elucidate fully the locus to three and four lines by describing which conic arises from which initial specification of the problem, and secondly, to understand the locus to five, six, or more lines. Descartes took up the challenge and solved it during five or six weeks in 1631–1632.

This is a remarkable testimony to the power of his new algebraic analysis. It is impressive that Descartes could find the locus to three or four lines. It is even more impressive that he could find the locus to five, six, seven, or any greater number of lines with almost equal ease — or rather, since it is clear that Descartes wanted his readers to be impressed, we should perhaps say that he wrote as though the problem was easy. 'I believe that I have ... completely accomplished what Pappus tells us the ancients sought to do', as the following jubilant passage from *La Géométrie* indicates:[43]

Descartes on Pappus's problem.

> This led me to try to find out whether, by my own method, I could go as far as they had gone. First, I discovered that if the question be proposed for only three, four, or five lines, the required points can be found by elementary geometry, that is — by the use of the straight-edge and compasses only, and the application of those principles that I have already explained, except in the case of five parallel lines. In this case, and in the cases where there are six, seven, eight, or nine given lines, the required points can always be found by means of the geometry of solid loci, that is, by using some one of the three conic sections. Here, again, there is an exception in the case of nine parallel lines. For this and the cases of ten, eleven, twelve, or thirteen given lines, the required points may be found by means of a curve of degree next higher than that of the conic sections. Again, the case of thirteen parallel lines must be excluded, for which, as well as for the cases of fourteen, fifteen, sixteen, and seventeen lines, a curve of degree next higher than the preceding must be used; and so on indefinitely.
>
> Next, I have found that when only three or four lines are given, the required points lie not only all on one of the conic sections but sometimes on the circumference of a circle or even on a straight line.
>
> When there are five, six, seven, or eight lines, the required points lie on a curve of degree next higher than the conic sections, and it is impossible to imagine such a curve that may not satisfy the conditions of the problem; but the required points may possibly lie on a conic section, a circle, or a straight line. If there are nine, ten, eleven, or twelve lines, the required curve is only one degree higher than the preceding, but any such curve may meet the requirements, and so on to infinity.
>
> Finally, the first and simplest curve after the conic sections is the one generated by the intersection of a parabola with a straight line in a way to be described presently. I believe that I have in this way completely accomplished what Pappus tells us the ancients sought to do, and I will try to give the demonstration in a few words, for I am already wearied by so much writing.

Descartes's idea of what constitutes a solution to a mathematical problem is an important and revealing aspect of his work, which we look at next.

What did Descartes understand by a 'solution'? Descartes's solution depends on how many given lines there are in the problem. If there are three, four, or five lines, then the solution can be obtained by straight-edge and compasses; if there are six to eight lines, the solution can be obtained by conic sections; for ten to twelve lines, the solution needs 'a curve of degree next higher'; and so on. Note that these curves are not the

---

[43]Descartes, *La Géométrie*, p. 26, and F&G 11.A6.

solutions as such, but are the curves that are needed to produce the solutions. The solution itself — that is, the required locus — for five to eight lines is a 'curve of degree next higher than the conic sections'; for nine to twelve lines, the locus is a curve one degree higher still; and so on.

Descartes not only found the solutions and classified them in terms of how many lines are given in the original problem, he also made a revealing distinction between the solution (the locus) and what needed to be done in order to construct it. As regards the solution curve, Descartes was careful to state that he was looking for points on the locus, not the whole curve itself — the phrase he used continually was 'the required points'. This is how, although a conic section is the solution to the three-line or four-line problem, it can be used for finding the solution up to the nine-line problem. The distinction between constructing points on a curve and the curve itself has important implications for Descartes's conception of curves, as we shall see shortly.

So how did Descartes reach these remarkable results, not previously attained, as he heavily implied, by any mathematician of ancient or his own times? Here is his own account, interspersed with our comments.[44]

> Let AB, AD, EF, GH, ... be any number of straight lines given in position, and let it be required to find a point C, from which straight lines CB, CD, CF, CH, ... can be drawn, making given angles $CBA, CDA, CFE, CHG, \ldots$ respectively, with the given lines, and such that the product of certain of them is equal to the product of the rest, or at least such that these two products shall have a given ratio, for this condition does not make the problem any more difficult.
>
> First, I suppose the thing done, and since so many lines are confusing, I may simplify matters by considering one of the given lines and one of those to be drawn (as, for example, AB and BC) as the principal lines, to which I shall try to refer all the others. Call the segment of the line AB between A and B, x, and call BC, y. Produce all the other given lines to meet these two (also produced if necessary) provided none is parallel to either of the principal lines. Thus, in the figure [Figure 13.12], the given lines cut AB in the points A, E, G, and cut BC in the points R, S, T.

'First, I suppose the thing done' — the opening gambit makes clear that this is a method of analysis. The 'thing' that Descartes supposed to have been done is the location of the point C, where C is a point lying on the locus — that is, a point satisfying the property laid down in the problem. Once the problem is solved, there are geometrical relations between the lines constituting the final diagram, so that those line-lengths can be expressed in terms of other line-lengths and ratios. In particular, Descartes showed that all the lengths in the problem can be given in terms of just two line segments, which he called x and y. It does not seem to matter particularly which segments are chosen; Descartes took the length of one of the lines from C to be y, and the length of some segment along one of the original lines to be x. All other lengths in the problem can then be written down in terms of these two lines.

This was a groundbreaking step, using two lines as a way to attach coordinates to points in the plane. It opened the way to the use of algebra to express the relationship between points, such as the points that satisfy a given condition (and form a locus), and it was to generate a huge transformation in people's ideas about geometry that has by no means exhausted its implications even today.

Descartes now argued as follows:

[44] Descartes, *La Géométrie*, pp. 26–34 and F&G 11.A6.

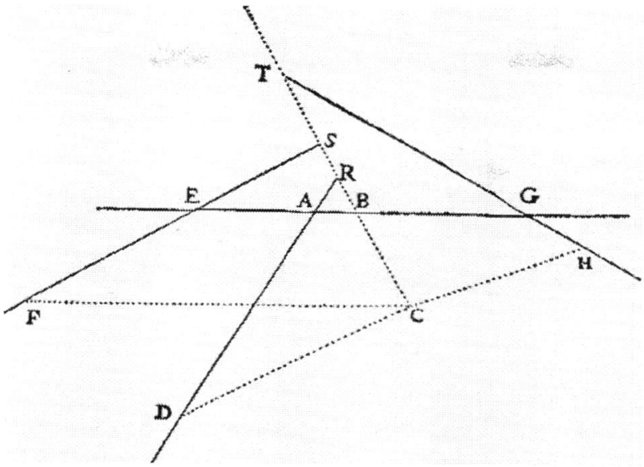

Figure 13.12. Descartes's figure

Now, since all the angles of the triangle ARB are known, the ratio between the sides AB and BR is known. If we let $AB : BR = z : b$, since $AB = x$, we have $RB = \frac{bx}{z}$; and since B lies between C and R, we have $CR = y - \frac{bx}{z}$. (When R lies between C and B, CR is equal to $y - \frac{bx}{z}$, and when C lies between B and R, CR is equal to $-y + \frac{bx}{z}$.)

Again, the three angles of the triangle DRC are known, and therefore the ratio between the sides CR and CD is determined. Calling this ratio $z : c$, since $CR = y + \frac{bx}{z}$ we have $CD = \frac{cy}{z} + \frac{bcx}{z^2}$. Then, since the lines AB, AD, and EF are given in position, the distance from A to E is known. If we call this distance k, then $EB = k + x$; although $EB = k - x$ when B lies between E and A, and $E = -k + x$ when E lies between A and B. Now the angles of the triangle ESB being given, the ratio of BE to BS is known. We may call this ratio $z : d$. Then $BS = \frac{dk+dx}{z}$ and $CS = \frac{zy+dk+dx}{z}$. When S lies between B and C we have $CS = \frac{zy-dk-dx}{z}$ and when C lies between B and S we have $CS = \frac{-zy+dk+dz}{z}$. The angles of the triangle FSC are known, and hence, also the ratio of CS to CF, or $z : e$. Therefore, $CF = \frac{ezy+dek+dex}{z^2}$.

Likewise, AG or l is given, and $BG = l - x$. Also, in triangle BGT, the ratio of BG to BT, or $z : f$, is known. Therefore, $BT = \frac{fl-fx}{z}$ and $CT = \frac{zy+fl-fx}{z}$. In triangle TCH, the ratio of TC to CH, or $z : g$, is known, whence $CH = \frac{gzy+fgl-fgx}{z^2}$.

It is one thing to express all lines in the problem in terms of x and y, and another to be able to cope with the resulting expressions. Fortunately for Descartes, as he observed, all the expressions are very straightforward: they are effectively all of the form $ax + by + c$, where a, b, and c are quantities that are known in terms of the problem. For instance, the line CH, the last whose length he deduced, can be written as

$$(-fg/z^2)x + (g/z)y + (fgl/z^2),$$

which is indeed of this form, with $a = -fg/z^2, b = g/z$ and $c = fgl/z^2$.

Descartes on the locus to any number of lines.

> And thus you see that, no matter how many lines are given in position, the length of any such line through C making given angles with these lines can always be expressed by three terms, one of which consists of the unknown quantity y multiplied or divided by some known quantity; another consisting of the unknown quantity x multiplied or divided by some other known quantity; and the third consisting of a known quantity. An exception must be made in the case where the given lines are parallel either to AB (when the term containing x vanishes), or to CB (when the term containing y vanishes). This case is too simple to require further explanation. The signs of the terms may be either $+$ or $-$ in every conceivable combination.
>
> You also see that in the product of any number of these lines the degree of any term containing x or y will not be greater than the number of lines (expressed by means of x and y) whose product is found. Thus, no term will be of degree higher than the second if two lines be multiplied together, nor of degree higher than the third, if there be three lines, and so on to infinity.
>
> Furthermore, to determine the point C, but one condition is needed, namely, that the product of a certain number of lines shall be equal to, or (what is quite as simple), shall bear a given ratio to the product of certain other lines. Since this condition can be expressed by a single equation in two unknown quantities, we may give any value we please to either x or y and find the value of the other from this equation. It is obvious that when not more than five lines are given, the quantity x, which is not used to express the first of the lines can never be of degree higher than the second.
>
> Assigning a value to y, we have $x^2 = \pm ax \pm b^2$, and therefore x can be found with ruler and compasses, by a method already explained. If then we should take successively an infinite number of different values for the line y, we should obtain an infinite number of values for the line x, and therefore an infinity of different points, such as C, by means of which the required curve could be drawn.

Thus, by calling certain lengths x and y, Descartes obtained an equation of degree 2 and so the required locus is a conic. This how geometry becomes algebra and the algebra turns back into geometry.

Because of its importance, it is worth spending time teasing out some details of his method. It is clear that Descartes's procedure differs from the system of 'Cartesian coordinates' that we now use, for his 'principal lines' are not necessarily at right angles and are not given in advance: they may be chosen in relation to the circumstances of the problem.

The analysis then seems to proceed in a straightforward way, but what has it to do with solving the original problem? Descartes first invoked the condition laid down in the problem. The locus property that C must satisfy is that the product of certain line-lengths, measured from C, is proportional to the product of certain other line-lengths. Descartes was then in a position to recast this geometrical locus property in terms of his algebraic analysis. Recall that, for the four-line locus, the defining locus property states that the product of two line-lengths is proportional to the product of two other line-lengths. Each side of this equation is a product of two terms of the form $ax + by + c$, multiplied by a constant.[45] When multiplied out, such a product may have terms in x^2, y^2, and xy, but no higher terms in x and y; it is at most a second-degree equation.

The aim of this procedure is to find some point C satisfying the property, and, as a result of the analysis, the property has now taken on the form of a second-degree equation in x and y. So, as Descartes observed, y can be given any value and the corresponding values of x can be constructed; this is because allocating some value to y leaves a second-degree equation in x — that is, an equation with x^2, x, and constant terms — and Descartes had shown at the start of *La Géométrie* how x can be constructed with straight-edge and compasses in such circumstances.

Descartes's conception of finding the locus curve was to find points on the curve, by repeatedly finding the values of x corresponding to different values of y. If we do this often enough, we effectively draw the curve. It seems that the possibility of being able to find (or construct) arbitrarily many points on the curve amounts to knowing the curve.

Descartes's argument here is evidently a general one. If we are given the problem of finding the locus to n lines, then the property turns out to be expressible as an equation in x and y, whose degree is about half the number of lines. One can then choose values of y at will, leaving an equation in x of degree $n/2$ or thereabouts, whose roots enable points on the locus curve to be determined. In summary, the solution to Pappus's problem is a matter of constructing the roots of equations. This is the perception to which Descartes's algebraic analysis led him. Notice that it does tell us whether the curve is an ellipse, a parabola, or a hyperbola.

Descartes's constructions. How are these roots to be found? Descartes had already shown how to construct a root of a quadratic equation by straight-edge and compasses, and thus his procedure solved the locus problem for three, four, or five lines. But the locus curve for more lines, whose defining property is a cubic, quartic, or yet higher-degree equation, generally needs some construction that goes beyond straight-edge and compasses.

Descartes needed to convince his readers that a whole array of curves of ever higher degree are just as fundamental and constructible as the basic line and circle. If he could show this, then his claim to be able to solve the Pappus problem for any number of lines would be justified, for some allowable curve could be introduced to make possible the construction of the roots of the higher-degree equation defining the locus. So Descartes's search for a complete solution to the problem led him to consider what curves could be considered as geometrical, and he gave a rather remarkable answer (see Figure 13.13).[46]

Descartes on curves.

> Probably the real explanation of the refusal of ancient geometers to accept curves more complex than the conic sections lies in the fact that the first curves to which their attention was attracted happened to be the spiral, the quadratrix, and similar curves, which really do belong only to mechanics, and are not among those curves that I think should be included here, since they must be conceived of as described by two separate movements whose relation does not admit of exact determination . . .
>
> Consider the lines AB, AD, AF, and so forth, which we may suppose to be described by means of the instrument YZ. This instrument consists

---

[45] It is a straightforward exercise in elementary coordinate geometry to show that the distance of a point (x_0, y_0) from the line $ax + by + c = 0$ is $\frac{ax_0+bx_0+c}{\sqrt{a^2+b^2}}$.

[46] Descartes, *La Géométrie*, pp. 44–48, 156, and F&G 11.A7.

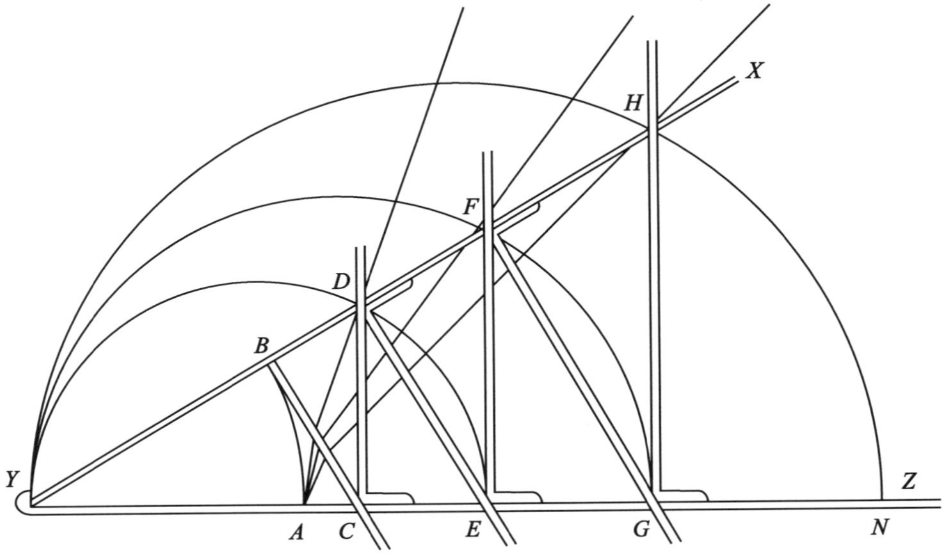

Figure 13.13. Descartes's machine (redrawn for clarity)

of several rulers hinged together in such a way that YZ being placed along the line AN the angle XYZ can be increased or decreased in size, and when its sides are together the points B, C, D, E, F, G, H, all coincide with A; but as the size of the angle is increased, the ruler BC, fastened at right angles to XY at the point B, pushes toward Z the ruler CD which slides along YZ always at right angles. In like manner, CD pushes DE which slides along YX always parallel to BC; DE pushes EF; EF pushes FG; FG pushes GH, and so on. Thus we may imagine an infinity of rulers, each pushing another, half of them making equal angles with YX and the rest with YZ.

Now as the angle XYZ is increased the point B describes the curve AB, which is a circle; while the intersections of the other rulers, namely, the points D,F, H describe other curves, AD, AF, AH, of which the latter are more complex than the first and this more complex than the circle. Nevertheless I see no reason why the description of the first cannot be conceived as clearly and distinctly as that of the circle, or at least as that of the conic sections; or why that of the second, third, or any other that can be thus described, cannot be as clearly conceived of as the first: and therefore I see no reason why they should not be used in the same way in the solution of geometric problems.

I could give here several other ways of tracing and conceiving a series of curved lines, each curve more complex than any preceding one, but I think the best way to group together all such curves and then classify them in order, is by recognizing the fact that all points of those curves which we may call 'geometric', that is, those which admit of precise and exact measurement, must bear a definite relation to all points of a straight line, and that this relation must be expressed by means of a single equation. If this equation contains no term of higher degree than the rectangle of two

> unknown quantities, or the square of one, the curve belongs to the first and simplest class, which contains only the circle, the parabola, the hyperbola, and the ellipse; but when the equation contains one or more terms of the third or fourth degree in one or both of the two unknown quantities (for it requires two unknown quantities to express the relation between two points) the curve belongs to the second class; and if the equation contains a term of the fifth or sixth degree in either or both of the unknown quantities the curve belongs to the third class, and so on indefinitely.
>
> ... But the curve AD is of the second class, while it is possible to find two mean proportionals by the use of the conic sections, which are curves of the first class. Again, four or six mean proportionals can be found by curves of lower classes than AF and AH respectively. It would therefore be a geometric error to use these curves. On the other hand, it would be a blunder to try vainly to construct a problem by means of a class of lines simpler than its nature allows.

To work out the criterion that Descartes used, we examine the words that he employed to justify his instrument. He was concerned with *clear* and *distinct* conceptions (as you would expect from an appendix to the *Discours*), and he required that geometrical curves should 'admit of precise and exact measurement'. His specification was somewhat imprecise until he spelled out what he took it to mean. We can infer this from understanding what the instrument does.

Although the figure looks forbidding, the curves that the instrument draws — the lines emanating from A — are geometrical, in the sense that each point is known exactly in its relation to A: in principle, the relationship of H (say) to A is no more complicated than the relationship between two lines in the earlier Pappus problem diagram. The feature that allows these curves to be geometrical in Descartes's understanding is that only one movement (rotating the bar XY about Y) generates and specifies all subsequent positions of all the points. This is quite different from the double movement that needs to be specified for the spiral, the quadratrix, and other curves that Descartes felt should not be considered as geometrical.

Descartes's instrument is better to hold in the mind than in one's hands — but that, in a way, was Descartes's point. It is a thought device for showing that constructions for curves more complicated than the circle can nevertheless be considered just as accurate, for the purposes of geometrical construction, as the straight-edge and compasses. This was not a prospectus for a precision scientific-instrument business, but a philosophical investigation into the foundations of geometrical truth.[47]

Allowable curves. The question of which curves are allowable in geometry — and indeed, what geometrical curves *are* — is a highly important one in *La Géométrie*, precisely because curves had two roles for Descartes: as solutions to problems, and as the means by which solutions are constructed. Thus, as Pappus knew, a conic is the solution to the locus problem for three or four lines. But a conic is also the *means by which* solutions are found to the locus problem for six to eight lines. The intersection of two conics, or a conic and a straight line, affords the construction of a point on the higher locus. And as the locus curves become more complicated, so it becomes important to know how to recognise a curve that yields a valid geometrical construction of the solution.

---

[47]John Aubrey provided anecdotal support for this bias in Descartes's thought: see F&G 9.G7.

Descartes decided that not every curve will do. His answers to the question of which curves are geometrical are surprising in several respects.[48] On the one hand, he wished to rule out certain curves, such as the quadratrix and the spiral, because their definitions could not be conceived in a sufficiently clear and distinct way; not even the authority of Archimedes was sufficient to validate the spiral for Descartes.

On the other hand, Descartes was quite relaxed about admitting curves constructed by means of pieces of string, and he regarded a *point-wise* construction of a curve as perfectly acceptable. He constructed the locus of points by solving the Pappus problem point by point, just as any modern computer does. Yet he must have placed some implicit restrictions on which curves could be constructed in this way, for it might seem that any curve (even the quadratrix) can be constructed point-wise if one tries hard enough. Everything hangs on what criteria a construction must satisfy to be acceptable, and suitable criteria are hard to provide. Certainly, the critical distinction of which curves are allowable in geometry does not emerge from Descartes's discussion in anything approaching a clear or distinct way. Most surprising of all, there is a clear and simple answer, implicit in Descartes's work, which he chose not to put forward: to allow into geometry precisely those curves that are represented in his analysis by polynomial equations.[49]

> But the fact that this method of tracing a curve by determining a number of its points taken at random applies only to curves that can be generated by a regular and continuous motion does not justify its exclusion from geometry. Nor should we reject the method in which a string or loop of thread is used to determine the equality or difference of two or more straight lines drawn from each point of the required curve to certain other points, or making fixed angles with certain other lines. We have used this method in *La Dioptrique* in the discussion of the ellipse and the hyperbola.
>
> On the other hand, geometry should not include lines that are like strings, in that they are sometimes straight and sometimes curved, since the ratios between straight and curved lines are not known, and I believe cannot be discovered by human minds, and therefore no conclusion based upon such ratios can be accepted as rigorous and exact. Nevertheless, since strings can be used in these constructions only to determine lines whose lengths are known, they need not be wholly excluded.

Why did Descartes not take as his criterion of geometrical acceptability the property that a curve should be *algebraic*, in the terminology of the next generation of mathematicians — that is, that it can be represented by a polynomial equation? This question, among others, has been cogently addressed by the historian Henk Bos, who argued that there were two main reasons for Descartes adopting the approach he did. The more technical one is that, for Descartes, algebraic curves were insufficiently geometrical, because they in turn would need a rule for their construction before they became intelligible. It was not clear to Descartes (and would have been difficult to show) that there was some continuous motion, of the kind he allowed, for the generation of all curves specified in terms of polynomial equations.[50]

The other reason is more general and philosophical. The whole Cartesian programme is an algebraic analysis of geometry (as the title of the book suggested), so the grounds of justification must be ultimately geometrical. An algebraic equation is not a geometrical

[48] See F&G 11.A7.

[49] F&G 11.A8.

[50] See (Bos 1981, 304–305 and 331–332) and F&G 11.A10.

object, and could not be taken as the defining property of one without sabotaging his entire enterprise. So Descartes's attitude towards the acceptability of algebraic formulations, as seen in *La Géométrie*, was ambivalent. He did not put forward an absolutely clear and consistent position. Bos has traced some of these ambiguities in Descartes's text back to the early 1630s, so it is fair to suppose that they reflect ambiguities in Descartes's mind at the time when when he first tackled the Pappus problem:[51]

> There is a conflict in the *Géométrie* between geometrical and algebraic methods of definition and criteria of acceptability. This conflict reflects a break in the development of Descartes' thought about geometry. In an early phase Descartes considered that the aim of geometry was to construct solutions of geometrical problems by means of curves traced by certain instruments; the instruments served as acceptable generalisations of straight-edge and compass. He tried to find new constructions in this way and to classify them.
>
> About 1630 that plan seemed to stagnate and Descartes also became fully aware of the power of algebraic methods. He then changed his programme. Algebra became the dominant tool, both for the solution of problems and for the classification of curves. But Descartes continued to believe in the principle of geometrical construction by means of curves traceable by instruments. As a result, there are conflicting elements in the *Géométrie*.

So it was a shift in what Descartes was attempting to achieve, in the historical development of his thought, that accounts for both the great significance and also the internal contradictions in *La Géométrie*. The algebraic language and thinking sit uneasily with the generalised compasses and the solution of problems by finding intersection points of curves.

Bos gives three reasons for Descartes's decision to go into print with what was, in fact, a contradictory programme. First, it is easy to understand what it means for curves to intersect, but nothing as so easy as accepting that pairs of algebraic equations define geometrical points — and Descartes's whole philosophy was based on the centrality of clear and distinct ideas. Second, and perhaps even more important, to define geometrical curves in such a way would virtually transform geometry into algebra — and Descartes's aim was to organise geometry, not to demolish it. Finally, though perhaps less convincingly, Bos suggests that the great emphasis that Descartes placed on finding the simplest solution methods is imperilled by going over to algebra. Indeed it is, but one might say: so what? Ever since Descartes's time, some mathematicians have followed an algebraic route while others have gone a more geometrical way, and they have differed over precisely this issue — Newton was to have very firm anti-Cartesian views on the matter.[52]

***La Géométrie*, Book II.** Having looked in some detail at Descartes's overall approach, mainly as it appeared in Book I of *La Géométrie*, we now look more briefly at the remaining two books.

In Book II, Descartes gave a thorough analysis of second-degree equations in two unknowns, showing that they represent conic sections, and in particular indicating when the locus in the Pappus problem is a parabola, hyperbola, or ellipse. He also indicated by which construction in Apollonius's *Conics* each of them could be realised — but that was as constructions involving a cone in three dimensions. He then took a special case of the

[51] See (Bos 1981, 298).

[52] As described in (Guicciardini 1999, 29).

problem of five lines, showed that the equation of the corresponding locus is

$$y - 2ay^2 - a^2y + 2a^3 = axy ,$$

for some constant a, and showed how to draw it by means of a movable parabola and a rotating line. This curve became known as the *Cartesian parabola*, and Descartes was to use it for constructing solutions to equations. Descartes then gave his method for finding normals to curves, which he regarded as very important; this is best considered in the context of the methods that other mathematicians proposed for finding tangents. He then constructed and analysed several quartic curves that he needed in his study of optics. Book II ended with some cryptic remarks about the study of curves in space.

***La Géométrie*, Book III.** In Book III, Descartes returned in detail to the problem to which his method always led: how to solve an equation geometrically. As we saw, a quadratic equation is solvable by a straight-edge and compass construction. Higher-degree equations, such as cubics and quartics, generally require conic sections, and he showed how to do this. A quadratic equation has a formula expressing its solutions, and you have seen how Descartes showed how to interpret the solutions geometrically and thus construct them. A cubic equation also has a solution given by a formula, and Descartes solved such equations geometrically, using a circle and a parabola; the same construction can also produce curves of degree 4.

Descartes then showed how to use the Cartesian parabola and the circle to solve the general equation of degree 6, and how this yields the solution of equations of degree 5, before he concluded *La Géométrie* with these words that succinctly capture his ambition and his achievement:[53]

> But it is not my purpose to write a large book. I am trying rather to include much in a few words, as will perhaps be inferred from what I have done, if it is considered that, while reducing to a single construction all the problems of one class, I have at the same time given a method of transforming them into an infinity of others, and thus of solving each in an infinite number of ways; that, furthermore, having constructed all plane problems by the cutting of a circle by a straight line, and all solid problems by the cutting of a circle by a parabola; and, finally, all that are but one degree more complex by cutting a circle by a curve but one degree higher than the parabola, it is only necessary to follow the same general method to construct all problems, more and more complex, ad infinitum; for in the case of a mathematical progression, whenever the first two or three terms are given, it is easy to find the rest.
>
> I hope that posterity will judge me kindly, not only as to the things which I have explained, but also as to those which I have intentionally omitted so as to leave to others the pleasure of discovery.

13.5 The Cartesian challenge to Euclid

As we saw from John Stuart Mill's comment in Section 13.3, many people make great claims for the significance of Descartes's *La Géométrie*: in fact, it has become a commonplace that Descartes accomplished something like a revolution in geometry. But this grievously over-simplifies a complicated and interesting story about the ways in which Descartes's ideas about geometry actually challenged those of Euclid, and the extent to which they became accepted.

[53]Descartes, *La Géométrie*, p. 240.

Novelties and continuities in Descartes's work. We can start with the apparent contrast between the synthetic Euclid's *Elements* and the analytic *La Géométrie*. Henk Bos[54] has summarised what is analytic in *La Géométrie* as follows:

> the method of the *Geometry* consisted of:
>
> A. An analytic part, using algebra to reduce any problem to an appropriate equation; and
>
> B. A synthetic part, finding the appropriate construction of the problem on the basis of the equation.

Bos added that Descartes's method of analysis provides constructions, at least for plane problems (those that lead to quadratic equations). So Descartes's analysis was followed by a process of synthesis that produced or realised the solution in a more geometrical way, by means of a construction.

Next, there is the marked contrast between how the concept of length was handled by Euclid and by Descartes. Numerical lengths, as such, do not enter into Euclid's account of geometry. Rather, Euclid spoke of two lines (meaning line-segments) being equal, meaning that one can be fitted exactly on the other; when this is not the case he spoke of one segment exceeding the other. He could compare ratios of segments, and when the segments are commensurable he could take one as the unit and assign the other a length (in terms of that unit). But, whether because of the difficulties with incommensurable segments or for some other reason, length is not a concept in Euclid's *Elements* in the way that it is in Cartesian geometry.

As we have seen, Descartes regarded the product of two lengths as another length. But even he regarded length as a primitive concept from which the concept of measuring numbers was derived: the approach that treats Cartesian coordinates as real numbers, and so explains lengths in terms of numbers, is a 19th-century one. Descartes's coordinates are to be understood as lengths — that is, as ratios of a certain segment to an arbitrary but fixed choice of unit. This became Newton's view of what measuring numbers are, as a ratio of a segment to a unit segment. It became the standard view and was used in applications every time that a ship builder, architect, or the like called for a twenty-foot long beam, two feet square. Descartes's concept of length greatly simplified and enhanced his algebra.

Descartes the revolutionary? Although there are marked differences between the work of Euclid and Descartes, there are considerable similarities, so it is no surprise that the claim that Descartes was a revolutionary in geometry has been made almost as often as it has been contested.

This debate has been analysed by the historian and philosopher Paolo Mancosu,[55] who found that the French mathematician Bernard le Bouvier de Fontenelle in 1696 was the first to speak of Descartes starting a revolt. Two generations later, in 1757, the French historian of mathematics Jean-Étienne Montucla called Descartes's algebraic analysis a 'happy revolution' and compared his impact with that of Plato. In 1837, in his famous and influential *Aperçu Historique*, the geometer and historian Michel Chasles spoke of Descartes achieving a complete revolution. The first cautionary note was sounded by Gaston Milhaud in his *Descartes Savant* (1921), who noted Fermat's geometrical work prior to *La Géométrie*. Another was struck by Belaval in his *Leibniz: Critique de Descartes* (1960), who argued that Descartes was an ancient, not a modern like Leibniz, because his mathematics could not allow the study of the infinite and the infinitesimal.

---

[54] See (Bos 2001, 287, 309).

[55] See Mancosu's chapter in (Gillies 1992, 105 and 107).

Inevitably, this disagreement hinges on the vexed issue of continuity (or evolution) versus revolution (or dramatic change). Must a revolution make a complete break with the past? How much can be left unchanged before milder terms are to be preferred (breakthrough, major development, ...)? Milhaud's argument is that Descartes and Fermat were simply continuing lines of enquiry raised by the Greeks, so there was

> neither a question of revolution nor a question of creation which radically transformed mathematics and renewed science. It is only a matter of normal development ...

Descartes would not have disagreed, inasmuch as it was part of his claim to particular merit that he had gone back to the Greeks and achieved what they had been unable to do. But we as historians might feel he was being short-changed: something about the way he presented his work suggests that he would not want to have been seen as chiefly making a return to the past.

The historian of science I.B. Cohen directly addressed the issue of what constitutes a revolution in science. He argued that a revolution has four characteristics:[56]

- the authors, or their contemporaries, must give it that status
- the relevant texts must bear such an interpretation
- competent historians must agree that a revolution has occurred
- and so must contemporary scientists or mathematicians.

On that basis, Mancosu found that the case for Descartes is not established. He noted that there is no evidence that Descartes's contemporaries spoke of a revolution and as we have seen, there is no consensus among historians. To this one might reply that the first characteristic may arise because the word 'revolution' was not used in the 17th century with the same suggestion of approbation as it is today, and the second characteristic may reflect the truth that complete agreement among historians is rather unlikely. Some historians delight in puncturing grand claims, either with a wealth of conflicting detail or with other equally grand claims, while others (often philosophers of mathematics who are eager for tidiness in their historical references) like sharp divisions.

When words are contentious, it can help to spell out how they are being used. We suppose there should not be too many revolutions. Standards must be high — but not too high, or there will be none. If we find that the evident continuities in the history are overwhelming, then it is inappropriate to speak of a revolution. Almost always in the history of mathematics and science the word 'revolution' is taken in a positive sense, which may not be what one thinks of revolutions in the political world, and one should be sensitive to that.

That said, how should we think about these issues? Everyone agrees that Descartes's significance derives from the way that he conducted analysis in geometry by means of algebra. Yes, Fermat had come to this idea before Descartes, but he applied it only to the study of conic sections; he did not present it as a general method, he had no dramatic examples of its use, and he confined it with the cumbersome notation of Viète. Yes, it was circulated in Paris in 1636, before Descartes's *La Géométrie* was published, but it was not *published* until it appeared in Fermat's *Varia Opera* of 1679. By then Descartes's *La Géométrie* had been out for forty years, had been translated into Latin, and had inspired helpful commentaries. In the Netherlands, England, and France, it was Descartes's analytic geometry that was promoted. Historians of science rightly speak of 17th-century Cartesians doing Cartesian mathematics.

[56] See (Cohen 1985, 41–45).

Descartes's challenge to Euclid. Can we say that Descartes's work posed a fundamental challenge to Euclid's *Elements*? Certainly, his method did not challenge the foundations of the *Elements*, rather it offered a new way of doing the same geometry. All of Euclid's theorems are true in Cartesian geometry. The real challenges to Euclid's *Elements* were presented by Descartes's style, by his choice of language and notation, and above all by his philosophy.

Descartes's *La Géométrie* was not presented in a form resembling the *Elements*, but in a discursive style, and this speeded up a process that was already under way. Newton's *Principia Mathematica* was written in a style that resembles the *Elements*, but there were few other examples for the next 250 years. By writing in French, Descartes encouraged mathematicians to write in the vernacular, and this undermined the claim that the *Elements* was a valuable part of the European heritage because it was in Greek and Latin. These observations suggest that when research is not finished, but ongoing and provisional, the time is not right for an axiomatic treatment. The Cartesian style was set up to tackle problems: it was adapted to work in progress and it proved to be well suited to the introduction of the calculus.

More threatening was Descartes's philosophy. Recall that *La Géométrie* was originally published as an appendix to his *Discourse on Method* of 1637. This *Discourse* in turn drew on unpublished ideas of Descartes that also came to magnify the impact of his philosophy: the *Regulae ad Directionem Ingenii* (*Rules for the Direction of the Mind*), first published in Dutch in 1684 and then in Latin in 1701. The *Regulae* will help us to elucidate what Descartes wanted to accomplish with his critique of geometry.

Descartes argued that thinking should start from ideas that were clear and distinct to the mind. But such ideas need not be those that Euclid had chosen. Knowledge was to be based on simple and distinct conceptions of the pure and attentive mind (Rule III) and these conceptions were to be amenable to logical deduction. In Rule IV Descartes wrote that arithmetic, geometry, and the new algebra, which replaces figures with numbers, are good examples of what can be done. But, he went on:[57]

> In neither discipline [arithmetic and geometry] did I happen to come upon writers who were altogether satisfactory ... they did not seem to show with sufficient intellectual clarity why these things were so and how they were discovered.

He began to suspect that:[58]

> what [the Ancients] recognized as mathematics was very different from that which people accept in our times ... [So] they preferred to show us in its place, as the product of their art, certain barren truths which they cleverly demonstrate deductively so that we should admire them, rather than teach us the method itself, which would indeed detract from the admiration.

Some hope might be found, he allowed, in recent work,[59]

> call[ed] by the barbarous name of algebra, which, if it could be freed from the vast array of numbers and inexplicable figures with which it is overrun, would no longer lack that clarity and great simplicity of application which we presume should characterize the true mathematics. [More generally] it finally became clear that only those subjects in which order or measure are considered are regarded

[57]See (Descartes 1964, 159).
[58]See (Descartes 1964, 160).
[59]See (Descartes 1964, 161).

> as mathematical, and it makes no difference whether such measure is sought in numbers, or figures, or stars, or sounds, or any other object whatever.
>
> It then follows that there must be a certain general science which explains everything which can be asked about order and measure, and which is concerned with no particular subject matter, and that this very thing is called '[universal] mathematics', not by an arbitrary appellation, but by a usage which is already accepted and of long standing, because in it is contained everything on account of which other sciences are called 'parts of mathematics'.

Descartes had high ambitions for a universal mathematics (*mathesis universalis*). It would be a general science of order and measure, greater in scope and simpler in method than Euclid's *Elements*. Indeed, by the standards he set himself, he failed. But what survived of this ambition in *La Géométrie* is more than enough to make it read more like an alternative to Euclid than an addition.

Problems with the *Regulae*. However, we must be cautious in dealing with these opinions of Descartes, because the *Regulae* was unfinished and abandoned. There is a considerable literature about when the *Regulae* was written. Gaukroger[60] believes that the part of Rule IV that mentions mathematics was written first, between March and November 1619. Much of the rest of the *Regulae*, including the rules up to Rule XI (except for some of Rule VIII) was written, he argues, in 1619–1620. They were then abandoned for almost a decade until they were resumed with a somewhat different aim in mind in 1628. On this interpretation, Descartes began to organise his ideas about the nature of knowledge with his attack on the method of Euclid's *Elements*. Analysis was promoted as a radical alternative to synthesis, one that is connected to a general problem-solving approach to all knowledge, with the ultimate aim of showing how all acquisition of knowledge can follow the example of a remodelled mathematics.

Gaukroger summarised his conclusions in this way:[61]

> This was a view that Descartes never abandoned, as we shall see, and it takes little reflection to realize that it is completely at odds with the notion that a mathematical model for knowledge commits one to the discovery of knowledge by means of a deduction from first principles. If one still has any doubt about this, one only needs to look at the *Géométrie*. None of the axioms, theorems, etc., of Classical mathematics can be found there, nor a deductive system of any kind.
>
> Descartes does, it is true, present a few synthetic proofs, but it is the analysis that carries one along; and after a few preliminaries the reader is thrown into one of the most difficult mathematical problems bequeathed by antiquity, Pappus's locus problem for four or more lines, which Descartes proceeds to solve analytically without further ado. Problem-solving is what mathematics is about for Descartes, not axiomatic demonstration. And problem-solving, in the guise of analysis, is what he is concerned to foster in his account of method.

Descartes resumed his *Regulae* in 1629, and here we see how his general thoughts about mathematical analysis as a universal method have now been sharpened and pushed towards algebra. Halfway through Rule XIV he wrote:[62]

---

[60] See (Gaukroger 1997, 111, 126).

[61] See (Gaukroger 1997, 126).

[62] See (Descartes 1964, 213–214).

> It is to be hoped at this point that the reader possesses a propensity for the study of arithmetic and geometry, even though I would prefer him never to have studied them, rather than to have studied them in the customary manner, for the use of the rules which I will give here in expounding these subjects, for which they are entirely sufficient, is much easier than is their use in any other kind of question. And the utility of this method in attaining deeper understanding is so great that I am not afraid to say that this part of our method has not been devised for the sake of solving mathematical problems, but rather that mathematics is principally to be studied for the sake of perfecting our skill in this method.
>
> In my treatment, I will therefore presuppose no previous knowledge of these disciplines, except perhaps of some things which are self-evident and obvious to everyone. For even though the knowledge of these things, as commonly held by our contemporaries, may not be corrupted by obvious errors, it is obscured by many imperfect and badly conceived principles; and these we will try to correct from time to time in the remainder of this book.

In an argument that continues through the remaining Rules, Descartes explained what he meant by extension and dimension, and how to use figures. This accurately foreshadows what he went on to do in *La Géométrie*, while continuing the radical theme of reshaping all of philosophy and science.

Descartes was never able to realise his grand vision, and had to content himself with partial successes, such as his solution to the Pappus problem of the locus to three and four lines, which he discovered around 1632. It appears in *La Géométrie* as a showcase of his method, much as his essays on optics and meteorology in his *Discours* stand as examples of what can be done in physics. They may be torsos, they may be deceptive, even intended to deceive, and *La Géométrie* may be as hostile to the infinitesimal as Descartes's critics observe, but they were examples of a general method that was openly anti-Euclidean — and successful.

13.6 Further reading

Bos, H.J.M. 1996. 'Tradition and modernity in early modern mathematics', Viète, Descartes and Fermat, in 1996. *L'Europe Mathématique, Mathematical Europe*, (ed. Goldstein, C., Gray, J.J., and Ritter, J.), Éditions de la Maison des Sciences de l'Homme, 183–204. An observant disentangling and comparison of the work of the three men most responsible for algebra in the early 17th century.

———. 2001. *Redefining Geometrical Exactness: Descartes's Transformation of the Early Modern Concept of Construction*, Sources and Studies in the History of Mathematics and Physical Sciences, Springer. This work rethinks the history of mathematics in the 17th and early 18th centuries by examining closely the ideas of problems and solutions (notably, constructions) held by the mathematicians of the time. It is not an easy read, but is likely to prove one of the few books to change our ideas about this vital period in the development of mathematics.

Cohen, I.B. 1985. *Revolution in Science*, Harvard University Press. A rewarding book by one of the leading Newton scholars of his generation.

Descartes, R. 1637. *Discours de la Méthode pour bien Conduire sa Raison et Chercher la Verité dans les Sciences. Plus la Dioptrique, Les Météores et la Géométrie, qui sont des Essais de cette Méthode*, Leyden. *La Géométrie* was translated as *The Geometry of René Descartes* by D.E. Smith and M.L. Latham, Open Court Publishing

Co., 1925, Dover reprint, 1954. The book carries a facsimile of the first, French, edition, and is a fascinating foundational document of modern mathematics; it is also a rewarding read.

Descartes, R. 1964. *Rules for the Direction of the Mind, Philosophical essays* L. J. Lafleur (ed. and transl.), The Library of the Liberal Arts, Bobbs-Merrill. A helpful guide to Descartes's evolving conversation with himself.

Gaukroger, S. (ed.) 1980. *Descartes: Philosophy, Mathematics and Physics*, Harvester Press. An interesting set of essays, including the one by Mahoney discussed in this chapter.

———. 1997. *Descartes: an Intellectual Biography*, Oxford University Press. A good introduction to Descartes which is unusual in discussing his mathematics.

Mahoney, M.S. 1994. *The Mathematical Career of Pierre de Fermat, 1601–1665* (second edition), Princeton University Press. Still the only attempt to survey all of Fermat's work in the context of his time.

Mancosu, P. 1992. Descartes's *Géométrie* and revolutions in mathematics, in *Revolutions in Mathematics* D. Gillies (ed.), Oxford University Press, pp. 83–116. A thoughtful and thought-provoking analysis assessing the magnitude of Descartes's achievement.

Sorell, T. 1987. *Descartes*, Oxford University Press. A good and accessible intellectual biography of Descartes, giving judicious consideration to the position of mathematics in Descartes's work as a whole.

Weil, A. 1983. *Number Theory: An Approach through History*, Birkhäuser. Written by a 20th-century master of the subject, this is not always an easy book, but is thorough, stimulating, and accurate. It spans the period from the Mesopotamian tablet Plimpton 322 up to Lagrange at the end of the 18th century; the discussion of Fermat is particularly lucid.

14

Concluding Remarks

Biographical information about mathematicians before the year 1650 may be slight, and in many cases non-existent, but we can nonetheless speculate usefully on their lives. We can begin by asking: What societies supported their efforts? What was demanded of them in return?

Whether we agree with the historian Otto Neugebauer that:[1]

> Of all the civilisations of antiquity, the Egyptian seems to me to have been the most pleasant. The excellent protection which desert and sea provide for the Nile valley prevented the excessive development of the spirit of heroism which must often have made life in Greece hell on earth.

or regard it as a provocative remark intended to shake his audience out of a shallow Hellenistic reverie, we may note that mathematics played a long and important role in these cultures, as it did also in Mesopotamia, India, and China. But were there *mathematicians* in these societies? It depends on what one means. Certainly there were numerate specialists with skills that most of their contemporaries (and quite possibly their rulers) did not possess, and whose jobs were to make complicated things happen. They kept the economy moving, they could answer important astrological questions, track and predict the seasons, provision an army, see that large-scale building works were carried out, and teach their successors. Many of their intellectual descendants in modern societies play similar roles, but we do not call them mathematicians: often they are economists, astronomers, architects, surveyors, and the like.

Invisible to us are those who reflected on their work and in so doing found ways to do it better, and who satisfied themselves and their successors that *this* method would always work or that *that* one was dubious. Even in ancient China the names we have are mostly those of talented commentators, not the originators. We can note only that the many sophisticated mathematical ideas that emerged over the centuries had to be invented by someone, polished by someone, and somehow made known before anyone could write the texts upon which our modern historical accounts are based.

Ancient Greece was different. Suddenly we have names, dialogues (however fictionalised), fragments, and eventually complete works that were written by someone whom we

[1] See (Neugebauer 1969, 71).

can name, or who wrote other works for people that we can also name. We can almost catch echoes of their conversations, and see the pictures that they drew upon the sand before the waves washed the diagrams away. We begin to get glimpses of their lives: what else they did, and even the manner of their deaths. The evidence about the earliest figures is so confused that it is possible to reject any and all of it as hopelessly unreliable, but one may also see Thales or Pythagoras as one of these oddly brilliant people for whom some ideas were unnaturally clear, and note that by the time of Aristophanes an interest in mathematics appears to have been something that a playwright could take for granted.

We cannot doubt that Plato gave mathematics a very high status indeed, and that he personally knew a number of people that we would call mathematicians, because for the first time we meet named characters with very unusual interests indeed. They cared about irrational magnitudes, abstract proof, problems such as the three classical problems that cannot be said to have any real practical significance, and novel curves such as the conic sections, the cissoid, and the conchoid. Historians, perhaps inspired by Aristotle's remarks that mathematics began in Egypt where a priestly class had the necessary leisure, or perhaps finding the apparently leisured lives of these Greeks reminiscent of 19th-century gentlemen, have emphasised the abstract, 'pure', quality of their work, and told us less about its practical character — the devices made by Archimedes and others, and the close links with astronomy. But these activities also required months and years of work, hours of focussed thought at a time, and a community (however small) that valued it and made it possible for people to participate at what had become a very high level.

We do not know what Dositheus did with the letters and lengthy manuscripts that he received from Archimedes. We do not know whether he found them easy to understand or barely intelligible. But we can surely allow ourselves to imagine him making the time to read them, finding somewhere quiet to do so, being prepared perhaps to copy out a diagram or an argument the better to understand it. We may suppose that he got up to stretch his legs, walk around to help himself sort out these ideas, and gaze at times over an olive grove and down to the sea, lost in thought.

In the sources we have, we get glimpses of a problem-solving culture. What else could have produced the sheer oddness of Diophantus's work, which hints at a tradition stretching back to pre-Greek times but about which we know almost nothing? And is it the siren of historical hindsight, or do we find some pathos in Pappus in 4th-century Alexandria, and Proclus in Athens, recording what they knew of an already ancient time as the Roman Empire steadily declined?

The same individual concerns recur wherever we look. When Āryabhaṭa wrote that:

> By the grace of Brahma, the sunken jewel of the best of true knowledge has been brought up by me out of the ocean of true and false knowledge by means of the boat of my intellect.

this was no arithmetical drudge grinding out useless volumes of calculation. His work on the 'Chinese' remainder theorem, and Bhāskara II's later work on indeterminate quadratic equations, was the work of people with a profound interest in numbers. When Chinese society flourished, our sources take us to practical men — the analogy here is with civil servants, not the gentry — but, as we saw, Dauben found it appropriate to end his survey with this remark by Xu Guangqi:

> In truth mathematics can be called the pleasure-garden of the myriad forms, the Erudite Ocean of the Hundred Schools of philosophy.

So too with the world of Islam. Islamic writers, often so varied in their interests that one cannot call them only mathematicians, took the measure of their Greek predecessors and surpassed them — not just with their 'algebra' and their 'algorithm', but with their careful reading of Euclid's *Elements*, and their understanding of Ptolemy's astronomy. Historians sometimes reflect on the lives of Europeans in what were once called the European 'Dark Ages', surrounded by crumbling Roman buildings on a scale no European then knew how to emulate, and wondering what knowledge had been lost. Another such window opened for Christians of the Middle Ages onto the Islamic world, when they saw it in southern Spain, in Sicily, or in the Middle East. From its grand architecture to its hand-sized astrolabes, the Islamic world was rich and the European one was poor.

The momentous reversal of fortunes that followed had many causes and will be long debated, but the study of Greek and Islamic mathematics formed part of it, as did a culture that valued such ideas. Now it was European writers who dedicated themselves to the lonely work of learning the appropriate foreign languages and writing the translations that others could use. Slowly, and interrupted by the Black Death, Europeans caught on to the power of mathematics. For the first time we find mathematicians (although the profession was still not named as such) in universities investigating motion, and others, in the commercial world, improving arithmetic and algebra. But it seems that after the Black Death the contributions of universities faded somewhat, and creative mathematics was done in the squabbling competitive cities and city-states of early modern Europe. Here an ability to tackle, and eventually solve, cubic equations was an advantage for some; Recorde attempted to educate an emerging class of merchants, surveyors, and the like; and Dee promoted mathematics, in part because of its use in navigation and the creation of an empire. It was a world of patrons, but patrons who would pay for work in astronomy and for scholarly editions of ancient texts. Here worked Harriot, Galileo, Kepler, and Viète.

Descartes's mathematics has proved to be his most lasting contribution. It would be provocative, but not wrong, to observe that his imperfect mathematics has only been improved by the contributions of later mathematicians, while philosophers have mostly dismantled Descartes's philosophy. Like some of his predecessors (Apollonius springs to mind) he wrote for a small elite, and, where Fermat, his equally gifted contemporary, was happy to revive Greek mathematics with remarkable ideas of his own, Descartes preferred to eclipse it with his own work. In this contrast these two figures represent the two halves of mathematics that make its history so interesting: its long but living past, and its original discoveries as new as the day.

15

Exercises

Advice on tackling the exercises

All our exercises take the form of essays, rather than exercises on the mathematics. We have not put a word length on these exercises, but you may wish to think in terms of 500–1000 words for each exercise in Part A, 1000–1500 words for each exercise in Part B, and 1500–2000 words for each exercise in Part C. Remember that keeping to the stated length is a skill that it is important to master.

When tackling a question from Part A, it is helpful to think in terms of the phrase 'content, context, and significance'. Being able to present the content of an extract, place it in a historical context, and then draw out its significance is a fundamental skill for a historian of mathematics. As a rough guide, you should allocate an equal amount of space to your discussion of each of the content, context, and significance passages in your answers, although sometimes you will find that there is rather more to say on each of the content and context than there is on the significance. There are also questions where the context and significance are quite distinct, and others where these categories merge into one another. A useful strategy is to go through the extract underlining all the words or phrases you intend to discuss (such as proper names, technical terms, etc.) and then, when you have finished writing your essay, go back and check that you have not omitted any of them.

The questions in Part B relate to a specific chapter (or chapters) in this book. Most of these questions consist of two equally weighted parts, and when answering the first part you should take care not to get so carried away that you forget to answer the second. You may, for example, be asked to describe a piece (or style) of mathematics, and then to comment in a particular way upon it.

The Part C questions are on more wide-ranging themes and require you to use material from several chapters of the book. With these questions you should present a variety of evidence before reaching a conclusion in which you balance the weight and merits of the different arguments.

Writing an essay.

Essays are sometimes arguments in favour of a judgement. One way to think of how you should approach an essay is to imagine that you are briefing your boss on the topic. It is your boss who will go into the meeting, who will present the arguments, and who will try to counter those on the other side. It is your job to tell your boss what he or she needs to win the arguments. Your boss will come out of the meeting and congratulate you or criticise you. You need to get it right, and remember, your boss cannot come out of the meeting halfway through to ask for clarification of what you wrote, or extra information. Everything has to be there in the essay, and it must not exceed the stated length (no-one wants to hear your boss droning on for far too long).

Or, to vary the metaphor, imagine that you are defending the teaching of the history of mathematics in a college. The syllabus can be filled many times over with topics, all of which have a legitimate claim on the students' attention. To argue for the subject is to give reasons why it matters (how will the students' education be enriched?), not to launch into an account of the life and work of Archimedes or Descartes, fascinating and important though they are. Of course, if the facts are relevant, as good accounts of the life and work of Archimedes or Descartes might well be, then they should be included, but the rule is: argument first, facts in support of the argument second.

So what does your boss want, and need, to know? First, what is the decision you are arguing for. Then, what are the good reasons for coming to that decision, what are the alternative decisions, and why are they not as good? Notice that it is a decision you are arguing for, a judgement that will help the company to perform better. We emphasise this point by including questions that explicitly call for a judgement. But it is not enough to state your opinion, however eloquently. You must have facts to back it up. With the facts clear in your mind, you can reach that judgement, discriminating between other judgements and refining your own in the process.

How to organise an essay. *Essays have a head, a long body, and a tail.*
There are many ways of writing an essay, but perhaps the main rules are these.

- Introduce the topic of your essay in a short paragraph at the start ('In this essay we argue that ...'; 'This essay describes ...').
- Fill the bulk of the essay with evidence in support of your initial claim.
- Each paragraph should make a single point.
- Organise your evidence clearly, and distinguish your position from others that have struck you as plausible but weaker than your own.
- In a short final paragraph, state your conclusion. This does not matter so much in a short essay, but in a lengthy article it is easy for the reader to lose the main point in a welter of detail.
- Diagrams/illustrations should be clearly captioned or referred to. If you discuss a diagram from a historical source and wish to refer to specific points on it, such as 'the point C on the line AB', a copy of the diagram should be included in your essay.
- If you transcribe historical mathematics into modern format/notation, make it clear that you are doing so and why.

Other rules are not so important, but they matter. Here are a few.

- Keep to the recommended length.
- Try to avoid writing in too personal a way, with an over-lavish use of the first person ('I') — you are supposed to be preparing a case that can be agreed by any reader.
- Use the past tense to refer to people who are now dead, but the present tense when writing about their surviving work, thus 'Descartes showed that ... ' but 'Descartes's *La Géométrie* establishes'. Of course, if the event happened in the past, then the past tense is required: 'Descartes's *La Géométrie* persuaded Mersenne ... '.

How to use sources. Always include citations for any information or opinion you use, so that readers can check them if they wish to. There are rules about using sources. If you want to quote you must say where the quotation come from. Citations that appear within the essay itself are given in many different forms in the literature. We suggest that they should be in one of these forms:

- author, date of publication, page number(s), e.g. (Robson 2008, 185–186)
- author, title of book or paper, page number(s), e.g. (Lyons, *The House of Wisdom*, p. 100).

Full references, including publisher etc., should be given in the bibliography at the end of your essay. If the item is in this book, however, it is enough to give the author's name, followed by BGW and the page number. If you want to summarise somebody else's opinion, give the author's name, followed by the date (or title of the book or paper) and page number.

These are the basic rules of evidence: in principle, any reader should be able to check that what you say is true. It follows that you must quote and summarise fairly, even if your source of information is one that you trust completely, or one that you have just pulled off a website. In this way you can also deal with errors and with conflicting opinions.

Examples of entries in a bibliography:

Article: Knorr, W.R. 1982. Observations on the early history of the conics, *Centaurus* 26, 1–24.

Book: Chace, A.B. 1927. *The Rhind Mathematical Papyrus*, Mathematical Association of America.

Book chapter: Mancosu, P. 1992. Descartes's *Géométrie* and revolutions in mathematics, in *Revolutions in Mathematics*, D. Gillies (ed.), Oxford University Press, pp. 83–116.

Website: Melville, D. *Old Babylonian mathematics*, http://it.stlawu.edu/ dmelvill/mesomath/index.html. Accessed 29 October 2018.

Plagiarism. *Using other people's words as your own is theft.*

You must not quote other people's work as if it were your own, and without giving proper reference to it. Presenting other people's words as your own is plagiarism. It is plagiarism if you send in a whole essay that somebody else wrote, and it is plagiarism if you just recycle a whole paragraph or a sentence without reference. Plagiarism is theft. It is stealing somebody else's work, and even when it is not illegal it is an insult to the academic community because it denies the reader a chance to check your sources or to give due credit to the original author. The person who suffers most is yourself, as you are depriving yourself of the opportunity to test your own growing understanding of the subject through constructing and expressing your own arguments, which is integral to your learning experience.

Nor is there any reason to do it. If it turns out that you have found an essay that answers your question perfectly, just say so. Write 'This essay takes the view of [author] who in [give publication details] argued that ...' and then re-state that case in your own words. It is almost certain that you will then find that you have extra information to bring to bear, or different shifts of emphasis to make. The result will be your work, and that is what the reader (and ultimately you) want.

Sample exercises. The following is not a set of model answers — we do not believe there can be such things — but is an indication of how you might proceed when confronted with various types of questions. They are sometimes longer than your answers should be, because we want to acquaint you with how you should proceed in general. They are intended to give hints about how to get started, and how you might present your answer. They are meant to be helpful, not a straitjacket.

Questions 1 and 2 are similar to Part A type questions, Question 3 is similar to a Part B type question, and Question 4 is similar to a Part C type question.

Question 1

A quantity and its $\frac{1}{2}$ *added together become 16. What is the quantity?*

Assume	2.
/1	2
/$\frac{1}{2}$	1
Total	3.

As many times as 3 must be multiplied to give 16, so many times 2 must be multiplied to give the required number.

/1	3
2	6
/4	12
$\frac{2}{3}$	2
$\frac{1}{3}$	1
Total	$5\frac{1}{3}$.
1	$5\frac{1}{3}$
/2	$10\frac{2}{3}$

Do it thus: The quantity is $10\frac{2}{3}$

$\frac{1}{2}$	$5\frac{1}{3}$
Total	16.

(The Rhind Papyrus)

Notes

You may like to structure your answer as follows.

- Explain what you take the problem to be — by putting it into more modern symbolism, for example.
- What do you think the problem (or the papyrus on which it occurs) was for?
- How characteristic of Egyptian mathematics do you consider the problem to be?

Content
What number, when added to half of itself, gives 16? In modern algebraic terms this can be written as

$$x + \tfrac{1}{2}x = 16,$$

which can be solved to give $x = 10\frac{2}{3}$.

Once you have explained what the problem is, you need to describe how the scribe (who did not have access to modern algebraic notation!) tackled it. The method given is not plugging some numbers into a formula and calculating the answer, but rather something more like a set of instructions, albeit rather cryptic ones from our perspective. You will need to consider the different techniques available to the scribe and note which ones he used (doubling and halving, using slanting lines to indicate which rows to add, calculating with unit fractions and $\frac{2}{3}$, etc.). Notice too that, on this occasion, the scribe asks the reader to check the answer ('Do it thus').

Context and significance
Whatever the extract, a good way to get started on its context and significance is ask yourself what you know about the author and the circumstances in which he or she (it is almost invariably a 'he') was working. In this case we know the author's name, and that he was copying from an earlier work. Next, what do you know about the source itself? What can you say about its age or date, rarity, typicality? What type of a source is it (book, letter, etc) and who was it written for? Is the extract the entire source or is it part of something larger? If the latter, is the extract from the source typical of the source as a whole? How have we (or those of us who cannot read hieratic script) been able to access its contents?

You're now in excellent shape to explain what the scribe did. You have also discovered more. It's unthinkable that the writing of this papyrus was a unique event in Egypt which, by some miracle, has come down to us. The generality of the method which is taken for granted here makes it clear that what is being described was well understood. Although we have very few primary sources for Egyptian mathematics, it is not unreasonable to deduce that there was a continuing tradition of mathematics at work, of which the Rhind Papyrus is but one example. Of course, your reading of Chapter 2 tells you that there was more going on.

What was the purpose of the Rhind Papyrus? We can't be sure, and historians (notably G.J. Toomer and A. Chace) have put forward differing ideas. Was it a tool for everyday life (in the way that an electronic calculator is today), or does it provide evidence that the Egyptians studied mathematics for its own sake? Your job is to weigh up what you know and arrive at a conclusion (which does not necessarily have to be definite, one way or the other).

If you now decide to write up your answer in the way that the Notes suggest, you can see that you are in a very good position to explain what the scribe did, using a modest amount of modern notation.

That's quite a journey, isn't it? From one papyrus we have gained a picture of Egyptian life, at least for some people, and a good deal of respect for the sort of things that they could do.

Question 2
Now the things peculiar to the science, the existence of which must be assumed, are the things with reference to which the science investigates the essential attributes, such as arithmetic with reference to units, and geometry with reference to points and lines. With these things it is assumed that they exist and that they are of such-and-such a nature. But, with regard to their essential properties, what is assumed is only the meaning of each

term employed: thus arithmetic assumes the answer to the question what is [meant by] 'odd' or 'even', 'a square' or 'a cube', and geometry to the question what is [meant by] 'the irrational' or 'deflection' or [the so-called] 'verging' [to a point]; but that there are such things is proved by means of the common principles and of what has already been demonstrated. Similarly with astronomy. For every demonstrative science has to do with three things, (1) the things which are assumed to exist, namely the genus [subject-matter] in each case, the essential properties of which the science investigates, (2) the common axioms so-called, which are the primary source of demonstration, and (3) the properties with regard to which all that is assumed is the meaning of the respective terms used.

(Aristotle, *Posterior Analytics*)

Notes

(i) You need to decide what is Aristotle's main concern in this passage, and what are his illustrative examples.

(ii) You might also consider what else you know about the technical matters that Aristotle raises by reading later writers, such as Euclid.

Content
Aristotle introduces three examples, presumably all leading to the same point. They were drawn from geometry, arithmetic, and astronomy. Each of these subjects is taken to be about some fundamental objects — which are assumed to exist — expressed in terms whose meanings we are supposed to know. These terms may be elementary or technical, but they are meaningful in virtue of some basic principles or of some detailed work. He then claimed all three subjects operate in the same way: they make some assumptions about what exists, they rely on some basic logical principles, and they invoke the essential meanings of the terms they use.

Context and significance
The context and the significance blur together here. Aristotle was writing before Euclid, so his comments on mathematics are particularly significant. We see the strong emphasis on proof that Euclid shared, and the separation of basic axioms from others, but note that some of Aristotle's examples are 'too advanced' for Euclid's *Elements*: 'verging', for example. But it feels somewhat odd to us to make assumptions about existence in astronomy. There, we like to feel we are talking about real things (but perhaps we should be more cautious, as some scientists and philosophers of science are). At all events, the easy transition from arithmetic and geometry to astronomy recalls Plato's views, which Aristotle seems to share.

Question 3
Briefly survey the range of Greek Mathematics that has come down to us, and assess how typical you judge Euclid's Elements *to have been. In the light of your assessment, explain why Euclid's* Elements *was so influential on mathematics in the West?*

Notes

(i) With a limitation on word length it is not possible to give more than a hint of what different Greek mathematicians did. It is therefore necessary to give a short accurate summary and make astute reference to Chapter 3.

(ii) Before making your assessment, you may find it helpful to list relevant items under a few headings, such as: topic, level of difficulty, method of treatment, use of examples, and problems.

(iii) The influence of Euclid's *Elements* is partly due to, and partly accounts for, the fact that it was transmitted much more forcefully than some other works, so your answer should consider this question of transmission.

You are asked to look at a lot of evidence, and locate Euclid's *Elements* in that spectrum. Since this could be a huge task, you were invited to zero in on a few key indicators (feel free to choose your own). The following experience of one of the authors of this book may reassure some readers.

> When I began to think about Greek mathematics my first experience was one of vertigo. I knew something about Euclid's *Elements* from my school education (but that proved to be misleading), and I had heard that there was harder stuff in Archimedes and Apollonius, which was indeed the case. Closer acquaintance revealed important similarities between these three (they all wrote geometry, one way or another) and important differences (Apollonius reads like a big leap beyond Euclid in the same direction, while Archimedes is altogether more diverse). There were also surprises in Euclid's *Elements*, such as the number theory and the difficulty of Book X. A much bigger surprise came with other people: the fragments of Hippocrates, the philosophical interpretations of Plato, the uses of mathematics by Ptolemy and Heron, and above all the work of Diophantus.

Looking over this material, you may still be torn. Was there a core to Greek mathematics, built upon a recognition of the key importance of proof and the power of geometrical reasoning? Or is that an artefact, built partly on accidents of transmission and partly on our prejudices for theory over application and tidy stories over messy ones? So your answer, and a very different one written by someone else with whom you disagree, may be equally valid and receive an equal mark. Disagreeing is part of being a historian!

But it is clear that Euclid, Archimedes, and Apollonius do fit together surprisingly well, and they fit with what little we know about what was going on a few years on either side of them. They placed a huge insistence on proving things, even (you might think) not very interesting things. Euclid, in his *Elements*, and Apollonius didn't spend many words (if any) on the uses of any of their work. In fact, words of any didactic or helpful kind are almost completely lacking, which is why some historians of mathematics suppose that the texts were accompanied by some oral tradition designed to motivate the student. Archimedes was different, but even he made a sharp distinction between discovery methods and proofs. And it is equally clear that, unless new evidence turns up, Diophantus will always have to be placed in a different tradition for which there is otherwise very little evidence in Greek times at all. For example, it was only Diophantus who set and solved problems — another way in which Euclid, Archimedes, and Apollonius were clearly not writing anything that we would call a modern textbook.

So you will have to make some choices in writing your answer. Do you want to raise the possibility of there being a history of Greek mathematics that does not place Euclid, Archimedes, and Apollonius centre stage? This would give you a perspective on the word 'typical' that is key to any answer to the question. But you might feel that the tradition is eminently sensible on this point (and in any case carries the bulk of the evidence) and there is nothing to say except that these three mathematicians define the core of Greek mathematics.

If you go with the first approach, you can still say that Euclid's *Elements* was typical, but is not paradigmatic (meaning that it is not the very essence of Greek mathematics). On the second approach, you might want to argue that it is paradigmatic. Whichever route you take, you must support your argument by considering the topic (geometry, mostly), the emphasis on proof, and the absence of problems and worked examples.

The second part of the question, asking you to assess the great influence of Euclid's *Elements* on mathematics in the West, may be difficult, since you might want to say that it calls for information you are going to learn later in the course. But it asks you to think about what aspects of Euclid's *Elements* could possibly contribute to its longevity. Were they part of what made it central to any tradition you have detected? Were they what made it distinctive and unusual? Several answers are possible here, and you should be looking for reasons that are rooted in what you have said about the *Elements* already, as well as those which are plausible. So take the opportunity to look back over the first part of your essay and ensure that the final part grows out of it, and isn't just stuck on, or (worse) that it contradicts it!

Question 4
To what extent did mathematics develop in the 16th and 17th centuries primarily as a means for responding to technological and scientific needs and to what extent was it undertaken for its own sake?

Notes
One way to begin your answer is to interpret some of the key words and phrases in the question, such as 'technological and scientific needs', 'its own sake'. Although you may have a clear idea about your answer to the question — 'a large amount', 'not at all', etc. — you need to produce evidence.

One strategy is to make a list of the mathematics developed within the time period and divide it into two, depending on which side of the argument the mathematics falls. In the first category you might cite *navigation* (related to development of instruments and maps); *astronomy* (theoretical and observational); *military purposes* (projectiles, surveying, fortification); *technological* (instruments, logarithms for ease of calculation); *educational* (Recorde's textbooks, founding of the Royal Society for promotion of experimental science). In the second category you might include the *solution of cubic equations*; the *rediscovery of Greek mathematics*; the *geometry of perspective* (related to practical surveying, but developed by Desargues and others as a subject in its own right); the increasing use of *algebra as a language* (especially in the posing and solution of geometrical problems); *number theory* (Fermat); *other influences* (religious standpoints, Commandino's search for 'certainty', social standing). You might further decide that there are certain topics that fall into both camps, such as the textbooks of Recorde which are practical and also helped with the spread of mathematical knowledge. As there is a lot of material to draw on, you will need to be disciplined in your answer in order to keep to the word limit. This is where judicious referencing to relevant passages in this book (or other sources) can be very helpful. Remember that you need to end with a conclusion, and that it must be supported by the evidence you have given.

Exercises: Part A

CHAPTER 2 Early Mathematics.

1. Find the volume of a cylindrical granary of diameter 9 and height 10. Take away 1/9 of 9, namely 1; the remainder is 8. Multiply 8 times 8; it makes 64. Multiply 64 times 10; it makes 640 cubed cubits.

   (Rhind Papyrus)

2. Example of dividing 100 loaves among 10 men, including a boatman, a foreman, and a door-keeper, who receive double portions. What is the share of each?

   The working out. Add to the number of the men 3 for those with double portions; it makes 13. Multiply 13 so as to get 100; the result is $7\frac{2}{3}\frac{1}{39}$. This then is the ration for seven of the men, the boatman, the foreman, and the door-keeper receiving double portions.

   For proof we add $7\frac{2}{3}\frac{1}{39}$ taken 7 times and $15\frac{1}{3}\frac{1}{26}\frac{1}{78}$ taken 3 times for the boatman, the foreman, and the door-keeper. The total is 100.

   (Rhind Papyrus)

3. [1] I found a stone, (but) did not weigh it; (after) I weighed (out) 6 times (its weight), (added) 2 gin, (and)

   [2] added one-third of one-seventh multiplied by 24,

   [3] I weighed (it): 1 ma-na. What was the origin(al weight) of the stone? The origin(al weight) of the stone was $4\frac{1}{3}$ gin.

   (From a Mesopotamian tablet)

4. I have added up the area and the side of my square: 0;45. You write down 1, the coefficient. You break off half of 1. 0; 30 and 0; 30 you multiply: 0; 15. You add 0; 15 to 0; 45: 1. This is the square of 1. From 1 you subtract 0; 30, which you multiplied. 0; 30 is the side of the square.

   (From a Mesopotamian tablet)

CHAPTER 3 Greek Mathematics: An Introduction.

1. The Pythagoreans considered all mathematical science to be divided into four parts: one half they marked off as concerned with quantity, the other half with magnitude; and each of these they posited as twofold. A quantity can be considered in regard to its character by itself or in its relation to another quantity, magnitudes as either stationary or in motion. Arithmetic, then, studies quantity as such, music the relations between quantities, geometry magnitude at rest, spherics magnitude inherently moving.

   (Proclus)

2. We can, then, properly lay it down that arithmetic shall be a subject for study by those who are told to hold positions of responsibility in our state; and we shall ask them not to be amateurish in their approach to it, but to pursue it till they come to understand, by pure thought, the nature of numbers — they aren't concerned with its usefulness for mere commercial calculation, but for war and for the easier conversion of the soul from the world of becoming to that of reality and truth.

   (Plato, *Republic*)

3. Everyone begins, as we said, with wondering that a certain thing should be so, as for example one does in the case of the puppet theatre (if one has not yet found out

the explanation), or with reference to the solstices or the incommensurability of the diagonal. For it must seem to everyone matter for wonder that there should exist a thing which is not measurable by the smallest possible measure. The fact is that we have to arrive in the end at the contrary and the better state, as the saying is. This is so in the cases just mentioned when we have learnt about them. A geometer, for instance, would wonder at nothing so much as that the diagonal should prove to be commensurable.

(Aristotle, *Posterior Analytics*)

4. *To bisect a given rectilineal angle.*
Let the angle BAC be the given rectilineal angle. Thus it is required to bisect it. Let a point D be taken at random on AB; let AE be cut off from AC equal to AD [I. 3]; let DE be joined, and on DE let the equilateral triangle DEF be constructed; let AF be joined. I say that the angle BAC has been bisected by the straight line AF. For, since AD is equal to AE, and AF is common, the two sides DA, AF are equal to the two sides EA, AF respectively. And the base DF is equal to the base EF; therefore the angle DAF is equal to the angle EAF [I. 8]. Therefore the given rectilineal angle BAC has been bisected by the straight line AF. Q.E.F.

(Euclid, *Elements*, Book I, Prop. 9)

CHAPTER 4 Greek Mathematics: Proofs and Problems.

1. *To two given straight lines to find a mean proportional.*
Let AB, BC be the two given straight lines; thus it is required to find a mean proportional to AB, BC. Let them be placed in a straight line, and let the semicircle ADC be described on AC; let BD be drawn from the point B at right angles to the straight line AC, and let AD, DC be joined. Since the angle ADC is an angle in a semicircle, it is right [III. 31]. And, since, in the right-angled triangle ADC, DB has been drawn from the right angle perpendicular to the base, therefore DB is a mean proportional between the segments of the base AB, BC [VI. 8, Por.]. Therefore to the two given straight lines AB, BC a mean proportional DB has been found. Q.E.F.

(Euclid, *Elements*, Book VI, Prop. 13)

2. *Circles are to one another as the squares on the diameters.*
Let $ABCD$, $EFGH$ be circles, and BD, FH their diameters; I say that, as the circle $ABCD$ is to the circle $EFGH$, so is the square on BD to the square on FH. For, if the square on BD is not to the square on FH as the circle $ABCD$ is to the circle $EFGH$, then, as the square on BD is to the square on FH, so will the circle $ABCD$ be either to some less area than the circle $EFGH$, or to a greater.

(Euclid, *Elements*, Book XII, Prop. 2)

3. It is my opinion that this problem is what led the ancients to attempt the squaring of the circle. For if a parallelogram can be found equal to any rectilinear figure, it is worth inquiring whether it is not possible to prove that a rectilinear figure is equal to a circular area.

(Proclus, *Commentary on Euclid's* Elements)

4. 'Reduction' is a transition from a problem or a theorem to another which, if known or constructed, will make the original proposition evident. For example, to solve the problem of doubling the cube geometers shifted their inquiry to another on which this depends, namely, the finding of two mean proportionals; and thenceforth they devoted their efforts to discovering how to find two means in continuous proportion

between two given straight lines. They say that the first to effect reduction of difficult constructions was Hippocrates of Chios, who also squared the lune and made many other discoveries in geometry, being a man of genius when it came to constructions, if there ever was one.

(Proclus)

CHAPTER 5 Greek Mathematics: Curves.

1. I set myself the task of communicating to you a certain geometrical theorem which had not been investigated before but has now been investigated by me, and which I first discovered by means of mechanics and then exhibited by means of geometry ... I am not aware that any one of my predecessors has attempted to square the segment bounded by a straight line and a section of a right-angled cone [a parabola], of which problem I have now discovered the solution. For it is here shown that every segment bounded by a straight line and a section of a right-angled cone [a parabola] is four-thirds of the triangle which has the same base and equal height with the segment, and for the demonstration of this property the following lemma is assumed: that the excess by which the greater of [two] unequal areas exceeds the less can, by being added to itself, be made to exceed any given finite area.

   (from a letter of Archimedes to Dositheus)

2. I say then that, even if a sphere were made up of the sand, as great as Aristarchus supposes the sphere of the fixed stars to be, I shall still prove that, of the numbers named in the Principles, some exceed in multitude the number of the sand which is equal in magnitude to the sphere referred to, provided that the following assumptions be made. The perimeter of the Earth is about 3,000,000 stadia and not greater. It is true that some have tried, as you are of course aware, to prove that the said perimeter is about 300,000 stadia. But I go further and, putting the magnitude of the Earth at ten times the size that my predecessors thought it, I suppose its perimeter to be about 3,000,000 stadia and not greater.

   (Archimedes)

3. Some time ago I expounded and sent to Eudemus of Pergamum the first three books of my conics which I have compiled in eight books, but, as he has passed away, I have resolved to dedicate the remaining books to you because of your earnest desire to possess my works. I am sending you on this occasion the fourth book. It contains a discussion of the question, in how many points at most it is possible for sections of cones to meet one another and the circumference of a circle, on the assumption that they do not coincide throughout, and further in how many points at most a section of a cone or the circumference of a circle can meet the hyperbola with two branches, [or two double-branch hyperbolas can meet one another]; and, besides these questions, the book considers a number of others of a similar kind.

   (Apollonius, *Conics*, Preface to Book IV)

4. If a straight line touching a parabola meets the diameter, the straight line drawn through the point of contact parallel to the diameter in the direction of the section bisects the straight lines drawn in the section parallel to the tangent.

   (Apollonius, *Conics*, Book I, Prop. 46)

CHAPTER 6 Greek Mathematics: Later Years.

1. *To find two rectangles such that the area of the first is three times the area of the second.*

I proceed thus: Take the cube of 3, making 27; double this, making 54. Now take away 1, leaving 53. Then let one side be 53 feet and the other 54 feet. As for the other rectangle, [I proceed] thus: Add together 53 and 54, making 107 feet: multiply this by 3, [making 321; take away 3], leaving 318. Then let one side be 318 feet and the other 3 feet. The area of one will be 954 feet and of the other 2862 feet.

(Heron, *Geometrica*)

2. *We wish to find two square numbers the sum of which is a cubic number.*
We put x^2 as the smaller square and $4x^2$ as the greater square. Then the sum of the two squares is $5x^2$, and this must be equal to a cubic number. Let us make its side any number of xs we please, say x again, so that the cube is x^3. Therefore, $5x^2$ is equal to x^3 . As the side which contains the xs is the lesser in degree, we divide the whole by x^2; hence x is equal to 5. Then, since we assumed the smaller square to be x^2, and since x^2 arises from the multiplication of x — which we found to be 5 — by itself, x^2 is 25. And, since we put for the greater square $4x^2$, it is 100. The sum of the two squares is 125, which is a cubic number with 5 as its side.

(Diophantus)

3. And so, in general, we have to state that the heavens are spherical and move spherically; that the Earth, in figure, is sensibly spherical also when taken as a whole; in position, lies right in the middle of the heavens, like a geometrical centre; in magnitude and distance, has the ratio of a point with respect to the sphere of the fixed stars, having itself no local motion at all.

(Ptolemy, *Almagest*)

4. Bees, then, know just this fact which is useful to them, that the hexagon is greater than the square and the triangle and will hold more honey for the same expenditure of material in constructing each. But we, claiming a greater share in wisdom than the bees, will investigate a somewhat wider problem, namely that, of all equilateral and equiangular plane figures having an equal perimeter, that which has the greater number of angles is always greater, and the greatest of them all is the circle having its perimeter equal to them.

(Pappus)

CHAPTER 7 Mathematics in India and China.

1. Since questions can scarcely be known without algebra, I shall therefore speak of algebra with examples. By knowing the pulverisers, zero, negative and positive quantities, unknowns, elimination of the middle terms, equations with one unknown, factum and square nature, one becomes the learned professor among the learned.

(Brahmagupta)

2. Analysis is certainly the innate intellect assisted by the various symbols, which for the instruction of duller intellects, has been expounded by the ancient sages who enlighten mathematicians as the sun irradiates the lotus, and that has now taken the name algebra ... The method of demonstration is always of two kinds: geometrical and algebraic.

(Bhāskara)

3. *(Given) a field of 1* mu*, how many (square) bu are there?*
(The answer) says: $15\frac{15}{31}$ (square) *bu*. The method says: a square 15 *bu* (on each side) is deficient by 15 (square) *bu*; a square of 16 *bu* (on each side) is in excess by 16 (square) *bu*. (The method) says: combine the excess and deficiency as the divisor;

(taking) the deficiency numerator multiplied by the excess denominator and the excess numerator times the deficiency denominator, combine them as the dividend. Repeat this, as in 'the method of finding the width'.

(Problem 53, *Book of Numbers and Computations*)

4. I read the *Nine chapters* as a boy, and studied it in full detail when I was older. [I] observed the division between the dual natures of Yin and Yang [the positive and negative aspects] which sum up the fundamentals of mathematics. Thorough investigation shows the truth therein, which allows me to collect my ideas and take the liberty of commenting on it. Things are known to belong to various classifications. Just as the branches of a tree are to its trunk, so are a multitude of things to an archetype. Therefore I have tried to explain the whole story as concisely as possible, with spatial forms shown in diagrams, so that the reader should have a reasonably good all-round understanding of it.

(Liu Hui, Preface to the *Nine Chapters*)

CHAPTER 8 Mathematics in the Islamic World.

1. For instance, "two squares and ten roots are equal to forty-eight dirhams"; that is to say, what must be the amount of two squares which, when summed up and added to ten times the root of one of them, make up a sum of forty-eight dirhams? You must first reduce the two squares to one, and you know that one square of the two is the half of both. Then reduce everything mentioned in the statement to its half, and it will be the same as if the question had been, a square and five roots of the same are equal to twenty-four dirhams; or, what must be the amount of a square which when added to five times its root, is equal to twenty-four dirhams? Now halve the number of the roots; the half is two and a half. Multiply that by itself; the product is six and a quarter. Add this to twenty-four; the sum is thirty dirhams and a quarter. Take the root of this; it is five and a half. Subtract from this the half of the number of the roots, that is two and a half; the remainder is three. This is the root of the square, and the square itself is nine.

(al-Khwārizmī)

2. We have said enough so far as numbers are concerned, about the six types of equations. Now, however, it is necessary that we should demonstrate geometrically the truth of the same problems which we have explained in numbers.

(al-Khwārizmī)

3. We say equations of first class in three places, that is, numbers, *objects* (x), and *squares* (x^2) have six forms; three of them are singletons and the other three are polynomials. Their unknowns can be obtained by Book II of the *Elements*. This has been explained in books of algebraists. But whenever *cubes* (x^3) come in, and among them and other places there is an equation, we need solid geometry, and especially conics and conic sections because a *cube* (x^3) is a solid.

(Omar Khayyam)

4. I have seen the book of Ibn Haytham, God bless his soul, called the solution of doubt. This postulate among other things was accepted without proof. There are many other things which are foreign to this field such as: If a straight line segment moves so that it remains perpendicular to a given line, and one end of it remains on the given line, then the other end of it draws a parallel. There are many things wrong here.

(Omar Khayyam)

CHAPTER 9 The mathematical awakening of Europe.

1. The point of special significance with respect to Adelard is that he stands at the meeting point of three intellectual movements, the traditional learning of the French schools, the Greek culture of southern Italy and the Arabic science of the East.
   (C.H. Haskins, *Adelard of Bath*, 1911)

2. Since each successive motion is proportionable to another with respect to speed, natural philosophy which studies motion, ought not to ignore the proportion of motions and their speeds, and, because an understanding of this is both necessary and extremely difficult, nor has as yet been treated fully in any branch of philosophy, we have accordingly composed the following work on the subject [*On the Ratios of Velocities in Motions*]. Since, moreover, (as Boethius points out in Book I of his Arithmetic), it is agreed that whoever omits mathematical studies has destroyed the whole of philosophic knowledge, we have commenced by setting forth the mathematics needed for the task at hand, in order to make the subject easier and more accessible to the student.
   (T. Bradwardine, *On the Ratios of Velocities in Motions*, 1328)

3. A lion would eat one sheep in four hours; and a leopard [would eat it] in 5 hours; and a bear [would eat it] in 6; we are asked, if a single sheep were to be thrown to them, how many hours would they take to devour it?
   You will do this: for 4 hours, in which the lion eats a sheep, put $\frac{1}{4}$; and for the 5 hours the leopard takes put $\frac{1}{5}$; and for the 6 hours the bear takes, put $\frac{1}{6}$: and because $\frac{1}{6}$, $\frac{1}{5}$ and $\frac{1}{4}$ are found [exactly] in 60, suppose that in 60 hours they will devour the sheep. Then consider how many sheep a lion would eat in the 60 hours: since in four hours it devours one sheep, it is obvious that it would eat 15 sheep in the 60 hours; and the leopard would eat 12 as a fifth of 60 is 12. Similarly the bear would eat 10; since 10 is of 60. Therefore in 60 hours they would eat 15 plus 12 plus 10 sheep, that is 37. So you will say: for the 60 hours, which I suppose, they will eat 37 sheep. What [time] should I suppose so that they will eat only one sheep? So multiply one by 60, and divide by 37, which gives $1\frac{23}{37}$. And in that number [of hours] they will have eaten up the sheep.
   (Leonardo of Pisa)

4. *If a given number is separated into two parts whose difference is known, then each of the parts can be found.*
   Since the lesser part and the difference equal the larger, the lesser with another equal to itself together with the difference make the given number. Subtracting therefore the difference from the total, what remains is twice the lesser. Having this yields the smaller and, consequently, the greater part.
   (Jordanus de Nemore)

CHAPTER 10 The Renaissance: Recovery and Innovation.

1. Having surveyed the basic authorities in mathematics, Maurolico examines the contributions of the moderns to the literature. He starts by saying that most modern books are to be passed over in silence, inasmuch as they are full of obscure curiosities, rather than of agreeable utility. Thus, he deems Luca Pacioli's *Summa* to be prolix and poor as Cardano has already pointed out. Cardano is one mathematician that Maurolico exempts from his condemnation of the moderns, even though his writings sometimes give more labour than delight. In general, says Maurolico, the moderns would do

much better if, instead of pouring out immaturely their own works, they were to examine carefully the books of the ancients.

(P. Rose, *The Italian Renaissance of Mathematics*, 1975)

2. The whole of this book is divided into five principal parts. In the first, numbers are discussed, in every way you would expect in simple and speculative practice ... And in this first part are also contained all commercial occurrences of problems and rules, that is by hundredweights, thousands, pounds, ounces, investments, sales, profits, losses, journeys or transportations of goods, weights, measures, and money from place to place.

(L. Pacioli, *Summa de Arithmetica*)

3. Find me two numbers such that when they are added together, they make as much as the cube of the lesser added to the product of its triple with the square of the greater; and the cube of the greater added to its triple times the square of the lesser makes 64 more than the sum of these two numbers.

(from a letter of L. Ferrari to N. Tartaglia)

4. Behold, the art which I present is new, but in truth so old, so spoiled and defiled by the barbarians, that I considered it necessary, in order to introduce an entirely new form into it, to think out and publish a new vocabulary, having gotten rid of all its pseudo-technical terms lest it should retain its filth and continue to stink in the old way, but since till now ears have been little accustomed to them, it will be hardly avoidable that many will be offended and frightened away at the very threshold.

(F. Viète, *Introduction to the Analytic Art*)

CHAPTER 11 The Renaissance of Mathematics in Britain.

1. Sith Merchauntes by shippes great riches do winne,
I may with good righte at their seate beginne.
The Shippes on the sea with Saile and with Ore,
Were first founde, and styll made, by Geometries lore.
Their Compas, their Carde, their Pulleis, their Ankers,
Were founde by the skill of witty Geometers.

(R. Recorde, *The Commodities of Geometry*)

2. The Elements of Geometrie of the most auncient Philosopher Euclide of Megara. Faithfully (now first) translated into the Englishe toung, by H. Billingsley, Citizen of London. Whereunto are annexed certaine Scholies, Annotations, and Inventions, of the best Mathematiciens, both of time past, and in this our age.

With a very fruitfull Praeface made by M. J. Dee, specifying the chiefe Mathematicall Sciences, what they are, and wherunto commodious: where, also, are disclosed certaine new Secrets Mathematicall and Mechanicall, untill these our daies, greatly missed.

(J. Dee)

3. Wonder not, that these logarithms are different from those which the excellent baron of Marchiston published in his Admirable Canon ... And as to the nature of the change, he thought it more expedient that 0 should be made the logarithm of 1; and 100 000 &c. the logarithm of the radius; which I could not but acknowledge was much better.

Therefore, rejecting those which I had before prepared, I proceeded, at his exhortation, to show him the principle of them; and should have been glad to do the same the third summer, if it had pleased God to spare him so long.

(H. Briggs)

4. And the better to distinguish upon the First view, what quantities were Known, and what Unknown, he [Oughtred] doth (usually) denote the Known to *Consonants*, and the Unknown by *Vowels*; as Vieta (for the same reason) had done before him ...
Mr Oughtred in his *Clavis*, contents himself (for the most part) with the solution of Quadratick Equations, without proceeding (or very sparingly) to Cubick Equations, and those of Higher Powers; having designed that Work for an *Introduction into Algebra* ... He contents himself likewise in Resolving Equations, to take notice of the *Affirmative* or *Positive Roots*; omitting the *Negative* or *Ablative Roots*, and such as are called *Imaginary* or *Impossible Roots*. And of those which, he calls *Ambiguous Equations* (as having more Affirmative Roots than one) he doth not (that I remember) any where take notice of more than Two Affirmative Roots; (Because in Quadratick Equations, which are those he handleth, there are indeed no more.) Whereas yet in Cubick Equations, there may be *Three*, and in those of Higher Powers, yet more. Which Vieta was well aware of, and mentioneth in some of his Writings; and of which Mr Oughtred could not be ignorant.

(J. Wallis on W. Oughtred's *Clavis Mathematicae*)

CHAPTER 12 The Astronomical Revolution.

1. Now that it has been shown that the Earth too has the form of a globe, I think we must see whether or not a movement follows upon its form and what the place of the Earth is in the universe. For without doing that it will not be possible to find a sure reason for the movements appearing in the heavens. Although there are so many authorities for saying that the Earth rests in the centre of the world that people think the contrary supposition inopinable and even ridiculous; if however we consider the thing attentively, we will see that the question has not yet been decided and accordingly is by no means to be scorned.

(N. Copernicus, *On the Revolutions of the Heavenly Spheres*)

2. Let the reader recall from my *Mysterium Cosmographicum*, which I published twenty-two years ago, that the number of the planets, or orbits about the Sun, was derived by the most wise Creator from the five solid figures, about which Euclid so many centuries ago wrote the book which, since it is made up of a series of propositions, is called *Elements*. That there cannot be more regular bodies, that regular plane figures, that is, cannot unite into a solid in more than five ways, was made clear in the second book of the present work.

[However] the cube and the octahedron enter somewhat their conjugate planetary orbits, the dodecahedron and the icosahedron do not quite reach their conjugate orbits, the tetrahedron just touches both orbits; in the first case there is a deficiency, in the second case an excess, in the last case an equality, in the distances of the planets. From these considerations it is apparent that the exact relations of the planetary distances were not derived from the regular figures alone; for the Creator, the very fountainhead of geometry, who, as Plato says, practises geometry eternally, does not deviate from his archetype.

(J. Kepler, *Harmonices Mundi*)

3. Why waste words? Geometry existed before the creation, is co-eternal with the mind of God, is God himself (what exists in God that is not God himself?); geometry provided God with a model for the Creation and was implanted into Man, together with God's own likeness-and not merely conveyed to his mind through the eyes.
(J. Kepler, *Harmonices Mundi*)

4. A projectile which is carried by a uniform horizontal motion compounded with a naturally accelerated vertical motion describes a path which is a semi-parabola.
(Galileo, *Two New Sciences*)

CHAPTER 13 European Mathematics in the Early 17th Century.

1. I have been assured that we shall have Mr. Gassendi here at the beginning of June. I am very pleased about this. He will see the most noble academy in the world which has recently been formed in this town. It no doubt will be the most noble for it is wholly mathematical. As for the names of the most excellent gentlemen who attend, they are: Etienne Pascal, ... Roberval, Desargues, and others. My good friend, René Descartes, does not attend. He will not commit himself to any group and is contemptuous of our deals. But I continue to correspond with him and he with us.
(from a letter of M. Mersenne)

2. There are infinitely many questions of this kind, but there are some others which demand new principles before the descent can be applied to them and the study of them is sometimes so difficult one cannot overcome without extreme effort. Such is the following question that Bachet said Diophantus had never been able to demonstrate, on the subject of which M. Descartes in one of his letters made the same declaration when he confessed that he found it so difficult that he could so no way of solving it. Every number is a square, or a sum of two, three, or four squares.
(from a letter of P. de Fermat to C. Huygens)

3. Let us apply it to the problem of finding two mean proportionals between the lines a and q. It is evident that if we represent one of the mean proportionals by z, then $a : z = z : \frac{z^2}{a} = \frac{z^2}{a} : \frac{z^3}{a^2}$. Thus we have an equation between q and $\frac{z^3}{a^2}$.
   Describe the parabola FAG with its axis along AC, and with AC equal to $\frac{1}{2}a$, that is, to half the latus rectum. Then erect CE equal to $\frac{1}{2}q$ and perpendicular to AC at C, and describe the circle AF about E as center, passing through A. Then FL and LA are the required mean proportionals.
(R. Descartes, *La Géométrie*)

4. I nowhere call Descartes a plagiarist; I would not appear so impolite. However this I say, the major part of his algebra (if not all) is found before him in other authors (notably in our Harriot) whom he does not designate by name. That algebra may be applied to geometry, and that it is in fact so applied, is nothing new. Passing the ancients in silence, we state that this has been done by Vieta, Ghetaldi, Oughtred, and others, before Descartes. They have resolved by algebra ... many geometrical problems ... But the question is not as to application of algebra to geometry (a thing quite old), but of the Cartesian algebra considered by itself.
(from a letter of J. Wallis to S. Morland)

Exercises: Part B

Chapter-based essay questions.

1. Outline briefly the main features of Egyptian and Mesopotamian mathematics. To what extent were their mathematical activities based on the desire to solve specific practical problems?

2. Aristotle wrote, in the mid-4th century BC, that 'the mathematical arts were first set up in Egypt; for there the priestly caste were allowed to enjoy leisure'.
   In the light of your knowledge of Mesopotamian and Egyptian mathematics, to what extent do you feel Aristotle's judgement was correct?

3. Was Plato's influence decisive in the development of Greek mathematics?

4. What do you consider to be the importance of Aristotle for the historian of mathematics?

5. Briefly characterise the main features of Euclid's *Elements*. How important do you judge the *Elements* to have been in the Greek study of mathematics?

6. Describe some of the main methods of proof used by Greek geometers. How important do you think proof was in Greek mathematics?

7. Explain the 'three classical problems' of Greek mathematics. In what ways did their study advance mathematics?

8. It is often said that Archimedes was the greatest of the Greek mathematicians. How valid do you consider this claim to be?

9. Were Archimedes and Apollonius engaged in the same tradition of geometrical research?

10. How important was the study of conic sections in the Greek study of mathematics?

11. Describe briefly one feature of Greek mathematics which seems to you to be of the highest significance, and indicate your reasons for making this judgement.

12. Choose a Greek mathematical result and a Mesopotamian one. Outline each briefly and explain how it was justified. What differences between Greek and Mesopotamian mathematical cultures do these examples seem to indicate?

13. Describe, with reference to examples, what seem to you to be the characteristic aims of the Egyptian, Mesopotamian, and Greek mathematical cultures. How did the Egyptians and Mesopotamians differ from the Greeks?

14. Outline briefly the main features of Indian and Chinese mathematics. To what extent were their mathematical activities based on the desire to solve specific practical problems?

15. Describe briefly the work of al-Khwārizmī on the solution of equations. Compare al-Khwārizmī's methods with those found in earlier mathematical cultures.

16. Choose a Chinese mathematical result, and an Islamic one. Outline each briefly and explain how it was justified. What differences between the Chinese and Islamic mathematical cultures do these examples indicate?

17. How did knowledge of the solution of cubic and quartic equations develop in 16th-century Italy?

18. Why was mathematics studied, and what was it used for, in 16th-century England?

19. Describe some of the main features of the mathematical works of Recorde and Dee. What would be the main advantages and disadvantages of learning mathematics from each of them?

20. Give a brief account of the achievements of the work of two of the following: Recorde, Oughtred, Dee, Harriot. To what extent were innovations in notation important in their work and in its reception?

21. Describe some of the translations of ancient Greek mathematical works that were made in the 16th century by Commandino, Maurolico, Dee, and others. How important do you judge these translations to have been, and why?

22. What were the contributions of Marin Mersenne to the mathematical and scientific culture of his time? In what ways did he advance the study of mathematics?

23. Fermat is sometimes described as being simultaneously 'ancient' and 'modern'. By making reference to examples of his work, decide how accurate you consider this description to be.

24. Describe the main contributions of Fermat to number theory. How successful was he as a mathematician in his own lifetime?

25. Descartes is sometimes said to be the inventor of modern algebra. In the light of your knowledge of the work of Viéte, Fermat, Harriot, and others, how fair do you consider this judgement to be?

26. Describe the main features of Descartes's work in mathematics. How did he strike a balance between geometry and algebra in *La Géométrie*?

27. To what extent did mathematics develop in the 16th and 17th centuries primarily as a means for responding to technological and scientific needs and to what extent was it undertaken for its own sake?

28. Is it fair to claim that mathematics developed as it did in the 16th and 17th centuries primarily as a result of the rediscovery of Greek mathematics?

29. How important do you judge advances in algebra to have been in the development of mathematics in the 16th and 17th centuries?

30. Describe how the relationship between algebra and geometry changed in the 16th and 17th centuries. How important do you judge these changes to have been?

Exercises: Part C

General essay questions.

1. How important has the role of proof been in the development of mathematics? Illustrate your answer with at least three examples from different periods, of which at least one illustrates where giving proofs seems to have been important to the mathematics being developed, and one where this was not a strong consideration.

2. Describe three mathematical achievements that, in your view, have a cultural significance reaching beyond the strict confines of the subject, and explain why they have been so important.

3. What are the main reasons why today we might still want to study the mathematics developed in ancient Greece? How do they differ from the reasons offered for studying Greek mathematics in the early 17th century?

4. How important has the role of applications been in the development of mathematics? Illustrate your answer with at least three examples showing differing degrees of importance.

5. For what reasons has mathematics been studied? Give examples from at least three different periods.

6. How have mathematical discoveries been disseminated at various periods in history? What were the important consequences of the introduction of printing?

7. Histories of mathematics often concentrate on the achievements of the great mathematicians. What has been the contribution to mathematics of the less than great? How important do you judge these to have been? Give examples from at least three different periods.

8. How important has good notation been in the development of mathematics? Give at least one example where good notation has been very important, and one where it has not.

9. It is sometimes said that the essential activity of mathematicians is proving theorems. Give three examples, chosen from different periods, to show that mathematics involves more than proof.

10. Some mathematicians have worked in almost complete isolation, other in large mathematical communities. By discussing at least three examples, consider how being part of a wider community helps or hinders a mathematician to lead a creative life.

Bibliography

Abū Kāmil, 1966. *The* Algebra *of abū Kāmil, in a commentary by Mordecai Finzi*, M. Levey (ed. and transl.), Wisconsin University Press.

Adamson, I.R. 1980. The administration of Gresham College and its fluctuating fortunes as a scientific institution in the seventeenth century, *History of Education* 9, 13–25.

Al-Khwārizmī, 1831. *The Algebra of Mohammed ben Musa*, F. A. Rosen (ed. and transl.), Kessinger Publishing, 2008.

———, 1930. *Robert of Chester's Latin Translation of the Algebra of al-Khowarizmi*, University of Michigan Press, reprinted in R. Calinger, *Classics in Mathematics*, Moore, 1982.

Amir-Moez, A.R. 1959. Omar Khayyam, *Discussions of difficulties in Euclid*, *Scripta Mathematica* 24, 275–303.

Anderson, M., Katz, V., and Wilson R.J. (eds.) 2004. *Sherlock Holmes in Babylon*, Mathematical Association of America.

Archimedes. 1953. *The Works of Archimedes*, Sir T.L. Heath (transl. and ed.), Cambridge University Press, Dover.

Aristophanes, 1978. *The Birds*, D. Barrett (transl.), Penguin.

Aristotle, 1998. *Politics*, E. Barker (transl.), Oxford University Press.

Artmann, B. 1999. *Euclid: The Creation of Mathematics*, Springer.

Āryabhaṭa, 1930. *The* Āryabhaṭīya *An Ancient Indian Work on Mathematics and Astronomy*, W.C. Clark (transl. and ed.), Chicago University Press.

Aubrey, J. 1949. *Brief Lives*, O.L. Dick (ed.), Secker and Warburg.

Baldi, B. 1859. *Versi e Prose di Bernadino Baldi*, F. Ugolino and F.-L. Polidori (eds.) Florence.

Banville, J. 1976. *Doctor Copernicus*, Secker and Warburg.

———. 1981. *Kepler*, Secker and Warburg.

Barrow, J.D. 1992. *Pi in the Sky: Counting, Thinking and Being*, Clarendon Press, Oxford.

Beer, A. and Beer, P. 1975. *Kepler: Four Hundred Years*, Pergamon Press.

Belaval, Y. 1960. *Leibniz, Critique de Descartes*, Gallimard, Paris.

Belyi, Yu. A. 1975. Johannes Kepler and the development of mathematics, in *Kepler: Four Hundred Years*, A. and P. Beer (eds,), Pergamon Press, 543–600.

Bennett, J.A. 1982. *The Mathematical Science of Christopher Wren*, Cambridge University Press.

———. 1986. The mechanics' philosophy and the mechanical philosophy, *History of Science* 24, 1–28.

Berggren, L.J. 1986. *Episodes in the Mathematics of Medieval Islam*, Springer.

Boas, M. 1970. *The Scientific Renaissance 1430–1630*, Fontana.

Boethius, 1983. *Boethian Number Theory: A Translation of the* De Institutione Arithmetica *with Introduction and Notes*, M. Masi (transl.), Rodopi.

Bombelli, R. 1966. *L'Algebra*, Feltrinelli.

Bos, H.J.M. 1981. On the representation of curves in Descartes's *Géométrie*, *Archive for History of Exact Sciences* 24, 295–338.

———. 1996. Tradition and modernity in early modern mathematics, Viète, Descartes and Fermat, in *L'Europe Mathématique, Mathematical Europe*, C. Goldstein, J.J. Gray, and J. Ritter (eds.), Éditions de la Maison des Sciences de l'Homme, 183–204.

———. 1997. The structure of Descartes's *Géométrie*, in *Lectures in the History of Mathematics*, American and London Mathematical Societies, 35–57.

———. 2001. *Redefining Geometrical Exactness: Descartes's Transformation of the Early Modern Concept of Construction*, Springer.

Bowen, A.C. 1982. Archytas, Fragment, *Ancient Philosophy* 2, 81–100.

Boyer, C.B. and Merzbach, U.C. 2011. *A History of Mathematics*, 3rd edn., Wiley.

Brecht, B. 1980. *The Life of Galileo*, H. Brenton (transl.), Eyre Methuen.

Brinsley, J. 1627. *Ludus Literarius; or, The Grammar Schoole* London.

Buck, R.C. 1980. Sherlock Holmes in Babylon, *American Mathematical Monthly* 87, 335–345, reprinted in Anderson, Katz, and Wilson (2004), 5–14.

Cardano, G. 1968. *Ars Magna*, T.R. Witmer (transl.), MIT Press, 1968.

Caspar, M. 1959. *Kepler*, Adelard–Schuman.

Chace, A.B. 1927. *The Rhind Mathematical Papyrus*, Mathematical Association of America.

Chartres, R. and Vermont, D. 1997. *A Brief History of Gresham College 1597–1997*, Gresham College, London.

Chasles, M. 1837. Aperçu Historique sur l'Origine et le Développement des Méthodes en Géométrie ... suivi d'un Mémoire de Géométrie, etc., Brussels, in *Mémoires sur les Questions Proposées par l'Académie Royale des Sciences et Belles-Lettres de Bruxelles* 11.

Chemla, K., Bray, F., Fu Daiwie, Huang Yi-Long, and Métailié, G. (eds.), 2001. La Scienza in Cina, in *Storia della scienza*, Vol. II. Sandro Petruccioli (ed.), Istituto della Enciclopedia Italiana, Roma.

Chuquet, N. 1985. *Nicolas Chuquet, Renaissance Mathematician: A Study with Extensive Translation of Chuquet's Mathematical Manuscript Completed in 1484*, H.G. Flegg, C.M. Hay, and B. Moss (eds. and transl.), Springer.

Clagett, M. 1964. *Archimedes in the Middle Ages, I: The Arabo–Latin Tradition*, University of Wisconsin Press.

Clavelin, M. 1974. *The Natural Philosophy of Galileo*, MIT Press.

Clucas, S. (ed.) 2006. *John Dee: Interdisciplinary Studies in English Renaissance Thought*, Springer.

Cohen, H.F. 1984. *Quantifying Music: The Science of Music at the First Stage of the Scientific Revolution, 1580–1650*, Reidel.

Cohen, I.B. 1985. *Revolution in Science*, Harvard University Press.

Cohen, M.R. and Drabkin, I.E. 1948. *A Source Book in Greek Science*, Harvard University Press.

Coolidge, J.L. 1968. *A History of the Conic Sections and Quadric Surfaces*, Oxford University Press, Dover.

Copernicus, N. 1543. *De Revolutionibus Orbium Coelestium*, Nuremberg. English transl. C.G. Wallis, Encyclopædia Britannica, 1952.

Crombie, A.C. 1953. *Robert Grosseteste and the Origins of Experimental Science*, Oxford University Press.

———. 1969. *Augustine to Galileo*, Penguin.

———. 1975. Mersenne, Marin. *Complete Dictionary of Scientific Biography*, C.C. Gillispie (ed.), 9, Charles Scribner's Sons, 316–322.

Cullen, C. 2004. *The Suan Shu Shu 'Writings on Reckoning'*, Needham Research Institute, No. 1, Cambridge.

Cuomo, S. 2001. *Ancient Mathematics*, Sciences of Antiquity, Routledge.

———. 2007. *Pappus of Alexandria and the Mathematics of Late Antiquity*, Cambridge University Press.

Dauben, J. 2008. *Suan Shu Shu.* A book on numbers and computations. English translation with commentary, *Archive for History of Exact Sciences* 62, 91–178.

Dee, J. 1661. Treatise of the division of superficies, etc. in *Euclid's Elements of Geometry*, London.

Desargues, G. 1986. *The Geometrical Work of Girard Desargues*, J.V. Field and J.J. Gray (transl. and eds.), Springer.

———. 1988. *L'Oeuvre Mathématique de G. Desargues: Textes Publiés et Commentés avec une Introduction Biographique et Historique*, 2nd edn., Institut Interdisciplinaire d'Études Epistemologiques.

Descartes, R. 1637. *Discours de la Méthode pour bien Conduire sa Raison et Chercher la Verité dans les Sciences. Plus la Dioptrique, Les Météores et la Géométrie, qui sont des Essais de cette Méthode*, Leyden; *La Géométrie* was translated as *The Geometry of René Descartes* by D.E. Smith and M.L. Latham, Open Court Publishing Co., 1925, Dover reprint, 1954.

———. 1964. *Rules for the Direction of the Mind, Philosophical Essays* L. J. Lafleur (transl. and ed.), Library of the Liberal Arts, Bobbs-Merrill.

Diophantus, 1621. *Arithmetica*, G.B. Bachet de Méziriac (ed.).

Doblhofer, E. 1973. *Voices in Stone*, Paladin.

Drake, S. 1957. *Discoveries and Opinions of Galileo*, Doubleday Anchor.

Edwards, C.H. Jr. 1979. *The Historical Development of the Calculus*, Springer.

Edwards, H.M. 1977. *Fermat's Last Theorem*, Springer.

Eisenstein, E.L. 1979. *The Printing Press as an Agent of Change: Communications and Cultural Transformations in Early-Modern Europe*, Cambridge University Press.

Euclid, 1925. *The Thirteen Books of Euclid's Elements*, T.L. Heath (ed.), Cambridge University Press.

Euler, L. 1984. *Elements of Algebra*, J. Hewlett (transl.), reprint of the 1840 edition, Springer.

Fauvel, J., Flood, R., and Wilson R.J. (eds.) 2013. *Oxford Figures: Eight Centuries of Oxford Mathematics*, 2nd edn., Oxford University Press.

Fauvel, J. and Gray, J.J. 1987. *The History of Mathematics: A Reader*, Macmillan, in association with the Open University.

Feingold, M. 1984a. The occult tradition in the English universities of the Renaissance: a reassessment, in B. Vickers (ed.), *Occult and Scientific Mentalities in the Renaissance*, 73–94, Cambridge University Press.

———. 1984b. *The Mathematicians' Apprenticeship*, Cambridge University Press.

Fermat, P. 1679. *Varia Opera Mathematica*, Toulouse.

———. 1891–1922. *Pierre de Fermat: Oeuvres*, 4 vols, Gauthier-Villars, Paris.

Fibonacci, 2002. *Fibonacci's* Liber Abaci*: A Translation into Modern English of Leonardo Pisano's* Book of Calculation, L.E. Sigler (transl. and ed.), Springer.

Field, J.V. 1988. *Kepler's Geometrical Cosmology*, Chicago University Press.

Flegg, G., Hay, C., and Moss, B. (eds.) 1985. *Nicolas Chuquet, Renaissance Mathematician*, Reidel.

Fowler, D.H. 1999. *The Mathematics of Plato's Academy*, 2nd edn., Oxford University Press.

Galilei, Galileo, 1957. *The Assayer*, Stillman Drake (transl.), Phildelphia.

———, 1967. *Dialogue Concerning the Two Chief World Systems*, Stillman Drake (transl.), University of California Press.

———, 1954. *Dialogues Concerning Two New Sciences*, H. Crew and A de Salvio (transl.), Dover.

Gandz, S. 1936. The sources of al-Khowarizmi's *Algebra*, *Osiris* 1, 263–277.

Gaukroger S. (ed.), 1980. *Descartes: Philosophy, Mathematics and Physics*, Harvester Press.

———. 1997. *Descartes: An Intellectual Biography*, Oxford University Press.

Gingerich, O. 2004. *The Book Nobody Read: Chasing the Revolutions of Nicolaus Copernicus*, Walker.

Goldstein, C., Gray, J.J., and Ritter, J. (eds.), 1996. *L'Europe Mathématique, Mathematical Europe*, Éditions de la Maison des Sciences de l'Homme, Paris.

Grant E.N. (ed.), 1971a. *Oresme and the Kinematics of Circular Motion*, University of Wisconsin Press.

———. 1971b. *Science in the Middle Ages*, Wiley.

Gray, J.J. 1989. *Ideas of Space, Euclidean, non-Euclidean and Relativistic*, Oxford University Press.

Guicciardini, N. 1999. *Reading the* Principia, Cambridge University Press.

Hallowes, D.M. 1961–1962. Henry Briggs, mathematician, *Transactions of the Halifax Antiquarian Society*, 8–81.

Hankel, H. 1874. *Zur Geschichte der Mathematik im Altertum und Mittelalter*, Teubner.

Harnack, A. and Mitchell, J.M. 1910–1911. Neo-Platonism, *Encyclopædia Britannica* (11th edn.).

Hannam, J. 2009. *God's Philosophers: How the Medieval World laid the Foundations of Modern Science*, Icon Books.

Hay, C. (ed.) 1987. *Mathematics from Manuscript to Print: 1300–1600*, Oxford University Press.

Heath, T.L. 1949. *Mathematics in Aristotle*, Oxford University Press.

———. 1981. *A History of Greek Mathematics*, 2 vols, Dover.

———. 2007. *Diophantos of Alexandria. A Study in the History of Greek Algebra*, Kessinger Publishing.

Heilbron, J. 2010. *Galileo, Oxford University Press.*

Heinzelin, J. de. 1962. Ishango, *Scientific American* 206, issue 6, 105–116.

Herodotus, 1996. *The Histories*, A. de Sélincourt (transl.), revised with an introduction by J. Marincola, Penguin.

Hill, C. 1982. *Intellectual Origins of the English Revolution*, Oxford University Press.

Hobbes, T. 1651. *Leviathan or The Matter, Forme and Power of a Common Wealth Ecclesiasticall and Civil*, Crooke, London.

Hogendijk, J. 1984. Greek and Arabic constructions of the regular heptagon, *Archive for History of Exact Sciences* 30, 197–330.

Howson, G. 1982. *A History of Mathematics Education in England*, Cambridge University Press.

Høyrup, J. 2002. *Lengths, Widths, Surfaces: A Portrait of Old Babylonian Algebra and its Kin*, Springer.

———. 2004. Conceptual divergence — canons and taboos — and critique: reflections on explanatory categories, *Historia Mathematica* 31, 129–147.

Huff, T.E. 1993. *The Rise of Early Modern Science*, Cambridge University Press.

Jaouiche, K. 1986. *La Théorie des Parallèles en Pays d'Islam*, Vrin.

Jenkyns, R. 19 June 1983. *The Sunday Times.*

Jones, A. 1991. The adaptation of Babylonian methods in Greek numerical astronomy, *Isis* 82, 441–453.

Jordanus, Nemorarius (de Nemore). 1981. *De Numeris Datis*. B.B. Hughes (transl. and ed.), University of California Press.

Katz, V. (ed.) 2007. *The Mathematics of Egypt, Mesopotamia, China, India, and Islam: A Sourcebook*, Princeton University Press.

Karpinski, L.C. 1915. *Robert of Chester's Latin Translation of the Algebra of Al-Khwarizmi*, New York.

Keller, A. 2006. *Expounding the Mathematical Seed, A Translation with Commentary of Bhāskara I on the Mathematics Chapter of the Āryabhaṭīya*, Birkhäuser.

Kepler, J. 1596. *Mysterium Cosmographicum*, Frankfurt.

———. 1609. *Astronomia Nova*, Prague.

———. 1619. *Harmonices Mundi*, Linz.

———. 1627. *Tabulae Rudolphinae*, Ulm.

———. 2015. *Astronomia Nova*, W. H. Donahue (ed. and transl.), Green Lion Press.

Khayyām, Omar. 1950. Quoted in H. J. J. Winter and W. Arafat. The algebra of Omar Khayyām, *Journal of the Royal Asiatic Society of Bengal: Science* 16, 27–78.

Klein, J. 1968. *Greek Mathematical Thought and the Origin of Algebra*, MIT Press.

Knorr, W.R. 1975. *The Evolution of the Euclidean Elements*, Reidel.

———. 1982. Observations on the early history of the conics, *Centaurus* 26, 1–24.

———. 1985. Untitled review of Books IV to VII of Diophantus' *Arithmetica: in the Arabic translation attributed to Qusta Ibn Luqa*, by Jacques Sesiano, *American Mathematical Monthly*, 91.2, 150–154.

———. 1993. *The Ancient Tradition of Geometric Problems*, Dover.

Knowles, D. 1962. *The Evolution of Mediaeval Thought*, Vintage Books.

Kuhn, T.S. 1957. *The Copernican Revolution: Planetary Astronomy in the Development of Western Thought*, Harvard University Press.

Lagrange, J.-L. 1768. Solution d'un problème d'arithmétique, *Miscellanea Taurinensia* IV, 1766–1769, in *Oeuvres* I, 671–731.

Leslie, J. 1820. *Philosophy of Arithmetic*, Edinburgh.

Li Yan and Dù Shírán. 1987. *Chinese Mathematics: A Concise History*, Clarendon Press, Oxford.

Libbrecht, U. 1973. *Chinese Mathematics in the 13th Century: The* Shu-Shu Chiu-Chang of Ch'in Chiu-Shao, MIT Press.

Lindberg D.C. 1978. *Science in the Middle Ages*, Chicago University Press.

———. 1981. *Theories of Vision from al-Kindi to Kepler*, Chicago University Press.

Lloyd, G.E.R. 1979. *Magic, Reason and Experience*, Cambridge University Press.

Lovell, K. 1961. *The Growth of Basic Mathematical and Scientific Concepts in Children*, London University Press.

Lyons, J. 2009. *The House of Wisdom*, Bloomsbury.

Mahoney, M.S. 1994. *The Mathematical Career of Pierre de Fermat, 1601–1665* (2nd edn.), Princeton University Press.

Mancosu, P. 1992. Descartes's *Géométrie* and revolutions in mathematics, in *Revolutions in Mathematics*, D. Gillies (ed.), Oxford University Press, 83–116.

Marshack, A. 1972. *The Roots of Civilization*, Weidenfeld and Nicolson.

Martzloff, J.-C. 1997. *A History of Chinese Mathematics* (2nd edn. 2006), Springer.

Menninger, K. 1969. *Number Words and Number Symbols*, MIT Press.

Mill, J.S. 1867. *An Examination of Sir W. Hamilton's Philosophy*, Spencer.

Milhaud, G. 1921. *Descartes Savant*, Alcan.

Molland, A.G. 1968. The geometrical background to the 'Merton School', *British Journal for the History of Science* 4, 108–125.

Morrow, G. 1970. *Commentary on the First Book of Euclid's* Elements, Princeton University Press.

Mosley, A. 2007. *Tycho Brahe and the Astronomical Community of the Late Sixteenth Century*, Cambridge University Press.

Murdoch, J.E. 1984. *Album of Science: Antiquity and the Middle Ages*, Charles Scribner's Sons.

Netz R. 1999. *The Shaping of Deduction in Greek Mathematics. A Study in Cognitive History*, Cambridge University Press.

Netz R. and Noel, W. 2007. *The Archimedes Codex*, Perseus.

Neugebauer, O. 1969. *The Exact Sciences in Antiquity*, Dover.

Neugebauer, O. and Sachs, A. 1945. *Mathematical Cuneiform Texts*, American Oriental Society.

Nicomachus of Gerasa, 1926. *Introduction to Arithmetic*, Book I, M. D'Ooge (transl.), Macmillan.

Nordgaard, M.A. 2004. Sidelights on the Cardan–Tartaglia controversy, in *Sherlock Holmes in Babylon* (ed. M. Anderson, V. Katz, and R.J. Wilson), Mathematical Association of America, 153–163.

O'Day, R. 1982. *Education and Society 1500–1800*, Longman.

Ong, W.J. 1958. *Ramus, Method, and the Decay of Dialogue*, Harvard University Press.

Ore, O. 1965. *Cardano, the Gambling Scholar*, Dover.

Osiander, A. 1543. Revolutions, Preface to Copernicus's *De Revolutionibus Orbium Coelestium.*

Oughtred, W. 1634. *To the English Gentrie, and All Others Studious of the Mathematicks ... The Just Apologie of Wil: Oughtred*, London: A. Mathewes, 1634(?).

———. 1647. *Clavis Mathematicae: The Key of the Mathematicks new Forged and Filed ...*, transl. R. Wood, published and printed by Tho. Harper for Rich. Whitaker.

Pascal, B. 1908. *Oeuvres Complètes de Blaise Pascal, Physique et Mathématiques,* Vol. 3, L. Brunschvicg and P. Boutroux (eds.), Hachette.

Pepper, J.V. 1968. Harriot's calculation of the meridional parts as logarithmic tangents, *Archive for History of Exact Sciences* 4, 359–413.

Phillips, D. 1985. Clerks of Oxenford: a new history of the University of Oxford, *Oxford Review of Education* 11, 105–109.

Pingree, D. 2003. The logic of non-Western science: mathematical discoveries in medieval India, *Daedalus* 132, 45–54.

Plato, 1892. *The Dialogues of Plato*, Vol. 2, B. Jowett (transl.), Oxford at the Clarendon Press.

———, 1970. *The Dialogues of Plato*, Vol. 3, B. Jowett (transl.) Sphere.

Plofker, K. 2009. *Mathematics in India*, Princeton University Press.

Plutarch, 1899. *Plutarch's Lives*, Vol. 2, A. Stewart and G. Long (transl.), George Bell and Sons.

Pomeroy, S.B. 1975. *Goddesses, Whores, Wives, and Slaves: Women in Classical Antiquity*, Schocken Books, New York.

Prior, M. 1718. *Alena, or the Progress of the Mind: in three cantos, Poems on Several Occasions*, London.

Proclus, 1970. *A Commentary on the First Book of Euclid's* Elements, G.R. Morrow (transl.), Princeton University Press.

Ptolemy 1952. *Almagest*, R. Catesby Taliaferro (transl.), Great Books of the Western World, Encyclopædia Britannica, Vol. 16.

Rabin, S. 2005. Nicolaus Copernicus, *The Stanford Encyclopedia of Philosophy*, E.N. Zalta (ed.): *http://plato.stanford.edu/archives/sum2005/entries/copernicus/.*

Ragep, F.J. 2007. Copernicus and his Islamic predecessors: some historical remarks. *History of Science* 45, 65–81.

Recorde, R. 1543. *The Grounde of Artes*, London, 2nd edn. 1552.

———. 1551. *The Pathway to Knowledg*, London.

———. 1557. *The Whetstone of Witte*, London.

Renfrew, C. 1987. *Archaeology and Language: The Puzzle of Indo-European Origins*, Penguin.

Roberts, G. and Smith, F. 2012. *Robert Recorde: The Life and Times of a Tudor Mathematician*, U. Wales Press.

Robins, G. and Shute, C. 1987. *The Rhind Mathematical Papyrus, An Ancient Egyptian Text*, British Museum Press.

Robson, E. 2001. Neither Sherlock Holmes nor Babylon: a reassessment of Plimpton 322, *Historia Mathematica* 28, 167–206.

———. 2002. Words and pictures: new light on Plimpton 322, *American Mathematical Monthly* 109, 105–120, reprinted in Anderson, Katz, and Wilson (2004), 14–26.

———. 2007. Mesopotamian Mathematics, in (Katz 2007, 58–186).

———. 2008. *Mathematics in Ancient Iraq*, Princeton University Press.

Rose, P.R. 1975. *The Italian Renaissance of Mathematics*, Droz.

Rosen, E. (ed.) 2004. *Three Copernican Treatises: The* Commentariolus *of Copernicus, the* Letter against Werner, *the* Narratio Prima *of Rheticus*, Dover.

———. 2008. Rheticus, *Complete Dictionary of Scientific Biography*, C.C. Gillispie (ed.), 11, 395–398, Charles Scribner's Sons.

Russell, B. 1918. *Mysticism and Logic*, Allen and Unwin.

Sarton, G. 1959. *Ancient Science and Modern Civilization*, Harper.

Savile, H. 1621. *Prælectiones Tresdecim in Principium Elementorum Euclidis, Oxonii Habitæ*, Oxford.

Shirley, J.W. 1983. *Thomas Harriot*, Oxford University Press.

Shukla K.S. and Sarma, K.V. 1976. *The Aryabhatiya of Aryabhata, Critically Edited with Translation and Notes*, Indian National Science Academy.

Smith, D.E. 1959. *A Source Book in Mathematics*, Dover.

Smith, D.E. and Latham M.L. (transl.) 1954. *The Geometry of René Descartes*, Dover Books.

Smith, F.R. 2008. The influence of Amatino Manucci and Luca Pacioli, *BSHM Bulletin* 23, 143–156.

Sorell, T. 1987. *Descartes*, Oxford University Press.

Stedall, J.A. 2003. *The Greate Invention of Algebra: Thomas Harriot's Treatise on Equations*, Oxford University Press.

———. 2011. *From Cardano's Great Art to Lagrange's Reflections: Filling a Gap in the History of Algebra*, European Mathematical Society: Heritage of European Mathematics.

Stevin, S. (1585). *L'Arithmétique*, Christophe Plantin.

Stone, E. 1723. Translator's preface to *The Construction and Principal Uses of Mathematical Instruments, translated from French of M. Bion*.

Stow, J. 1615. *The Annales, or Generall Chronicle of England*.

Straffin, P.D. 1998. Liu Hui and the first golden age of Chinese mathematics, *Mathematics Magazine* 71, 163–181, reprinted in Anderson, Katz, and Wilson (2004), 69–82.

Sude, B.H. 1974. *Ibn al-Haytham's Commentary on the Premises of Euclid's* Elements, PhD thesis, Princeton University Press.

Taliaferro, R.C. (transl.) 1952. Apollonius *Conics*, in Great Books of the Western World, Encyclopædia Britannica.

Tartaglia, N. 1546. *Quesiti et Inventioni Divese*, A. Masotti (ed.), Brescia, 1959.

Taylor, E.G.R. 1954. *The Mathematical Practitioners of Tudor & Stuart England*, Cambridge University Press.

Taylor, R. and Wiles, A. 1995. Ring-theoretic properties of certain Hecke algebras, *Annals of Mathematics* (2), 141, 553–572.

Thomaidis, Y. 2005. A framework for defining the generality of Diophantos' methods in *Arithmetica*, *Archive for History of Exact Sciences* 59, 591–640.

Thomas, I. 1980. *Selections Illustrating the History of Greek Mathematics*, Heinemann.

Toomer, G.J. 1971. Mathematics and Astronomy, in *The Legacy of Egypt*, J. R. Harris (ed.), 27–54, Oxford University Press.

———. 1975. Ptolemy, *Complete Dictionary of Scientific Biography*, C. C. Gillispie (ed.), 11, Charles Scribner's Sons.

———. (transl.) 1976. *Diocles on Burning Mirrors*, Springer.

———. 1988. Hipparchus and Babylonian astronomy in *A Scientific Humanist: Studies in Memory of Abraham Sachs*, E. Leichty, M. de J. Ellis, and P. Gerardi (eds), The University Museum, Philadelphia.

Tuplin, C.J. and Rihll, T.E. (eds). 2002. *Science and Mathematics in Ancient Greek Culture*, Oxford University Press.

Urquhart T. (transl.) 1653. *The First Book of the Works of Mr Francis Rabelais*, R. Baddeley, London.

Van Brummelen, G. 2009. *The Mathematics of the Heavens and the Earth: The Early History of Trigonometry*, Princeton University Press.

———. 2013. *Heavenly Mathematics: The Forgotten Art of Spherical Trigonometry*, Princeton University Press.

Van der Waerden, B.L. 1961. *Science Awakening*, A. Dresden (transl.), Oxford University Press.

Vieta, F. 1646. *Opera Mathematica . . . Recognita Opera*, F. Schooten (ed.), Lugduni Batavorum.

Voelkel, J.R. 2001. *The Composition of Kepler's* Astronomia Nova, Princeton University Press.

Waterhouse, W. 1972. The discovery of the regular solids, *Archive for History of Exact Sciences* 9, 212–221.

Weil, A. 1983. *Number Theory: An Approach through History*, Birkhäuser.

Weisheipl, J.A. 1959. The place of John Dumbleton in the Merton School, *Isis* 50, 439–454.

Westfall, R.S. 1977. *The Construction of Modern Science*, Cambridge University Press.

Whitehead, A.N. 1933. *Adventures of Ideas*, Cambridge University Press.

Wiles, A. 1995. Modular elliptic curves and Fermat's last theorem. *Annals of Mathematics* (2), 141, 443–551.

Wood, A. à. 1792. *The History and Antiquities of the University of Oxford*, 2 vols., J. Gutch (ed.), Oxford.

Yates, F.A. 1939. Giordano Bruno's conflict with Oxford, *Journal of the Warburg Institute* 2, 227–242.

Index

Published Titles in This Series

47 **Przemyslaw Bogacki,** Linear Algebra, 2019

45 **June Barrow-Green, Jeremy Gray, and Robin Wilson,** The History of Mathematics: A Source-Based Approach, 2019

44 **Maureen T. Carroll and Elyn Rykken,** Geometry: The Line and the Circle, 2018

43 **Virginia W. Noonburg,** Differential Equations: From Calculus to Dynamical Systems, Second Edition, 2019

41 **Owen D. Byer, Deirdre L. Smeltzer, and Kenneth L. Wantz,** Journey into Discrete Mathematics, 2018

40 **Zbigniew Nitecki,** Calculus in 3D, 2018

39 **Duff Campbell,** An Open Door to Number Theory, 2018